T0313868

Energy Landscapes, Inherent Structures, and Condensed-Matter Phenomena

Energy Landscapes, Inherent Structures, and Condensed-Matter Phenomena

Frank H. Stillinger

Princeton University Press
Princeton and Oxford

Library of Congress Cataloging-in-Publication Data

Stillinger, F. H., author.
  Energy landscapes, inherent structures, and condensed-matter phenomena / Frank H. Stillinger.
    pages cm
  Includes bibliographical references and index.
  ISBN 978-0-691-16680-3 (hardback : alk. paper) 1. Condensed matter. 2. Nuclear physics. I. Title.

  QC173.457.S7S75 2015
  530.4'1—dc23   2015004169

British Library Cataloging-in-Publication Data is available

This book has been composed in Minion Pro and Ocean Sans

Printed on acid-free paper. ∞

Printed in the United States of America

10 9 8 7 6 5 4 3 2 1

Contents

Preface

Condensed matter consists of particles interacting simultaneously with many surrounding neighbors. The measurable physical and chemical properties exhibited by such matter consequently arise as intrinsically cooperative phenomena determined by those simultaneous and continuing interactions. These phenomena include attributes that involve time dependence and irreversibility, as well as features that pertain to reversible states of thermal equilibrium. The primary purpose of this volume is to present and to apply a set of conceptual and analytical tools for statistical mechanics that are a natural choice for theoretical study of a wide range of condensed-matter properties. A secondary purpose is to connect these tools and techniques to other more traditional approaches, including those that are applicable even to rarified forms of matter, such as dilute gases and isolated clusters. The unifying concept that underlies this development is the material system's multidimensional potential energy hypersurface, and how its "landscape" topography controls both kinetic and thermodynamic characteristics of any many-body system of interest. As a historical note, the first explicit suggestion that a multidimensional potential energy landscape view might be an insightful approach to condensed-matter phenomena can be found in papers by Martin Goldstein [Goldstein, 1969, 1977] devoted to liquid viscosity and glass formation. Subsequent developments have significantly broadened the range of applicability of the approach, while strengthening it with mathematical rigor and detail [e.g., Frauenfelder et al., 1997; Wales, 2003].

Unless ample scientific motivation is present, simply generating and advocating a "new" representation for statistical mechanics is little more than sterile pedagogy. Consequently, an effort has been exercised in the following text to identify concepts and related calculations that otherwise might remain unexplored in the traditional statistical mechanical formalism but which arise as natural consequences of this potential energy hypersurface "landscape paradigm." There is no intention to replace more traditional analytical methods, but rather to add to them with the hope of strengthening the discipline of statistical mechanics and its wide range of condensed-matter applications. Whereas the present approach has demonstrated some significant advantages, it is by no means a scientifically exhausted subject. The material presented in this text can serve as the starting point for a wide range of future research projects.

The remarkable advances in computer power experienced over the past half century have presented condensed-matter science with an equally remarkable enabling capacity to penetrate quantitatively into basic aspects of that subject. Specifically, accurate computations are now attainable for quantum mechanical potential energy surfaces, at least for modest-sized atomic and molecular systems. Furthermore, these computations can feed into rational fitting procedures that bootstrap those results to credible modeling of more extensive collections of particles. As a consequence of these advances, classical molecular dynamics and Monte Carlo simulations with large numbers of particles (10^4–10^6) have become almost routine. Quantum simulations are computationally more demanding and thus limited to smaller many-particle systems but are also growing in feasible size.

A developing realization over the years of the exquisite power offered by computer simulation generated at least one independent rediscovery of the multidimensional landscape paradigm [Stillinger and Weber, 1981, 1982]. The theoretical methods developed in this volume are the result of contributions from many researchers and constitute a suite of useful (and occasionally unorthodox) ways to analyze and interpret results both from laboratory experiments as well as from computer simulations. The approach also offers the benefit of strengthening intellectual ties between the liquid-state and the solid-state research communities.

The entire scientific literature devoted to condensed-matter phenomena and the statistical mechanics describing those phenomena is enormous. No attempt has been made to cover all the relevant parts of that literature in a historically complete manner. Instead, the motivation has been to identify key references that represent historic milestones in the development of the subject, as well as those general works that can serve as guides to a wider range of published literature. The availability of search engines on the Internet nowadays also provides access to that wider range in an effective manner that was not possible in the past.

The primary readership to which this work has been directed includes graduate students and their mentors in condensed-matter physics, chemical physics, and materials science, as well as senior researchers in those fields. Prerequisites include basic familiarity with quantum theory and the principles of statistical mechanics, although the following text briefly reviews those parts of these disciplines that underlie the approach to be developed. Some of the specific applications of the general energy landscape/inherent structure formalism that have been included penetrate into traditional chemistry and molecular biology. Although such examples in this text are modest in scope, their inclusion was largely motivated by the widely held belief that all natural sciences can benefit from transfer of concepts and techniques across traditional disciplinary boundaries.

While working for several decades in the area covered by this volume, I have benefited enormously from scientific contacts and collaborations with a large group of scientific colleagues. These interactions have occurred for the most part at the former AT&T Bell Laboratories in Murray Hill, New Jersey, and at Princeton University. A complete and balanced list of those contacts still grows and would be impractical to record at this late stage. However, among those colleagues who played an early formative role in developing this subject, special mention deserves to be bestowed on Thomas A. Weber and Randall A. LaViolette. Furthermore, I am deeply indebted to my wife, Dorothea Keller Stillinger, for research collaborations, for advice concerning writing style, and for assistance in producing figures for this book.

Frank H. Stillinger
October 2014

Energy Landscapes, Inherent Structures, and Condensed-Matter Phenomena

I.

Potential Energy Functions

Potential energy functions that control nuclear motions in various forms of matter provide a central object of theoretical attention in the formalism to be developed in subsequent chapters. The quantum mechanics of electrons interacting with the nuclei provides the source of those potential energy functions. It is natural, then, to begin this volume that is devoted to multidimensional potential energy landscapes and their implications by reviewing the relevant underlying quantum mechanical fundamentals. Understanding these fundamentals is a necessary prerequisite for selecting many-body models that are typically used for analytical theory, for computer simulation, and for interpreting experimental observations of a wide range of material systems, both in and out of thermal equilibrium. This chapter includes a survey of some of the commonly encountered model potentials and discusses their principal characteristics for representing both "simple" and "complex" substances.

A complete review of quantum mechanical basics would include relativistic effects that become important when heavy elements with high nuclear charges are present. However, the development to follow is limited to the nonrelativistic regime that is described by the Schrödinger equation. For present purposes, this is a justifiable simplification because it is quantitatively accurate for many applications, and even when relativistic effects exist, this nonrelativistic approximation can still qualitatively capture the majority of properties needed for the development of the energy landscape/inherent structure formalism.

A. Quantum Mechanical Basis

The natural starting point for consideration of condensed-matter properties is the quantum mechanics of the constituent electrons and nuclei, regarded as point particles with fixed masses and electrostatic charges. The dynamical evolution of collections of these point particles is described by the time-dependent Schrödinger equation [Pauling and Wilson, 1935; Schiff, 1968]:

$$\mathbf{H}\Psi = -(\hbar/i)(\partial\Psi/\partial t) \tag{I.1}$$

This basic partial differential equation is to be solved subject to applicable initial and boundary conditions, where the system volume in general may be either finite or infinite. Here the Hermitian operator $\mathbf{H}$ is the spin-independent Hamiltonian operator for the collection of electrons and nuclei, Ψ is the time-dependent wave function for that collection, and $\hbar$ is Planck's constant h divided by 2π. In addition to time t, Ψ also has as its variables the set of electronic and nuclear coordinates, to be denoted respectively by $\{\mathbf{x}_i'\} \equiv \{\mathbf{r}_i', s_i'\}$ and $\{\mathbf{x}_j\} \equiv \{\mathbf{r}_j, s_j\}$, where these coordinates include both spatial positions ($\mathbf{r}'$, $\mathbf{r}$), and spins if any (s', s). Because of the linearity of the Schrödinger equation, Ψ can be resolved into a linear combination of components corresponding to each of the eigenfunctions of the time-independent wave equation.

$$\Psi(\{\mathbf{x}_i'\}, \{\mathbf{x}_j\}, t) = \sum_n A_n \psi_n(\{\mathbf{x}_i'\}, \{\mathbf{x}_j\}) \exp(-iE_n t/\hbar) \tag{I.2}$$

$$[\mathbf{H} - E_n]\psi_n(\{\mathbf{x}_i'\}, \{\mathbf{x}_j\}) = 0$$

In these expressions, n indexes the time-independent spin and space eigenfunctions ψ_n with respective energy eigenvalues E_n. The constants A_n are determined by the initial state of the system under consideration. Because $\mathbf{H}$ is spin independent, each ψ_n can also be chosen simultaneously as an eigenfunction of total spin-squared ($\mathbf{S}^2$) and directionally projected spin operators ($\mathbf{S}_Z$) for the electrons and each nuclear species present. The ψ_n are obliged to exhibit the exchange symmetries required for the particle types that are present (symmetric for identical bosons, antisymmetric for identical fermions).

A basic simplification upon which the formalism to be developed in this volume rests is the separation of electronic and nuclear degrees of freedom in the full wave mechanics, using the Born-Oppenheimer approximation [Born and Oppenheimer, 1927]. This approximation is motivated and justified by the wide discrepancy between the mass m_e of the light electrons on one side and the masses M_j of the much heavier nuclei on the other. In the hypothetical limit for which all ratios m_e/M_j of electron to nuclear masses approach zero, the eigenfunction and eigenvalue errors committed by the Born-Oppenheimer approximation would also asymptotically approach zero [Takahashi and Takatsuka, 2006]. In most applications to be considered in this text, the actual errors attributable to the Born-Oppenheimer approximation with nonzero mass ratios are in fact negligibly small.

To apply the Born-Oppenheimer approximation, the full Hamiltonian operator $\mathbf{H}$ is first separated into two parts,

$$\mathbf{H} = \mathbf{H}_I + \mathbf{H}_{II}, \tag{I.3}$$

where $\mathbf{H}_I$ refers to motion of electrons in the presence of fixed nuclei, and $\mathbf{H}_{II}$ collects all remaining terms (nuclear kinetic energy, and nuclear pair Coulomb interactions). Because the full Schrödinger Hamiltonian $\mathbf{H}$ is spin-independent, and if no external forces are present, the specific forms are

$$\mathbf{H}_I = -(\hbar^2/2m_e)\sum_i \nabla_{\mathbf{r}_i}^2 - e^2\sum_i\sum_k Z_k/|\mathbf{r}_i'-\mathbf{r}_k| + e^2 \sum_{i<j} 1/r_{ij}', \tag{I.4}$$

and

$$\mathbf{H}_{II} = -\hbar^2\sum_j(\nabla_{\mathbf{r}_j}^2/2M_j) + e^2\sum_{j<k}Z_jZ_k/r_{jk}. \tag{I.5}$$

As indicated earlier, the $\mathbf{r}_i'$ in these equations denotes spatial positions of the electrons, and the $\mathbf{r}_j$ does the same for spatial positions of the nuclei. Following normal convention, e represents the fundamental Coulomb charge of a proton, and the nuclei bear respective charges $Z_j e$. This separation permits the full quantum mechanical problem to be resolved into an ordered sequence of two simpler problems. In the first stage, eigenfunctions and eigenvalues are obtained just for the operator $\mathbf{H}_I$, i.e., for the electrons moving in the static Coulomb field supplied by the nuclei at fixed positions, with those electronic eigenfunctions subject to the necessary antisymmetry conditions [Pauling and Wilson, 1935; Schiff, 1968]:

$$[\mathbf{H}_I - E_{n'}^{(el)}(\{\mathbf{x}_j\})]\psi_{n'}^{(el)}(\{\mathbf{x}_i'\} \mid \{\mathbf{x}_j\}) = 0. \tag{I.6}$$

As the vertical-bar notation for the wavefunctions emphasizes, the entire set of electronic eigenfunctions and eigenvalues depends parametrically on the nuclear coordinates. For each electron-subsystem indexing quantum number n' there exists a well-defined limit for the eigenvalue as all nuclei recede from one another to infinity:

$$E_{n'}^{(el)}(\infty) = \lim_{\{r_{kl}\} \to \infty} E_{n'}^{(el)}(\{\mathbf{x}_j\}). \tag{I.7}$$

This limiting energy consists of a sum of independent contributions from the individual atoms and/or ions that are formed as the electrons become partitioned and localized around the widely separated nuclei.

The second stage of the solution sequence addresses the nuclear quantum mechanics. This involves the operator $\mathbf{H}_{II}$ augmented by the relevant position-dependent energy eigenvalue from the electron subsystem. Consequently, the effective nuclear Hamiltonian becomes

$$\mathbf{H}^{(nuc)} = -\hbar^2 \sum_j (\nabla_{\mathbf{r}_j}^2 / 2M_j) + \Phi(\{\mathbf{r}_j\} \mid n'). \tag{I.8}$$

Here $\Phi(\{\mathbf{r}_j\}|n')$ is the Born-Oppenheimer potential energy (hyper)surface on which the nuclei move:

$$\Phi(\{\mathbf{r}_j\} \mid n') = e^2 \sum_{j<k} Z_j Z_k / r_{jk} + E_{n'}^{(el)}(\{\mathbf{x}_j\}) - E_{n'}^{(el)}(\infty). \tag{I.9}$$

Notice that $E_{n'}^{(el)}(\infty)$ has been inserted into Φ to provide a convenient energy origin for the nuclear subsystem. In other words, this convention implies that $\Phi = 0$ when all nuclei are infinitely far from one another. Nuclear eigenfunctions and eigenvalues (indexed by n'') are subsequently to be determined by solving the nuclear Schrödinger equation:

$$[\mathbf{H}^{(nuc)} - E_{n''}^{(nuc)}(n')]\psi_{n''}^{(nuc)}(\{\mathbf{x}_j\} \mid n') = 0, \tag{I.10}$$

subject to given boundary conditions, and to symmetry requirements imposed by the presence of identical nuclei with identical spin components (symmetric under exchange for bosons, antisymmetric under exchange for fermions) [Pauling and Wilson, 1935, Chapter XIV; Landau and Lifshitz, 1958a, Chapter IX].

As a consequence of this sequential solution process, the Born-Oppenheimer approximations to the full-system eigenfunctions and eigenvalues have the following forms:

$$\psi_n(\{\mathbf{x}_i'\}, \{\mathbf{x}_j\}) \cong \psi_{n'}^{(el)}(\{\mathbf{x}_i'\} \mid \{\mathbf{x}_j\})\psi_{n''}^{(nuc)}(\{\mathbf{x}_j\} \mid n'); \tag{I.11}$$

$$E_n \cong E_{n'}^{(el)}(\infty) + E_{n''}^{(nuc)}(n'). \tag{I.12}$$

The index ("quantum number") n for the initial problem has been replaced by the pair n', n'' for electronic and nuclear problems, respectively. It is possible and usually convenient to take the electronic wave functions to be real and orthonormal among themselves for any nuclear configuration.

Because of the presence of unbound electrons in highly excited states, this method requires that the system be at least temporarily confined to a finite volume V, which can be allowed to pass to infinity at a later stage. Consequently one can write

$$\int \psi_{m'}^{(el)}(\{\mathbf{x}_i'\} \mid \{\mathbf{x}_j\}) \psi_{n'}^{(el)}(\{\mathbf{x}_i'\} \mid \{\mathbf{x}_j\}) d\{\mathbf{x}_i'\} = \delta(m', n'). \tag{I.13}$$

The same can be assumed true for the nuclear eigenfunctions for any given electronic quantum state n':

$$\int \psi_{m''}^{(nuc)}(\{\mathbf{x}_j\} \mid n') \psi_{n''}^{(nuc)}(\{\mathbf{x}_j\} \mid n') d\{\mathbf{x}_j\} = \delta(m'', n''). \tag{I.14}$$

The "integrations" indicated in these last two equations implicitly include spin summations where necessary. The notation $\delta(j, k)$ stands for the Kronecker delta function (equal to unity for $j = k$, zero otherwise).

In parallel with Eq. (I.2), the time dependence of the nuclear quantum dynamics on the n' electronic potential surface can then be expressed as follows:

$$\Psi^{(nuc)}(\{\mathbf{x}_j\}, t \mid n') = \sum_{n''} B_{n''} \psi_{n''}^{(nuc)}(\{\mathbf{x}_j\} \mid n') \exp[-iE_{n''}(n')t/\hbar], \tag{I.15}$$

with numerical coefficients $B_{n''}$ determined by initial conditions. In circumstances where very heavy nuclei are involved, moving through regions of the configuration space where potential Φ varies slowly with position, quantum dynamics often can be replaced by its classical limit. The corresponding Newtonian equations of motion for the N nuclei then describe the nuclear configurational dynamics [Goldstein, 1953]:

$$M_j d^2 \mathbf{r}_j(t)/dt^2 = -\nabla_{\mathbf{r}_j} \Phi(\mathbf{r}_1 \ldots \mathbf{r}_N \mid n') \qquad (1 \leq j \leq N). \tag{I.16}$$

One might legitimately question whether the "bare" nuclear masses M_j that appear in Eqs. (I.8) and (I.16) are appropriate or whether a set of effective masses M_j^* that include contributions from the electron subsystem would lead to a more precise description. If the electrons are in their lowest energy (ground) state, or in a low-lying excited state, each nucleus would tend to have bound to it a number of electrons equal, or nearly equal, to its atomic number Z_j. Because these bound electrons would preferentially move with that nucleus, their mass should in principle be added to the bare nuclear mass. Thus, if the physical system of interest corresponds to well-separated electrostatically neutral atoms (e.g., noble gas atoms), the effective masses would be

$$M_j^* = M_j + Z_j m_e, \tag{I.17}$$

where m_e is the electron mass. If the nuclei exist within the system as ions with fixed oxidation states (electrostatic charges), then expression (I.17) would have to be modified accordingly to account for the deficit or surfeit of electrons compared to the atomic number Z_j. Chemical bonding to produce molecular species or extended covalent networks within which included nuclei execute vibrational motions present a more complicated situation; detailed calculations would be necessary to reveal the extent to which electron mass follows the vibrating nuclei, and thus to assign appropriate effective masses. In any case, these effective mass corrections to the Born-Oppenheimer approximation are small in absolute magnitude but should be carefully considered when high-accuracy theory and/or calculations are contemplated.

Each of the Born-Oppenheimer potential functions $\Phi(\mathbf{r}_1 \ldots \mathbf{r}_N \mid n')$ can be viewed as defining a hypersurface embedded in a $(3N + 1)$-dimensional space that is generated by the $3N$ Cartesian coordinates of nuclear position supplemented by an energy axis. Because many of the basic interests of condensed matter science require N to be roughly comparable to Avogadro's number ($\approx 6.022 \times 10^{23}$), these hypersurfaces are enormously complicated. Nevertheless, mathematical description at least of a statistical sort is feasible for these hypersurfaces, and it is facilitated to some extent by analogies to the topographies of familiar three-dimensional landscapes. A substantial portion of the analyses presented in subsequent chapters exploits this viewpoint.

Although it proves to be an excellent description for many problems in condensed-matter physics and chemical physics, the Born-Oppenheimer approximation is inaccurate and thus inappropriate in some special circumstances. In addition to the effective-mass corrections mentioned above, one of these cases involves nuclear dynamics where close approach (in energy), or even intersection, of two or more potential energy hypersurfaces occurs [Yarkony, 1996, 2001; Domcke and Yarkony, 2012]. Another important case concerns superconductivity arising from the BCS mechanism of electron coupling through phonons [Bardeen, Cooper, and Schrieffer, 1957], and in fact the relevant deviations from the Born-Oppenheimer description underlie the nuclear isotopic-mass dependence of superconducting transition temperatures [Kittel, 1963, Chapter 8; Ashcroft and Mermin, 1976, Chapter 34].

B. Properties of Electronic Ground and Excited States

Although the number of electronic eigenstates indexed by n' in the Born-Oppenheimer approximation is infinite for all material systems of interest and includes states of unbounded excitation energy above the ground state, the principal focus of attention in the present volume is on the ground state itself, and the low-lying excited states. If the atoms comprised in a many-body system are geometrically well isolated from one another, the lowest lying electronic states are nearly degenerate (i.e., confined to very narrow bands) and have energies substantially determined by those of the separate atoms. Bound excited states for isolated atoms are limited above by that atom's ionization energy. Table I.1 presents the ionization energies for individual atoms of several elements, i.e., the energy difference between the electronic ground state and the lowest lying ionized continuum state for those atoms. The substantial variation among entries for the elements is an indication of their distinctive chemical properties.

As a configuration of many identical atoms, initially widely separated from one another, is uniformly compressed, the narrow bands of nearly degenerate energy levels of the many-atom system tend to broaden in a manner strongly dependent on the specific atomic elements involved. When brought to normal ambient-pressure crystal densities and configurations, different elements

TABLE I.1. Ionization energies (in eV) for individual atoms of several elements[a]

Element	H	He	Li	C	O	Na	Si
Ionization energy	13.59844	24.58741	5.39172	11.26030	13.61806	5.13908	8.15169

[a]Lide, 2003, p. **10**–178.

can become electronic insulators (large remaining band gap between ground and lowest lying excited electronic state), semiconductors (small but nonzero band gap), or metals (no gap) [Harrison, 1980]. But it is well to keep in mind that even atomic species that are normally regarded as insulators (e.g., the noble gases) eventually become metallic materials under sufficiently strong compression. A case in point is the element xenon, which has been reported to transform to a metallic ground state at a pressure of 132(5) GPa [Goettal et al., 1989].

Electronic excitation affects the occurrence and nature of potential energy minima. This is a phenomenon present even in the simplest molecular systems and is illustrated by the elementary example of molecular hydrogen. Figure I.1 shows a plot of the single-molecule potential energy curves for H_2, as a function of internuclear separation, for the $^1\Sigma_g^+$ ground state (electron spin singlet) and the lowest lying $^3\Sigma_u^+$ (electron spin triplet) excited state. This pair of curves vividly demonstrates the disappearance of a well-developed covalent chemical bond minimum upon excitation from the ground state, with replacement by a repulsive potential curve. For strict accuracy, it should be mentioned that even the repulsive $^3\Sigma_u^+$ curve possesses an extremely shallow minimum at very large internuclear separation because of very weak dispersion attraction between well-separated hydrogen atoms [Landau and Lifshitz, 1958a, p. 270; Kołos and Wolniewicz, 1965]. The two-electron example illustrated in Figure I.1 is the simplest case of a more general result for a spin-independent Schrödinger Hamiltonian, specifically that the lowest energy eigenvalue occurs when the net electron spin is at its lowest because of the Pauli exclusion principle

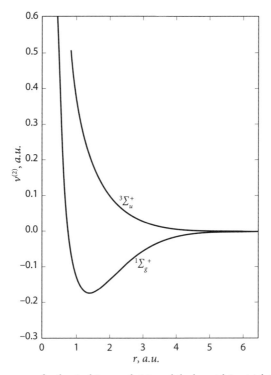

FIGURE I.1. Potential energy curves for the singlet ground state and the lowest lying triplet excited state of molecular hydrogen [Kołos and Wolniewicz, 1965]. The scales involve atomic units ("a.u."), where for distance 1 a.u. equals the Bohr radius 0.5292 Å, and for energy 1 a.u. equals 27.21 eV.

(Pauli, 1925). In most applications to be considered later in this text, that is the electronic state involved.

The Hellmann-Feynman theorem, which is applicable to any electronic eigenstate, can be used to provide valuable insight into the nature of interparticle interactions [Hellmann, 1937; Feynman, 1939; Weissbluth, 1978]. This theorem concerns the change in an electronic eigenvalue as a result of a differential perturbation in the electronic Hamiltonian operator $\mathbf{H}_I$ (Eq. (I.4)). Let λ represent the strength of that perturbation. The Hellmann-Feynman theorem states the following:

$$dE_{n'}^{(el)}(\{\mathbf{x}_j\})/d\lambda = \int \psi_{n'}^{(el)}(\{\mathbf{x}_i'\} \mid \{\mathbf{x}_j\})[d\mathbf{H}_I/d\lambda]\psi_{n'}^{(el)}(\{\mathbf{x}_i'\} \mid \{\mathbf{x}_j\})d\{\mathbf{x}_i'\}. \qquad (I.18)$$

That is, the rate of change with λ of the n' electronic eigenvalue is precisely equal to the diagonal matrix element in that eigenstate of the corresponding rate of change of the electronic Hamiltonian operator $\mathbf{H}_I$. The specific application of this result that is useful in the present context identifies the perturbation as a spatial shift in one of the nuclear positions, say $\mathbf{r}_l$ for nucleus l. As a result, Eq. (I.18) implies the following result:

$$\nabla_{\mathbf{r}_l} E_{n'}^{(el)}(\{\mathbf{x}_j\}) = \int [\psi_{n'}^{(el)}(\{\mathbf{x}_i'\} \mid \{\mathbf{x}_j\})]^2 \left[-Z_l e^2 \nabla_{\mathbf{r}_l} \sum_i 1/\mid \mathbf{r}_i' - \mathbf{r}_l \mid\right] d\{\mathbf{x}_i'\} \equiv -\mathbf{F}_l^{(Coul)}(n'). \qquad (I.19)$$

This equation declares that the rate of change of the electronic eigenvalue $E_{n'}^{(el)}$, with respect to a shift in position of nucleus l, is precisely the negative of the coulombic force exerted on nucleus l by the charge distribution of the electrons in that quantum state n'. In this respect, the electronic contribution to Born-Oppenheimer potential energy surfaces has a simple interpretation in terms of classical electrostatics. As a special case, if all nuclei are at positions of mechanical equilibrium, the net electrostatic force on each of those nuclei caused by the electron density and the other nuclei must vanish identically. More generally, in the case of a nearby pair of atoms engaging in a covalent chemical bond, whether or not those atoms are at positions of mechanical equilibrium, Eq. (I.19) places a constraint on the distribution of electron density in the vicinity of that pair.

The quantum mechanical virial theorem [Born, Heisenberg, and Jordan, 1925–1926; Slater, 1933; Hirschfelder, Curtiss, and Bird, 1954] for a collection of electrons and nuclei establishes a rigorous connection between the overall kinetic energy operator

$$\mathbf{K} = -(\hbar^2/2m_e)\sum_i \nabla_{\mathbf{r}_i}^2 - \hbar^2\sum_k (\nabla_{\mathbf{r}_k}^2/2M_k), \qquad (I.20)$$

and the virial operator

$$\mathbf{V} = \sum_i \mathbf{r}_i' \cdot \mathbf{F}_i' + \sum_k \mathbf{r}_k \cdot \mathbf{F}_k. \qquad (I.21)$$

Here, $\mathbf{F}_i'$ and $\mathbf{F}_k$, respectively, are the vector forces experienced by electron i and by nucleus k for any given spatial configuration of all electrons and nuclei. Presuming that this system is confined to the interior of a container with reflecting walls, and is in overall quantum state n, the theorem declares

$$\int \psi_n^* \mathbf{K} \psi_n d\{\mathbf{x}_i'\}d\{\mathbf{x}_k\} = -(1/2)\int \psi_n^* \mathbf{V} \psi_n d\{\mathbf{x}'\}d\{\mathbf{x}_k\}. \qquad (I.22)$$

Note that this theorem does not rely upon the Born-Oppenheimer approximation but is a general statement about eigenstates for the combined quantum mechanics of electrons plus nuclei. If the quantum state denoted by n is a bound state inside an arbitrarily large container and has vanishing center-of-mass motion, then the only forces present are those between pairs of electrons and/or nuclei. Because those forces arise exclusively from Coulomb interactions, the virial operator is equivalent to the sum of all of those Coulomb terms:

$$\mathbf{V} \rightarrow e^2 \sum_{i<j} 1/r_{ij}' - e^2 \sum_i \sum_k Z_k / |\mathbf{r}_i' - \mathbf{r}_k| + e^2 \sum_{k<l} Z_k Z_l / r_{kl} \tag{I.23}$$

$$\equiv \mathbf{V}_{Coul}.$$

In this circumstance, Eq. (I.22) states that the expectation value of the total kinetic energy is minus one-half that of the total Coulomb interaction:

$$\int \psi_n^* \mathbf{K} \psi_n d\{\mathbf{x}_i'\} d\{\mathbf{x}_k\} = -(1/2) \int \psi_n^* \mathbf{V}_{Coul} \psi_n d\{\mathbf{x}_i'\} d\{\mathbf{x}_k\}. \tag{I.24}$$

In the formal limit where masses of all the nuclei approach infinity, those nuclei can remain localized at vanishing-force mechanical equilibrium points of the Born-Oppenheimer potential energy surface $\Phi(\{\mathbf{r}_k\}|n')$, and they would have no kinetic energy. As a result, the left side of Eq. (I.24) involves only electron kinetic energy, whereas the right side contains electron–electron, electron–nucleus, and nucleus–nucleus contributions. Alternatively, suppose that the nuclei were constrained at positions not necessarily those of mechanical equilibrium on the Born-Oppenheimer potential energy surface but just acted as fixed force centers for the electron quantum mechanics. This is the situation for the first stage of the Born-Oppenheimer approximation outlined in Section I.A of this chapter. With the resulting nonvanishing net forces on the nuclei, the virial theorem then states the following for the electronic degrees of freedom:

$$\sum_k \mathbf{r}_k \cdot \nabla_{\mathbf{r}_k} \Phi(\{\mathbf{r}_l\}|n') = \int \psi_{n'}^{(el)*} \left\{ \sum_i [-(\hbar^2/m_e)\nabla_{\mathbf{r}_i}^2 + \mathbf{r}_i' \cdot \mathbf{F}_i'] \right\} \psi_{n'}^{(el)} d\{\mathbf{x}_j'\}. \tag{I.25}$$

This equation indicates that the first spatial moment of the forces on the nuclei is proportional to the imbalance in kinetic and virial-force contributions of the electronic degrees of freedom.

A final application of the quantum virial theorem concerns the case of a macroscopic system confined to a small enough region V that it exerts nonvanishing forces on the constraining walls. These wall forces can be interpreted as a normal outward pressure p, and their presence requires that the virial operator $\mathbf{V}$ be separated into a part internal to the system of electrons and nuclei (*Coul*), and a part involving the external-wall interactions (*ex*):

$$\mathbf{V} = \mathbf{V}_{Coul} + \mathbf{V}_{ex}. \tag{I.26}$$

As a result, the theorem leads to a formal expression for the pressure–volume product pV for the system in quantum state n:

$$3pV = \int \psi_n^* [2\mathbf{K} + \mathbf{V}_{Coul}] \psi_n d\{\mathbf{x}_i'\} d\{\mathbf{x}_k\}. \tag{I.27}$$

This is another result whose validity does not rely on the Born-Oppenheimer approximation.

C. Mathematical Attributes of Potential Energy Functions

Whether it is the electronic ground state or any of the electronic excited states that is under consideration, it is important to recognize the general mathematical properties that must be obeyed by the corresponding Born-Oppenheimer potential energy functions that control nuclear motions. These general properties in turn influence the statistical behavior of the corresponding condensed-matter systems. Here is a list of the most basic of those general properties. Some are obvious, others more subtle, but all provide a necessary background for the theoretical developments of the next two sections, I.D and I.E, and of subsequent chapters.

(1) For any quantum state, the potential energy function Φ is a single-valued function of the nuclear position coordinates $\mathbf{r}_1 \ldots \mathbf{r}_N$.

(2) Any geometrically isolated set of nuclei in free space (i.e., remote from container walls) can be translated and/or rotated arbitrarily without changing the value of the potential energy function Φ.

(3) Permutation of the positions of any pair of nuclei with equal atomic numbers ($Z_j = Z_l$, but possibly different isotopes) leaves the potential energy function Φ unchanged. This invariance remains valid even in the presence of wall forces.

(4) The electrostatic charges $Z_j e$ carried by the nuclei cause coulombic divergences in the $\Phi(\mathbf{r}_1 \ldots \mathbf{r}_N | n')$ whenever the separation between two nuclei tends to zero. Specifically,

$$\Phi(r_{jk} \to 0 \,|\, n') \sim Z_j Z_k / r_{jk} + A(\mathbf{r}_1 \ldots \mathbf{r}_N | n'), \tag{I.28}$$

where the function A remains finite in that nuclear confluence limit. For virtually all condensed-matter applications of interest, the instances of r_{jk} remain large enough even at their smallest occurrences that these nuclear Coulomb singularities are obscured by powerful (but bounded) electron-cloud overlap repulsions. The latter are implicit in the A function.

(5) Aside from rare nuclear configurations producing electronic degeneracies such as "conical intersections" [Yarkony, 1996, 2001; Domcke and Yarkony, 2012], Φ is continuous and at least twice differentiable in nuclear position coordinates $\mathbf{r}_1 \ldots \mathbf{r}_N$ away from any nuclear coincidences. In particular, this phenomenon implies that relative or absolute minima of Φ are locally at least quadratic in those nuclear coordinates. Such minima are mechanically stable configurations, that is, forces on all nuclei vanish. These Φ minima are henceforth called "inherent structures" throughout the remainder of this text.

(6) Φ obeys a global stability criterion [Fisher and Ruelle, 1966; Ruelle, 1969, p. 33]. There exists some $K > 0$ that is independent of N, of $\mathbf{r}_1 \ldots \mathbf{r}_N$, of boundary conditions, and of quantum number n' such that

$$\Phi > -KN. \tag{I.29}$$

This property stands in accordance with the common observation that cohesive energy of a macroscopic material sample at fixed number density is an extensive quantity, i.e., proportional to sample size.

(7) If the system is constrained to overall charge neutrality, and if all pairs of nuclei are constrained to be more widely separated than some minimum distance $r_0 > 0$, then in the $N \to \infty$ asymptotic regime, Φ must possess the extensivity property, i.e., it can increase in magnitude no more rapidly than proportionally to N. This condition is consistent with, but independent of, the previous condition (6).

(8) In a large system of N nuclei, configurational rearrangements involving only a small local subset of $O(1)$ nuclei change Φ only by $O(1)$. Removal to infinity of $O(1)$ nuclei is one example of this limited change in Φ.

A useful context in which to view model potentials includes the general resolution of a many-body potential function $\Phi(\mathbf{r}_1 \ldots \mathbf{r}_N \,|\, n')$ into one-body, two-body, three-body, ..., N-body contributions:

$$\Phi(\mathbf{r}_1 \ldots \mathbf{r}_N \,|\, n') \equiv \sum_{i=1}^{N} v^{(1)}(\mathbf{r}_i) + \sum_{i=2}^{N} \sum_{j=1}^{i-1} v^{(2)}(\mathbf{r}_i, \mathbf{r}_j) + \sum_{i=3}^{N} \sum_{j=2}^{i-1} \sum_{k=1}^{i-1} v^{(3)}(\mathbf{r}_i, \mathbf{r}_j, \mathbf{r}_k) + \ldots + v^{(N)}(\mathbf{r}_1 \ldots \mathbf{r}_N). \qquad (I.30)$$

For notational simplicity, this expression assumes that all nuclei are the same species, but the multicomponent extension is straightforward. By requiring Eq. (I.30) to be an identity successively for $N = 1, 2, 3, \ldots$ the $v^{(1)}, v^{(2)}, v^{(3)}, \ldots$ in turn each receive precise definitions. Specifically each $v^{(l)}$ for a set of l nuclei corrects the estimate of the potential energy for that set provided by the collection of lower order $v^{(j)}$ functions. Because the coordinates $\mathbf{r}_i$ refer to single nuclei, the single-particle contributions $v^{(1)}$ would describe interactions with external force fields, such as those associated with containment vessel walls or with a gravitational field. If the circumstances are appropriate, the series of terms in identity (I.30) may decline in magnitude sufficiently rapidly with superscript order that truncation may be a reasonable approximation at some fixed order $l_{max} \ll N$. In some cases, it suffices to choose $l_{max} = 2$ or $l_{max} = 3$. Although Eq. (I.30) applies in principle to any electronic excited state n', virtually all applications to be considered later involve the ground state $n' = 0$.

D. Model Potential Functions: Simple Particles

Direct solution of the electronic Schrödinger equation is a formidable task even for small numbers of electrons and nuclei, let alone for the very large numbers of both that are present in extended condensed phases. Although a wide variety of approximate solution methods for the underlying quantum problem exists [Parr, 1963; Torrens, 1972], these methods can be computationally demanding, their results may suffer from uncontrolled errors, and they may by themselves be uninformative from the standpoint of analytical statistical mechanics. This situation creates legitimacy for the introduction of model potential energy functions that mimic in at least some respects the experimental and quantum-computational information known about systems of interest. The basic desirable features of such models are conceptual clarity and mathematical simplicity. In addition, they can frequently offer computational speed advantage for numerical simulation when large numbers of particles are involved. The illustrative examples discussed in this and in Section I.E include specific cases as well as generic families of model potentials. They have been selected for a variety of reasons. Some have been widely used, some have unusual mathematical features, and some have been included to illustrate the wide variety of substances that can be described at least approximately by relatively simple modeling. These examples are all intended to represent qualitatively the electronic ground state potential energy "landscapes" for nuclear motion (i.e., $n' = 0$).

The cases discussed in this section all involve structureless, spherically symmetrical particles that (aside from possible external forces) by assumption engage only in pairwise additive interactions. The following section (Section I.E) extends the discussion to include more complex particles with internal degrees of freedom, such as polyatomic molecules and cases requiring nonpairwise interactions. It should be kept in mind for all of these illustrations that model potentials need not, and typically do not, adhere strictly to all of the attributes mentioned in Section I.C. Nevertheless, their virtue is in representing some selected nontrivial many-body phenomena, without the "baggage" of distracting complication.

(1) INVERSE-POWER POTENTIALS

A particularly simple and transparent family of model many-particle potential energy functions consists of a sum of pair terms that are inverse powers of scalar distances. If only a single particle species is present, this model involves a potential energy function of the following form:

$$\Phi(\mathbf{r}_1 \ldots \mathbf{r}_N) = \varepsilon \sum_{i<j} (\sigma/r_{ij})^{\bar{n}}, \tag{I.31}$$

where the sum includes all pairs of particles that are present. Here, ε and σ are positive parameters with units of energy and length, respectively, which establish the scales for those basic properties. When this inverse-power form is applied to three-dimensional modeling, the exponent $\bar{n}$ must obey the inequality $\bar{n} > 3$ in order to ensure that Φ is an extensive quantity [attribute (7) in Section I.C]. This case focuses attention exclusively on the strong repulsive forces that atoms or molecules experience when their electron clouds overlap, as when the many-particle system is subject to high external pressure, or when high-energy particle collisions are a dominant kinetic feature.

For a macroscopic collection of particles at fixed density, interacting according to this inverse-power potential, the configuration producing the absolute minimum value of Φ is a periodic face-centered cubic crystal structure [Dubin and Dewitt, 1994]; see Chapter IV. However, this is not the only mechanically stable arrangement (inherent structure) for the particles. In particular, the body-centered cubic lattice is also a local Φ minimum, though lying higher in energy for all $\bar{n} > 3$.

The fact that expression (I.31) is a homogeneous function of coordinates with degree $-\bar{n}$ has an obvious but mathematically useful consequence. Assume that periodic boundary conditions apply and that the N-particle system, initially confined to a region with volume V, has all of its coordinates scaled by a common factor $\zeta > 0$:

$$\mathbf{r}_1 \ldots \mathbf{r}_N \rightarrow \zeta \mathbf{r}_1 \ldots \zeta \mathbf{r}_N. \tag{I.32}$$

The volume then becomes $\zeta^3 V$, and the initial potential energy $\Phi(\mathbf{r}_1 \ldots \zeta \mathbf{r}_N)$ changes by a factor $\zeta^{-\bar{n}}$. In addition, this transformation has the property that any N-particle configuration that initially was a local Φ minimum (inherent structure) remains a local Φ minimum after scaling; and a similar remark applies to Φ extrema of other types (saddle points and maxima, if any). In particular, the number of mechanically stable potential energy minima, and their relative ordering in energy, are both invariant to the scaling. The same is true for the saddle points of Φ of various orders.

As exponent $\bar{n}$ increases, the repelling pair interactions become steeper and steeper. The infinite-$\bar{n}$ limit in Eq. (I.31) consequently produces the venerable hard-sphere model, a staple of statistical mechanical theory [Hill, 1956; McQuarrie, 1976; Hansen and McDonald, 1986]. Length parameter σ then plays the role of the collision diameter for the hard spheres. Specifically, one has

$$\lim_{\bar{n}\to\infty} \Phi(\mathbf{r}_1\ldots\mathbf{r}_N) = \sum_{i<j} \mathrm{v}_{hs}(r_{ij}), \tag{I.33}$$

where

$$\mathrm{v}_{hs}(r) = +\infty \qquad (r \le \sigma), \tag{I.34}$$

$$= 0 \qquad (\sigma < r).$$

Because of the singular nature of the limit operation, ε does not explicitly appear in this result but nevertheless still manages formally to confer the dimension of energy on $\mathrm{v}_{hs}(r)$.

Pairwise additive inverse power potentials for finite $\bar{n}$ (often called "soft core models") have provided the basis for several published studies of both crystal-phase and dense fluid-phase properties [Hoover, Young, and Grover, 1972; Martin and Singer, 1991; Vieira and Lacks, 2003].

(2) LENNARD-JONES 12,6 POTENTIAL

The Lennard-Jones pair potential model was originally proposed [Jones, 1924] to approximate the interactions between neutral particles in the gas phase, including both attractive and repulsive forces. It is most appropriately viewed as representing the electronic ground states of the noble gases helium, neon, argon, krypton, and xenon. Once again, single-atom interactions that might be caused by boundary confinement are put aside, and this model also truncates the identity (I.30) at just the atom pair level. Specifically, it postulates the following form for the total potential energy experienced by N atoms, all of the same species:

$$\Phi(\mathbf{r}_1\ldots\mathbf{r}_N \,|\, 0) = \varepsilon \sum_{i<j} \mathrm{v}_{LJ}(r_{ij}/\sigma) \qquad (\varepsilon, \sigma > 0), \tag{I.35}$$

where again the sum includes all particle pairs, r_{ij} is the scalar distance between particles i and j, and $\mathrm{v}_{LJ}(x)$ is a unit-depth reduced interaction defined as follows for $x > 0$:

$$\mathrm{v}_{LJ}(x) = C(\bar{n}, \bar{m})(x^{-\bar{n}} - x^{-\bar{m}}) \qquad (\bar{n} > \bar{m} > 3), \tag{I.36}$$

$$C(\bar{n}, \bar{m}) = \left(\frac{\bar{n}}{\bar{m}}\right)^{\bar{m}/(\bar{n}-\bar{m})}\left(\frac{\bar{n}}{\bar{n}-\bar{m}}\right).$$

This equation constitutes an obvious extension of the preceding inverse-power pair potential. Although exponents $\bar{n}$ and $\bar{m}$ were originally treated as adjustable [Jones, 1924], it has become traditional more recently to refer to the "Lennard-Jones model" as specifically involving the choice $\bar{n}, \bar{m} = 12,6$:

$$\mathrm{v}_{LJ}(x) = 4(x^{-12} - x^{-6}). \tag{I.37}$$

The positive parameters ε and σ, respectively, once again set the energy and length scales for the pair interactions, as appropriate for the specific substance of interest. The x^{-12} term produces strong interparticle repulsion at small separation, whereas the attractive x^{-6} term dominates at large separation and possesses the correct distance dependence expected for long-range (nonrelativistic) dispersion interactions [Atkins, 1970]. It should be noted in passing that the Hellmann-Feynman relation (I.19) requires that neutral particles participating in attractive dispersion attractions at moderate to large distance from one another must at the same time display distortions of their isolated-particle electron distributions to be electrostatically consistent with those attractions.

The reduced Lennard-Jones 12,6 pair potential $v_{LJ}(x)$ in Eq. (I.37) has a single zero (at $x = 1$), and a single minimum (at $x = x_{min} = 2^{1/6} \cong 1.1225$), for positive x:

$$v_{LJ}(1) = 0, \tag{I.38}$$

$$v_{LJ}(x_{min}) = -1. $$

The curvature at x_{min} is the following:

$$v_{LJ}''(x_{min}) = 72 \cdot 2^{-1/3} \cong 57.146. \tag{I.39}$$

Furthermore, this function possesses a single inflection point at

$$x_{infl} = (26/7)^{1/6} \cong 1.2445, \tag{I.40}$$

$$v_{LJ}(x_{infl}) = -133/169 \cong -0.78698. $$

By optimizing fits to experimental gas-phase data of simple substances, numerical values can be assigned to the Lennard-Jones scale parameters ε and σ. Measured equation of state virial coefficients, or alternatively measured viscosities, provide the type of input required. Table I.2 contains values that have been suggested as appropriate for the noble gases [Hirschfelder, Curtiss, and Bird, 1954]. Both parameters increase steadily with increasing atomic number Z because of the growing number of electrons bound in the neutral atoms. The result is increased overlap repulsion at small separation and stronger dispersion attraction at large separation.

The scientific literature contains a large number of publications in which the pairwise-additive Lennard-Jones 12,6 potential plays a central role. Representative examples of those publications can be cited in which this model potential has been used to represent gas-phase clusters [Wales and Berry, 1990; Tsai and Jordan, 1993b], the liquid state [Hansen and McDonald, 1986; Malandro and Lacks, 1998], crystalline solids [Somasi et al., 2000; Stillinger, 2001], and amorphous solids [Fox and Andersen, 1984; Vollmayr et al., 1996], thus approximating each of these states exhibited by the heavier noble gases. The special cases of the light helium isotopes involving

TABLE I.2. Energy and length scale parameters for the noble gases represented by the Lennard–Jones 12,6 pair potential[a]

Species	He	Ne	Ar	Kr	Xe
Atomic no. (Z)	2	10	18	36	54
ε/k_B, K	10.22	35.60	119.8	171	221
σ, nm	0.2556	0.2749	0.3405	0.360	0.410

[a]Taken from Hirschfelder, Curtiss, and Bird, 1954, p. 1110.

strong atomic quantum effects also incorporate Lennard-Jones-like interactions but require separate discussion, which is presented in Chapter VIII.

When mixtures of different substances are present, several distinct pairs of Lennard-Jones energy and length scale parameters are required. It can then be impractical to determine all such pairs experimentally. In that situation, it has been traditional to invoke empirical combining rules that use the parameters for the pure substances. For two species a and b, the cross-species Lennard-Jones energy and length parameters conventionally are set equal respectively to the geometric and arithmetic means of those for the pure substances [Hirschfelder, Curtiss, and Bird, 1954, p. 168]:

$$\varepsilon_{ab} = (\varepsilon_{aa}\varepsilon_{bb})^{1/2}, \tag{I.41}$$

$$\sigma_{ab} = (\sigma_{aa} + \sigma_{bb})/2. $$

(3) MORSE POTENTIAL

The Lennard-Jones 12,6 pair potential contains only two adjustable parameters, fixing energy and length scales. For some applications, it is desirable to have greater functional flexibility. The Morse pair potential offers three adjustable parameters (ε, σ, r_e) [Morse, 1929]. It has the following parameterized form:

$$\varepsilon v_M[(r - r_e)/\sigma], \tag{I.42}$$

where

$$v_M(x) = \exp(-2x) - 2\exp(-x). \tag{I.43}$$

As was the case with the Lennard-Jones pair potential, ε denotes the depth of the Morse potential at its single minimum, and σ controls its length scale. The position of the minimum (i.e., the mechanical equilibrium pair distance) is independently set by r_e. Although v_M has neither the capacity to produce a divergence at small pair separation nor a long-range dispersion attraction varying as r^{-6}, these attributes may be inconsequential for some applications.

The properties of v_M to be compared with those shown in Eqs. (I.38)–(I.40) for v_{LJ} are as follows ($x_{min} = 0$):

$$v_M(-\ln 2) = 0, \tag{I.44}$$

$$v_M(x_{min}) = -1;$$

$$v_M''(x_{min}) = 2; \tag{I.45}$$

$$x_{infl} = \ln 2, \tag{I.46}$$

$$v_M(x_{infl}) = -3/4.$$

One obvious advantage of v_M over v_{LJ} is its ability to provide via σ independent control of the curvature at the minimum of the rescaled pair potential (I.42) for fixed equilibrium pair bond length r_e and depth ε.

In the large-N limit, a many-body potential energy function composed of Morse pair interactions, to describe an electronic ground-state situation,

$$\Phi(\mathbf{r}_1 \ldots \mathbf{r}_N \mid 0) = \varepsilon \sum_{i<j} v_M[(r_{ij} - r_e)/\sigma], \tag{I.47}$$

must conform to the lower-bound criterion stated in Eq. (I.29). This criterion places a constraint on the acceptable values of r_e and σ. If the N particles are arranged so that essentially equal subsets of $N/4$ are placed precisely on top of one another at the vertices of a regular tetrahedron with sides equal to r_e, then the requirement that Φ be bounded below by $-KN$, $K > 0$, leads to the inequality

$$\exp(r_e/\sigma) \geq 3. \tag{I.48}$$

(4) GAUSSIAN CORE MODEL POTENTIAL

Several unusual mathematical properties create an intrinsic interest in the model potential energy function consisting of a pairwise sum of suitably scaled Gaussian terms:

$$\Phi(\mathbf{r}_1 \ldots \mathbf{r}_N) = \varepsilon \sum_{i<j} \exp[-(r_{ij}/\sigma)^2] \qquad (\varepsilon, \sigma > 0). \tag{I.49}$$

Historically, this form could be viewed as originating in polymer solution theory, where the individual Gaussian terms approximated interactions between randomly coiling linear polymers [Flory, 1953]; see Chapter X. However, the simple form in Eq. (I.49) is best interpreted as corresponding to structureless, spherically symmetric "soft" particles.

In its remote tail region, each Gaussian pair interaction approaches zero more and more rapidly in relative terms as r increases. This attribute is measured by the logarithmic rate of change quantity

$$d\ln\{\varepsilon \exp[-(r/\sigma)^2]\}/dr = -2r/\sigma^2, \tag{I.50}$$

indicating that the relative steepness increases in direct proportion to the separation r. Of course, the pair interaction itself is very small at large r. But if two Gaussian core particles (in isolation from all others), initially widely separated, approach one another with very low relative kinetic energy E, they experience virtually a rigid sphere collision at an effective collision diameter:

$$r_{coll}(E) \approx \sigma[\ln(\varepsilon/E)]^{1/2}. \tag{I.51}$$

This observation underlies the low-temperature, low-density asymptotic reduction of the Gaussian core model to the hard-sphere model discussed in Eqs. (I.33)–(I.34) [Stillinger and Stillinger, 1997].

The spatial arrangement of Gaussian core particles in a macroscopic system that attains the minimum potential energy is a face-centered cubic lattice at low density but a body-centered cubic arrangement at high density [Stillinger, 1976]. The corresponding lattice energies Φ_{fcc} and Φ_{bcc} obey an exact duality relation [Stillinger, 1979; Stillinger and Stillinger, 1997]. To exhibit this connection, define

$$I_{fcc}(\rho) = \varepsilon + \lim_{N\to\infty} [2\Phi_{fcc}(\rho)/N], \tag{I.52}$$

$$I_{bcc}(\rho') = \varepsilon + \lim_{N\to\infty} [2\Phi_{bcc}(\rho')/N],$$

where ρ and ρ' are the respective values of the particle number density N/V. If these number densities satisfy the following identity:

$$\rho\rho' = \pi^{-3}\sigma^{-6}, \tag{I.53}$$

then it follows from the self-similarity of the Gaussian function under Fourier transformation that

$$(\rho)^{-1/2}I_{fcc}(\rho) = (\rho')^{-1/2}I_{bcc}(\rho'). \tag{I.54}$$

This duality relation has the useful property that lattice energy at high density, where many neighbors interact, can be reduced to evaluation of a low-density lattice energy, where few neighbors are close enough to interact to a significant extent. The lattice energies per particle for the two structures are equal at the self-dual density $\rho\sigma^3 = \rho'\sigma^3 = \pi^{-3/2}$.

Let $f(\mathbf{R})$ be a function defined in the $3N$-dimensional configuration space $\mathbf{R} \equiv \mathbf{r}_1\ldots\mathbf{r}_N$, and let $\mathbf{L}(\lambda)$ be a convolution operator that entails smoothing over this space with a Gaussian kernel possessing width λ:

$$\mathbf{L}(\lambda)*f(\mathbf{R}) = (\pi^{1/2}\lambda)^{-3N}\int \exp[-(\mathbf{R} - \mathbf{R}')^2/\lambda^2]f(\mathbf{R}')d\mathbf{R}'. \tag{I.55}$$

In particular, applying this operator to a Gaussian function results in another Gaussian function with increased width but with a diminished maximum value. For the N-particle potential energy function in Eq. (I.49) consisting of a sum of Gaussians, one finds [Stillinger and Stillinger, 1997]:

$$\Phi[\mathbf{R}/(1+\delta)] = (1+\delta)^3\mathbf{L}\{[(\delta^2/2) + \delta]^{1/2}\sigma\}*\Phi(\mathbf{R}), \qquad \delta \geq 0. \tag{I.56}$$

Consequently, shrinking all particle coordinates by factor $(1+\delta)^{-1}$, or equivalently increasing the number density by factor $(1+\delta)^3$, is equivalent to applying both the smoothing operator and a magnitude renormalization to Φ. In view of the fact that smoothing tends to eliminate maxima and minima, though the absolute minimum of the Gaussian core model potential at high density is the body-centered cubic lattice, this lattice evidently is the only mechanically stable structure that survives in the asymptotic high-density limit.

(5) YUKAWA POTENTIAL

Another frequently cited pair interaction is the Yukawa potential. For a single-component many-body system, the potential energy function (exclusive of single-particle terms $v^{(1)}$) has the form

$$\Phi(\mathbf{r}_1\ldots\mathbf{r}_N) = \varepsilon\sum_{i<j}v_Y(r_{ij}/\sigma), \tag{I.57}$$

where

$$v_Y(x) = \exp(-x)/x. \tag{I.58}$$

This elementary function retains the name attributable historically to its appearance in the early theory of meson-mediated nucleon interactions [Yukawa, 1935]. More relevant to the general subject currently under consideration is that it has the form of a shielded Coulomb pair interaction, as appears in Debye-Hückel electrolyte theory [Debye and Hückel, 1923; also see Section V.G in Chapter V]. In order for the potential to satisfy the global stability criterion (I.29), the energy scaling parameter ε cannot be negative.

(6) COLLECTIVE DENSITY VARIABLE REPRESENTATION

In the event that the potential energy function Φ consists only of a sum of spherically symmetric pair interactions for identical particles, an alternative representation of that function may be available that is based on collective density variables. For some purposes, this alternative offers conceptual and mathematical advantages.

Let the particle system be contained within a rectangular solid region with linear dimensions L_x, L_y, L_z, and volume $V = L_x L_y L_z$, and furthermore assume that periodic boundary conditions apply to this region. This geometry allows introduction of an infinite set of wave vectors $\mathbf{k}$, each one of which has Cartesian components:

$$\mathbf{k} = (2\pi m_x/L_x, \, 2\pi m_y/L_y, \, 2\pi m_z/L_z), \tag{I.59}$$

$$m_x, m_y, m_z = 0, \pm 1, \pm 2, \ldots .$$

Collective density variables $\rho(\mathbf{k})$ are then defined for each wave vector by sums over all particle positions $\mathbf{r}_j$ within V:

$$\rho(\mathbf{k}) = \sum_{j=1}^{N} \exp(i\mathbf{k}\cdot\mathbf{r}_j). \tag{I.60}$$

As a result of this definition, one has the following obvious properties:

$$\rho*(\mathbf{k}) = \rho(-\mathbf{k}),$$

$$\rho(0) = N, \tag{I.61}$$

$$0 \le |\rho(\mathbf{k})| \le N \qquad (\mathbf{k} \ne 0).$$

The spherically symmetric pair potential acting between particles is denoted by $v^{(2)}(r)$, i.e.:

$$\Phi = \sum_{j<l} v^{(2)}(r_{jl}). \tag{I.62}$$

Assuming that it exists, the Fourier transform of $v^{(2)}(r)$, defined for the wave vector set (I.59), involves an integral over the finite system volume V:

$$V^{(2)}(\mathbf{k}) = \int_V v^{(2)}(r) \exp(i\mathbf{k}\cdot\mathbf{r}) d\mathbf{r}. \tag{I.63}$$

The inverse transform requires a sum over all instances of $\mathbf{k}$:

$$v^{(2)}(r) = V^{-1} \sum_{\mathbf{k}} V^{(2)}(\mathbf{k}) \exp(-i\mathbf{k}\cdot\mathbf{r}). \tag{I.64}$$

Each pair potential term in the right side of Eq. (I.62) may be replaced by this last inverse transform sum, with the result

$$\Phi = V^{-1} \sum_{j<l} \sum_{\mathbf{k}} V^{(2)}(\mathbf{k}) \exp[-i\mathbf{k}\cdot(\mathbf{r}_j - \mathbf{r}_l)]$$

$$= (2V)^{-1} \sum_{j \ne l=1}^{N} \sum_{\mathbf{k}} V^{(2)}(\mathbf{k}) \exp[-i\mathbf{k}\cdot(\mathbf{r}_j - \mathbf{r}_l)]$$

$$= (2V)^{-1}\sum_{\mathbf{k}}V^{(2)}(\mathbf{k})\left\{\sum_{j,l=1}^{N}\exp[-i\mathbf{k}\cdot(\mathbf{r}_j - \mathbf{r}_l)] - N\right\}$$

$$= (2V)^{-1}\sum_{\mathbf{k}}V^{(2)}(\mathbf{k})[\rho(\mathbf{k})\rho(-\mathbf{k}) - N]. \tag{I.65}$$

Although this set of equations formally expresses the total potential as a quadratic form in the collective density variables, it is important to keep in mind that these variables are not independent, and as a result they are subject to complicated nonlinear constraints [Fan et al., 1991; Uche et al., 2004].

The Fourier transform $V^{(2)}(\mathbf{k})$ does not exist for the Lennard-Jones pair potentials, Eqs. (I.36)–(I.37). However, it does exist for the more shape-flexible Morse potential, Eqs. (I.42)–(I.43). In the limit where the linear dimensions of the system volume V far exceed the range of that pair function, one has

$$V^{(2)}(\mathbf{k}) = \varepsilon\int v_M[(r - r_e)/\sigma]\exp(i\mathbf{k}\cdot\mathbf{r})d\mathbf{r}$$

$$= 16\pi\varepsilon\sigma^3\left[\frac{\exp(2r_e/\sigma)}{(4 + \sigma^2 k^2)^2} - \frac{\exp(r_e/\sigma)}{(1 + \sigma^2 k^2)^2}\right]. \tag{I.66}$$

Another case of interest involves a repelling Gaussian pair potential that generates the Gaussian core model mentioned above [Stillinger and Stillinger, 1997]. This is one of the simplest circumstances in which the pair potential and its Fourier transform are self-similar:

$$v^{(2)}(r) = \varepsilon\exp[-(r/\sigma)^2]; \tag{I.67}$$

$$V^{(2)}(\mathbf{k}) = \pi^{3/2}\varepsilon\sigma^3\exp(-\sigma^2 k^2/4), \tag{I.68}$$

a property that underlies the duality identities for this model mentioned above, Eqs. (I.52)–(I.53). Finally note that the Yukawa pair potential $\varepsilon v_Y(r/\sigma)$ has the following Fourier transform:

$$V^{(2)}(k) = \frac{4\pi\varepsilon\sigma^3}{1 + \sigma^2 k^2}, \tag{I.69}$$

exhibiting a long algebraic falloff with increasing k because of the coulombic divergence of the potential at the origin in $\mathbf{r}$ space.

An unusual feature can arise if the pair potential $v^{(2)}(r)$ has a non-negative Fourier transform that vanishes identically beyond some radius in $\mathbf{k}$ space:

$$V^{(2)}(\mathbf{k}) > 0 \qquad (0 \leq |\mathbf{k}| \leq K) \tag{I.70}$$

$$= 0 \qquad (K < |\mathbf{k}|).$$

If K is sufficiently small that fewer than $3N$ wave vectors $\mathbf{k}$ fall within that radius about the origin, then the absolute minimum for Φ is attained by having all of the corresponding $|\rho(\mathbf{k})| = 0$. However, this condition does not exhaust all of the system's degrees of freedom. Consequently, the absolute Φ minimum remains configurationally highly degenerate, and for small enough K this situation can produce amorphous classical ground states in which the N particles are free to reconfigure along a large set of configuration space pathways [Uche et al., 2006].

E. Model Potential Functions: Complex Particles

(1) SATURATING CHEMICAL BONDS

The behavior of a wide range of chemical substances involves formation and dissociation of strong chemical bonds between the constituent atoms. Obvious examples include monovalent elements that tend to be present as diatomic molecules, such as hydrogen (H_2), oxygen (O_2), nitrogen (N_2), and chlorine (Cl_2). Elemental sulfur (S) is divalent and consequently forms closed polygons or chains of bonded atoms [Donohue, 1982]. Silicon (Si) is tetravalent, forming at low temperature a diamond-structure crystal with local chemical bonding geometry that connects each atom to four equivalent neighbors located at the vertices of a surrounding regular tetrahedron. These chemically preferred valences, and the local patterns of bond angles at each participating atom, cannot be described accurately in terms of spherically symmetric pair interactions $v^{(2)}(r_{ij})$ alone. However, at least a rudimentary description of chemical valency in the electronic ground state can be achieved by combining pair potentials with a proper choice of three-atom functions $v^{(3)}(\mathbf{r}_i, \mathbf{r}_j, \mathbf{r}_k)$:

$$\Phi(\mathbf{r}_1 \ldots \mathbf{r}_N \mid 0) \cong \sum_{i<j} v^{(2)}(r_{ij}) + \sum_{i<j<k} v^{(3)}(\mathbf{r}_i, \mathbf{r}_j, \mathbf{r}_k). \tag{I.71}$$

Here the role of the pair terms $v^{(2)}$ is to create chemical bonds individually with appropriate strength and length scales. The atom-triplet terms $v^{(3)}$ can then be chosen to discourage formation of more than the number of chemical bonds allowed by the valence of the element under consideration and to establish the preferred angles between bonds at any atom when the valence of that atom is 2 or greater.

One example suffices to illustrate the fixed-valency concept. This example involves the ground electronic state for elemental fluorine (F), which is monovalent and which consequently can exist as a covalently bonded diatomic molecule, F_2. A combination of two-atom and three-atom interactions of the type in Eq. (I.71) has been constructed and used in computer simulation [Harris and Stillinger, 1990]. In the interests of computational simplicity, both the two-atom and three-atom interactions were taken to vanish identically beyond a modest distance range on the molecular scale. The pair function was chosen to be

$$v^{(2)}(r) = \varepsilon_{FF} u^{(2)}(r/\sigma_{FF}), \tag{I.72}$$

in which

$$\varepsilon_{FF} = 160.2263 \text{ kJ/mol}, \tag{I.73}$$

$$\sigma_{FF} = 0.12141 \text{ nm}, $$

and

$$u^{(2)}(x) = A(x^{-8} - x^{-4})E(x - 3.6). \tag{I.74}$$

Here the function $E(y)$ has the following definition:

$$E(y) = \exp(y^{-1}) \qquad (y < 0), \tag{I.75}$$

$$= 0 \qquad (y \geq 0),$$

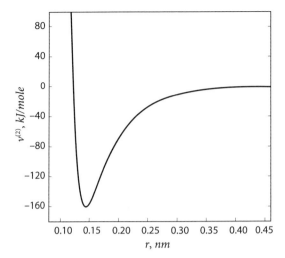

FIGURE I.2. Model pair potential for elemental fluorine, Eq. (I.72) [Harris and Stillinger, 1990].

so that $E(y)$ is continuous and has continuous derivatives of all orders for all real y. The dimensionless constant A was set equal to $6.052\,463\,017$ so that $u^{(2)}(x)$ possesses a unit-depth minimum. With these choices, the implied equilibrium bond length for the isolated F_2 molecule is 0.1435 nm, the harmonic vibrational frequency is 896.9 cm^{-1}, and the dissociation energy is 160.2263 kJ/mol, all close to the corresponding spectroscopically measured values (0.1435 nm, 892 cm^{-1} [Herzberg, 1950], and 160.0508 kJ/mol [Huber and Herzberg, 1979]). The pair potential presented in Eq. (I.72), as well as all its distance derivatives, continuously approach zero and remain vanishing, as r increases to and exceeds the cutoff distance 0.437076 nm. Figure I.2 shows a plot of this fluorine pair potential.

Besides enforcing the monovalency of the model fluorine atoms, the three-body contributions $v^{(3)}$ can be chosen to yield a moderately realistic description of the chemical exchange reaction $F + F_2 \rightarrow F_2 + F$, avoiding the pitfall of too large a transition-state barrier. A choice matched to the above pair potential in Eq. (I.72) has the following form that is necessarily symmetrical under atom permutations [Harris and Stillinger, 1990]:

$$v^{(3)}(\mathbf{r}_1, \mathbf{r}_2, \mathbf{r}_3) = \varepsilon_{FF}[\,f(r_{12}/\sigma_{FF}, r_{13}/\sigma_{FF}, \theta_1) + f(r_{12}/\sigma_{FF}, r_{23}/\sigma_{FF}, \theta_2), \qquad (I.76)$$

$$+ f(r_{13}/\sigma_{FF}, r_{23}/\sigma_{FF}, \theta_3)],$$

where $0 \le \theta_i \le \pi$ is the triangle vertex angle at nucleus i, and

$$f(x, y, \theta) = 8.4(xy)^{-4}E(x - 3.6)E(y - 3.6) \qquad (I.77)$$

$$+ (55 - 25\cos^2\theta)E[(x - 2.80)/3]E[(y - 2.80)/3].$$

With these two-body and three-body interaction choices, the transition state for three fluorine atoms is a symmetric linear arrangement, and at that state the potential energy possesses negative curvature along the direction of asymmetric stretch, the reaction coordinate. Figure I.3 presents a

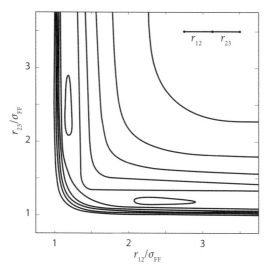

FIGURE I.3. Contour plot of the reduced potential energy for three fluorine atoms in a linear arrangement, shown vs. reduced distances. The two asymmetric minima occur within the closed contours at which the reduced energy $\Phi/\varepsilon_{FF} = -1.01$. Contours of successively increasing energy, moving outward from the minima, are at reduced energies -1.01, -0.85, -0.70, -0.50, -0.30, and -0.10.

contour plot of this model's potential energy for three atoms in the linear-configuration subspace containing the transition state. An equivalent pair of asymmetrical minima occur at $r_{12}/\sigma_{FF} = 1.1887$, $r_{23}/\sigma_{FF} = 2.3322$ and the reverse, at which the reduced interaction energy is $\Phi/\varepsilon_{FF} = -1.03895$. The symmetrical transition state at $r_{12}/\sigma_{FF} = r_{23}/\sigma_{FF} = 1.3649$ has $\Phi/\varepsilon_{FF} = -0.85997$, which is 28.677 kJ/mol above the minima.

Analogously structured combinations of two-atom and three-atom interaction functions can be selected to describe substances exhibiting higher valence chemical bonding. Divalent elemental sulfur (S) has been modeled in this manner in order to simulate temperature effects in its liquid phase [Stillinger, Weber, and LaViolette, 1986; Stillinger and Weber, 1987]. The divalency of this element is known experimentally to lead to preferred angles between bonds at an atom and thus to formation of octameric molecular rings (S_8) at low temperature (specifically below the melting temperature at low pressure). But this divalency also leads to rings of other sizes containing both even and odd numbers of atoms and to long linear polymers at elevated temperature [MacKnight and Tobolsky, 1965; Steudel, 1984].

The same kinds of two-atom plus three-atom model potentials have also been constructed to describe and to simulate the tetravalent semiconductors silicon (Si) [Stillinger and Weber, 1985a; Biswas and Hamann, 1987] and germanium (Ge) [Ding and Andersen, 1986]. As was the case for the monovalency and divalency cases just mentioned, the role of the three-body nonadditive interactions is to cause saturation of chemical bond formation and to control the angles between bonds simultaneously impinging on any given atom.

It should be noted in passing that whenever three-body interactions are present, the number of terms to be considered in a large N-atom system nominally requires arduous triple summation over all particles. However, if the $v^{(3)}$'s have the forms shown in Eq. (I.76), with f's that are products of radial functions and polynomials in angle sines and cosines as in Eq. (I.77), then in certain cases it is possible to reduce triple sums to double sums [Biswas and Hamann, 1987; Weber and Stillinger, 1993]. Such reductions can substantially speed up computer calculations based on that class of models.

(2) INTACT MOLECULES

Many applications in the study of condensed-matter phenomena concern neutral molecules or polyatomic ions whose chemical structures remain stable in the physical circumstances of interest (temperature, pressure, irreversible flows). Nuclei continue to be chemically bonded within the same molecular or ionic groupings throughout the observation "time window." For these applications, it is unnecessary, even undesirable, to consider the entire Born-Oppenheimer potential energy hypersurface throughout its full configuration space. Instead, the relevant chemical subspace of the full configuration space is identified, and model potential functions are then generated for just that subspace. In doing so, the model functions should be chosen to ensure that the desired chemical bonding patterns cannot be violated by chance nuclear excursions. In these circumstances, it is natural to subtract the intramolecular ground-state minimum energies for the molecules or ions from Φ, which then measures only the combination of internal distortion energies of the stable molecules or ions, the potentials of interaction they have with any walls present, and the interactions between neighboring intact molecules or ions.

Water offers a representative and obviously important example. Its condensed-phase properties and their molecular interpretations form the subject of Chapter IX. A system comprising N oxygen nuclei and $2N$ hydrogen nuclei, in the presence of $10N$ electrons so as to be electrostatically neutral overall, can form N H_2O molecules. The Born-Oppenheimer electronic ground state for one such molecule has C_{2v} symmetry at its own potential energy minimum, with a structure illustrated in Figure I.4. The dipole moment for the molecule in isolation points along the direction of the symmetry axis and has magnitude (in Debye) 1.855 D [Dyke and Muenter, 1973]. Its three vibrational normal modes (with light hydrogens ^{1}H) have frequencies equal to 3,656.65 cm^{-1} (symmetric stretch), 3,755.79 cm^{-1} (asymmetric stretch), and 1,594.59 cm^{-1} (symmetric bend)

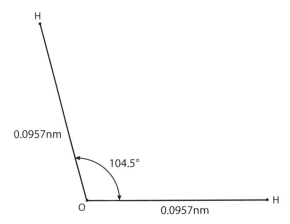

FIGURE I.4. Ground-state structure of the water molecule [Eisenberg and Kauzmann, 1969].

[Eisenberg and Kauzmann, 1969, p. 7]. Needless to say, each of these mode frequencies decline upon isotopic substitution of the light hydrogens by heavier deuterium.

Water molecules are internally quite stiff, i.e., their internal restoring forces resisting bond bending and bond stretching are quite large, as indicated by the cited mode frequencies. Consequently in liquid water at room temperature and pressure or in ice at lower temperature and similar pressure the molecules retain nearly their isolated-molecule structure. Furthermore, only about 5 molecules in 10^8 are found to have dissociated into solvated H^+ and OH^- ions in the room-temperature liquid [Eisenberg and Kauzmann, 1969, Table 4.9], although this fraction rises considerably with increasing temperature and pressure [Tödheide, 1972]. As a result, it is reasonable to construct and exploit water models that postulate permanently intact molecules and that do not involve exchange of atoms between those molecules. Chapter IX discusses several such cases. Without any restriction to permanently intact H_2O molecules, the full Born-Oppenheimer potential function Φ for N oxygens and $2N$ hydrogens would have each of its geometrically distinct mechanically stable minima (inherent structures) belonging to a set of $N!(2N)!$ permutation-equivalent inherent structures. However, the intact-molecule restraint reduces this equivalence number to the much smaller value $N!2^N$ because then the N molecules can be permuted among themselves, but each of the N molecules has only two equivalent configurations that differ by intramolecular hydrogen exchange through rotation by angle π about its symmetry axis.

The generalization of this reduction in number of geometrically equivalent inherent structures is straightforward when considering stiff-molecule or stiff-ion models for other substances. Consider the case of N molecules, each of which contains $v(i)$ atoms of species i. Therefore, the system of interest has the corresponding composition: precisely $v(i)N$ atoms for each of the species i. Then without the restriction to permanently intact stiff molecules, each Φ inherent structure would belong to an equivalence set comprising

$$\prod_i [v(i)N]! \tag{I.78}$$

members. By contrast, N permanently intact stiff molecules that each have symmetry number $\bar{\sigma}$ for reorientation among their own equivalent positions would be able to attain only

$$N!\bar{\sigma}^N \tag{I.79}$$

indistinguishable inherent structures on the multidimensional Φ landscape.

The potential energy function for intact molecules confined to their own portion of the full configuration space has a resolution into one, two, three, $\dots$, N molecule contributions analogous to that exhibited in Eq. (I.30) for individual atoms. In the case of a single species, let $\mathbf{y}_i$ stand for the collection of coordinates required to specify the location, orientation, and internal deformation of molecule i, $1 \leq i \leq N$. Then one has the following expression for the many-stable-molecule potential energy function (implicitly assumed to refer to the electronic ground state):

$$\Phi(\mathbf{y}_1 \dots \mathbf{y}_N) \equiv \sum_{i=1}^{N} v^{(1)}(\mathbf{y}_i) + \sum_{i=2}^{N} \sum_{j=1}^{i-1} v^{(2)}(\mathbf{y}_i, \mathbf{y}_j) + \sum_{i=3}^{N} \sum_{j=2}^{i-1} \sum_{k=1}^{j-1} v^{(3)}(\mathbf{y}_i, \mathbf{y}_j, \mathbf{y}_k) + \dots + v^{(N)}(\mathbf{y}_1 \dots \mathbf{y}_N) \tag{I.80}$$

where as before the individual $v^{(l)}$ functions, $1 \leq l \leq N$, are to be determined sequentially from interactions of increasing numbers of molecules. The generalization of this expression to the case of systems containing several intact molecular and/or ionic species is straightforward. The leading contributions $v^{(1)}(\mathbf{y}_i)$ now contain intramolecular interactions for flexible molecules, as well as any external force contributions operating on the individual molecules.

(3) CHIRAL MOLECULE ENANTIOMERS

The tetrahedrally radiating pattern of four single covalent bonds emanating from a carbon atom is a rather rigid feature in a molecule. Under ambient temperature and pressure conditions, only minor angular displacements in this pattern would occur. If the four portions of the molecule attached to such a carbon atom by those covalent bonds are chemically distinguishable, then that carbon is identified as a "stereocenter." This name implies that a molecule containing such a specific tetrahedral arrangement of the bonds to those four portions cannot be superimposed on its mirror-image molecule by any combination of translation and rotation. If only one stereocenter is present, then the molecule and its mirror image are distinguishable chiral species, called "enantiomers." Figure I.5 illustrates this situation, generically labeling the four distinct attachments A_1, A_2, A_3, A_4. As an example, if these attachments are specifically chosen to be a hydrogen atom (–H), a fluorine atom (–F), a methyl group (–CH_3), and an ethyl group (–CH_2–CH_3), the result would be one of the two enantiomers of a 2-fluorobutane molecule. As is discussed in Chapter XI, the overwhelming majority of amino acids that form the building blocks of protein molecules in terrestrial living organisms are chiral species of just one enantiomer.

Enantiomers are experimentally easily distinguishable from one another because they are optically active, that is, they rotate the plane of transmitted polarized light in opposite directions. A mixture containing equal amounts of the two enantiomers is called a "racemic" mixture, and it has no net rotational effect on the plane of transmitted polarized light.

In principle, the full Born-Oppenheimer ground-state potential function Φ would not discriminate between enantiomers. In other words, when Φ displays an inherent structure minimum for one enantiomer of a chiral molecule, it must also exhibit an equivalent mirror-image inherent structure minimum for the other enantiomer. But for substances such as 2-fluorobutane and the

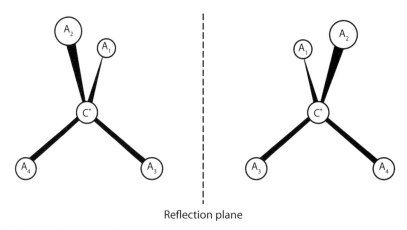

Reflection plane

FIGURE I.5. Schematic illustration of the geometric structures of a chiral molecule and its mirror image, each of which contains a stereocenter carbon atom C^*. The four covalently attached structures denoted by A_1, A_2, A_3, A_4 must be chemically distinguishable.

chiral amino acids, these inherent structures must be separated by high energy barriers in the Φ landscape. This is necessary to be consistent with the experimental fact that the enantiomers can be treated as distinct molecular species whose numbers remain unchanged over time if the ambient conditions are not extreme. This barrier height requirement would place restrictions on any approximation used for the intramolecular interaction portion of $v^{(1)}(\mathbf{y}_i)$ in Eq. (I.80).

(4) RIGID BOND LENGTHS

Because of the relative nondeformability of many small molecular species such as water, a useful level of simplification often invoked treats each molecule as a strictly rigid object with fixed bond lengths and in some cases fixed bond angles. This method reduces the relevant configuration space even further than as discussed in Section I.E(2). Rigid linear molecules [examples are diatomic nitrogen (N_2) and triatomic carbon dioxide (CO_2)] would then require only five configurational coordinates, the center position coordinates and two polar angles. Rigid nonlinear molecules [such as water (H_2O), and benzene (C_6H_6)] would require six configurational coordinates, the center position coordinates and three Euler angles, to specify spatial orientation [Goldstein, 1953].

Realistically modeling the interactions of some classes of intact but nonrigid molecules requires explicit inclusion of some internal conformational degrees of freedom. This is the case, for example, with the saturated linear hydrocarbons (normal alkanes) that have chemical compositions C_nH_{2n+2} for $n \geq 4$. Even with fixed chemical bond lengths and fixed angles between vertex-sharing chemical bond pairs emanating from its carbon atoms, $n - 3$ dihedral angles are still needed to specify the conformation of the carbon-atom backbone chain of the alkane molecule, as well as two additional angles for rotation of its two terminal methyl groups ($-CH_3$). This process requires that a model potential energy function Φ for normal alkanes, even when simplified by the fixed-bond-length and fixed-bond-angle assumptions, must include within its single-molecule $v^{(1)}$ contributions at least a portion depending explicitly on those $n-1$ remaining angles.

(5) FORCE CENTER ASSIGNMENTS

The interactions that actually operate in solid and fluid condensed phases of intact molecules may involve a substantial contribution from nonpair terms $v^{(j)}(j > 2)$ in Φ, when resolved as in Eq. (I.80) into one, two, three, …, molecule portions. In particular, this is the case for collections of the polar, and polarizable, water molecules. It is also a significant issue when ions are present. The calculation of induced dipoles in a condensed-phase configuration involves solving $O(N)$ simultaneous linear equations, often an impractical computational situation. Consequently, these and other nonpair effects need to be incorporated at least approximately in any model that successfully represents these condensed-phase situations. One strategy to account for nonpair terms is to devise an "effective pair potential," chosen to fit a set of measured properties for the substance of interest, then applied to the study of other properties [Stillinger, 1970; Stillinger et al., 2002]. Because the specific choice of fitting criteria would typically depend on an investigator's special interests, and consequently vary from case to case, there can be no generally agreed upon "uniquely best" effective pair potential [see Box IX.1 for details]. However, this non-uniqueness does not invalidate the approach.

The most straightforward yet flexible strategy for creating an effective intermolecular pair potential involves use of an appropriately chosen number of spherically symmetric force centers within each molecule or ion. These centers are fixed with respect to local subsets of the nuclei and chemical bonds of each molecule and can interact with force centers attached to neighboring molecules via radial functions of the distance between those force centers. Consequently, the effective molecule–molecule pair interaction would have the following form:

$$v_{eff}^{(2)}(i,j) = \sum_{\eta}\sum_{\zeta} b_{\eta\zeta}(|\mathbf{R}_j + \mathbf{u}_\eta - \mathbf{R}_i - \mathbf{u}_\zeta|). \qquad (I.81)$$

In this expression, single "origin-center" locations embedded in each of molecules i and j have respectively been denoted by $\mathbf{R}_i$ and $\mathbf{R}_j$. Indices η and ζ label the various force centers in i and j, with displacements $\mathbf{u}_\eta$ and $\mathbf{u}_\zeta$ from $\mathbf{R}_i$ and $\mathbf{R}_j$, respectively. The individual force centers in a molecule need not be located precisely at its nuclei nor be equal in number to those nuclei. The radial functions $b_{\eta\zeta}$ could then be chosen from a suitable function set to minimize error in prediction of a weighted sum of experimental properties for the molecular substance under consideration. Such selection should be carried out so as to satisfy, to the extent possible, the general mathematical properties of potential functions discussed in Section I.C. For some applications, the functional forms exhibited by the pair potentials described in Section I.D may serve as useful force-center interaction potentials for complex molecules [e.g., the Lennard-Jones 12,6 function, Eq. (I.37)].

Choosing $b_{\eta\zeta}$'s for homonuclear diatomic molecules such as oxygen (O_2), nitrogen (N_2), or chlorine (Cl_2) can rely on short-range functions because no molecular dipole moments are involved. Although the positions of the nuclei could serve as the force centers in the event that two centers per diatomic molecule are to be used, this may not be the optimal choice. It may be more appropriate to displace the force centers inward toward the molecular center. This would be consistent with the accumulation of electron density between the chemically bonded atoms, as described by the adapted Hellmann-Feynman theorem, Eq. (I.19).

Molecules containing nonequivalent nuclei, such as water, can have net electrostatic charges attributed to each of the constituent atoms and assigned in turn to the force centers. This concept justifies inclusion of long-range coulombic tails in the force-center functions:

$$b_{\eta\zeta}(r) \sim q_\eta q_\zeta / r. \qquad (I.82)$$

If the molecular species under consideration is electrically neutral, the sum of force-center charges for each molecule of course must vanish:

$$\sum_{\eta} q_\eta = 0. \qquad (I.83)$$

Formally, the dipole moment magnitude due to force-center charges for a molecule is

$$\mu = |\sum_{\eta} q_\eta \mathbf{u}_\eta|. \qquad (I.84)$$

In creating an appropriate $v_{eff}^{(2)}$ for a strongly polar molecule, this dipole magnitude may differ from that of the isolated molecule. In the case of water, the value required for optimal condensed-phase modeling may be substantially larger than that of the isolated species because of mutual polarization effects between each molecule and those comprising its surroundings. Various forms of evidence suggest that μ should increase from the dilute gas value 1.855 D [Dyke and Muenter, 1973] to approximately 2.95 ± 0.20 D in the liquid-water phase [Gubskaya and Kusalik, 2002]. Similar enhancements should be expected for other polar molecules embedded in condensed phases.

If stable ionic species such as the ammonium cation (NH_4^+) or sulfate anion (SO_4^{2-}) are present, the sums $\sum q_\eta$ for all such species should be integer multiples of a "unit charge." However, the optimal choice for the force-center functions $b_{\eta\zeta}(r)$ may not set this "unit charge" equal to the fundamental proton charge e because of the need to account for dielectric polarization effects.

A comparative discussion of effective pair potentials for water, consisting of rigid molecules with simple force centers, may be found in Jorgensen et al. (1983). An example of force-center assignment for the normal-alkane homologous series appears in a publication by Errington and Panagiotopoulos (1999). Other chemical substances for which simple force-center approximations to intermolecular interactions have been proposed are the linear carbon disulfide molecule (CS_2) [Ciccotti et al., 1982], acetone (CH_3COCH_3) and chloroform ($CHCl_3$) [Rickardi et al., 1998], and various aromatic hydrocarbons [Williams, 1966].

The force-center functions devised for effective pair potentials can also be useful for modeling some aspects of intramolecular interactions comprised in $v^{(1)}$. This method becomes especially relevant when high-molecular-weight species such as long-chain alkanes, nonbiological polymers, or proteins are under consideration. Physically attainable conformations for those species can bring remote portions of the same molecule with their assigned force centers into close contact, in which case the resulting local contribution to the potential energy would be substantially indistinguishable from a corresponding intermolecular contribution that involves the same kind of local groupings of atoms.

II.

Statistical Mechanical Basics

T he formalism of statistical mechanics provides the logical connection between atomic-level structure and interactions on one side and the collective properties of arbitrary-size groups of atoms, ions, and molecules on the other. These properties include those exhibited by systems at thermodynamic equilibrium, by systems in the process of relaxing to equilibrium, and even by systems driven essentially permanently into metastable states. This chapter sketches the fundamentals of the statistical mechanical theory that are necessary to develop a general energy landscape/inherent structure formalism. Later chapters build upon those fundamentals with specific technical elaborations as needed for the various applications to be examined. The coverage of statistical mechanical basics in this chapter is necessarily abbreviated. More detailed presentations of the complex subject of statistical mechanics are available in a large number of specialized texts [e.g., Landau and Lifshitz, 1958b; Huang, 1963; Feynman, 1972; Kubo et al., 1991; Toda et al., 1992].

A. Density Matrices

Although a more general development is possible, in the interest of simplicity this section confines attention to the quantum behavior of a set of N nuclei, moving on a Born-Oppenheimer potential surface Φ for the electronic ground state. Intact molecular or ionic groupings may or may not be present on this potential surface, depending on which nuclei are involved as well as the details of their dynamics on that surface. Following the notation used in Chapter I, the collection of nuclear coordinates is denoted by $\{\mathbf{x}_j\}$, comprising both spatial positions $\{\mathbf{r}_j\}$ and spin variables $\{s_j\}$, if any. The Hamiltonian operator for this dynamical system of N nuclei, assumed to be spin independent, is [cf. Eq. (I.8)]:

$$\mathbf{H} = -\hbar^2 \sum_j (\nabla_j^2 / 2M_j) + \Phi(\mathbf{r}_1 \ldots \mathbf{r}_N), \qquad (\text{II.1})$$

where M_j is the mass of the jth nucleus. Here and in the following, the superscript "(*nuc*)" that was introduced in Chapter I, Eqs. (I.8)–(I.12), has been dropped as unnecessary because only nuclear motions are explicitly considered. The time-dependent wave function Ψ for such a system is generally a normalized linear combination of terms for each of the eigenfunctions of the nuclear Schrödinger equation:

$$\Psi(\{\mathbf{x}_j\}, t) = \sum_a C_a \psi_a(\{\mathbf{x}_j\}) \exp[-iE_a t/\hbar]. \qquad (\text{II.2})$$

Here a indexes the nuclear quantum states for which the corresponding ψ_a and E_a are respectively the normalized time-independent eigenfunction and energy eigenvalue that are appropriate for the finite system volume V and associated boundary conditions. Spin independence of $\mathbf{H}$ implies that one can choose each ψ_a to be an eigenfunction of the nuclear total-spin operators $\mathbf{S}^2$ and $\mathbf{S}_z$ for each species of nucleus, an option that is now exercised. The (generally complex) coefficients C_a appearing in Eq. (II.2) are such that the real numbers $C_a * C_a$ represent the probabilities that the time-dependent system would be found in each state a. Consequently, they are subject to the normalization condition

$$\sum_a C_a * C_a = 1. \tag{II.3}$$

The entire set of measurable properties for the time-dependent system can conveniently be generated with the help of the Hermitian density matrix operator $\boldsymbol{\rho}^{(N)}$. When the system has expression (II.2) as its wave function, $\boldsymbol{\rho}^{(N)}$ has matrix elements of the following form, in the basis of the complete set of eigenfunctions $\{\psi_a\}$ of the Hamiltonian operator $\mathbf{H}$:

$$\langle a \,|\, \boldsymbol{\rho}^{(N)} \,|\, b \rangle = \int \psi_a^*(\{\mathbf{x}_j\}) \boldsymbol{\rho}^{(N)} \psi_b(\{\mathbf{x}_j\}) d\{\mathbf{x}_j\} \tag{II.4}$$

$$= C_a^* C_b \exp[i(E_a - E_b)t/\hbar],$$

where the simplifying Dirac bracket notation [Dirac, 1947] has been used to represent the integrals. The "integration" indicated in this last equation should be interpreted as including a sum over nuclear spin variables. As an alternative to the form in Eq. (II.4), where some other complete set of normalized basis functions (φ_ζ) was chosen to represent the density matrix, the corresponding matrix elements would be obtained by the transformation:

$$\langle \zeta \,|\, \boldsymbol{\rho}^{(N)} \,|\, \eta \rangle = \sum_{a,b} \langle \zeta \,|\, a \rangle \langle a \,|\, \boldsymbol{\rho}^{(N)} \,|\, b \rangle \langle b \,|\, \eta \rangle, \tag{II.5}$$

where

$$\langle \zeta \,|\, a \rangle = \int \varphi_\zeta^*(\{\mathbf{x}_j\}) \psi_a(\{\mathbf{x}_j\}) d\{\mathbf{x}_j\}, \tag{II.6}$$

$$\langle b \,|\, \eta \rangle = \int \psi_b^*(\{\mathbf{x}_j\}) \varphi_\eta(\{\mathbf{x}_j\}) d\{\mathbf{x}_j\}.$$

Let $\mathbf{O}$ represent the Hermitian operator that corresponds to some observable property O. The time-dependent measurable expectation value for this operator involves calculating the trace of the product of this operator with the density matrix:

$$\langle O(t) \rangle = Tr(\mathbf{O}\boldsymbol{\rho}^{(N)}) \tag{II.7}$$

$$= Tr(\boldsymbol{\rho}^{(N)}\mathbf{O}) .$$

Because the trace is independent of the basis in which it is computed, either the eigenfunction set $\{\psi_a\}$ or an alternative basis $\{\varphi_\zeta\}$ could be used:

$$\langle O(t) \rangle = \sum_{a,b} \langle a \,|\, \mathbf{O} \,|\, b \rangle \langle b \,|\, \boldsymbol{\rho}^{(N)} \,|\, a \rangle \tag{II.8}$$

$$= \sum_{\zeta,\mu} \langle \zeta \,|\, \mathbf{O} \,|\, \eta \rangle \langle \eta \,|\, \boldsymbol{\rho}^{(N)} \,|\, \zeta \rangle.$$

Because of the fact that energy is conserved in a closed system, the expectation value of total energy E must be independent of time:

$$\langle E \rangle = \sum_{a,b} \langle a | \mathbf{H} | b \rangle \langle b | \boldsymbol{\rho}^{(N)} | a \rangle$$
$$= \sum_{a} E_a \langle a | \boldsymbol{\rho}^{(N)} | a \rangle \tag{II.9}$$
$$= \sum_{a} E_a C_a^* C_a,$$

which transparently illustrates the earlier identification of the real quantity $C_a^* C_a$ as the weight (probability) of quantum state a in the time-dependent wave function $\Psi(t)$, Eq. (II.2). By contrast, the expectation value of some nonconserved quantity, such as the potential energy function $\Phi(\mathbf{r}_1 \dots \mathbf{r}_N)$, which constitutes just part of the Hamiltonian, is not time invariant in the mixed state $\Psi(t)$.

The time-dependent configurational probability for the system of N nuclei is obtained by summing the product $\Psi^*(t)\Psi(t)$ over nuclear spin degrees of freedom, if any. This probability, defined over the $3N$-dimensional configuration space of position coordinates $\{\mathbf{r}_j\}$, is denoted by P_N:

$$P_N(\{\mathbf{r}_j\}, t) = \sum_{\{s_j\}} \Psi^*(\{\mathbf{x}_j\}, t)\Psi(\{\mathbf{x}_j\}, t) \tag{II.10}$$
$$= \sum_{a,b} \exp[i(E_a - E_b)t/\hbar] C_a^* C_b \sum_{s_j} \psi_a^*(\{\mathbf{x}_j\})\psi_b(\{\mathbf{x}_j\}).$$

The same quantity can also be expressed in terms of the density matrix. Suppose in Eq. (II.5) that $\langle \zeta |$ and $| \eta \rangle$ are chosen to represent basis vectors that are localized in the configuration and spin coordinates, so that we write in their place

$$\langle \zeta | \rightarrow \langle \{\mathbf{x}_j\} |, \tag{II.11}$$

$$| \eta \rangle \rightarrow | \{\mathbf{x}_j'\} \rangle .$$

This expression implies that

$$\langle \{\mathbf{x}_j\} | a \rangle = \psi_a^*(\{\mathbf{x}_j\}), \tag{II.12}$$

$$\langle \{\mathbf{x}_j'\} | b \rangle = \psi_b(\{\mathbf{x}_j'\}),$$

so that

$$\langle \{\mathbf{x}_j\} | \boldsymbol{\rho}^{(N)} | \{\mathbf{x}_j'\} \rangle = \sum_{a,b} \psi_a^*(\{\mathbf{x}_j\}) \langle a | \boldsymbol{\rho}^{(N)} | b \rangle \psi_b(\{\mathbf{x}_j\}). \tag{II.13}$$

In conjunction with Eqs. (II.4) and (II.10), this equation leads immediately to the result

$$P_N(\{\mathbf{r}_j\}, t) = \sum_{\{s_j\}} \langle \{\mathbf{x}_j\} | \boldsymbol{\rho}^{(N)} | \{\mathbf{x}_j\} \rangle. \tag{II.14}$$

That is, the diagonal elements of the density matrix in a configuration-localized basis, summed over nuclear spin variables, provide the $3N$-dimensional configuration space probability function.

Consider next the case of the system of nuclei in a pure quantum state, say a, so that

$$\Psi(\{\mathbf{x}_j\}, t) = C_a \exp(-iE_a t/\hbar)\psi_a(\{\mathbf{x}_j\}). \tag{II.15}$$

In many cases of interest, there exists a nonempty set Θ of nuclear configurations $\{\mathbf{r}_j\}$ for which the potential energy exceeds the eigenvalue:

$$\Phi(\{\mathbf{r}_j\}) - E_a > 0, \qquad (\{\mathbf{r}_j\} \in \Theta). \tag{II.16}$$

In this circumstance, it is well known that the phenomenon of quantum mechanical tunneling permits the probability function $P_N(\{\mathbf{r}_j\})$ to be greater than zero within the classically forbidden set Θ. Rather less well known is the fact that the Schrödinger equation also allows for the existence of a phenomenon that might be regarded as the inverse of tunneling. This is a feature whose possibility was realized first in the early years of the development of quantum mechanics [von Neumann and Wigner, 1929]. It concerns the existence of square-integrable (i.e., spatially localized) eigenfunctions with energy eigenvalues larger than the maximum of the potential energy Φ. In a system occupying a large volume, this discrete energy is located within a dense band of delocalized eigenfunctions and so is often called a "bound state in the continuum." This phenomenon illustrates the possibility for quantum mechanics to confine particle motion at an energy for which only unconfined motion would exist in classical mechanics subject to the same potential. The existence of bound states in the continuum depends sensitively on the geometric details of the potential energy surface defined by $\Phi(\mathbf{r}_1 \ldots \mathbf{r}_N)$. Many theoretical examples of bound states in the continuum have been constructed subsequent to the original von Neumann-Wigner observation [Moses and Tuan, 1959; Weidmann, 1967; Stillinger and Herrick, 1975]. These examples have led in turn to experimental realization of bound states in the continuum by suitably crafted semiconductor composites [Capasso et al., 1992].

B. Thermal Equilibrium

A system of N nuclei left to evolve in isolation from the rest of the physical universe adheres to the time-dependent wave function displayed in Eq. (II.2). The eigenstate probabilities $C_a {}^* C_a$ are constants that are set by initial conditions and would not change with the passage of time. By contrast, the condition of thermal equilibrium requires breaking this isolation, normally by bringing the system into contact (i.e., into weak interaction) with a thermal reservoir whose state is specified by a measurable absolute temperature T. As a result of this coupling, the eigenstate probabilities can slowly change with time, and the nuclear system's energy slowly drifts and fluctuates during this period of weak interaction. The thermal reservoir must be much larger than the system of interest itself, so that its initial temperature is not significantly altered by the interaction. We assume in this section that contact with the reservoir does not affect the nuclear spin degrees of freedom within the system.

Although it is the role of the thermal reservoir to induce perturbations affecting the eigenstate occupation probabilities, one can assume that the system-reservoir interactions are sufficiently weak that final average values of any system property can be evaluated without direct inclusion of that weak reservoir interaction. Thermal equilibrium then can be associated with the long-time-average properties of the nuclear system alone. In particular, the matrix elements shown in Eq. (II.4) for the density matrix reduce to diagonal form under such time averaging:

$$\lim_{t_0 \to \infty} t_0^{-1} \int_0^{t_0} \langle a \,|\, \boldsymbol{\rho}^{(N)} \,|\, b \rangle \, dt = 0 \qquad (a \neq b), \tag{II.17}$$

$$= \langle C_a {}^* C_a \rangle \qquad (a = b).$$

Furthermore, these averaged diagonal elements are not arbitrary, but as a result of thermal equilibration with the reservoir they are driven to adopt the form of a Boltzmann factor for the prevailing reservoir temperature T [Toda et al., 1992, Sections 2.4.1 and 2.4.2]:

$$\langle C_a^* C_a \rangle = [Q_N(\beta)]^{-1} \exp(-\beta E_a). \tag{II.18}$$

Here $\beta = (k_B T)^{-1}$ with k_B standing for Boltzmann's constant, and the normalization factor $Q_N(\beta)$ is the "canonical partition function":

$$Q_N(\beta) = \sum_a \exp(-\beta E_a) \tag{II.19}$$

$$= Tr[\exp(-\beta \mathbf{H})].$$

As a consequence of this reduction to diagonal form, the time-averaged density matrix itself (indicated by an overbar) can simply be expressed in terms of the nuclear Hamiltonian operator $\mathbf{H}$, Eq. (II.1):

$$\bar{\boldsymbol{\rho}}^{(N)} = [Q_N(\beta)]^{-1} \exp(-\beta \mathbf{H}). \tag{II.20}$$

The thermal equilibrium form of the N-particle configurational probability function follows from Eqs. (II.14) and (II.20):

$$\bar{P}_N(\{\mathbf{r}_j\}, \beta) = \sum_{\{s_j\}} \langle \{\mathbf{x}_j\} \mid \bar{\boldsymbol{\rho}}^{(N)} \mid \{\mathbf{x}_j\} \rangle \tag{II.21}$$

$$= [Q_N(\beta)]^{-1} \sum_{\{s_j\}} \langle \{\mathbf{x}_j\} \mid \exp(-\beta \mathbf{H}) \mid \{\mathbf{x}_j\} \rangle.$$

Although the thermally averaged density matrix has just been described as a long-time average for a single dynamical system in weak contact with a large thermal reservoir, an alternative (and more conventional) view is equally valid. Specifically, it is possible in principle to imagine creating a collection, or ensemble, of equivalent systems that have been prepared so as to have indistinguishable macroscopic histories, but without control over the fine details concerning the set of eigenfunction coefficients C_a. If each of the ensemble members after initial preparation is then placed in long-time contact with the same temperature-T reservoir, the time average shown in Eq. (II.17) can be replaced by, and is equivalent to, a subsequent average at fixed time over members of this "canonical ensemble." The canonical partition function expression (II.19), the diagonal form (II.20) for the averaged density matrix $\bar{\boldsymbol{\rho}}^{(N)}$, and the N-particle configurational probability function (II.21) are still applicable for this alternative interpretation, whereby they represent ensemble averages.

Connection with thermodynamics for the N-nucleus system occurs through the Helmholtz free energy F, related directly to the canonical partition function:

$$\beta F = -\ln Q_N. \tag{II.22}$$

Standard thermodynamic identities then in principle provide access to the mean energy $\bar{E}$, pressure p, and entropy S, all as functions of temperature:

$$\bar{E} = \left(\frac{\partial \beta F}{\partial \beta} \right)_V, \tag{II.23}$$

$$p = -\left(\frac{\partial F}{\partial V}\right)_T, \tag{II.24}$$

$$\frac{S}{k_B} = \beta^2\left(\frac{\partial F}{\partial \beta}\right)_V. \tag{II.25}$$

Here, V is the volume to which the system of nuclei is confined. In most cases of interest for which these thermodynamic connections would be invoked, the large-system asymptotic regime would apply, with $\bar{E}$ and S being strictly extensive $O(N)$ quantities, and p a strictly intensive $O(1)$ quantity. However, if interest includes surface and/or phase interface properties, or even the properties of small isolated clusters, corrections to these asymptotic limits must explicitly be considered.

A variant of the canonical ensemble is possible, to represent the physical circumstance that includes fluctuations of the volume V, subject to the condition of constant externally applied pressure p. In this case, one or more walls of the confining container act as a piston, with constant applied force equal to the product of piston-face area and pressure. This variant is called the "isothermal–isobaric ensemble." It treats the variable volume V as a dynamical degree of freedom (determined for example by a piston displacement), nominally on the same footing as the nuclear position coordinates, though typically with a much larger mass M_V. Let a caret (^) signify this isothermal–isobaric option. The correspondingly extended Hamiltonian is [cf. Eq. (II.1)]:

$$\hat{\mathbf{H}} = -\hbar^2\sum_{j=1}^{N}(\nabla_j^2/2M_j) - (\hbar^2/2M_V)\nabla_V^2 + \Phi(\mathbf{r}_1\ldots\mathbf{r}_N) + pV. \tag{II.26}$$

Here the nuclear potential energy Φ has been augmented by the pressure–volume product, the combination of which amounts to a "potential enthalpy" function. The "isothermal–isobaric partition function" is

$$\hat{Q}_N(\beta, p) = Tr[\exp(-\beta\hat{\mathbf{H}})] \tag{II.27}$$

where the indicated trace operation includes V as one of the now $3N + 1$ configurational degrees of freedom. Extension of the N-particle configurational probability function to the isothermal–isobaric case is formally effected by the simple modification of Eq. (II.21), which replaces Q_N, $\mathbf{H}$ by $\hat{Q}_N$, $\hat{\mathbf{H}}$:

$$\hat{P}_N(\{\mathbf{r}_j\}, V; \beta, p) = [\hat{Q}_N(\beta, p)]^{-1}\sum_{\{s_j\}}\langle\{\mathbf{x}_j\}, V\,|\exp(-\beta\hat{\mathbf{H}})\,|\,\{\mathbf{x}_j\}, V\rangle. \tag{II.28}$$

The thermodynamic connection occurs through the Gibbs free energy G, provided directly by the isothermal–isobaric partition function:

$$\beta G = -\ln\hat{Q}_N, \tag{II.29}$$

from which other thermodynamic quantities can be obtained by means of standard thermodynamic manipulations.

C. Open Systems

Whereas the member systems of the canonical ensemble can exchange energy with the thermal reservoir to establish an equilibrium distribution, the number N of nuclei is the same for each member and does not vary. However, physical circumstances may often involve matter exchange with the reservoir as well as energy exchange. Therefore, it is desirable to supplement the statistical mechanical formalism of the closed canonical ensemble with an open "grand canonical ensemble" in which N can fluctuate. To begin, we shall assume that all nuclei are the same species. However, the multispecies generalization is straightforward, and its format is indicated below. With this extended role for the reservoir, it is necessary to supplement the temperature control parameter T with a matter control parameter, the chemical potential μ. The appropriate partition function for an equilibrated open system, the "grand partition function" $\mathcal{Q}(\beta, \mu)$, involves a sum over canonical partition functions $Q_N(\beta)$ for different N with weighting determined by the chemical potential:

$$\mathcal{Q}(\beta, \mu) = \sum_{N=0}^{\infty} \exp(\beta\mu N) Q_N(\beta). \tag{II.30}$$

For notational convenience, $Q_0(\beta)$ has been defined to be unity in this last expression. The individual terms in the grand partition function sum are proportional to the probability that the open system contains exactly N nuclei, so the mean value of N can be obtained as follows:

$$\bar{N}(\beta, \mu) = [\mathcal{Q}(\beta, \mu)]^{-1} \sum_{N=0}^{\infty} N \exp(\beta\mu N) Q_N(\beta) \tag{II.31}$$

$$= (\partial \ln \mathcal{Q} / \partial \beta\mu)_\beta.$$

In the large-system limit, where the effects of boundaries and of interfaces between coexisting phases are negligible relative to bulk properties of the system interior, the grand partition function has a straightforward thermodynamic identification in terms of the pressure–volume product:

$$\ln \mathcal{Q}(\beta, \mu) = \beta p V. \tag{II.32}$$

It should be noted that no straightforward isobaric extension for the grand canonical ensemble exists using the $\hat{Q}_N(\beta, p)$ in place of the $Q_N(\beta)$.

Passing now to the case where several nuclear species are present, separate chemical potentials must be introduced to describe the matter exchanges of the open system with the reservoir. At any instant, suppose that the system contains exactly $1, \ldots, N_v$ nuclei of species v. These species numbers can conveniently be abbreviated as a v-component vector:

$$\mathbf{N} \equiv (N_1, \ldots, N_v), \tag{II.33}$$

and the corresponding set of chemical potentials by the vector

$$\boldsymbol{\mu} = (\mu_1, \ldots, \mu_v). \tag{II.34}$$

Then the multispecies generalization of the grand partition function is the following transparent extension of Eq. (II.30):

$$\mathcal{Q}(\beta, \boldsymbol{\mu}) = \sum_{\mathbf{N}} \exp(\beta\boldsymbol{\mu}\cdot\mathbf{N}) Q_{\mathbf{N}}(\beta), \tag{II.35}$$

where $Q_{\mathbf{N}}(\beta)$ represents a multicomponent canonical partition function, and the summation carries each of the species numbers independently from zero to infinity. The large-system-limit thermodynamic connection to the pressure, Eq. (II.32), remains equally valid for the multispecies generalization.

Many cases of physical and chemical interest involve only intact molecules, as discussed in Chapter I, where each such unit consists of a fixed integer number of nuclei of different kinds. Familiar examples include the molecular species water (H_2O), methane (CH_4), sulfur hexafluoride (SF_6), and glycerol [$HOCH_2CH(OH)CH_2OH$], provided the temperature and pressure are not so high as to produce dissociation or chemical rearrangements of some kind. These intact molecules enter or leave the system to or from the reservoir, not the individual nuclei, so that the stoichiometric (molecular composition) ratios within the system remain fixed. In this circumstance, the chemical potentials of the nuclear species that combine to form the molecules can be treated in combinations corresponding to those molecules. Suppose that the system of interest contained just a single molecular species and that its stable molecule of interest consisted of n_1 nuclei of species 1, …, n_ξ nuclei of species ξ. Then the chemical potential of the molecular species formally would be

$$\mu_{mol} = n_1\mu_1 + \ldots + n_\xi\mu_\xi. \tag{II.36}$$

The grand partition function would reduce to a single sum over the number N of intact molecules, as in Eq. (II.30), with chemical potential μ_{mol}, and the canonical partition functions $Q_N(\beta)$ would be determined by the N-molecule Hamiltonian for that molecular species.

In cases involving ionic substances such as sodium chloride (Na^+Cl^-) or magnesium carbonate ($Mg^{2+}CO_3^{2-}$), the corresponding grand ensembles can formally treat the numbers of separate ionic species as independently variable, each with its own chemical potential parameter. However, a macroscopic system comprising such substances must remain close to electric neutrality because the net Coulomb potential would otherwise become prohibitively large and positive. Consequently, the number variations permitted by reservoir coupling are statistically strongly correlated. So far as macroscopic bulk properties are concerned, it then becomes appropriate in the grand partition function to treat the positive (cationic) and negative (anionic) species as stoichiometrically fully coupled, with a single chemical potential analogous to that for a single intact polyatomic molecular species, Eq. (II.36).

D. Classical Limit

By taking any N-particle system to sufficiently high temperature, under constant volume conditions, most properties identifiable as intrinsically quantum mechanical disappear. As a result, classical statistical mechanics becomes an appropriate (and simpler) description. Quantum mechanical attributes that may be significant in the low-temperature regime but not at high temperature include Bose-Einstein and Fermi-Dirac statistics distinctions arising from wave function symmetry constraints, as well as wave-mechanical delocalization and diffraction effects. The next objective is to document how this reduction from quantum to classical statistics arises in the thermal equilibrium partition functions and to show what role the potential energy function Φ plays in that transformation.

The canonical partition function $Q_N(\beta)$, Eq. (II.19), serves to illustrate this reduction. In the interests of conceptual transparency, the analysis is first presented for a single particle species, where all N particles have identical spin states. As a result, spin variables $s_1 \ldots s_N$ can be temporarily suppressed, since they do not occur explicitly in the nuclear Hamiltonian operator $\mathbf{H}$. The more general case, with distinct species possessing distinct spin states, is discussed later.

The trace required by Eq. (II.19) is independent of the basis set used for its evaluation, provided that the basis set is complete. For present purposes, it is useful to employ a basis of properly symmetrized plane wave functions. In the interests of notational brevity, let $\mathbf{r} \equiv (\mathbf{r}_1 \ldots \mathbf{r}_N)$ stand for the full set of spatial coordinates for the N particles. The basis functions to be used are the following:

$$\varphi^{(N)}(\mathbf{r}, \mathbf{p}) = (N! V^N)^{-1/2} \sum_P (\pm 1)^{|P|} \exp\left[(i/\hbar) \sum_{j=1}^{N} \mathbf{r}_j \cdot P\mathbf{p}_j\right]. \qquad (II.37)$$

Here P is a permutation operator, and as indicated it is summed over all $N!$ particle permutations. The parity indicator of any one such permutation has been designated by $|P|$ and is zero for even permutations, unity for odd permutations. The upper sign in Eq. (II.37) applies to bosons, the lower sign to fermions, and as stated above for both of these cases, the N particles have identical (parallel) spins. The $\mathbf{p}_j$ appearing in the plane waves are particle momenta, denoted collectively by $\mathbf{p}$. Assuming that the system is subject to periodic boundary conditions and is contained in a cubical box with volume V, the allowed values of the momenta are

$$\mathbf{p}_j \equiv (hn_x/V^{1/3}, hn_y/V^{1/3}, hn_z/V^{1/3}), \qquad (II.38)$$

in which $h = 2\pi\hbar$ is Planck's constant, and n_x, n_y, n_z are positive or negative integers, or zero. For fermions, to avoid identical vanishing of the right side of Eq. (II.37), it is necessary for all $\mathbf{p}_j$ in the basis function $\varphi^{(N)}(\mathbf{r}, \mathbf{p})$ to be distinct. Using this basis for either bosons or fermions, we have

$$Q_N(\beta) = \sum_{\mathbf{p}} \int \varphi^{(N)*}(\mathbf{r}, \mathbf{p}) \exp(-\beta\mathbf{H}) \varphi^{(N)}(\mathbf{r}, \mathbf{p}) d^{3N}\mathbf{r} \qquad (II.39)$$

$$\rightarrow (N! h^{3N})^{-1} V^N \int d^{3N}\mathbf{p} \int d^{3N}\mathbf{r} \varphi^{(N)*}(\mathbf{r}, \mathbf{p}) \exp(-\beta\mathbf{H}) \varphi^{(N)}(\mathbf{r}, \mathbf{p}),$$

where for the large-system case the sum over momenta has been replaced by an integral, with $N!$ as a denominator factor to compensate for redundancy arising from the $\mathbf{p}$ integration, as appropriate for the large-system limit [Kirkwood, 1933]. The particle position integrals span the interior of the confinement volume V.

Both basis function occurrences in the latter form of Eq. (II.39) contain permutation operator sums, as exhibited in Eq. (II.37). However, one of these summations can be regarded as unnecessary because integration yields results dependent only on the difference between the two permutations involved. Consequently it is proper to rewrite the canonical partition function trace in the following simplified version:

$$Q_N(\beta) = (V^N/N!)^{1/2} h^{-3N} \int d^{3N}\mathbf{p} \int d^{3N}\mathbf{r} \varphi^{(N)*}(\mathbf{r}, \mathbf{p}) \exp(-\beta\mathbf{H}) \exp(i\mathbf{p}\cdot\mathbf{r}/\hbar). \qquad (II.40)$$

The exponential operator appearing in the integrand has the formal expansion

$$\exp(-\beta\mathbf{H}) = \sum_{l=0}^{\infty} [(-\beta)^l/l!] \mathbf{H}^l. \qquad (II.41)$$

One can easily verify the following identity for any differentiable function $\Lambda(\mathbf{r}, \mathbf{p})$:

$$\mathbf{H} \exp(i\mathbf{r}\cdot\mathbf{p}/\hbar) \Lambda(\mathbf{r}, \mathbf{p}) = \exp(i\mathbf{r}\cdot\mathbf{p}/\hbar) \mathbf{H}' \Lambda(\mathbf{r}, \mathbf{p}), \qquad (II.42)$$

where $\mathbf{H}'$ is a modified form of the Hamiltonian operator $\mathbf{H}$ [Stillinger and Kirkwood, 1960]:

$$\mathbf{H}' = (2m)^{-1} \sum_{j=1}^{N} (\mathbf{p}_j - i\hbar \nabla_{\mathbf{r}_j})^2 + \Phi(\mathbf{r}_1 \ldots \mathbf{r}_N), \tag{II.43}$$

and m is the common mass of the particles. Using this identity repeatedly, it is possible to move the last factor in the integrand of Eq. (II.40) from right to left, through each term of the operator power series (II.41), to yield

$$Q_N(\beta) = (N! h^{3N})^{-1} \int d^{3N}\mathbf{p} \int d^{3N}\mathbf{r} \, \sigma_{\pm}^{(N)}(\mathbf{r}, \mathbf{p}) F^{(N)}(\mathbf{r}, \mathbf{p}, \beta), \tag{II.44}$$

where

$$\sigma_{\pm}^{(N)}(\mathbf{r}, \mathbf{p}) = \sum_{P} (\pm 1)^{|P|} \exp\left[(i/\hbar) \sum_{j=1}^{N} \mathbf{r}_j \cdot (\mathbf{p}_j - P\mathbf{p}_j) \right], \tag{II.45}$$

and

$$F^{(N)}(\mathbf{r}, \mathbf{p}, \beta) = \exp(-i\mathbf{p} \cdot \mathbf{r}/\hbar) \exp(-\beta \mathbf{H}) \exp(i\mathbf{p} \cdot \mathbf{r}/\hbar) \tag{II.46}$$

$$= \exp(-\beta \mathbf{H}') \cdot 1.$$

Planck's constant h (equivalently $\hbar$) is the only fundamental physical constant introduced by Schrödinger's wave equation that is not explicitly present in classical mechanics. Thus, Planck's constant may be viewed as controlling the presence of nonrelativistic quantum phenomena. With this in mind, we now proceed to examine leading terms in a formal series expansion for $F^{(N)}(\mathbf{r}, \mathbf{p}, \beta)$ in ascending powers of $\hbar$. Starting with the operator expansion of definition (II.46) for $F^{(N)}$, one has

$$F^{(N)}(\mathbf{r}, \mathbf{p}, \beta) = \{ 1 - \beta \mathbf{H}' + (\beta \mathbf{H}')^2/2 - (\beta \mathbf{H}')^3/6 + \ldots \} \cdot 1$$

$$= \{ 1 - \beta[(\mathbf{p} - i\hbar\nabla_{\mathbf{r}})^2/2m + \Phi(\mathbf{r})] + (\beta^2/2)[(\mathbf{p} - i\hbar\nabla_{\mathbf{r}})^2 + \Phi(\mathbf{r})]^2$$

$$- (\beta^3/6)[(\mathbf{p} - i\hbar\nabla_{\mathbf{r}})/(2m) + \Phi(\mathbf{r})]^3 + \ldots \} \cdot 1$$

$$= 1 - \beta[(p^2/2m) + \Phi(\mathbf{r})] + O(\beta^2)$$

$$= \exp\{ -\beta[p^2/2m + \Phi(\mathbf{r})] + O(\beta^2) \}. \tag{II.47}$$

That is, $F^{(N)}$ asymptotically reduces to just the Boltzmann factor for the classical Hamiltonian in the high-temperature ($\beta \to 0$) limit. For any temperature, $F^{(N)}$ obeys a Bloch equation [Bloch, 1932], as inspection of Eq. (II.46) immediately verifies:

$$\frac{\partial F^{(N)}}{\partial \beta} = -\mathbf{H}' F^{(N)}(\mathbf{r}, \mathbf{p}, \beta). \tag{II.48}$$

One should note in passing that this is formally equivalent to the Schrödinger equation for the modified Hamiltonian operator $\mathbf{H}'$ but where β plays the role of imaginary time $it/\hbar$.

One may assume that quantum corrections to the limiting classical form of $F^{(N)}$ at high temperature have the form of a power series in $\hbar$:

$$F^{(N)}(\mathbf{r}, \mathbf{p}, \beta) = \exp\{ -\beta[p^2/2m + \Phi(\mathbf{r})] + \sum_{n=1}^{\infty} C_n(\mathbf{r}, \mathbf{p}, \beta)\hbar^n \}. \tag{II.49}$$

In order to have the proper high-temperature behavior, the C_n must satisfy the conditions ($n \geq 1$):

$$\lim_{\beta \to 0} C_n(\mathbf{r}, \mathbf{p}, \beta) = 0. \tag{II.50}$$

Upon substituting expression (II.49) into the Bloch equation (II.48), and then equating coefficients of the same power of $\hbar$, one finds the following for the first two orders:

$$\frac{\partial C_1}{\partial \beta} = -(i\beta/m)\mathbf{p} \cdot \nabla_\mathbf{r}\Phi, \tag{II.51}$$

$$\frac{\partial C_2}{\partial \beta} = (\beta^2/2m^2)\mathbf{pp} : \nabla_\mathbf{r}\nabla_\mathbf{r}\Phi + (\beta^2/2m)(\nabla_\mathbf{r}\Phi)^2 - (\beta/2m)\nabla^2\Phi,$$

These differential equations can be integrated in turn, subject to the boundary values (II.50), to yield

$$C_1(\mathbf{r}, \mathbf{p}, \beta) = -(i\beta^2/2m)\mathbf{p} \cdot \nabla_\mathbf{r}\Phi, \tag{II.52}$$

$$C_2(\mathbf{r}, \mathbf{p}, \beta) = (\beta^3/6m^2)\mathbf{pp} : \nabla_\mathbf{r}\nabla_\mathbf{r}\Phi + (\beta^3/6m)(\nabla_\mathbf{r}\Phi)^2 - (\beta^2/4m)\nabla^2\Phi.$$

Higher order C_n's can be sequentially determined in the same way, but they contain higher spatial derivatives of the potential function, and except for special circumstances (see Box II.1), rapidly become unwieldy.

Returning to the general canonical partition function in Eq. (II.44), it is also necessary to analyze the role of the exchange factor $\sigma_\pm^{(N)}$, Eq. (II.45), in passing to the classical limit. This factor contains $N!$ terms, and with the exception of the one term corresponding to the identity permutation, all are dependent on Planck's constant. To see this in detail, consider for the moment any one permutation P contributing to the sum in $\sigma_\pm^{(N)}$. This P can be resolved uniquely into sets of n-particle cyclic interchanges ($1 \leq n \leq N$), each one of which involves a succession of exchanges of indexed particles, thus:

$$j_1 \to j_2 \to j_3 \to \cdots \to j_n \to j_1.$$

The trivial cases with $n = 1$ are identities ("self-exchanges"), $j_1 \to j_1$. If the chosen P consists of l_n cycles of length n, then $|P|$, the parity of P, is 0 or 1 according to whether the following integer is even or odd, respectively:

$$\sum_{n=1}^{N}(n-1)l_n \equiv N - \sum_{n=1}^{N} l_n. \tag{II.53}$$

Aside from algebraic sign, each of the $N!$ terms in $\sigma_\pm^{(N)}$ consists of a product of factors, one for each exchange cycle, of the type

$$\exp\left[(i/\hbar)\sum_{u=1}^{n} \mathbf{r}_{j_u} \cdot (\mathbf{p}_{j_u} - \mathbf{p}_{j_{u+1}})\right], \tag{II.54}$$

in which the index $j_{n+1} \equiv j_1$. Unless the exchange cycle involved is the identity, $n = 1$, factors of the type in expression (II.54) oscillate arbitrarily rapidly in the classical $\hbar \to 0$ limit, with variations in either particle positions or momenta. Consequently, in that limit their contribution to the

BOX II.1. Example of a Particle in a Harmonic Well

By way of elementary illustration, consider the simple case for which a single particle is confined by a symmetric harmonic well in three dimensions. The harmonic well is centered at the coordinate origin, and its force constant is K in all directions. The potential energy function for this particle is

$$\Phi(\mathbf{r}) = Kr^2/2$$

$$= (K/2)(x^2 + y^2 + z^2).$$

It can be verified by direct substitution for this case that the expression

$$F^{(1)}(\mathbf{r}, \mathbf{p}, \beta) = \exp[3W(\beta) + X(\beta)p^2 + Y(\beta)\mathbf{p}\cdot\mathbf{r} + Z(\beta)r^2]$$

satisfies both the Bloch equation (II.48) and the proper classical limit, provided that one sets

$$W(\beta) = -\frac{1}{2}\ln\cosh\left[\left(\frac{K}{m}\right)^{1/2}\beta\hbar\right],$$

$$X(\beta) = -\frac{1}{2(mK)^{1/2}\hbar}\tanh\left[\left(\frac{K}{m}\right)^{1/2}\beta\hbar\right],$$

$$Y(\beta) = \frac{i}{\hbar}\left\{\text{sech}\left[\left(\frac{K}{m}\right)^{1/2}\beta\hbar\right]-1\right\},$$

$$Z(\beta) = -\frac{(mK)^{1/2}}{2\hbar}\tanh\left[\left(\frac{K}{m}\right)^{1/2}\beta\hbar\right].$$

Each of these four functions has a convergent power series in $\hbar$, so that for this simple illustrative example it is possible to exhibit the coefficient functions $C_n(\mathbf{r}, \mathbf{p}, \beta)$ for the expansion shown in Eq. (II.49) in terms of Bernoulli and Euler numbers that arise from the hyperbolic functions [Abramowitz and Stegun, 1964].

For this case of a single particle inhabiting a spherically symmetric harmonic well in three dimensions, the partition function expression (II.44) can be immediately integrated, with the result

$$Q_1(\beta) = \left[\frac{\exp(-\beta\hbar\omega/2)}{1 - \exp(-\beta\hbar\omega)}\right]^3,$$

where

$$\omega = (K/m)^{1/2}$$

is the angular frequency of harmonic oscillation.

The result for $F^{(1)}(\mathbf{r}, \mathbf{p}, \beta)$ can be directly extended to the case of an anisotropic harmonic well, where each principle direction has its own force constant. The correspondingly generalized $F^{(1)}(\mathbf{r}, \mathbf{p}, \beta)$ then consists of a product of terms for each of those principle directions, with coefficients $W(\beta), X(\beta), Y(\beta), Z(\beta)$ containing the respective force constants. Analogously, for an *N*-particle ideal gas confined by a harmonic potential, the multidimensional solution $F^{(N)}(\mathbf{r}, \mathbf{p}, \beta)$ to the Bloch equation (II.48) consists of a product of the same kind of factors for each independent degree of freedom.

With a bit of extra effort, the partition function expression Eq. (II.44) can be evaluated for two identical ideal gas particles confined to an isotropic harmonic well. In that case, a pair of permutation operations (identity and pair exchange) contribute to the result. One obtains the following expression:

$$Q_2(\beta) = \frac{1}{2}\left\{\left[\frac{\exp(-\beta\hbar\omega/2)}{1-\exp(-\beta\hbar\omega)}\right]^2 \pm \frac{\exp(-\beta\hbar\omega)}{1-\exp(-2\beta\hbar\omega)}\right\}^3,$$

in which the upper sign refers to bosons, the lower sign to fermions.

canonical partition function integral, Eq. (II.44), is negligible. The only significant contributor is the term with all unit-length exchange cycles, i.e., the identity permutation.

Combining this last fact with the previous classical limit for $F^{(N)}$, we obtain the following classical limit for the canonical partition function:

$$Q_N(\beta) = (N!h^{3N})^{-1}\!\int d^{3N}\mathbf{r}\!\int d^{3N}\mathbf{p} \exp\{-\beta[(p^2/2m) + \Phi(\mathbf{r})]\} \qquad \text{(classical limit).} \qquad \text{(II.55)}$$

That is, the classical canonical partition function entails the multidimensional position and momentum integral of the Boltzmann factor for the classical Hamiltonian, and the distinction between Bose-Einstein and Fermi-Dirac statistics has vanished. Because the kinetic and potential energy terms are separated in the integrand exponent of this last expression, momentum integrations can be performed explicitly to yield

$$Q_N(\beta) = (N!\lambda_T^{3N})^{-1}\!\int d^{3N}\mathbf{r} \exp[-\beta\Phi(\mathbf{r})], \qquad \text{(II.56)}$$

in which

$$\lambda_T = \hbar(2\pi\beta/m)^{1/2} \equiv h/(2\pi m k_B T)^{1/2} \qquad \text{(II.57)}$$

is a length parameter conventionally called the mean thermal de Broglie wavelength [Hill, 1956; Hansen and McDonald, 1986]. Notice that although Planck's constant h is an intrinsic indicator of the quantum regime, it nevertheless remains via λ_T in a normalizing factor for the classical partition function Q_N.

In the case of the isothermal–isobaric partition functions $\hat{Q}_N$ in their classical limit, the extra degree of freedom associated with volume variations also contributes a mean thermal de Broglie wavelength to the denominator. However, this wavelength is normally a much smaller length than that produced by each of the particle momentum component integrations because of the much larger mass $M_v \gg m$ involved. Inclusion of this additional denominator factor renders each $\hat{Q}_N$ dimensionless, as are the instances of Q_N.

Recall that the grand partition function $\mathfrak{Q}$, Eq. (II.30), consists of a chemical-potential-weighted sum of canonical partition functions Q_N. The classical limit for the former simply requires inserting the classical limit of the latter, Eq. (II.56), for each term in that sum.

Essentially the same considerations just applied to the canonical partition function can also be brought to bear on the N-particle distribution function $\bar{P}_N$ for the canonical ensemble. Upon using the symmetrized plane-wave basis functions $\varphi^{(N)}(\mathbf{r}, \mathbf{p})$, Eq. (II.37), to evaluate the diagonal matrix element in Eq. (II.21), one has

$$\bar{P}_N(\mathbf{r}, \beta) = [Q_N(\beta)N!h^{3N}]^{-1}\int d^{3N}\mathbf{p}\,\sigma_{\pm}^{(N)}(\mathbf{r}, \mathbf{p})F^{(N)}(\mathbf{r}, \mathbf{p}, \beta), \qquad (\text{II.58})$$

where the integrand factors have been defined in Eqs. (II.45) and (II.46). The classical limit for this probability function requires use of the classical limits for $Q_N(\beta)$ Eq. (II.56), and for $F^{(N)}$ from Eq. (II.49) with $\hbar = 0$, as well as retention of only the identity permutation in $\sigma_{\pm}^{(N)}$. The momentum integrals can then be eliminated to leave the following elementary classical probability:

$$\bar{P}_N(\mathbf{r}, \beta) = \exp[-\beta\Phi(\mathbf{r})]/\int d^{3N}\mathbf{r}\exp[-\beta\Phi(\mathbf{r})] \qquad \text{(classical limit).} \qquad (\text{II.59})$$

The corresponding classical probability function for the isothermal–isobaric equilibrium ensemble is just an obvious extension of this last form:

$$\bar{P}_N(\mathbf{r}, V; \beta, p) = \exp\{-\beta[\Phi(\mathbf{r}) + pV]\}/\int dV\int d^{3N}\mathbf{r}\exp\{-\beta[\Phi(\mathbf{r}) + pV]\}$$
$$\text{(classical limit).} \qquad (\text{II.60})$$

In the event that the system contains a mixture of v particle species with distinct spins, the symmetrized basis functions $\varphi^{(N)}$ that were introduced in Eq. (II.37) need to be generalized. If all members of each of those species possess identical (i.e., parallel) spins, the necessary extension of the form in Eq. (II.37) must include separate permutation operators $P\gamma$ for every species γ:

$$\varphi^{(N)}(\{\mathbf{r}\}, \{\mathbf{p}\}) = \left[V^{N_\gamma}\prod_{\gamma=1}^{v}N_\gamma!\right]^{-1/2}\sum_{P_1\ldots P_v}\left[\prod_{\gamma=1}^{v}(\pm 1)^{|P_\gamma|}\right]\exp\left[(i/\hbar)\sum_{\gamma=1}^{v}\sum_{j=1}^{N_\gamma}\mathbf{r}_{\gamma j}\cdot P_\gamma\mathbf{p}_{\gamma j}\right]. \qquad (\text{II.61})$$

If each of the mixture species includes particles with nonaligned spins, however, a more elaborate representation is required, with intermixed configurational and spin coordinates so as to satisfy symmetry requirements. Such technical details for the general case are not required in the following development but are covered in special cases as they arise in subsequent chapters. However, the classical-limit partition function expression (II.55) again appears, except for an additional nuclear-spin-degeneracy multiplier Δ_{spin} that is independent of temperature and volume.

The passage from the quantum regime to the classical limit in a finite volume V involves a change from a set of discrete energy eigenvalues E_a for the Schrödinger Hamiltonian operator $\mathbf{H}$ to a continuum of energies that can be exhibited by the classical Hamiltonian. In that classical limit, it is then possible to define a "microcanonical ensemble" of systems, all constrained to possess the same energy E but subject to that constraint uniformly distributed over the multidimensional space of particle configurations and momenta [ter Haar, 1954, p. 123]; see Box II.2. This alternative ensemble, its partition function and various statistical attributes, plays only a subsidiary role in the developments of subsequent chapters.

E. Reduced Density Matrices and Distribution Functions

Unless the number of particles N is very small, the density matrix elements $\langle\{\mathbf{x}_j\}|\boldsymbol{\rho}^{(N)}|\{\mathbf{x}_j'\}\rangle$ would contain an overwhelming amount of detailed information. This is certainly obvious when a many-particle system under consideration is macroscopic, i.e., when N is comparable to Avogadro's

BOX II.2. Microcanonical Ensemble

The classical limit for the canonical partition function, shown in Eq. (II.55), can formally be written in terms of a Laplace transform with respect to the continuous energy E:

$$Q_N(\beta) = (N!h^{3N})^{-1} \int_{-\infty}^{+\infty} \exp(-\beta E)\Gamma(E)dE.$$

Here the notation $\Gamma(E)$ stands for hyperarea measure of the classically available constant-E hypersurface in the $6N$-dimensional $\mathbf{r}, \mathbf{p}$ phase space. This measure vanishes for $E \leq \Phi_{min}$, the global minimum of the N-particle potential energy function, but remains positive for all larger E.

Because E is a constant of the motion for isolated-system classical dynamics, a phase point that lies on the $\Gamma(E)$ hypersurface at time t remains on that hypersurface thereafter. Dynamical systems whose phase-space trajectories in the long-time limit visit the entirety of $\Gamma(E)$ are called "ergodic" [ter Haar, 1954, Section 5.7]. However, it is important to keep in mind that this property is not attained in a variety of circumstances. At very low E values, $\Gamma(E)$ may become disconnected, so the N-particle system may then be trapped around a local potential energy minimum and thus unable to surmount local barriers to explore the full set of configurations with $\Phi(\mathbf{r}) < E$. In other cases, the dynamical system may possess other constants of motion besides E, one example of which would be the center of mass momentum when periodic boundary conditions permit free translation of the entire system. When these circumstances apply, the non-ergodic system's dynamical trajectory is confined to a proper subset of $\Gamma(E)$.

In the long-time limit, an ergodic system covers its energy hypersurface uniformly. That is, if its phase space position $\mathbf{r}, \mathbf{p}$ is recorded repeatedly at a fixed time interval over an arbitrarily long observation period, the corresponding occupancy probability would be found to approach a constant over hypersurface $\Gamma(E)$. This uniform distribution of phase space positions is time independent and could equally well be viewed as the distribution of an ensemble of distinct systems viewed at a fixed time. Such a collection of equal energy systems is conventionally called a "microcanonical ensemble." It can be endowed with a "microcanonical partition function":

$$Q^{(micro)}(E) = (N!h^{3N})^{-1}\Gamma(E)\delta E,$$

where δE is an arbitrary energy of order unity included to make the result dimensionless in conformity with other ensemble distribution functions.

In case the system of interest is not dynamically ergodic, a collection of initial conditions can in principle be selected so that the hypersurface is again uniformly populated by a time-independent probability distribution.

The Laplace integral in the classical canonical partition function can be evaluated asymptotically for large N by identifying the single large integrand maximum at $E^*(\beta)$. This maximum is located by the solution to

$$[d \ln \Gamma(E)/dE]_{E=E^*} = \beta.$$

Consequently, the logarithm of the classical canonical partition function has the asymptotic value

$$\ln Q_N(\beta) = -\beta E^*(\beta) + \ln \Gamma[E^*(\beta)] + O(N/\ln N),$$

where the last term on the right is negligible in the large system limit, compared to the $O(N)$ terms that precede it. In that limit, $E^*(\beta)$ can be identified as the equilibrium energy in either the canonical or microcanonical ensemble. Then because Eq. (II.22) allows the left side of this last equation to be identified in terms of Helmholtz free energy F, one immediately connects the hypersurface measure to the thermodynamic entropy at inverse temperature β:

$$\ln \Gamma[E^*(\beta)] \sim S(\beta)/k_B,$$

to leading order in N.

number $N_A \cong 6.022 \times 10^{23}$. But it is even so when only a few dozen to a few hundred interacting particles are present, as has often been the case in computer simulations of modest scope. Consequently, a contracted description is desirable, and for that purpose it is convenient to introduce a set of reduced density matrices. For the remainder of this section, we shall leave aside the isobaric variant of the canonical ensemble and just consider the fixed-volume canonical ensemble and the grand ensemble to which its Q_N contribute.

Assuming that interest focuses on configurational degrees of freedom, not spin, the reduced density matrices of orders $1 \leq n < N$ in a system comprising a single species, at time t, are defined by means of the following partial traces:

$$\rho_{n,N}(\{\mathbf{r}_i\} \mid \{\mathbf{r}_i'\}, t) = [N!/(N-n)!] \sum_{\{s_j\}} \int d\mathbf{r}_{n+1} \ldots \int d\mathbf{r}_N \langle \{\mathbf{x}_j\} \mid \boldsymbol{\rho}^{(N)}(t) \mid \{\mathbf{x}_j'\} \rangle^{**}, \qquad (II.62)$$

where the "**" superscript appended to the matrix element serving as the integrand implies setting all $s_j = s_j'$, as well as setting $\mathbf{r}_l = \mathbf{r}_l'$ for $n + 1 \leq l \leq N$. The combinatorial factor in the right side of Eq. (II.62) has been chosen to confer a simple and useful interpretation on diagonal elements of the reduced density matrices. Specifically, the probability that differential volume elements $d\mathbf{r}_1 \ldots d\mathbf{r}_n$ simultaneously contain any n of the N particles at time t is just

$$\rho_{n,N}(\{\mathbf{r}_i\} \mid \{\mathbf{r}_i\}, t) d\mathbf{r}_1 \ldots d\mathbf{r}_n. \qquad (II.63)$$

so that the diagonal portions of the reduced density matrices are n-particle configurational distribution functions.

Introduction of the reduced density matrices simplifies expressions for averages of many operators. As an example, suppose that spin-independent operator $\mathbf{O}$ for property O consists of a sum of symmetric n-body operators:

$$\mathbf{O} = \sum_{i_1 \ldots i_n} \mathbf{O}^{(n)}(i_1 \ldots i_n), \qquad (II.64)$$

with one term in the right member for each of the $N!/(N-n)!n!$ particle n-tuples within the system. Then the t-dependent expectation value of property O can be written in the following way:

$$\langle O(t) \rangle = (n!)^{-1} \int d\{\mathbf{r}_i\} \int d\{\mathbf{r}_i'\} \langle \{\mathbf{r}_i\} \mid \mathbf{O}^{(n)} \mid \{\mathbf{r}_i'\} \rangle \rho_{n,N}(\{\mathbf{r}_i'\} \mid \{\mathbf{r}_i\}, t). \qquad (II.65)$$

In the event that $\mathbf{O}^{(n)}$ is diagonal in the position representation [i.e., a function $O^{(n)}(\{\mathbf{r}_i\})$], this last expression undergoes the obvious simplification

$$\langle O(t) \rangle = (n!)^{-1} \int d\{\mathbf{r}_i\} O^{(n)}(\{\mathbf{r}_i\}) \rho_{n,N}(\{\mathbf{r}_i\} \mid \{\mathbf{r}_i\}, t). \qquad (II.66)$$

If the N-particle system exists in a pure quantum state or is represented by an equilibrium ensemble, the reduced density matrices are independent of time t, in which case that variable may be dropped from the notation. Even with this simplification, it is important to realize that the $\rho_{n,N}$ are sensitive to the boundary conditions that apply to the N-particle system. In particular, this is true for the singlet density matrix $\rho_{1,N}(\mathbf{r}_1|\mathbf{r}_1')$. If a canonical ensemble is under consideration for which periodic boundary conditions are present, and with no externally applied potential to inhibit free translation across those periodic boundaries, then $\rho_{1,N}$ remains unchanged under the replacement $\mathbf{r}_1, \mathbf{r}_1' \to \mathbf{r}_1 + \mathbf{u}, \mathbf{r}_1' + \mathbf{u}$, where $\mathbf{u}$ is an arbitrary displacement. Furthermore, the diagonal elements of this singlet density matrix are then equal to the overall particle number density:

$$\rho_{1,N}(\mathbf{r}_1|\mathbf{r}_1) = N/V \equiv \rho$$
(periodic boundary conditions, no external potential). (II.67)

This translational invariance property with periodic boundary conditions is true whether crystal, liquid, or vapor phases, or even their coexisting combinations, are present. The pair ($n = 2$) and higher order ($n \geq 3$) reduced density matrices under the same periodic boundary conditions depend configurationally only on relative particle positions, which could be taken for example as displacements relative to the "mean" location $\bar{\mathbf{r}}_1 = (\mathbf{r}_1 + \mathbf{r}_1')/2$ of the first particle:

$$\rho_{n,N}(\{\mathbf{r}_j\}|\{\mathbf{r}_j'\}) \to \rho_{n,N}(\{\mathbf{r}_j - \bar{\mathbf{r}}_1\}|\{\mathbf{r}_j' - \bar{\mathbf{r}}_1\})$$
(periodic boundary conditions, no external potential). (II.68)

The presence of impenetrable confining walls destroys translational invariance. The singlet density for a fluid phase deviates from the overall number density ρ most strongly in the immediate vicinity of such walls. Details of those deviations are affected by whether or not the particle-wall interactions are purely repulsive or involve adsorbing attractions, a distinction that can affect whether or not a fluid phase wets the wall surfaces. Far from the wall in a large system, a fluid-phase singlet density becomes substantially position independent. But because of the identity for a volume-V closed system

$$\int_V d\mathbf{r}_1 \rho_{1,N}(\mathbf{r}_1|\mathbf{r}_1) = N,$$
(II.69)

a wall-region accumulation or depletion inevitably influences the magnitude of that singlet density within the interior of the system. Because the presence of walls breaks the translational symmetry of periodic boundary conditions, the higher order particle distribution functions for a fluid phase would no longer exhibit everywhere the configurational reduction shown in Eq. (II.68).

If the prevailing equilibrium conditions cause the system to exist in a single crystalline phase, then the wall interactions could essentially clamp that crystal both translationally and orientationally. If that were the case, then (aside from localized wall-induced perturbations), the singlet particle density $\rho_{1,N}(\mathbf{r}_1|\mathbf{r}_1)$ would become a periodic function of position $\mathbf{r}_1$, the details of which depend on crystal symmetry (cf. Chapter IV). However, this ideal situation requires near-perfect fitting of the crystal surfaces against the confining walls in a single orientation. The full set of possible wall interactions can violate this requirement to varying degrees, and so the amplitude of periodic oscillation displayed by the singlet density for a crystal can vary in magnitude anywhere between zero and the ideal-case maximum.

Let V_0 be a spatially fixed volume entirely contained within the interior of a volume-V equilibrium system represented by a canonical ensemble. The number of particles N_0 contained within V_0 fluctuates with the passage of time, and from one ensemble member to another, but its average is determined by the singlet density function within V_0:

$$\langle N_0 \rangle = \int_{V_0} d\mathbf{r}_1 \rho_{1,N}(\mathbf{r}_1 | \mathbf{r}_1). \tag{II.70}$$

Likewise, the number of particle pairs inside V_0, $N_0(N_0 - 1)/2$, also fluctuates. But by virtue of the fact that the pair density function $\rho_{2,N}$ has been defined to measure the simultaneous occurrence frequency of particle pairs at specified locations, we also have

$$\langle N_0(N_0 - 1)/2 \rangle = (2!)^{-1} \int_{V_0} d\mathbf{r}_1 \int_{V_0} d\mathbf{r}_2 \rho_{2,N}(\mathbf{r}_1, \mathbf{r}_2 | \mathbf{r}_1, \mathbf{r}_2). \tag{II.71}$$

The numerical factor preceding the integral is present to avoid double-counting of particle pairs. By combining results (II.70) and (II.71), one obtains a conventional second-moment measure of the particle number fluctuation within V_0:

$$\langle (N_0 - \langle N_0 \rangle)^2 \rangle = \int_{V_0} d\mathbf{r}_1 \int_{V_0} d\mathbf{r}_2 [\rho_{2,N}(\mathbf{r}_1, \mathbf{r}_2 | \mathbf{r}_1, \mathbf{r}_2) - \rho_{1,N}(\mathbf{r}_1 | \mathbf{r}_1) \rho_{1,N}(\mathbf{r}_2 | \mathbf{r}_2)]. \tag{II.72}$$

Reduced density matrices and their spatial distribution functions in the grand ensemble are equal to weighted averages of those for the contributing canonical ensembles, where the weighting represents the occurrence probabilities for presence of exactly N particles in the open system. These occurrence probabilities are proportional to the separate terms in the grand partition function $\mathcal{Q}$, Eq. (II.30), and after normalization the corresponding probabilities (for a single-species system) become

$$\exp(\beta\mu N) Q_N(\beta)/\mathcal{Q}(\beta, \mu). \tag{II.73}$$

Consequently, the grand ensemble reduced density matrices are defined as follows:

$$\rho_n(\mathbf{r}_1 \dots \mathbf{r}_n | \mathbf{r}_1' \dots \mathbf{r}_n') = \sum_{N=n}^{\infty} \exp(\beta\mu N) Q_N(\beta) \rho_{n,N}(\mathbf{r}_1 \dots \mathbf{r}_n | \mathbf{r}_1' \dots \mathbf{r}_n')/\mathcal{Q}(\beta, \mu), \tag{II.74}$$

where it is to be understood implicitly that the result depends both on β and μ.

F. Thermodynamic Properties

As a supplement to the basic thermodynamic identities in Eqs. (II.23)–(II.25), the reduced density matrices introduced in Section II.E can be used to provide formal expressions for several thermodynamic properties in the event that equilibrium conditions apply. An obvious example is the mean energy $\bar{E}$, consisting of a sum of kinetic (K) and potential (Φ) portions:

$$\bar{E} = \langle K \rangle + \langle \Phi \rangle. \tag{II.75}$$

As Eq. (II.1) indicates, the kinetic energy operator involves a sum of single-particle operators, and in the case of a single-species system with particle mass m in the position representation, it is

$$\mathbf{K} = -(\hbar^2/2m) \sum_{j=1}^{N} \nabla_j^2. \tag{II.76}$$

Furthermore, suppose that Φ for the same single-species system includes only isotropic particle–pair interactions:

$$\Phi = \sum_{i=1}^{N-1} \sum_{j=i+1}^{N} v_2(r_{ij}), \qquad (\text{II.77})$$

$$r_{ij} = |\mathbf{r}_{ij}| = |\mathbf{r}_j - \mathbf{r}_i|.$$

Then in the manner indicated in Eq. (II.65) for a closed system, these two operator forms allow $\langle K \rangle$ and $\langle \Phi \rangle$, respectively, to be expressed in terms of the singlet and pair reduced density matrices. For the former average, one has

$$\langle K \rangle = -N(\hbar^2/2m)\int d\mathbf{r}_1 [\nabla_{\mathbf{r}_1}{}^2 \rho_{1,N}(\mathbf{r}_1 | \mathbf{r}_1')]_{\mathbf{r}_1 = \mathbf{r}_1'}, \qquad (\text{II.78})$$

which depends on the way that $\rho_{1,N}$ decays initially away from its diagonal elements. The latter average reduces to the following:

$$\langle \Phi \rangle = (1/2)\int d\mathbf{r}_1 \int d\mathbf{r}_2 v_2(r_{12}) \rho_{2,N}(\mathbf{r}_1, \mathbf{r}_2 | \mathbf{r}_1, \mathbf{r}_2). \qquad (\text{II.79})$$

In both Eqs. (II.78) and (II.79), the integrals span the system volume V.

An expression for the pressure p can be obtained from the virial theorem, Eq. (I.22), adapted to the Born-Oppenheimer approximation for nuclear motion. Restricting attention once again to the single-component case, with pairwise additive interactions as shown in Eq. (II.77), this leads to the result [Feynman, 1972, p. 57]:

$$pV = 2\langle K \rangle/3 - (1/6)\int d\mathbf{r}_1 \int d\mathbf{r}_2 [r_{12}v_2'(r_{12})] \rho_{2,N}(\mathbf{r}_1, \mathbf{r}_2 | \mathbf{r}_1, \mathbf{r}_2). \qquad (\text{II.80})$$

As before, the position integrals cover system volume V.

In the classical limit, the average kinetic energy becomes simply proportional to the absolute temperature:

$$\langle K \rangle \to 3Nk_B T/2 \qquad \text{(classical limit)}. \qquad (\text{II.81})$$

In that limit, the reduced density matrices collapse toward diagonal forms that are the n-particle distribution functions. These classical distribution functions at temperature $T = 1/k_B\beta$ can be obtained in principle by integrating the classical canonical N-body probability function shown in Eq. (II.59):

$$\rho_{n,N}(\mathbf{r}_1 \ldots \mathbf{r}_n, \beta) = [N!/(N-n)!]\int d\mathbf{r}_{n+1} \ldots \int d\mathbf{r}_N \bar{P}_N(\mathbf{r}_1 \ldots \mathbf{r}_N, \beta). \qquad (\text{II.82})$$

Consequently, the quantum mechanical formulas for system energy and pressure reduce in the classical limit to the following:

$$\bar{E}/N = (3k_B T/2) + (1/2N)\int d\mathbf{r}_1 \int d\mathbf{r}_2 v_2(r_{12}) \rho_{2,N}(\mathbf{r}_1, \mathbf{r}_2, \beta); \qquad (\text{II.83})$$

$$pV/N = k_B T - (1/6N)\int d\mathbf{r}_1 \int d\mathbf{r}_2 [r_{12}v_2'(r_{12})] \rho_{2,N}(\mathbf{r}_1, \mathbf{r}_2, \beta). \qquad (\text{II.84})$$

Under the assumptions that the N-particle system is homogeneous and macroscopically large and that the pair potential v_2 (and thus its derivative v_2') are spatially short ranged, the classical-limit canonical pair distribution function in these last two equations may be replaced by its grand ensemble version $\rho_2(\mathbf{r}_1, \mathbf{r}_2, \beta)$ without incurring significant error. It should be noted in passing that the grand ensemble averaging, Eq. (II.74), similarly converts each set of classical canonical distribution functions $\rho_{n,N}(\mathbf{r}_1 \ldots \mathbf{r}_n, \beta)$ to the corresponding open system function $\rho_n(\mathbf{r}_1 \ldots \mathbf{r}_n, \beta)$.

Next we consider the number fluctuation formula (II.72) for a convex region with volume V_0. Suppose that V_0 is macroscopically large in each dimension but is still much smaller than the volume V of the full system, i.e., $V_0 \ll V$ and $\langle N_0 \rangle \ll N$. Suppose also that V_0 is embedded well within V so as to be free of boundary effects if any are present. Because the boundary of V_0 is only a mathematical surface fully penetrable by particles, the exterior region $V - V_0$ acts as a particle reservoir, and the interior region V_0 can be described as an open system represented by its own grand partition function even if the large embedding system contains a fixed number N of particles. The magnitude of the number fluctuation $\langle (N_0 - \langle N_0 \rangle)^2 \rangle$ in this macroscopic subvolume is controlled by the isothermal compressibility κ_T of the material at the prevailing conditions of temperature and density (or equivalently temperature and pressure):

$$\kappa_T = \left(\frac{\partial \ln \rho}{\partial p} \right)_T. \tag{II.85}$$

Grand ensemble theory establishes the connection to local order and leads to the expression [Hill, 1956, p. 236]:

$$\rho^2 k_B T \kappa_T = \rho + (1/V_0) \int_{V_0} d\mathbf{r}_1 \int_{V_0} d\mathbf{r}_2 [\rho_2(\mathbf{r}_1, \mathbf{r}_2 \,|\, \mathbf{r}_1, \mathbf{r}_2; \beta) - \rho_1(\mathbf{r}_1 \,|\, \mathbf{r}_1; \beta) \rho_1(\mathbf{r}_2 \,|\, \mathbf{r}_2; \beta)]. \tag{II.86}$$

If the overall system permits free translation so that the pair distribution function depends only on relative displacement and so that the singlet densities become constants, Eq. (II.86) reduces to the following:

$$\rho^2 k_B T \kappa_T = \rho + (1/V_0) \int_{V_0} d\mathbf{r}_1 \int_{V_0} d\mathbf{r}_2 [\rho_2(\mathbf{r}_{12} \,|\, \mathbf{r}_{12}; \beta) - \rho^2]. \tag{II.87}$$

The entropy divided by Boltzmann's constant, S/k_B, can be extracted from the canonical distribution function $Q_N(\beta)$ for the quantum system, Eq. (II.19), via the Helmholtz free energy $F(\beta)$, Eq. (II.22), and the thermodynamic relation Eq. (II.25). This extraction leads to the following expression in terms of the energy eigenvalues for the system:

$$\frac{S}{k_B} = \frac{\sum_a \beta E_a \exp(-\beta E_a)}{\sum_b \exp(-\beta E_b)} + \ln(\sum_c \exp(-\beta E_c)]. \tag{II.88}$$

An alternative form follows upon re-expressing this last equation in terms of the normalized canonical ensemble probabilities p_a to find the system in eigenstate a [cf. Eq. (II.18)]:

$$p_a = \frac{\exp(-\beta E_a)}{\sum_b \exp(-\beta E_b)}. \tag{II.89}$$

As a result, one has the following entropy expression:

$$S/k_B = -\sum_a p_a \ln p_a. \tag{II.90}$$

This last relation reduces to zero if all of the $p_a = 0$ except for that of the ground state, otherwise $S/k_B > 0$.

In contrast to Eqs. (II.83), (II.84), and (II.86) for $\bar{E}$, p, and κ_T, respectively, the entropy cannot be expressed compactly in terms of low-order distribution functions for the thermodynamic state of interest.

G. Transport Phenomena

The response of a many-body system that is initially in an equilibrium state, then subjected to a weak external perturbation, can be described in terms of linear relaxation behavior or by transport coefficients. The perturbations can be electrical or mechanical disturbances, or local deviations from equilibrium values of temperature or species concentrations. Familiar examples of properties that describe the response of the system include shear viscosity, thermal and electrical conductivities, frequency-dependent dielectric response, and diffusion constants. This section provides a sketch of the theory and a few of its results for those properties that are examined in subsequent chapters. The general approach to linear response behavior can be described in terms of the so-called "fluctuation-dissipation theorem" [Callen and Welton, 1951; Callen and Greene, 1952; Greene and Callen, 1952]. Zwanzig has written a convenient and useful review of this subject [Zwanzig, 1965] as well as a general text [Zwanzig, 2001]. A fundamental characteristic of these linear response properties is that they can be expressed as fluctuation attributes of the perturbation-free equilibrium ensemble.

For clarity of presentation, the theory is first developed within the regime of the classical canonical ensemble. Modifications required by extension into the quantum regime are specified at the end of this section.

The first property to be considered is the self-diffusion constant D for a single-component system at equilibrium. It measures the long-time rate of increase of the mean-square displacement of an arbitrarily chosen particle i from its position at some chosen time origin:

$$\langle [\mathbf{r}_i(t) - \mathbf{r}_i(0)]^2 \rangle \sim 6Dt. \tag{II.91}$$

In this context, the "external perturbation" is the identification, or labeling, of the specific particle i. Here the operation denoted by $\langle \ldots \rangle$ is an average over a canonical ensemble of identical systems during a common time interval t for the thermodynamic state of interest. It should be noted here that for polyatomic species this definition of D is invariant to the choice of molecular "center" that vector $\mathbf{r}_i$ locates, although conventionally the choice would be the center of mass or the intramolecular position of a chosen nucleus. In the infinite system limit, it is legitimate to write the last equation in the more precise form

$$D = \lim_{t \to \infty} (6t)^{-1} \langle [\mathbf{r}_i(t) - \mathbf{r}_i(0)]^2 \rangle. \tag{II.92}$$

Within the classical regime under temporary consideration, the Newtonian equations of motion describe the dynamical evolution of the many-particle system, including both intramolecular motions and molecular diffusive translations. The time-rate-of-change of position vector $\mathbf{r}_i$ defines its velocity $\mathbf{v}_i$, so consequently Eq. (II.92) is equivalent to

$$D = \lim_{t \to \infty} (6t)^{-1} \int_0^t ds \int_0^t ds' \langle \mathbf{v}_i(s) \cdot \mathbf{v}_i(s') \rangle. \tag{II.93}$$

The thermal equilibrium average denoted here and above by $\langle \dots \rangle$, for the case of a canonical ensemble, implies that the mean value of the velocity scalar product that serves as the integrand can depend only on the time difference $s - s'$, and furthermore it must be an even function of that difference. Consequently, a change of integration variables converts Eq. (II.93) to

$$D = \lim_{t \to \infty} (3t)^{-1} \int_0^t (t - |u|) \langle \mathbf{v}_i(0) \cdot \mathbf{v}_i(u) \rangle du. \tag{II.94}$$

Considering the fact that in a large system dynamical processes cause the particles relatively quickly to "lose memory" of their initial particle velocities, Eq. (II.94) reduces even further to the frequently cited velocity autocorrelation function expression for D:

$$D = (1/3) \int_0^\infty \langle \mathbf{v}_i(0) \cdot \mathbf{v}_i(u) \rangle du. \tag{II.95}$$

Anticipating results below for other transport coefficients, it is useful to generate an alternative expression for the self-diffusion constant D. In particular, this alternative involves displaying D in terms of an autocorrelation function for a current-flow quantity that refers explicitly to all particles in the system, not just particle i. Start by defining a time-dependent "parity current" vector $\mathbf{J}_\pm(t)$ for the N-particle system as a whole:

$$\mathbf{J}_\pm(t) = N^{-1/2} \sum_{i=1}^N (-1)^i \mathbf{v}_i(t). \tag{II.96}$$

This parity current divides the particles essentially equally, but arbitrarily, into subsets with even and odd labels and weights their contributions with corresponding signs. Next, form the canonical ensemble autocorrelation function for this parity current:

$$\langle \mathbf{J}_\pm(0) \cdot \mathbf{J}_\pm(t) \rangle = N^{-1} \sum_{i=1}^N \sum_{j=1}^N (-1)^{i+j} \langle \mathbf{v}_i(0) \cdot \mathbf{v}_j(t) \rangle \tag{II.97}$$

$$= \langle \mathbf{v}_i(0) \cdot \mathbf{v}_i(t) \rangle + O(N^{-1}).$$

This result is based on the recognition, first, that all N terms with $i = j$ are equal to one another, and second, that to within a negligible $O(N^{-1})$ correction, the arbitrary assignment of relative signs to terms with $i \neq j$ causes them to cancel in the canonical average. Therefore, in view of Eq. (II.95), the self-diffusion constant in the large-system limit can be presented as a current autocorrelation function expression:

$$D = (1/3) \int_0^\infty \langle \mathbf{J}_\pm(0) \cdot \mathbf{J}_\pm(t) \rangle dt. \tag{II.98}$$

If the system of interest contains several species of particles, the mean-square displacement definition (II.91) applies individually to each and supplies the diffusion constant D_v when particle i has been chosen as a representative of species v. Then by straightforward extension of the preceding transformation, the velocity autocorrelation function expression (II.95) for particle i also yields D_v. But in order for D_v to be expressed as a current autocorrelation function integral of the type shown in Eq. (II.98), the parity current double sum in Eq. (II.97) should include only those particles i and j of species v, and their total number N_v must be much larger than unity.

Next, consider the shear viscosity η, specifically for an isotropic system. This property measures the linear flow-response rate of the N-particle system to an imposed shear stress. An equilibrium ensemble autocorrelation function expression for η analogous to that just displayed in Eq. (II.98) for D was first derived by Green using the Fokker-Planck equation [Green, 1954]. More recent derivations of the same expression have been based on the full Newtonian dynamics of the equilibrated many-particle system [Zwanzig, 1965; McQuarrie, 1976; Kubo et al., 1991]. The result uses the systemwide quantity $J^{xy}(t)$, the x,y component of a symmetric momentum flux tensor:

$$J^{xy}(t) = \sum_{j=1}^{N} \{p_{jx}(t)p_{jy}(t)/m + (1/2)[x_j(t)F_{jy}(t) + y_j(t)F_{jx}(t)]\}. \tag{II.99}$$

Here p_{jx} and p_{jy} are Cartesian components of the momentum of particle j, while x_j, y_j and F_{jx}, F_{jy} are the x and y components of its position and of the net force exerted on it by other particles, respectively. The expression shown in Eq. (II.99) thus includes spatial transfer of momentum by particle drift (first summand term) and by force exerted over the length scale of intermolecular interactions (second summand term). The shear viscosity at absolute temperature T is then provided by the following time integral:

$$\eta = (1/k_B T V) \int_0^\infty \langle J^{xy}(0) J^{xy}(t) \rangle dt, \tag{II.100}$$

where as usual V is the volume of the system. This identity connects η to the temporal persistence of shear stress fluctuations that arise spontaneously in the equilibrium ensemble. Low-viscosity media relax those fluctuations rapidly, and high-viscosity media relax them slowly.

Many-particle systems that contain two or more species obviously exhibit only a single collectively determined shear viscosity, not individual values for each of those species. Equation (II.100) retains validity for that multicomponent generalization, for which the momentum flux summation in Eq. (II.99) then includes all particles of all species present. The first summation term, of course, must reflect the appropriate mass m_v for each species-v particle.

If the medium possesses hydrodynamic isotropy, the choice of x and y is arbitrary as the pair of perpendicular directions to use for expression (II.100). Any other perpendicular pair would serve equally well. However, if that isotropy is not present, the shear viscosity is not a scalar quantity but a tensor, for which Eq. (II.100) gives the one element η_{xy}. In that event, the proper hydrodynamic description requires all elements of that tensor, so all choices of directions for autocorrelation function expression (II.100) must be separately evaluated.

Shear viscosity η measures the energy dissipated when a many-body system is forced to change its shape at constant volume. An analogous quantity, the volume (or bulk) viscosity η_V, provides a corresponding energy dissipation measure for the case where the system is forced to change its volume but not its shape, i.e., uniform compression or expansion at a nonzero rate [Graves and Agrow, 1999]. In comparison with the shear viscosity, the experimental determination of η_V is much more challenging [Gillis et al., 2005]. A corresponding autocorrelation function expression is available. However, η_V does not play a significant role in the material to be developed in later chapters and so is left aside. Interested readers should consult the available literature for details concerning the theory for η_V [Zwanzig, 1965, 2001; Hansen and McDonald, 1976, Chapter 8].

The coefficient of thermal conductivity λ in an isotropic medium measures heat flow in response to a temperature gradient ∇T. The autocorrelation function expression for λ requires a systemwide heat current vector $\mathbf{J}_H$ [Zwanzig, 1965, 2001; Hansen and McDonald, 1976, Chapter 8]. This vector takes the following form:

$$\mathbf{J}_H(t) = (d/dt)\sum_{i=1}^{N}[E_i(t) - h_0]\mathbf{r}_i(t). \tag{II.101}$$

After dividing the n-particle potential energy contributions $v^{(n)}$ equally among the n particle participants and adding them to the respective kinetic energies, the resulting energies assigned to each particle are denoted here by E_i, $1 \le i \le N$. The overall average enthalpy per particle has been denoted by h_0; it is present in expression (II.101) to make the final result independent of the choice of origin for the particle positions $\mathbf{r}_i$. Then λ has the following representation as an equilibrium ensemble autocorrelation function expression:

$$\lambda = (3Vk_BT^2)^{-1}\int_0^\infty \langle \mathbf{J}_H(0)\cdot\mathbf{J}_H(t)\rangle dt. \tag{II.102}$$

If the interaction potential Φ has caused the nuclei to assemble into molecules, this general expression incorporates heat flow both within individual molecules as well as between neighboring molecules.

Heat conductivity in anisotropic media (e.g., oriented low-symmetry crystals) is not simply a constant, as shown in Eq. (II.102), but depends on direction. In such cases, the scalar λ must be replaced by a tensor analogous to that required for shear viscosity in anisotropic media. The separate elements of that heat-conductivity tensor are represented by integrals analogous to those in Eq. (II.102) but where the averaged product becomes those for the individual components of $\mathbf{J}_H$ and in general must include cross terms for different-direction components.

The autocorrelation function expressions shown above for D, η, and λ have used classical statistical mechanics. However, with appropriate reinterpretation of the quantities involved, all such expressions can be extended into the quantum regime [Zwanzig, 1965]. Each of the classical ensemble properties D, η, and λ have involved current–current autocorrelation function time integrals of the type

$$\int_0^\infty \langle \dot{\mathbf{A}}(0)\cdot\dot{\mathbf{A}}(t)\rangle dt, \tag{II.103}$$

where in this form the current has been represented as the time derivative $\dot{\mathbf{A}}$ of a vector quantity $\mathbf{A}$, an example of which appears in Eq. (II.101). The following three rules properly transcribe those classical expressions to the quantum regime. First, the classical canonical ensemble average of a dynamical quantity B that has been denoted by $\langle B \rangle$ must be interpreted as the trace of the product of the density matrix for the system and the proper operator form for B:

$$\langle B \rangle \to Tr(\mathbf{B}\boldsymbol{\rho}^{(N)}). \tag{II.104}$$

Second, the time dependence of an operator $\mathbf{C}(t)$ [such as that corresponding to $\dot{\mathbf{A}}(t)$ in Eq. (II.103)] must be represented by

$$\mathbf{C}(t) = \exp(it\mathbf{H}/\hbar)\mathbf{C}(0)\exp(-it\mathbf{H}/\hbar), \tag{II.105}$$

where as before, $\mathbf{H}$ is the nuclear system's Hamiltonian operator. Finally, the operator $\dot{\mathbf{A}}(0)$ corresponding to the first of the pair of systemwide currents needs to be replaced by its "Kubo transform":

$$\dot{\mathbf{A}}(0) \rightarrow \tilde{\dot{\mathbf{A}}}(0) = k_B T \int_0^{1/k_B T} \exp(\xi \mathbf{H}) \dot{\mathbf{A}}(0) \exp(-\xi \mathbf{H}) d\xi. \tag{II.106}$$

These modifications in principle allow evaluation of transport coefficients for systems within the quantum regime and include the effects of quantum diffraction, of tunneling, and of Bose-Einstein or Fermi-Dirac statistics.

H. Phase Transitions—A Preview

In thermodynamic equilibrium states of macroscopically large systems, with just a single phase present, the reduced density matrices and their associated distribution functions display simplification when the configuration variables for the particles involved are widely separated. This method is especially simple for the grand ensemble, where one has

$$\rho_n(\{\mathbf{r}_i\} \mid \{\mathbf{r}_i'\}) \rightarrow \prod_{i=1}^n \rho_1(\mathbf{r}_i \mid \mathbf{r}_i'). \tag{II.107}$$

That is, spatial separation removes correlation that might exist between directly or indirectly interacting particles that are neighbors on the atomic or molecular length scale. This reduction in Eq. (II.107) implies that the integrands in the isothermal compressibility relations, Eqs. (II.86) and (II.87), are short ranged, thus contributing to the integral only when the two particles involved are close on the atomic or molecular length scale, and thus yielding an isothermal compressibility that is an intensive $[O(1)]$ quantity.

The situation changes qualitatively when the system of interest possesses one or more first-order phase transitions, and the thermodynamic control parameters (temperature, mean density) locate the state point within a phase coexistence region. For illustration, suppose that two macroscopic coexisting uniform phases δ and ε have distinct particle number densities ρ_δ and ρ_ε at the prevailing temperature, respectively. Coexistence means that the system's overall number density ρ lies between these limits:

$$\rho_\delta < \rho < \rho_\varepsilon. \tag{II.108}$$

This then implies that the macroscopic volumes V_δ and $V_\varepsilon = V - V_\delta$ occupied, respectively, by the two phases are given by

$$\frac{V_\delta(\rho)}{V} = \frac{\rho_\varepsilon - \rho}{\rho_\varepsilon - \rho_\delta}, \tag{II.109}$$

$$\frac{V_\varepsilon(\rho)}{V} = \frac{\rho - \rho_\delta}{\rho_\varepsilon - \rho_\delta}.$$

Under normal phase coexistence conditions, positive surface free energy (surface tension for fluids) at the interface between distinct phases drives the separate phase volumes into connected macroscopic shapes, with linear dimensions that are themselves macroscopic, not atomic or molecular. As a result, when periodic boundary conditions are present to permit free translation, reduced density matrices and the associated distribution functions for coexistence reduce in the large system limit to linear combinations of those for the separate phases, provided that the inter-particle separations are small compared to the linear dimensions of the separate phases. Specifically, one has the following form for two-phase coexistence [Mayer, 1947; Uhlenbeck et al., 1963]:

$$\rho_{n,N}(\{\mathbf{r}_i\}, \{\mathbf{r}_i'\}) = (V_\delta/V)\rho_{n,N;\delta}(\{\mathbf{r}_i\}, \{\mathbf{r}_i'\}) + (V_\varepsilon/V)\rho_{n,N;\varepsilon}(\{\mathbf{r}_i\}, \{\mathbf{r}_i'\}), \tag{II.110}$$

where the reduced density matrices on the right side are those for the homogeneous δ and ε phases at the phase transition state. In conjunction with Eqs. (II.83) and (II.84), this linear combination ensures that the mean energy $\bar{E}$ varies linearly with system density ρ as it sweeps across its coexistence interval $\rho_\delta \leq \rho \leq \rho_\varepsilon$, while the pressure p remains constant over that same interval.

Equation (II.110) implies that the integrand of an isothermal compressibility relation (II.87) in a coexistence state at fixed N takes on the form

$$\rho_{2,N}(\mathbf{r}_{12}|\mathbf{r}_{12}) - \rho^2 \to \left(\frac{V_\delta}{V}\right)\rho_{2,N;\delta}(\mathbf{r}_{12}|\mathbf{r}_{12}) + \left(\frac{V_\varepsilon}{V}\right)\rho_{2,N;\varepsilon}(\mathbf{r}_{12}|\mathbf{r}_{12}) - \left[\left(\frac{V_\delta}{V}\right)\rho_\delta + \left(\frac{V_\varepsilon}{V}\right)\rho_\varepsilon\right]^2 \tag{II.111}$$

For values of the particle pair separation $\mathbf{r}_{12}$ large enough on the atomic or molecular length scale to eliminate correlation effects within homogeneous phases [as indicated in Eq. (II.107)] but still small on the macroscopic length scale set by $V_0^{1/3}$ [cf. Eq. (II.86)], expression (II.111) undergoes a further simplification to

$$\rho_{2,N}(\mathbf{r}_{12}|\mathbf{r}_{12}) - \rho^2 \to \left(\frac{V_\delta V_\varepsilon}{V^2}\right)[\rho_\varepsilon - \rho_\delta]^2 \tag{II.112}$$

$$> 0.$$

Consequently, the isothermal compressibility integral in Eq. (II.87) becomes unbounded in the large system limit and thus exhibits consistency with the fact that pressure must remain constant across the coexistence interval of density.

The remarks in this section so far have been concerned with two-phase coexistence. However, larger numbers of phases separated by first-order transitions can also coexist. Perhaps the most prominent example is the "triple point" for pure substances, at which crystal, liquid, and vapor phases can simultaneously be present in a fixed temperature and pressure state of thermodynamic equilibrium. Suppose that the three phases δ, ε, and ζ at a triple point of coexistence possess the respective pure-phase number densities:

$$\rho_\delta < \rho_\varepsilon < \rho_\zeta. \tag{II.113}$$

In order to have triple phase coexistence, the overall density ρ must lie between the extremes of these three, but that parameter alone is insufficient to specify the relative amounts of the three phases present. The specific amounts of each of these three phases are controlled both by the number of particles present and also by the independently variable system average energy $\bar{E}$. In any case, the total macroscopic volume V would be divided into three distinct portions:

$$V = V_\delta + V_\varepsilon + V_\zeta. \tag{II.114}$$

The reduced density matrices for atomic or molecular scale separations then become linear combinations of those for the pure phases, generalizing Eq. (II.110):

$$\rho_{n,N}(\{\mathbf{r}_i\}, \{\mathbf{r}_i'\}) = (V_\delta/V)\rho_{n,N;\delta}(\{\mathbf{r}_i\}, \{\mathbf{r}_i'\}) + (V_\varepsilon/V)\rho_{n,N;\varepsilon}(\{\mathbf{r}_i\}, \{\mathbf{r}_i'\}) \qquad \text{(II.115)}$$

$$+ (V_\zeta/V)\rho_{n,N;\zeta}(\{\mathbf{r}_i\}, \{\mathbf{r}_i'\}).$$

Analogous considerations apply for multicomponent systems that can exhibit more than three simultaneously coexisting phases.

III.

Basins, Saddles, and Configuration-Space Mapping

The two preceding chapters have established the basic properties of potential energy hypersurfaces and the way that they control the kinetics and thermodynamics of interacting many-particle systems. The logical next step involves detailed analysis of the topographic features of those hypersurfaces. For that purpose, some general mathematical procedures need to be defined and applied. Special attention is focused on the extrema of the relevant N-body potential energy function $\Phi(\mathbf{r}_1 \ldots \mathbf{r}_N)$, including both minima, and saddle points of various orders. The electronic quantum number, previously denoted by n', refers to the ground state for most applications but in any case is temporarily suppressed in the rest of the chapter for notational simplicity.

A. Steepest Descent Mapping

In order to implement the landscape paradigm for potential energy functions describing systems at constant volume, the first step is to create a $(3N + 1)$-dimensional Euclidean space whose generators are the nuclear position variables $\mathbf{r}_1 \ldots \mathbf{r}_N$, and the value of Φ itself. The potential energy function represents a relation between all $3N + 1$ coordinates that is satisfied on a $3N$-dimensional hypersurface inhabiting this space. Except at a subset of the nuclear coordinates that correspond to vanishing of one or more internuclear distances r_{ij}, this hypersurface is continuous and differentiable in those nuclear coordinates. In addition, its Φ coordinate is everywhere bounded below by $-BN$, where B is some positive number. These are general properties of the potential energy function that were discussed earlier (Chapter I, Section 1.C).

Local minima of this Φ hypersurface with respect to variations in the nuclear coordinates $\mathbf{r}_1 \ldots \mathbf{r}_N$ are mechanically stable configurations of the full set of N nuclei that play a special role in the following developments. These stable nuclear configurations are denoted as "inherent structures." Provided that impenetrable walls are present to prevent free translation or rotation of the system as a whole, each inherent structure corresponds to an isolated point on the Φ hypersurface. However, when periodic boundary conditions are present, free translation of the system as a whole stretches each inherent structure into a three-dimensional manifold of constant-Φ configurations within the full $(3N+1)$-dimensional Euclidean space.

If the many-body system of interest consists of identical nuclear species, the potential function Φ has full permutational symmetry. That is, Φ remains unchanged under exchange of the positions of any pair of particles (and therefore any sequence of such pair exchanges):

$$\Phi(\mathbf{r}_1 \ldots \mathbf{r}_i \ldots \mathbf{r}_j \ldots \mathbf{r}_N) \equiv \Phi(\mathbf{r}_1 \ldots \mathbf{r}_j \ldots \mathbf{r}_i \ldots \mathbf{r}_N) \qquad \text{(all } i, j\text{)}. \tag{III.1}$$

When impenetrable walls are present, this implies that any inherent structure belongs to a family of $N!$ permutationally equivalent inherent structures. However, when periodic boundary conditions are present, the possibility of free translations for the system as a whole can reduce this multiplicity to $(N - 1)!$ for inherent structures that exhibit translational symmetry in three-dimensional space (e.g., crystal structures).

The properties just outlined for the set of inherent structures apply to virtually all potential functions that are designed realistically to represent known atomic and molecular materials. However, it is important to recognize that mathematical models can be constructed that exhibit significant deviations from those properties. One such anomalous class of model potentials involves short-range positive interactions in Fourier space and was mentioned at the end of Chapter I, Section 1.D [see Eq. (I.70)]. These models possess highly degenerate classical ground states, which is to say that their ground-state inherent structures are high-dimension manifolds on the Φ hypersurface. This unusual class of models is not included in the analysis of the remainder of this chapter.

In the interest of notational simplicity, let $\mathbf{R} \equiv (\mathbf{r}_1 \ldots \mathbf{r}_N)$ now stand for the set of configurational coordinates. The $3N$-dimensional $\mathbf{R}$ space can be divided into nonoverlapping "basins of attraction," one for each inherent structure. This division is accomplished by means of steepest descent mapping on the Φ hypersurface, to be carried out while maintaining the given boundary conditions. Steepest descent paths are generated by the equation

$$d\mathbf{R}(s)/ds = -\nabla_{\mathbf{R}} \Phi[\mathbf{R}(s)], \tag{III.2}$$

where $s \geq 0$ is a virtual "time" of descent. Starting at an essentially arbitrary initial configuration $\mathbf{R}(0)$, the trajectory defined by Eq. (III.2) slides downward on the Φ hypersurface until at $s = +\infty$ it is trapped at a nearby inherent structure, to be given an identifying label α. As indicated above, the final configuration of the steepest descent process is unique if container walls are present to prevent free translation and/or rotation of the many-body system as a whole, so in such circumstances it is proper to write

$$\lim_{s \to \infty} \mathbf{R}(s) = \mathbf{R}_\alpha \tag{III.3}$$

for that unique inherent-structure configuration. However, if the boundary conditions permit free translation and/or rotation, or if the potential possesses a highly degenerate classical ground state, then the inherent structures are not unique configurations. In such circumstances, the collection of configurations generated by the steepest descent operation, Eq. (III.2), becomes a set the members of which can be identified uniquely by the initial condition, i.e., $\mathbf{R}_\alpha[\mathbf{R}(0)]$.

An equivalent description of the steepest descent mapping, Eq. (III.2), is that all N nuclei move in directions and at proportional rates determined by the forces they individually experience, until all forces simultaneously vanish at a configuration of mechanical equilibrium, the inherent structure.

The set of all nuclear configurations that converge onto inherent structure α by means of this steepest descent mapping constitute the basin $\mathbf{B}_\alpha$ belonging to that inherent structure. By construction, every basin is a connected set of configurations, but in principle basins need not be simply connected. With the exception of a zero-measure set of starting configurations, the collection of all basins exhaustively cover ("tile") $\mathbf{R}$ space.

The shared boundary between a pair of neighboring basins is a $(3N-1)$-dimensional hypersurface. The points contained in such boundaries are not assigned to either one of the contiguous basin pair by the steepest descent mapping. Instead, applying the mapping operation, Eq. (III.2), to these boundary configurations as initial conditions generates trajectories that converge onto a vanishing-gradient saddle point contained in the boundary. Such saddle points are "transition states" separating the respective inherent structures, and the trajectories terminate because $\nabla_{\mathbf{R}}\Phi$ vanishes there. Circumstances could also exist in which a basin B_α contains an interior saddle point, and steepest descent mapping cannot connect such a point to an inherent structure. However, there is no problem declaring that, by convention, any such interior saddle should always be assigned to the basin that encloses it.

The steepest descent mapping just described refers to constant-volume ("isochoric") boundary conditions. This method could involve a volume defined either by the location of confining walls surrounding the system of interest or by the positions of periodic boundary conditions. However, many laboratory experiments are performed under constant-pressure ("isobaric") conditions, for which a variant of the steepest descent mapping may become appropriate. If constant-pressure (p) circumstances apply, the system volume V fluctuates, so it is natural to treat V itself as an additional system configurational variable on a basis similar to that of the nuclear configurational coordinates. Then, as indicated in Section II.B, Eq. (II.26), it is useful to define a "potential enthalpy" function as follows:

$$\hat{\Phi}(\mathbf{r}_1\ldots\mathbf{r}_N, V) = \Phi(\mathbf{r}_1\ldots\mathbf{r}_N) + pV, \tag{III.4}$$

where here and in the remainder of the book, the isobaric condition is indicated by a circumflex. The augmented configuration space $(\mathbf{r}_1\ldots\mathbf{r}_N, V)$ now has dimension $3N+1$, and thus one is naturally led to examine the entire potential enthalpy landscape for the hypersurface in $3N+2$ dimensions defined by the expression in Eq. (III.4). If $\mathbf{R}$ now refers to the augmented set of configurational coordinates, an isobaric version of the steepest descent mapping can be defined by the following analog of Eq. (III.2):

$$d\mathbf{R}(s)/ds = -\nabla_R\hat{\Phi}[\mathbf{R}(s)] \qquad (s \geq 0). \tag{III.5}$$

This procedure identifies isobaric inherent structures, and their surrounding basins, as the virtual time parameter $s \to +\infty$.

It needs to be emphasized that even if the premapping collection of configurations comes from an isobaric ensemble, the steepest descent operation can optionally be carried out while holding the initial volume constant. In other words, the mapping can be performed either on the Φ or the $\hat{\Phi}$ multidimensional landscape. Which choice should be elected depends on the specific physical insights sought. Indeed, some circumstances might justify carrying out both mapping procedures and examining the differences in results obtained. In the various applications examined in the remainder of this monograph, it is clear whether the constant-volume (isochoric) or the constant-pressure (isobaric) version of steepest descent mapping has been used.

The strategic advantage of identifying inherent structures and their embedding basins in either constant-volume or constant-pressure circumstances is that a natural separation thereby occurs between pure structural aspects of the many-body system and intrabasin vibrational motions taking the system locally away from the reference inherent structure. This separation is done without forcing any unnatural theoretical framework on the system, such as the introduction of arbitrary length or energy parameters. Instead, through the steepest descent mapping, the procedure relies strictly on the topography of the potential energy or the potential enthalpy "rugged landscape" itself to effect the separation. Implications of that separation appear in later sections of this chapter and in subsequent chapters.

B. Hypersurface Curvatures

The shape of the hypersurface corresponding to a potential energy function $\Phi(\mathbf{R})$ can at least locally be described by a Taylor expansion:

$$\Phi(\mathbf{R} + \Delta\mathbf{R}) = \Phi(\mathbf{R}) + [\nabla_{\mathbf{R}}\Phi]\cdot\Delta\mathbf{R} + \frac{1}{2}[\nabla_{\mathbf{R}}\nabla_{\mathbf{R}}\Phi]:\Delta\mathbf{R}\Delta\mathbf{R} + O[\Delta\mathbf{R}]^3. \qquad (III.6)$$

The gradient vector $\nabla_{\mathbf{R}}\Phi$ vanishes at each inherent structure, at any other intrabasin extremum, and at the saddle points (transition states) occurring on the shared boundary between contiguous basins. The symmetric matrix of second derivatives of Φ contains information about the curvatures of the hypersurface at configuration $\mathbf{R}$. This matrix is often called the Hessian matrix [Hertz et al., 1991] and is denoted here simply by

$$\mathbf{K} = \nabla_{\mathbf{R}}\nabla_{\mathbf{R}}\Phi(\mathbf{R}). \qquad (III.7)$$

A rotational (unit-Jacobian) transformation of coordinates always exists to render $\mathbf{K}$ into diagonal form, with each of the diagonal elements representing a principal curvature of the hypersurface at location $\mathbf{R}$. The eigenvectors for each of these diagonal elements define the directions of the principal curvatures. In this diagonal representation, the quadratic term in the Taylor expansion [Eq. (III.6)] has the following appearance:

$$\frac{1}{2}\sum_{i=1}^{3N} K_i u_i^2, \qquad (III.8)$$

where the u_i are the transformed coordinates and the K_i are the respective harmonic force constants for those coordinates. If the system were to obey periodic boundary conditions, three of the directions u_i would correspond to translation of the system as a whole, and the resulting translational invariance of Φ would require vanishing hypersurface curvatures ($K_i = 0$) in those directions.

As mentioned, at any inherent-structure configuration $\mathbf{R}_\alpha$ the linear terms in Taylor expansion (III.6) vanish. Furthermore, in the immediate neighborhood of $\mathbf{R}_\alpha$, no force constant K_i is negative. However, moving away from the inherent structure within its basin may cause one or more of the K_i to become negative, corresponding to directions of downward hypersurface curvature. In particular, the vanishing-gradient saddle points have one negative curvature direction if they are "simple," but higher order saddle points also exist and can be classified by the number of negative curvatures that they display. Changes with position of the values of the K_i's within a basin provide one way to characterize basin anharmonicity. The number of negative K_i's at any system configuration is uniquely defined and so could form the basis for a subdivision of every inherent-structure basin B_α.

The harmonic approximation to the potential energy function Φ in the immediate vicinity of any given inherent structure α may be written

$$\Phi(\mathbf{R}) \cong \Phi_\alpha + \frac{1}{2}\sum_{i=1}^{3N} K_{\alpha i} u_{\alpha i}^2, \tag{III.9}$$

$$\Phi_\alpha \equiv \Phi(\mathbf{R}_\alpha).$$

Subscript α has been appended to the force constants and coordinates because these variables differ fundamentally from one basin to another (except those corresponding merely to permutations of identical nuclei). If the system contains only a single nuclear species and that species has mass m, then the independent harmonic normal modes in basin α have angular frequencies given by

$$\omega_{\alpha i} = (K_{\alpha i}/m)^{1/2}. \tag{III.10}$$

In preparation for later analysis, it is now appropriate to mention a variant of the steepest descent mapping that in principle treats all extrema on an equal footing. Define a scalar function $\Gamma(\mathbf{R})$ in terms of the squared gradient of Φ:

$$\Gamma(\mathbf{R}) = [\nabla_{\mathbf{R}}\Phi(\mathbf{R})]^2. \tag{III.11}$$

This auxilliary function vanishes if and only if $\mathbf{R}$ is the location of an extremum of Φ. These extrema include, but are not restricted to, inherent structures. The function $\Gamma(\mathbf{R})$ also vanishes at saddle points of all orders, and at local maxima, if any. Steepest descent on the Γ hypersurface, defined by the analog to Eq. (III.2):

$$d\mathbf{R}(s)/ds = -\nabla_{\mathbf{R}}\Gamma[\mathbf{R}(s)] \qquad (s \geq 0), \tag{III.12}$$

generate paths, some of which converge to zeros of Γ as $s \to +\infty$. Others may converge to relative minima with $\Gamma > 0$ and have less significance in the present context. Nevertheless, this auxiliary function approach has been useful for locating saddle points in computer simulations of classical many-body systems [Stillinger and Weber, 1984b; Angelani et al., 2002].

Although Eqs. (III.6)–(III.12) have referred to the constant-volume potential function Φ, minor modifications for the most part permit immediate application to the constant-pressure potential enthalpy function $\hat{\Phi}$ defined in Eq. (III.4). The exception concerns the normal mode vibrational spectrum because of the "mass" that must be attributed to the volume coordinate V. To agree with normal experimental practice, this attribute might be assigned a macroscopic magnitude (e.g., a piston mass). Calculation of normal mode frequencies then would require diagonalizing a matrix of mass-weighted Hessian matrix elements.

C. Inherent-Structure Enumeration

It has already been noted that permutational symmetry for the nuclei comprised in the system implies, for a single-component case with confining walls, that any inherent structure and its surrounding basin is just one of $N!$ equivalent inherent structures and basins [periodic boundary conditions can reduce this to $(N-1)!$ for structurally perfect crystals]. However, this phenomenon leaves open the question of how many distinguishable (geometrically distinct) inherent structures that are not simply related by permutation exist for a given system of N nuclei. Even for small numbers of nuclei, the distinguishable inherent structures appear to rise rapidly with N. This characteristic has become obvious from efforts to identify and describe the inherent structures in free space for small numbers of particles ($N < 14$) in the Lennard-Jones [Hoare, 1979; Tsai and Jordan, 1993a] and Morse [Hoare and McInnes, 1976, 1983] potential models. In this section, attention concentrates just on the topography of the electronic ground-state potential hypersurface.

Consider the case of a nonfluctuating macroscopic volume V_0 that contains N_0 identical particles comprising a uniform condensed phase. Suppose that this system possesses some number $\Omega_d(N_0)$ of distinguishable inherent structures. Now for some integer $n > 1$ expand the system size to volume nV_0 and nN_0 particles by placing the original system and $n-1$ copies together and removing any walls between them. Because the original and each of the copies are macroscopically large, they can be virtually independently rearranged internally to distinct mechanically stable structures, with a relative error of surface-to-volume character that vanishes in the $V_0, N_0 \to \infty$ limit. Consequently, one estimates that the count of distinguishable inherent structures for the combined system is the product of those for the subsystems:

$$\Omega_d(nN_0) \cong [\Omega_d(N_0)]^n. \tag{III.13}$$

An obviously equivalent mathematical statement is that

$$\ln[\Omega_d(nN_0)] \sim n\ln[\Omega_d(N_0)], \tag{III.14}$$

that is, the logarithm of the number of distinguishable inherent structures rises linearly with the number of particles, holding the number density within the system fixed. This linear asymptotic relationship may be put into an equivalent but more useful form, describing the growth rate with N, of the count of distinguishable inherent structures:

$$\ln[\Omega_d(N)] \sim \sigma_\infty(\rho)N. \tag{III.15}$$

In this expression, $\sigma_\infty > 0$ generally depends on the particle number density ρ as indicated, as well as on the species of the particles that comprise the system. Leading corrections to the linear term shown in Eq. (III.15) depend on the boundary conditions, and in the event that confining walls are present, those corrections would be expected to consist of surface contributions proportional to surface area, i.e., proportional to $N^{2/3}$. For the case of periodic boundary conditions, one expects the corrections to have yet lower order in N, provided that the many-body system consists of a single homogeneous phase.

A relatively minor technical point may be worth noting at this stage. If the system's containing volume has a high symmetry (e.g., cubic), any given inherent structure can be reoriented by a discrete amount within that volume without changing its potential energy or basic structure. The same would be true upon replacing that inherent structure, if it were chiral, by its mirror image.

The results of these symmetry operations would formally be counted as distinct inherent structures. However, the resulting multiplicity, or degeneracy, of those inherent structures is only an order-unity integer and consequently would have no effect on asymptotic expressions such as that in Eq. (III.15).

Besides exhibiting differences of geometric structure, distinguishable inherent structures possess a significant range of potential energy values. In the large-system regime, potential energy is an extensive quantity (i.e., proportional to N). Consequently, the span of inherent-structure potential energies, from the lowest (Φ_{min}) to the highest (Φ_{max}), likewise is asymptotically proportional to N. The "best" spatial arrangement of particles necessary to attain Φ_{min} would in virtually all cases amount to the most nearly perfect version of the most stable crystal structure that could be formed with the given boundary conditions and integer number of particles. The "worst" spatial arrangement that still meets the criterion of mechanical stability, exhibiting potential energy Φ_{max}, likely depends sensitively on the details of the potential energy function Φ. It is not clear a priori whether this configuration corresponds to an irregular amorphous arrangement of particles or whether it involves an unusual periodic structure. In any case, one might reasonably anticipate that this worst inherent structure possesses a very small basin, its configuration-space region of marginal local stability.

Because of the great diversity of mechanically stable particle arrangements that are possible, with corresponding spread of inherent-structure potential energies Φ_α, it becomes relevant and useful to inquire about the distribution of basin depths. For this purpose, introduce the intensive "order parameter" φ that generically represents the potential energy per particle over the entire range of inherent structures α:

$$\varphi = \Phi_\alpha/N, \tag{III.16}$$

$$\varphi_{min} \leq \varphi \leq \varphi_{max}.$$

In view of the asymptotic growth rate in expression (III.15), the same asymptotic behavior must apply to the distribution of distinguishable inherent structures when classified by depth per particle. Specifically, for large-system size the expected number of inherent structures with intensive depth parameters in a narrow range $\varphi \pm \delta\varphi/2$ can be expressed in the form

$$C\exp[N\sigma(\varphi,\rho)]\delta\varphi, \tag{III.17}$$

$$\sigma(\varphi,\rho) \geq 0,$$

where C is a positive scale factor with dimension (energy)$^{-1}$. The total number of distinguishable inherent structures can be obtained by integrating this last expression with respect to φ, with a result that must be consistent with Eq. (III.15).

$$\ln\left\{C\int_{\varphi_{min}}^{\varphi_{max}} \exp[N\sigma(\varphi,\rho)]d\varphi\right\} \sim N\sigma_\infty(\rho). \tag{III.18}$$

The fact that N is very large implies that the integral in this last expression is dominated by the integrand evaluated in the immediate vicinity of its maximum. This implication leads one to conclude that

$$\max_{(\varphi)}[\sigma(\varphi,\rho)] = \sigma_\infty(\rho). \tag{III.19}$$

Inherent structures with potential energies near the lower or upper limit must be structurally rather special and therefore rare among all inherent structures; most inherent structures are found to lie between these limits. In other words, the maximum of σ indicated in Eq. (III.19) should occur well above φ_{min} and well below φ_{max}.

The derivations just presented refer to isochoric (constant-volume) circumstances. Exactly parallel considerations apply for isobaric (constant-pressure) circumstances. The total number of distinguishable inherent structures on the $\hat{\Phi}$ hypersurface again rises exponentially with particle number N. The appropriate intensive order parameter for classifying isobaric inherent structures and their basins is $\hat{\varphi}$, the potential enthalpy per particle. The distribution expression corresponding to Eq. (III.17) takes the form

$$\hat{C}\exp[N\hat{\sigma}(\hat{\varphi}, p)]\delta\hat{\varphi}, \tag{III.20}$$

$$\hat{\sigma}(\hat{\varphi}, p) \geq 0,$$

where use of the circumflex extends to the isobaric version of basin enumeration.

Deriving quantitative results for the enumeration functions σ and $\hat{\sigma}$ for specific substances is a nontrivial task. Later chapters consider this issue in detail.

D. Partition Function Transformations

The introduction of inherent structures and their basins into the statistical mechanical description of many-body systems illuminates several aspects of those systems' cooperative phenomena that might tend otherwise to remain obscure. This illumination becomes apparent at least partly through the natural transformations that this inherent-structure representation permits for the partition functions that describe equilibrium properties. The transformations are most straightforward for single-component systems in the classical limit, so that is the situation to be examined first.

The canonical partition function for that single-component, classical-limit case was introduced in Section II.D, Eq. (II.56). Its form is repeated here:

$$Q_N(\beta, V) = \exp(-\beta F_N) = (\lambda_T^{3N}N!)^{-1}\int\exp[-\beta\Phi(\mathbf{R})]d\mathbf{R}, \tag{III.21}$$

where $\beta = 1/k_BT$, F_N is the Helmholtz free energy for the N-particle system, and the mean thermal de Broglie wavelength is given by

$$\lambda_T = h(\beta/2\pi m)^{1/2}. \tag{III.22}$$

The N-body configuration vector $\mathbf{R}$ has Euclidean components that in the partition function (III.21) are integrated over the interior of the container volume V.

The $3N$-dimensional configuration integral in Q_N is the sum of partial contributions from each of the inherent-structure basins. Because there are $N!$ equivalent basins (for repelling-wall boundaries) caused by permutational symmetry, Eq. (III.21) can be recast into the following form:

$$Q_N(\beta, V) = \lambda_T^{-3N}\sum_{\alpha}\exp(-\beta\Phi_\alpha)\int_{B_\alpha}\exp[-\beta\Delta_\alpha\Phi(\mathbf{R})]d\mathbf{R}. \tag{III.23}$$

The α summation covers every distinguishable inherent-structure type, and $\Delta_\alpha \Phi(\mathbf{R})$ is the rise in potential energy in the connected basin B_α above that of its embedded inherent structure:

$$\Delta_\alpha \Phi(\mathbf{R}) = \Phi(\mathbf{R}) - \Phi_\alpha. \tag{III.24}$$

The configurational integrals over each of the basins B_α express the contributions of intrabasin vibrational displacements away from the inherent-structure configurations. At low temperature, the mean vibrational amplitudes are small, so the approximation of harmonic vibrations becomes appropriate. This is the circumstance for which the Hessian diagonalization, Eqs. (III.8)–(III.10), is relevant for identifying normal mode frequencies $\omega_{\alpha i}$ for small excursions from inherent structure α. In this harmonic normal mode approximation, the classical intrabasin contribution to the canonical partition function is

$$\lambda_T^{-3N} \int_{B_\alpha} \exp[-\beta \Delta_\alpha \Phi(\mathbf{R})] d\mathbf{R} \cong \prod_{i=1}^{3N} (\hbar \omega_{\alpha i} \beta)^{-1} \qquad \text{(low temperature)}. \tag{III.25}$$

As temperature rises, vibrational excursions away from the inherent structure become larger on average, so that anharmonic corrections to this harmonic result become significant. These larger amplitude intrabasin vibrations eventually begin to probe regions of negative hypersurface curvature and to encounter and cross the basin boundaries.

As noted earlier, use of periodic boundary conditions causes three normal mode frequencies to disappear as such and to be replaced by free translational motions. In that circumstance, Eq. (III.25) needs to be modified. After taking due account of the unit-Jacobian transformation to harmonic normal mode coordinates, the right side of Eq. (III.25) then becomes

$$\mathrm{N}^{-1/2} V \prod_{i=1}^{3N-3} (\hbar \omega_{\alpha i} \beta)^{-1}; \tag{III.26}$$

however, this expression causes no change to the asymptotic large-system form of the free energy.

In order to take advantage of the intensive order parameter φ to simplify the partition function in the large-system regime, it is useful to introduce an average vibrational partition function for the basins of all inherent structures whose depths fall in the narrow range $\varphi \pm \delta\varphi/2$. This introduction is justified by the huge number of inherent structures implied by the exponentially rising distribution, Eq. (III.17), provided that φ is neither at its lower or upper limits where the enumeration function $\sigma(\varphi, \rho)$ may vanish. Therefore, set

$$\exp[-N\beta f_{vib}(\varphi, \beta, \rho)] = \lambda_T^{-3N} \left\langle \int_{B_\alpha} \exp[-\beta \Delta_\alpha \Phi(\mathbf{R}) d\mathbf{R} \right\rangle_{\varphi \pm \delta\varphi/2} \tag{III.27}$$

where $\langle \dots \rangle_{\varphi \pm \delta\varphi/2}$ denotes the requisite average. As thus defined, f_{vib} represents a mean vibrational free energy per particle, including all anharmonic effects, for that subset of basins lying within a narrow range about the chosen depth parameter φ. If the N-body system exists as a single-phase spatially uniform state, the vast majority of the basins within the narrow depth limits used for the average in the last equation have vibrational properties narrowly clustered about one another and therefore are close to the average itself.

The α summation in Eq. (III.23) may now be converted to an integral over the intensive depth parameter φ, using the density expression Eq. (III.17) and the vibrational free-energy expression (III.27).

$$\exp(-\beta F_N) = C \int_{\varphi_{\min}}^{\varphi_{\max}} \exp\{N[\sigma(\varphi, \rho) - \beta\varphi - \beta f_{vib}(\varphi, \beta, \rho)]\}d\varphi. \tag{III.28}$$

The presence of the large parameter N in the integrand exponent implies that in order to evaluate the Helmholtz free energy per particle, it is only necessary to evaluate that integrand at its maximum; the rest of the integrand is irrelevant for that large-N asymptotic limit. Therefore, define $\varphi^*(\beta, \rho)$ to be the value of the depth parameter φ that maximizes the combination $\sigma - \beta\varphi - \beta f_{vib}$ at the prevailing temperature and density and which therefore locates the basins of preferred occupancy under those conditions. As a consequence of this φ^* property, the Helmholtz free energy per particle adopts the simple form

$$\beta F_N/N = \beta\{\varphi^*(\beta, \rho) + f_{vib}[\varphi^*(\beta, \rho), \beta, \rho]\} - \sigma[\varphi^*(\beta, \rho)]. \tag{III.29}$$

A conceptual advantage of this expression is that it fully separates purely configurational-enumeration (σ) aspects of the statistical mechanics from vibrational (f_{vib}) aspects.

Figure III.1 offers a schematic graphical version of the variational φ^* determination, which involves setting the φ derivative of $\sigma - \beta\varphi - \beta f_{vib}$ equal to zero. The figure separately displays the enumeration function σ and a family of curves for $-\beta(\varphi + f_{vib})$ over a wide temperature range, all as functions of the intensive depth parameter φ. The distinguished quantity φ^* at any given temperature (i.e., at any given β) is identified as the position at which slopes of the σ curve and the relevant $-\beta(\varphi + f_{vib})$ curve have equal magnitudes but opposite signs. These positions are identified in the figure by vertical dashed lines and move monotonically to lower φ as temperature declines (i.e., as β increases). In the high-temperature limit, φ^* is close to the value that produces the

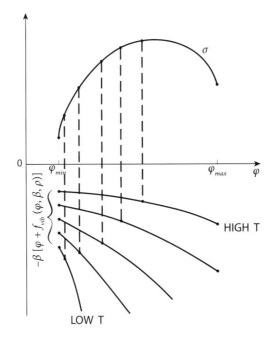

FIGURE III.1. Graphical determination of the dominant inherent structure depth $\varphi*(\beta, \rho)$. The vertical dashed lines, shifting with temperature, locate $\varphi*$ as the position of equal magnitude but opposite signs of slopes for the σ and the $-\beta(\varphi + f_{vib})$ curves.

maximum of σ, while in the low-temperature limit, φ^* should converge to φ_{min}. It should be clear from this construction that most inherent structures on the high-φ side of the σ maximum would not be accessed in states of thermal equilibrium; however, a variety of nonequilibrium preparation methods (such as cold working, radiation damage, or irreversible deposition from the vapor) can direct the system into at least some of the high-lying basins of those inherent structures. In this connection, it is worth recalling the earlier remark that the highest lying basins (those with φ nearly equal to φ_{max}) tend to have small extensions in the multidimensional configuration space because their inherent structures are close to being unstable, and so their vibrational partition functions are correspondingly reduced.

The isobaric variant of Eq. (III.29) expresses the Gibbs free energy per particle in a similarly separated manner:

$$\beta G_N/N = \beta\{\hat{\varphi}^*(\beta, p) + \hat{f}_{vib}[\hat{\varphi}^*(\beta, p), \beta, p]\} - \hat{\sigma}[\hat{\varphi}^*(\beta, p)]. \tag{III.30}$$

By analogy, $\hat{\varphi}^*$ represents the enthalpy per particle for those isobaric basins that dominate the equilibrium statistical mechanics at the prevailing temperature and pressure, and $\hat{f}_{vib}$ is the depth-dependent mean vibrational free energy for this variable-volume scenario. This variant emerges naturally after extending the set of configurational variables to include volume V, using this extended set for steepest descent identification of isobaric basins, then carrying out the same types of partition function transformations as have been applied to the constant-volume canonical partition function. An isobaric version of the graphical construction indicated in Figure III.1 would contain a curve for $\hat{\sigma}(\hat{\varphi})$ and a family of curves $\beta(\hat{\varphi} + \hat{f}_{vib})$ for varying temperature, plotted against the intensive enthalpy parameter $\hat{\varphi}$.

Whether constant-volume or constant-pressure conditions apply, the extension of free-energy expressions (III.29) and (III.30) to multicomponent mixtures is formally straightforward. Let N stand for the total number of particles of all species:

$$N = N_1 + \ldots + N_\nu. \tag{III.31}$$

Then of course, the enumeration functions σ and $\hat{\sigma}$ depend on the respective mole fractions and count the larger numbers of inherent structures arising from the distributions of the distinct species among possible sites. In the average vibrational partition function expression (III.27), the principal format change required is to include the necessary product of mean thermal de Broglie wavelengths λ_{Ti}. Thus, for the constant-volume case, one has

$$\exp[-N\beta f_{vib}(\varphi, \beta, \rho_1 \ldots \rho_\nu)] = \left[\prod_{i=1}^{\nu} \lambda_{Ti}^{-3N_i}\right]\left\langle \int_{B_\alpha} \exp[-\beta\Delta_\alpha \Phi(\mathbf{R})] d\mathbf{R} \right\rangle_{\varphi \pm \delta\varphi/2}. \tag{III.32}$$

Naturally, the boundaries of the basins B_α and the landscape contours within them depend upon the mixture species present. After implementing these straightforward generalizations, Eqs. (III.29) and (III.30) are still applicable for the system Helmholtz and Gibbs free energies, respectively.

Although the fundamental free-energy expressions (III.29) and (III.30) have been derived under the assumption that the classical limit of statistical mechanics applied, the same expressions are still valid in the general quantum-statistical-mechanical regime, provided that the vibrational free-energy functions f_{vib} and $\hat{f}_{vib}$ are properly redefined in terms of averages of intrabasin quantum partition functions. The enumeration functions σ and $\hat{\sigma}$ require no modification because the landscapes that they describe are the same whether they support classical or quantum dynamics for the particles present in the system.

The quantum mechanical canonical partition function (II.19) may be transparently rewritten in the following way:

$$Q_N(\beta) = \sum_a \exp(-\beta E_a) \int \psi_a^*(\mathbf{R})\psi_a(\mathbf{R})d\mathbf{R}, \tag{III.33}$$

where $\psi_a(\mathbf{R})$ is the normalized eigenfunction corresponding to energy E_a. The $\mathbf{R}$ integration that spans the entire $3N$-dimensional configuration space can be resolved into separate contributions from each of the $N!$-member basin families (for impenetrable walls), so Eq. (III.33) may be transformed to

$$Q_N(\beta) = \sum_a \exp(-\beta\Phi_\alpha) \int_{B_\alpha} \sum_a \exp[-\beta(E_a - \Phi_\alpha)][N!\psi_a^*(\mathbf{R})\psi_a(\mathbf{R})]d\mathbf{R}. \tag{III.34}$$

Here the $N!$ appearing as an integrand factor can be viewed as compensating for the fact that the normalized wave functions are spread over all $N!$ permutation replicas of basin B_α. The integral shown is the proper identification of the intrabasin vibrational partition function for basin B_α. As before in analysis of the classical canonical partition function, it makes sense for the large-system asymptotic limit to classify basins by φ, the inherent structure's potential energy per particle, and consequently to define the mean vibrational free energy by lumping together all basins within a narrow φ range. Consequently, one has the following quantum version of the classical Eq. (III.27):

$$\exp[-N\beta f_{vib}(\varphi, \beta, \rho)] = \left\langle \int_{B_\alpha} \sum_a \exp[-\beta(E_a - \Phi_\alpha)][N!\psi_a^*(\mathbf{R})\psi_a(\mathbf{R})]d\mathbf{R} \right\rangle_{\varphi \pm \delta\varphi/2}. \tag{III.35}$$

By utilizing this definition for f_{vib}, the mean vibrational free energy per particle, the earlier Helmholtz free energy expression (III.29) continues to be valid, along with relevance of the graphical construction indicated in Figure III.1.

Suppose that basin B_α is deep and accurately harmonic, even for large displacements from its inherent structure. Suppose further that this basin is configurationally well isolated so as not to be significantly inhabited by tunneling states lower in energy E_a than its bottom at Φ_α. Then this basin's vibrational partition function becomes a simple result [Hill, 1960, Chapter 5]:

$$\prod_{i=1}^{3N} \exp(\hbar\omega_{\alpha i}\beta/2)/[\exp(\hbar\omega_{\alpha i}\beta) - 1], \tag{III.36}$$

which reduces to the classical result shown in Eq. (III.25) in the asymptotic regime where all $\hbar\omega_{\alpha i}\beta$ approach zero.

The applications of the quantum extension to isobaric circumstances and to mixtures both in isochoric and isobaric ensembles are sufficiently straightforward in principle, given the analysis just described, that they do not require additional discussion.

E. Phase Transitions—An Overview

In a state of thermal equilibrium for which a single phase spans the entire system volume, any small continuous change in control parameters (temperature, and volume or pressure) only causes corresponding small continuous changes in thermodynamic functions and in molecular distribu-

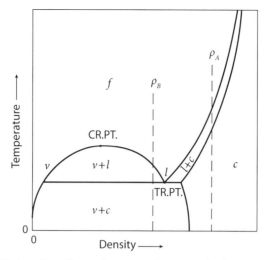

FIGURE III.2. Generic equilibrium-phase diagram for a pure substance, in the density–temperature plane, illustrating liquid–vapor, liquid–crystal, and crystal–vapor transitions. Here liquid, crystal, and vapor regions are denoted by *l*, *c*, and *v*, and the triple point and critical point are abbreviated TR.PT. and CR.PT. The letter *f* encompasses the subcritical liquid and vapor regions, as well as the supercritical fluid region. It has been implicitly assumed that only a single isotropic fluid phase *f* arises, i.e., no anisotropic liquid crystal phases.

tion functions. Indeed these changes would be arbitrarily-many-times differentiable as functions of those control variables. The quantities φ^*, $\sigma(\varphi^*)$, and $f_{vib}(\varphi^*, \beta, \rho)$ appearing as separate components of the Helmholtz free energy, Eq. (III.29), or equivalently $\hat{\varphi}^*$, $\hat{\sigma}(\hat{\varphi}^*)$, and $\hat{f}_{vib}(\hat{\varphi}^*, \beta, p)$ for the Gibbs free energy, Eq. (III.30), likewise would be arbitrarily-many-times differentiable in the control variables.

However, such smooth behavior is interrupted upon encountering a phase transition. The schematic Figure III.2 shows a typical equilibrium-phase diagram for a single-component system in the density–temperature plane and serves to illustrate what qualitative features would be involved in its first-order phase transitions. It relates specifically to substances for which melting of the crystal phase under constant-pressure conditions results in a volume increase. Two constant-density loci, at ρ_A and ρ_B, have been explicitly designated. The former, at the higher density ρ_A, begins at high temperature within the single phase region of the supercritical fluid, but as temperature declines, this locus encounters the fluid freezing point for that density, $T_{ff}(\rho_A)$. Further cooling causes a downward traversal of the fluid–crystal coexistence region, until the system fully crystallizes at the lower temperature for crystal melting at density ρ_A, $T_{cm}(\rho_A)$. Subsequent temperature reduction to absolute zero leaves the system in a volume-spanning uniform crystal phase.

Let $\rho_{ff}(T)$ and $\rho_{cm}(T)$ represent the number densities, respectively, along the fluid-freezing and crystal-melting curves that bound the liquid–crystal coexistence region in Figure III.2. When the temperature is such that the system inhabits this coexistence region while confined to the ρ_A locus, one has

$$\rho_{ff}(T) \le \rho_A \le \rho_{cm}(T). \tag{III.37}$$

It is then easy to see [cf. Eqs. (II.109)] that the macroscopic fluid phase occupies a portion V_f of the entire volume V, while the coexisting crystal occupies the remainder $V_c = V - V_f$, where

$$\frac{V_f(T)}{V} = \frac{\rho_{cm}(T) - \rho_A}{\rho_{cm}(T) - \rho_{ff}(T)}, \tag{III.38}$$

$$\frac{V_c(T)}{V} = \frac{\rho_A - \rho_{ff}(T)}{\rho_{cm}(T) - \rho_{ff}(T)}.$$

The existence of a positive interfacial free energy between contacting fluid and crystal phases can be inferred from the fact that the coexisting phases remain geometrically distinct [Landau and Lifshitz, 1958b], and the requirement that the total interfacial free energy be at a minimum implies that the macroscopic phase regions with volumes V_f and V_c would be connected and would have appropriate shapes. In view of the fact that crystal–fluid interfacial free energies in general depend on crystal-plane direction [Herring, 1951], inferring the shapes of those minimal surfaces can be a nontrivial problem.

The lower density locus in Figure III.2, at ρ_B, involves similar considerations, but three distinct phases come into play. As temperature declines along this locus from the supercritical regime, liquid–vapor coexistence is the first phase transition encountered. Between the critical-point temperature and the triple-point temperature, let $\rho_{lb}(T)$ stand for the equilibrium liquid-boiling curve. Also, let its equilibrium vapor-condensation counterpart be denoted by $\rho_{vc}(T)$. Then for that portion of the ρ_B locus that falls within the liquid–vapor coexistence area, one has

$$\rho_{vc}(T) \leq \rho_B \leq \rho_{lb}(T). \tag{III.39}$$

Furthermore, analogs to the Eqs. (III.38) specify the relative portions of the full-volume V that are occupied by the vapor (v) and liquid (l) phases:

$$\frac{V_v(T)}{V} = \frac{\rho_{lb}(T) - \rho_B}{\rho_{lb}(T) - \rho_{vc}(T)}, \tag{III.40}$$

$$\frac{V_l(T)}{V} = \frac{\rho_B - \rho_{vc}(T)}{\rho_{lb}(T) - \rho_{vc}(T)}.$$

Because liquid–vapor interfacial free energy (surface tension) has no direction dependence, in contrast with the cases involving a crystal phase, determining the minimal surfaces that separate the liquid and vapor regions is straightforward. If the overall volume V is cubical, and if periodic boundary conditions are present, the interface adopts one of three possible shapes. The interface is spherical if either the liquid or the vapor comprises a relatively small portion of the entire volume. Specifically, this phenomenon occurs when

$$0 < V_v/V, \ V_l/V < 4\pi/81 \cong 0.1551 \qquad \text{(sphere)}. \tag{III.41}$$

However, the minimal-surface geometry converts to a circular cylinder parallel to a cube edge when

$$4\pi/81 < V_v/V, \ V_l/V < 1/\pi = 0.3183 \qquad \text{(cylinder)}. \tag{III.42}$$

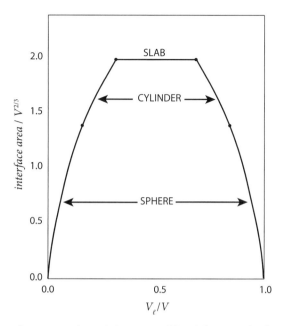

FIGURE III.3. Interfacial area for macroscopic coexisting vapor and liquid phases, confined to a cubical box with periodic boundary conditions, plotted against the fraction of the total volume occupied by the liquid. Separate regimes of minimal surface geometry have been delineated by bold dots.

Finally, a slab geometry with planar interfaces parallel to one of the cube faces prevails when

$$1/\pi < V_v/V,\ V_l/V < 1 - 1/\pi \qquad \text{(slab)}. \tag{III.43}$$

Consequences of these alternatives are illustrated in Figure III.3, where the reduced interface area has been plotted vs. the volume fraction occupied by the liquid phase.

For that portion of the ρ_B locus in Figure III.2 lying below the triple-point temperature, vapor and crystal phases coexist. The respective volume fractions can be expressed in the same manner as shown in Eqs. (III.38) and (III.40):

$$\frac{V_v(T)}{V} = \frac{\rho_{cs}(T) - \rho_B}{\rho_{cs}(T) - \rho_{vc}(T)}, \tag{III.44}$$

$$\frac{V_c(T)}{V} = \frac{\rho_B - \rho_{vc}(T)}{\rho_{cs}(T) - \rho_{vc}(T)}.$$

Here $\rho_{cs}(T)$ represents the equilibrium crystal sublimation curve, and $\rho_{vc}(T)$ represents the extension (not necessarily an analytic continuation!) of the equilibrium vapor condensation curve to the range below the triple-point temperature. As in the liquid–crystal coexistence case, the anisotropy of crystal interface free energy complicates determination of the minimal interface surface shapes [Herring, 1951].

When the temperature is that of the triple point, T_{tp}, and ρ_B is located as shown in Figure III.2, all three phases, vapor, liquid, and crystal, can simultaneously coexist. The overall number density

ρ_B is then not sufficient to determine what fractions of the system volume V would be occupied by each of the three phases. Another intensive thermodynamic variable needs to be specified. In particular, the overall system energy per particle can serve in this capacity. Because each of the three coexisting phases at T_{tp} possesses its own distinct value of energy per particle, the system's overall energy per particle depends nontrivially on the relative amounts of those phases present. For a given assignment of the phase volumes so determined, the shapes of the three coexisting phase regions and how they contact one another depend on the three respective interfacial free-energy functions.

Chapter VII examines the formation and detailed characteristics of inherent structures for dilute vapor under constant-volume steepest descent mapping. These inherent structures typically fill the volume occupied by the premapped vapor but as sparse "sponges" or "aerogels." Such open structures can be multiply connected networks of fragile filamentary bridges. In the case of equilibrium states that involve vapor as one of two or more coexisting phases, one expects constant-volume steepest descent mapping to generate these kinds of open architectures within the macroscopic vapor regions.

A technical issue arises concerning steepest descent mapping when the system exhibits phase coexistence. Inherent structures that underlie phase-coexistence states, and that are identified by isochoric (constant V) steepest descent mapping, retain the large-scale spatial patterns of the initial phase shapes, at least in a qualitative manner. However, one should keep in mind that the mapping operation may induce modest changes in those region shapes belonging to the separate phases, including small relative changes from the premapping volumes of those regions, as specified by Eqs. (III.38), (III.40), and (III.44). Such changes do not strictly conform to either isochoric (constant V) or isobaric (constant p) mappings for the homogeneous single-phase cases, thus resulting in small shifts in the enumeration functions and the related vibrational free-energy functions. However, these small shifts compensate one another so far as the large-system-limit free energies are concerned. In that respect, it suffices simply to interpolate the homogeneous phase functions and values with weightings linearly proportional to the relative amounts of the two phases in the coexistence state. In particular, this means that for coexisting phases γ and δ, respectively containing N_γ and $N_\delta = N - N_\gamma$ particles, the system's enumeration function at its free-energy-maximizing value can be assigned the form

$$\sigma(\varphi^*, \rho) = (N_\gamma/N)\sigma_\gamma(\varphi_\gamma^*, \rho_\gamma) + (N_\delta/N)\sigma_\delta(\varphi_\delta^*, \rho_\delta), \tag{III.45}$$

$$\varphi^* = (N_\gamma/N)\varphi_\gamma^* + (N_\delta/N)\varphi_\delta^*,$$

and the vibrational free-energy function is then assigned the corresponding form

$$f_{vib}(\varphi^*, \beta, \rho) = (N_\gamma/N)f_{vib,\gamma}(\varphi_\gamma^*, \beta, \rho_\gamma) + (N_\delta/N)f_{vib,\delta}(\varphi_\delta^*, \beta, \rho_\delta). \tag{III.46}$$

In these expressions, the intensive quantities $\sigma_\gamma, \sigma_\delta, f_{vib,\gamma}, f_{vib,\delta}, \varphi_\gamma^*, \varphi_\delta^*, \rho_\gamma, \rho_\delta$ are the functions and values that pertain to the pure phases at the transition. Use of Eqs. (III.45) and (III.46) in the canonical partition function leads to identification of an effective system basin depth function $\varphi^*(\beta, \rho)$ that as a function of overall system density ρ interpolates linearly between its values for the single phases.

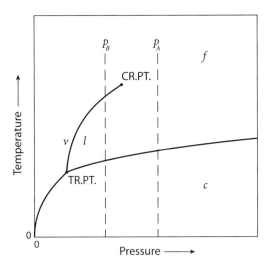

FIGURE III.4. Generic phase diagram for a pure substance in the pressure–temperature plane. Identifying symbols are the same as in Figure III.2.

A more detailed analysis of coexistence mapping is required in order to extract properties of the interfacial free energy and structure. Some aspects of these higher order corrections are analyzed in Chapter V, Section V.H.

Figure III.4 presents a generic phase diagram analogous to that in Figure III.2 but plotted in the pressure–temperature plane. This alternative is relevant to steepest descent mapping (including V as a configuration variable) on the potential enthalpy hypersurface [$\hat{\Phi}$, Eq. (III.4)] under constant-pressure (isobaric) conditions. The vertical locus shown for pressure p_A passes through the first-order transition between the supercritical fluid at high temperature and the crystal phase. The lower pressure p_B locus, upon cooling, passes first through the subcritical vapor–liquid condensation, then through the liquid–crystal freezing point, also first-order transitions.

In contrast to the previously considered constant-volume circumstance, the constant-pressure condition does not produce phase coexistence with determinate volumes occupied by each of the distinguishable phases. Just at the intersection point of the p_A locus with the fluid–crystal transition curve, the system exhibits distinctly bimodal character, virtually always displaying just one of the two phases at a time, with the system volume essentially equal, respectively, to that required by just pure fluid or by just pure crystal. Similarly, the intersections of the p_B locus with the vapor–liquid condensation curve and with the liquid–crystal freezing–melting curve entail bimodal distributions for system volumes and configurations for just one at a time of the two phases in thermodynamic equilibrium. The reason for this bimodality stems from the positive interfacial free energy that necessarily accompanies phase coexistence, which would be associated with interfacial area such as that plotted in Figure III.3. Under these constant-pressure conditions, that interfacial free-energy cost, of the order $N^{2/3}$, strongly diminishes the occurrence probability of system states with both macroscopic phases present. Of course, these statements refer to the large-system-limit behavior; in a sufficiently small system, intermediate states with apparent phase coexistence would occasionally appear.

F. Interbasin Transitions

With the exception of the lowest temperature extreme, dynamical motions in a many-body system inevitably involve transitions between neighboring basins on the potential energy, or potential enthalpy, landscape. This is true whether the system is in a state of thermal equilibrium, out of equilibrium and relaxing toward equilibrium, or even trapped more or less permanently in a macroscopic metastable state such as a vitreous solid. Therefore, any comprehensive analysis of many-body landscapes inevitably must include a detailed characterization of interbasin transitions.

By definition, interbasin transitions are crossings of shared basin boundaries. In the simple case of N structureless particles, these boundaries are $(3N - 1)$-dimensional hypersurfaces for constant-volume conditions within containing walls, $3N$-dimensional hypersurfaces for corresponding constant-pressure conditions. If the total energy of the system is sufficiently high, a dynamical transition could in principle penetrate a basin-bounding surface at any one of its points. However, the threshold energy for any given interbasin transition is determined by the height of the lowest simple saddle point embedded in the shared boundary. Furthermore, the dominant transition paths for energies just above that threshold of necessity cross the shared boundary in the immediate vicinity of that simple saddle point. Consequently, it is natural to begin examination of interbasin transitions with analysis of saddle points themselves.

Saddle points of all orders (as well as inherent structures) can be located on any potential energy or potential enthalpy hypersurface by finding absolute minima (equal to zero) of the isochoric gradient-squared function $\Gamma(\mathbf{R})$, Eq. (III.11), or of its isobaric analog. But one must keep in mind that saddle points need to be distinguished by whether they reside on shared basin boundaries or whether they occur strictly within the interior of a single basin. All steepest descent trajectories passing near the latter converge to the single inherent structure of the host basin. By contrast, steepest descent trajectories passing near a boundary-embedded saddle point fall into distinct subsets, according to the specific inherent structures to which they converge. A simple saddle point (one negative curvature) embedded in a shared boundary has only two distinct trajectory subsets, corresponding to the pair of inherent structures that share the boundary and that are dynamically connected by transitions through the neighborhood of that saddle point.

Elementary transitions between two boundary-sharing basins can be usefully classified by a "participation ratio." Let $\Delta\mathbf{r}_i$ denote the difference in respective locations of particle i in the two inherent structures α and α' for those contiguous basins:

$$\Delta\mathbf{r}_i(\alpha \rightarrow \alpha') = \mathbf{r}_i(\alpha') - \mathbf{r}_i(\alpha). \tag{III.47}$$

[*Note*: If periodic boundary conditions are in effect, the inherent structures should be translated relative to one another so that their centroids coincide.] The participation ratio is defined by the following combination of mean-square and mean-fourth-power values of the inherent displacements for all N particles [Stillinger and Weber, 1983b; Weber and Stillinger, 1985a]:

$$P(\alpha, \alpha') \equiv P(\alpha', \alpha)$$

$$= \left[N^{-1} \sum_{i=1}^{N} (\Delta\mathbf{r}_i)^4 \right] / \left[N^{-1} \sum_{i=1}^{N} (\Delta\mathbf{r}_i)^2 \right]^2$$

$$\equiv \overline{(\Delta\mathbf{r}_i)^4} / [\overline{(\Delta\mathbf{r}_i)^2}]^2. \tag{III.48}$$

In the hypothetical situation in which all N particles experienced displacements with exactly the same magnitude, the participation ratio would equal unity. In the hypothetical opposite extreme in which only a single particle is displaced, all others remaining fixed, the participation ratio would equal N. If the displacement vectors for all N particles conformed to a single three-dimensional Gaussian distribution, $P(\alpha, \alpha')$ would be 5/3.

Various simulations of single-phase condensed matter systems at low to moderate temperatures reveal that the typical participation ratios tend to be substantially larger than unity [Weber and Stillinger, 1985a; Wales and Uppenbrink, 1994]. Consequently, the basic interbasin transitions can be described as local rearrangements of only $O(1)$ particles that leave most of the N particles virtually undisturbed. The change in potential energy required to surmount a transition state between two inherent structures would likewise be an $O(1)$ quantity. In a very large system of particles, these localized rearrangements would tend to be isolated from one another, and thus independent. This implies that if some modest number, say $1 < n \ll N$, of these localized rearrangements were to occur simultaneously, the configuration space description would involve surmounting an nth-order saddle, with its n negative curvatures respectively describing the independent reordering processes. Carrying the estimate a bit further, it suggests that the mean potential energy rise necessary to surmount saddles is proportional to the number of negative Hessian eigenvalues exhibited at those saddles. Indeed, just this proportionality has been reported in a simulation study [Broderix et al., 2000].

The localization of the elementary interbasin transitions has further important consequences for characterizing the multidimensional landscape and the dynamics it supports. Because the number of locales in an N-body system at which $O(1)$ particle rearrangements can occur is itself $O(N)$, so too are the expected number of simple transition states in any basin's boundary $O(N)$. Within the regime of classical statistical mechanics, this fact in turn implies that residence lifetime in any given basin, except at the very lowest temperatures, is proportional to N^{-1} [Stillinger and Weber, 1983b].

G. Metastable Phases

An important experimental fact is that strict thermodynamic equilibrium can often be frustrated at first-order phase transitions. Because of the sluggishness of nucleation processes, metastable extensions of one phase can often be followed into what for strict thermodynamic equilibrium under isochoric conditions should be a coexistence region for two phases, or a completed phase change under isobaric conditions. The metastable phases frequently encountered in this scenario include undercooled liquids, overheated solids and liquids, overcompressed and supercooled vapors, and crystal phases that are not those of lowest free energy. In each of these cases, the homogeneous single metastable phase can exhibit experimentally reproducible properties that are smooth extensions of those in the stable phase region. This section presents a conceptually straightforward modification of the strict equilibrium theory, again based on inherent structures and the basin tiling of the configuration space, which is specifically designed to describe metastable phases.

A simple chemical example offers a vivid example of metastability and how it should be described in terms of the potential energy or potential enthalpy landscape. At room temperature or below, and at ambient pressure, the molecular gases hydrogen (H_2) and oxygen (O_2) can be mixed, and the properties of that unreacted mixture can be measured reproducibly. In this binary mixture, there are high potential barriers to disruption of the covalent chemical bonds within the diatomic molecules, rendering such chemical processes unmeasurably slow. However, it is obvious that a far lower energy in this system would be obtained by catalyzing the exothermic reaction between these substituents to form water molecules (H_2O) to the extent possible and then cooling the result. In order to attain a statistical mechanical description of the unreacted mixture, it is clear that one should focus just on those inherent structures and their steepest descent basins in the multidimensional landscape that exhibit only intact diatomic hydrogen and oxygen molecules. In other words, all basins failing to meet this strict criterion should be projected out of consideration. The relevant statistical mechanical description would then be confined to the surviving subspace of the full configuration space, whether constant volume or constant pressure conditions were involved.

The general approach to be followed to describe metastable phases associated with first-order transitions adopts the same kind of strategy as for the preceding $H_2 + O_2$ example. It involves discriminating inherent structures and their surrounding basins according to their structural relevance for the metastable phase of interest [Stillinger, 1988a]. In particular, the geometric patterns presented by the inherent structures are undisturbed by vibrational distortions and therefore should be the clearest configurational identifiers of legitimate contributions to the homogeneous phase whose metastable extension is under consideration. In the case of an undercooled liquid (see Chapter VI), one eliminates basins whose inherent structures exhibit significant clusters or domains of particles arranged in the "undesirable" crystal patterns. For an overheated liquid, any inherent structure containing a substantial void, or bubble, would be eliminated, while the reverse case of an undercooled vapor would require elimination of inherent structures displaying large compact high-density regions. In all of these cases, it must be assumed that a relevant pattern-recognizing capacity is available to project out of consideration the undesirable inherent structures and their basins. For all such metastable phase extensions, the geometric boundary between the allowed and disallowed portions of the multidimensional configuration space has been constructed to consist of parts of basin boundaries, specifically those parts shared by an allowed and a disallowed basin. The retained basin subset is where the metastable-phase system kinetics resides during the reproducible measurement duration.

Let subscript m denote the metastable phase of interest. A consequence of the projection operation is that the isochoric inherent-structure depth order parameter φ for the accepted basins may have a shortened interval of existence compared to that for the full set of basins [Eq. (III.16)]:

$$\varphi_{\min} \leq \varphi_{\min,m} \leq \varphi \leq \varphi_{\max,m} \leq \varphi_{\max}. \tag{III.49}$$

Over this possibly shortened interval, the basin enumeration is also subject to the obvious inequality:

$$0 \leq \sigma_m(\varphi) \leq \sigma(\varphi). \tag{III.50}$$

The vibrational free-energy function for the restricted basin set $f_{vib,m}(\varphi, \beta, \rho)$ in general does not obey an inequality with that for all basins $f_{vib}(\varphi, \beta, \rho)$. However, just as before, it is an average over basin partition functions for the specified depth, now within just the restricted basin set.

The Helmholtz free energy per particle has the same format as in the equilibrium situation, shown in Eq. (III.29), but with the projected-basin-set quantities

$$\beta F_{N,m}/N = \beta\{\varphi_m{}^*(\beta, \rho) + f_{vib,m}[\varphi_m{}^*(\beta, \rho), \beta, \rho]\} - \sigma_m[\varphi_m{}^*(\beta, \rho)]. \tag{III.51}$$

The quantity $\varphi_m{}^*$ is the free-energy minimizing order parameter for this metastable-phase projected case. It should be realized that in the equilibrium single-phase portion of the β, ρ plane, each of the m-subscripted quantities in expression (III.51) reduces almost exactly to the conventional equilibrium values that have no subscript; that is, when only a single homogeneous phase is thermodynamically stable, the projection operation is substantially irrelevant. The projection permits smooth extrapolation beyond the formal phase transition point into the regime of metastability.

The isobaric version of metastable-phase Helmholtz free energy in expression (III.51) involves a corresponding expression for the Gibbs free energy per particle. The metastable-phase version of Eq. (III.30) incorporates the following straightforward modification paralleling Eq. (III.51):

$$\beta G_{N,m}/N = \beta\{\hat{\varphi}_m{}^*(\beta, p) + \hat{f}_{vib,m}[\hat{\varphi}_m{}^*(\beta, p), \beta, p]\} - \hat{\sigma}_m[\hat{\varphi}_m{}^*(\beta, p)]. \tag{III.52}$$

Before closing this section, two related points need to be mentioned. First, the specific projection criterion to be implemented for a metastable state extension involves some arbitrariness. As one example of this feature, specifically in the case of a supercooled liquid, this would concern choice of the lower size limit and of the structural perfection of crystalline clusters or domains to be excluded in the inherent structures. For an overheated liquid, this would involve the choice of maximum allowable vapor bubble size and shape. Nevertheless, when the system of interest is known to resist nucleation strongly at least for moderate penetration of the metastable phase regime, it is reasonable to suppose that the predictions of metastable phase behavior are quite insensitive to modest and reasonable variations in these parameters.

The second point is primarily mathematical. It arises from the fact that statistical mechanical theory of strict equilibrium behavior predicts that in principle metastable extensions are prevented by the existence of essential singularities in all infinite-system-limit thermodynamic functions that are located precisely at the first-order phase transition [Andreev, 1964; Fisher, 1967b; Fisher and Felderhof, 1970]. Such singularities have very small numerical influence on the thermodynamic properties in the equilibrium range, even at the transition itself, but nevertheless in principle prevent analytic continuation (in density, temperature, or other intensive control parameters) of the properties from the equilibrium range into the metastable range. The source of these essential singularities is the presence of very low concentrations of spontaneously forming and decaying very small local particle clusters of the "other" phase, even before the experimental control parameters reach values allowing that phase kinetically to nucleate and become macroscopically dominant. Imposition of the basin projection strategy outlined above necessarily truncates the spontaneous cluster distribution, eliminating its large-cluster tail. Such truncation eliminates essential singularities. A justification is that experimental protocols applied in the study of metastable phases have a similar truncating effect, by discarding any experimental run that spontaneously nucleates, and at least for modest intrusions beyond the thermodynamic transition generate reproducible measurements.

H. Distribution Function Mapping

The removal of intrabasin vibrational displacements that results from steepest descent mapping on potential energy or potential enthalpy hypersurfaces inevitably affects the various n-particle distribution functions. In simple qualitative terms, the vibrational displacements have a smoothing or smearing effect on the geometric order that is present in the many-body system under consideration. Examination of the distribution functions following the steepest descent mapping restores image detail for inherent structural order conveyed by those functions. Some technical details of distribution function behavior under that mapping are now examined.

The canonical-ensemble-averaged configurational distribution function for an N-particle system confined to a fixed volume V, denoted by $\bar{P}_N(\mathbf{R}, \beta)$, has been introduced earlier [Eq. (II.21)]. It describes systems subject to quantum effects, as well as those in the classical limit. Here $\mathbf{R}$ stands for the full set of $3N$ configurational coordinates, and as usual $\beta = 1/k_B T$. As a result of steepest descent mapping on the $\Phi(\mathbf{R})$ hypersurface, this normalized distribution deforms into a vibrationally quenched distribution [identified by a superscripted (q)] that is also normalized:

$$\bar{P}_N(\mathbf{R}, \beta) \xrightarrow[\text{mapping}]{} \bar{P}_N^{(q)}(\mathbf{R}, \beta). \tag{III.53}$$

More specifically, steepest descent mapping does not change the occupancy probability in basin B_α for any inherent structure α:

$$\int_{B_\alpha} \bar{P}_N(\mathbf{R}, \beta) d\mathbf{R} = \int_{B_\alpha} \bar{P}_N^{(q)}(\mathbf{R}, \beta) d\mathbf{R}. \tag{III.54}$$

But whereas $\bar{P}_N$ represents probability spread over the interior of each occupied basin B_α, $\bar{P}_N^{(q)}$ is concentrated at the inherent-structure configuration as a suitably weighted Dirac delta function. These remarks for the isochoric canonical ensemble can be straightforwardly reinterpreted for an isobaric ensemble, with premapping and postmapping distributions $\hat{P}_N(\mathbf{R}, \beta)$ and $\hat{P}_N^{(q)}(\mathbf{R}, \beta)$ and symbol $\mathbf{R}$ then representing the particle configuration coordinates augmented by system volume V.

Reduced distribution functions for the canonical ensemble were introduced in Chapter II, Section II.E. For a single-component system (including spins, if any, that are in identical states), with $1 \leq n \leq N - 1$, the definition of the n-point distribution function is the following:

$$\rho_{n,N}(\mathbf{r}_1 \ldots \mathbf{r}_n) = [N!/(N - n)!] \int d\mathbf{r}_{n+1} \ldots \int d\mathbf{r}_N \bar{P}_N(\mathbf{r}_1 \ldots \mathbf{r}_N, \beta), \tag{III.55}$$

where it is understood implicitly that this function depends on temperature and density. The motivation for this definition rests upon the fact that the probability of simultaneous occurrence of any n particles in the differential volume elements $d\mathbf{r}_1 \ldots d\mathbf{r}_n$ respectively at $\mathbf{r}_1 \ldots \mathbf{r}_n$ is $\rho_{n,N}(\mathbf{r}_1 \ldots \mathbf{r}_n) d\mathbf{r}_1 \ldots d\mathbf{r}_n$. Precisely parallel to this definition is that for postmapping n-point distribution functions:

$$\rho_{n,N}^{(q)}(\mathbf{r}_1 \ldots \mathbf{r}_n) = [N!/(N - n)!] \int d\mathbf{r}_{n+1} \ldots \int d\mathbf{r}_N \bar{P}_N^{(q)}(\mathbf{r}_1 \ldots \mathbf{r}_N, \beta). \tag{III.56}$$

The simplest of these premapping and postmapping reduced distributions are those for $n = 1$, describing the single-particle density distribution throughout the system volume V. When periodic boundary conditions are present, allowing free translation of the N-particle system, then for structureless particles one has

$$\rho_{1,N}(\mathbf{r}_1) = \rho_{1,N}^{(q)}(\mathbf{r}_1) = N/V \equiv \rho. \tag{III.57}$$

However, with impenetrable walls surrounding the system, this translation invariance is broken. In that case, even when a nominally uniform fluid phase fills the system, the singlet density is not constant near such walls, and application of the steepest descent mapping amplifies the local wall-region variations in singlet density from the macroscopic expectation value ρ. Alternatively, if the thermodynamic equilibrium phase is crystalline, and if N, the shape of volume V, and the wall forces are conducive to retaining that crystal in a fixed orientation, then throughout the major portion of V away from the walls $\rho_{1,N}(\mathbf{r}_1)$ is a periodic function of position $\mathbf{r}_1$. The vibration-quenched singlet density function $\rho_{1,N}^{(q)}(\mathbf{r}_1)$ exhibits the same spatial periodicity, but with enhanced amplitude, assuming that the boundary conditions continue to clamp the crystal in the same fixed orientation and position in contact with the walls. If the crystalline state under study contains a small concentration of structural defects, their presence and geometric characteristics would be more obvious in $\rho_{1,N}^{(q)}(\mathbf{r}_1)$ than in $\rho_{1,N}(\mathbf{r}_1)$.

The premapped and postmapped pair distribution functions $\rho_{2,N}(\mathbf{r}_1, \mathbf{r}_2)$ and $\rho_{2,N}^{(q)}(\mathbf{r}_1, \mathbf{r}_2)$ play an especially important role in characterizing isotropic fluid states. Away from any confining walls, these functions for structureless particles would depend configurationally only on scalar pair distance r_{12}. For large systems, they would then approach the same uncorrelated asymptotic limit at large separation:

$$\rho_{2,N}(r_{12}), \rho_{2,N}^{(q)}(r_{12}) \sim \rho^2. \tag{III.58}$$

The fluid state's short-range order is at least partially revealed by the small-r_{12} behavior of these functions, and the steepest descent mapping process enhances that short-range order and should aid in its statistical interpretation.

Numerical studies of several classical models in their equilibrium liquid states have compared pair distribution functions $\rho_{2,N}$ and $\rho_{2,N}^{(q)}$ computed, respectively, for the systems' configurational distributions before and after application of the steepest descent operation. These investigations have demonstrated a dramatic sharpening of the image of short-range order in simulations devoted to models of various monatomic substances [Stillinger and Weber, 1984a; Weber and Stillinger, 1984; Stillinger and Weber, 1985b; Qi et al., 1992]. Under constant density conditions for systems in equilibrium thermodynamic states, comparison of the premapped and postmapped functions demonstrates that most of the temperature dependence of $\rho_{2,N}$ for these simple models (spherically symmetric additive pair interactions) can be attributed just to the temperature dependence of intrabasin vibrational motions, not to variation in the depth population of basins occupied. For simulations involving polyatomic molecules, the quenching of vibrational smearing can allow otherwise-unattainable resolution of intramolecular ordering; one clear example arose from the investigation of a united-atom model for liquid cyclohexane (C_6H_{12}) [Harris and Stillinger, 1991].

For simple monatomic substances, the pair distribution functions before and after mapping provide the distribution of pair separations. However, knowing the distributions of those pair separations is insufficient in principle to determine the corresponding distributions of triangles of different shapes and sizes that are present. The extra information required of course resides in the triplet distribution functions $\rho_{3,N}(\mathbf{r}_1, \mathbf{r}_2, \mathbf{r}_3)$ and $\rho_{3,N}^{(q)}(\mathbf{r}_1, \mathbf{r}_2, \mathbf{r}_3)$. An analogous situation concerns additional information required for occurrence probabilities for particle tetrahedra; in particular, $\rho_{4,N}$ and $\rho_{4,N}^{(q)}$ would be required to distinguish relative occurrence probabilities for left- and right-handed chiral mirror-image tetrahedra in the premapped and postmapped ensembles.

IV.

Crystal Phases

The most easily understood and mathematically well characterized portions of a potential energy landscape are those corresponding to spatial periodic order of the constituent particles, i.e., crystals. Crystal symmetry simplifies analysis and renders feasible many otherwise formidable condensed-phase calculations. Examples of such feasible calculations include identification of absolute minimum potential energy configurations, construction of phonon spectra, and enumeration and quantitative study of the low-lying excited-state basins (point defect configurations) that flank the basins for the absolute minima. This chapter aims to provide an overview of the energy landscape/inherent-structure representation of crystal phases, including temperature effects up to the thermodynamic melting point. Consideration of strong quantum effects in nuclear motions is for the most part reserved for consideration in Chapter VIII.

Solid-state physics is a vast subject that deals with an enormous collection of fascinating and technologically important phenomena. The energy landscape/inherent-structure representation provides a useful framework within which to analyze many of those phenomena. This chapter necessarily has modest scope, only touching a few basics for the crystalline state. This leaves for detailed examination elsewhere such engaging topics as dislocation formation and motion, martensitic transformations and shape-memory materials, surface roughening transitions, and materials displaying negative thermal expansion or auxetic (negative Poisson ratio) behavior.

A. Bravais Lattices

The defining characteristic of crystals is their spatially periodic structure. These are structures into which the constituent particles spontaneously arrange themselves, given proper circumstances. Details of those periodic arrangements are especially vivid when viewing the inherent structures for the crystal, and they reflect the nature of the interactions that are present. The smallest periodic unit defines a primitive cell whose shape is a parallelepiped. Three edges of that cell, emanating from one of its vertices, can be specified by a set of three linearly independent vectors $\mathbf{a}_1$, $\mathbf{a}_2$, $\mathbf{a}_3$. The volume v_0 of the primitive unit is the magnitude of the scalar triple product,

$$v_0 = |\mathbf{a}_1 \cdot (\mathbf{a}_2 \times \mathbf{a}_3)|. \tag{IV.1}$$

All possible linear combinations of these vectors with coefficients that are positive or negative integers, or zero, define an infinite "Bravais lattice" [Ashcroft and Mermin, 1976, Chapter 4]:

$$\mathbf{A}(n_1, n_2, n_3) = n_1\mathbf{a}_1 + n_2\mathbf{a}_2 + n_3\mathbf{a}_3, \qquad \text{(IV.2)}$$

$$n_1, n_2, n_3 = 0, \pm 1, \pm 2, \dots .$$

This lattice of vectors $\mathbf{A}$ represents the entire set of vertices for the primitive cell and all its periodic images that tile unbounded three-dimensional space.

The primitive cell and all its images, for a structurally perfect crystal, each contain an integer number W of particle positions. We take these to be the ideal positions of the nuclei, whether the particles are distinct atoms, ions, or chemically bound molecules. The locations of those nuclei within any periodic unit (cell) can be expressed in the following way:

$$\mathbf{r}_i(n_1, n_2, n_3) = \mathbf{A}(n_1, n_2, n_3) + \mathbf{w}_i \qquad (1 \leq i \leq W), \qquad \text{(IV.3)}$$

where the displacement vectors $\mathbf{w}_i$ must lie within the same cell, one of whose vertices is at $\mathbf{A}$. As a result of this restriction, one writes

$$\mathbf{w}_i = \xi_{i1}\mathbf{a}_1 + \xi_{i2}\mathbf{a}_2 + \xi_{i3}\mathbf{a}_3, \qquad \text{(IV.4)}$$

subject to the inequalities

$$0 \leq \xi_{i1}, \xi_{i2}, \xi_{i3} < 1. \qquad \text{(IV.5)}$$

The intracell location vectors $\mathbf{w}_i$ are called the "basis" for the crystal structure of interest. Provided that inequalities (IV.5) continue to be obeyed, an arbitrary common translation can be applied to the basis vectors. In particular, a possible choice would be to make one of the basis vectors vanish, say $\mathbf{w}_1 = 0$, in which case every Bravais lattice point $\mathbf{A}(n_1, n_2, n_3)$ would be occupied by a nucleus.

Based on their symmetry properties, there are 14 distinct types of Bravais lattices in three dimensions [Ashcroft and Mermin, 1976, Chapter 7; Kittel, 1996, Chapter 1]. Among the simplest of these are (1) the simple cubic, (2) the body-centered cubic, and (3) the face-centered cubic lattices. The first of these can be produced by the following generating vector choice:

$$\mathbf{a}_1 = l\mathbf{u}_x, \mathbf{a}_2 = l\mathbf{u}_y, \mathbf{a}_3 = l\mathbf{u}_z, \qquad \text{(simple cubic)} \qquad \text{(IV.6)}$$

where l denotes the nearest neighbor distance between pairs of points in the Bravais lattice, and $\mathbf{u}_x, \mathbf{u}_y, \mathbf{u}_z$ are unit vectors along the axes of a Cartesian coordinate system. For the body-centered cubic case, one can choose

$$\mathbf{a}_1 = (2/3^{1/2})l\mathbf{u}_x, \mathbf{a}_2 = (2/3^{1/2})l\mathbf{u}_y, \mathbf{a}_3 = 3^{-1/2}l(\mathbf{u}_x + \mathbf{u}_y + \mathbf{u}_z), \qquad \text{(body-centered cubic).} \qquad \text{(IV.7)}$$

The face-centered cubic Bravais lattice is generated by the vectors

$$\mathbf{a}_1 = 2^{-1/2}l(\mathbf{u}_x + \mathbf{u}_y), \mathbf{a}_2 = 2^{-1/2}l(\mathbf{u}_x + \mathbf{u}_z), \mathbf{a}_3 = 2^{-1/2}l(\mathbf{u}_y + \mathbf{u}_z), \qquad \text{(face-centered cubic).} \qquad \text{(IV.8)}$$

The remaining eleven Bravais lattices fall into six groups: "triclinic" (one case), "monoclinic" (two cases), "orthorhombic" (four cases), "tetragonal" (two cases), "trigonal" (one case), and "hexagonal" (one case).

In the simple cubic lattice, the angles between pairs of vectors from any lattice point to two of its six nearest neighbors can only be 90° or 180°. The corresponding angles to two of eight nearest neighbors in the body-centered cubic lattice are $\cos^{-1}(1/3) \cong 70.53°$, the so-called tetrahedral angle:

$$\theta_t = \cos^{-1}(-1/3) \cong 109.47°, \tag{IV.9}$$

and 180°. The face-centered cubic lattice, with 12 nearest neighbors, displays nearest neighbor pair angles 60°, 90°, 120°, and 180°.

Representatives of these three cubic Bravais lattices can be found among the crystal phases of the elements, i.e., crystal structures with a basis of one atom per periodic repeat cell. The simple cubic structure appears for α-polonium (Po), that element's stable low temperature, low pressure form [Donohue, 1982]. Furthermore, calcium (Ca) has been reported to exhibit a simple cubic crystal at elevated pressure [Olijnyk and Holzapfel, 1984], although this structure is apparently subject to weak symmetry-breaking distortions [Mao et al., 2010]. The elements sodium (Na), iron (Fe), and tungsten (W) adopt the body-centered cubic form at low pressure, whereas argon (Ar), calcium (Ca), and copper (Cu) have face-centered cubic crystals at low pressure [Donohue, 1982].

The diamond crystal structure is not a Bravais lattice. However, it involves a face-centered cubic Bravais lattice with a basis of two atoms. This structure amounts to a pair of interpenetrating face-centered cubic lattices. Figure IV.1 shows the corresponding spatial arrangement, with dashed connecting line segments ("bonds") that indicate the tetrahedral disposition of four nearest neighbors around each atom. The local geometries of the two atoms in each periodic repeat cell are mirror (rotated) images of one another but are otherwise equivalent. In addition to the diamond form of elemental carbon (C), silicon (Si) and germanium (Ge) also exhibit this structure.

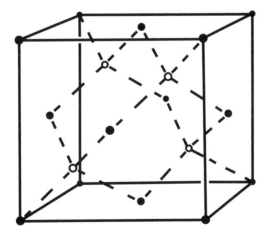

FIGURE IV.1. Spatial arrangement of particles comprising the diamond lattice. This structure involves a face-centered cubic lattice with a basis of two particles, distinguished by black and white circles.

Along with the generators in expression (IV.8) for the face-centered cubic Bravais lattice, the diamond crystal structure utilizes the following basis vectors:

$$\mathbf{w}_1 = 0, \tag{IV.10}$$

$$\mathbf{w}_2 = 2^{-3/2}l(\mathbf{u}_x + \mathbf{u}_y + \mathbf{u}_z).$$

The angle between any two bonds from a particle position to two of its nearest neighbors is the tetrahedral angle θ_t, Eq. (IV.9). The distance r_1 that separates nearest neighbor particle pairs in the diamond structure is given by

$$r_1 = |\mathbf{w}_2| = (3/8)^{1/2}l, \tag{IV.11}$$

in which l stands as before for the nearest neighbor distance within either of the two interpenetrating face-centered cubic Bravais lattices.

Ideal crystal structures can be at least partly characterized by their local coordination shell radii and occupancies. These local geometric properties are the same for all particles in each of the four cases thus far considered: simple cubic, body-centered cubic, face-centered cubic, and diamond. Moving radially outward from any chosen particle, the coordination shells of first, second, third, ..., neighbors occur, respectively, at distances that can be denoted by $r_1, r_2, r_3, \ldots$, and the numbers of particles residing in these spherical shells can be denoted by the positive integers Z_1, $Z_2, Z_3, \ldots$. Tables IV.1 to IV.4 present the first few coordination shell distances and occupancies for these four lattices. These tables also indicate the respective relations between the first-shell distance r_1 and the number density ρ for each case, which can be expressed in terms of unit-cell volume v_0 and the required number W of basis vectors:

$$\rho = W/v_0. \tag{IV.12}$$

In the event that the interaction potential Φ for the many-particle system consists only of spherically symmetric pair interactions $v_2(r_{ij})$, this type of coordination shell data is immediately applicable

TABLE IV.1. Coordination structure for the ideal simple-cubic lattice. As a function of the number density ρ, the nearest neighbor distance is $r_1 = \rho^{-1/3}$

v	$(r_v/r_1)^2$	Z_v
1	1	6
2	2	12
3	3	8
4	4	6
5	5	24
6	6	24
7	8	12
8	9	30
9	10	24
10	11	24

TABLE IV.2. Coordination shell structure for the ideal body-centered cubic lattice. As a function of the number density ρ, the nearest neighbor distance is $r_1 = 2^{-2/3}3^{1/2}\rho^{-1/3}$

v	$(r_v/r_1)^2$	Z_v
1	1	8
2	4/3	6
3	8/3	12
4	11/3	24
5	4	8
6	16/3	6
7	19/3	24
8	20/3	24
9	8	24
10	9	32

TABLE IV.3. Coordination shell structure for the ideal face-centered cubic lattice. As a function of the number density ρ, the nearest neighbor distance is $r_1 = 2^{1/6}\rho^{-1/3}$

v	$(r_v/r_1)^2$	Z_v
1	1	12
2	2	6
3	3	24
4	4	12
5	5	24
6	6	8
7	7	48
8	8	6
9	9	36
10	10	24

TABLE IV.4. Coordination structure for the ideal cubic diamond lattice. This non-Bravais structure has two particles per unit cell, both of which possess the same local coordination shell enumeration. As a function of the number density ρ, the nearest neighbor distance is $r_1 = 2^{-1}3^{1/2}\rho^{-1/3}$

v	$(r_v/r_1)^2$	Z_v
1	1	4
2	8/3	12
3	11/3	12
4	16/3	6
5	19/3	12
6	8	24
7	9	16
8	32/3	12
9	35/3	24

to calculation of the ideal crystal structure's potential energy per particle, that is, for its inherent structure:

$$\lim_{N\to\infty} \Phi/N = (1/2)\sum_{\nu=1}^{\infty} Z_\nu v_2(r_\nu). \tag{IV.13}$$

For notational simplicity, one can denote the position-dependent singlet density for an arbitrarily large single-component crystal, under thermal equilibrium conditions, by $\rho_1(\mathbf{r})$. This density distribution exhibits the same periodicity as the relevant Bravais lattice, i.e.

$$\rho_1(\mathbf{r} + \mathbf{A}) = \rho_1(\mathbf{r}). \tag{IV.14}$$

Here $\mathbf{A}$ is any one of the vectors in Eq. (IV.2). The amplitude of the spatial periodicity depends sensitively on the boundary conditions. If periodic boundary conditions are present, permitting free translation of the many-particle system, the resulting configurational averaging reduces that amplitude to zero, i.e., $\rho_1(\mathbf{r}) \equiv \rho$, the overall number density. However, other boundary condition choices can effectively clamp a crystal in position and orientation, revealing the underlying periodicity that distinguishes crystal phases from fluids. In this clamped circumstance, $\rho_1(\mathbf{r})$ is finite and continuous, peaked around the nominal crystal locations for particles but with peak heights and widths determined by quantum zero-point motions and/or thermal vibrations. Under ideal circumstances, these peak broadening effects are removed by the steepest descent operation that produces inherent structures for the many-particle system, so that at least for structurally perfect crystals the finite-width peaks sharpen into zero-width Dirac delta functions.

If the crystal of interest is free of defects (discussed in Section IV.D), the net occupancy of a primitive cell v_0 equals the basis number W. Thus the singlet density would satisfy the relation

$$\int_{v_0}\rho_1(\mathbf{r})d\mathbf{r} = W. \tag{IV.15}$$

However, a distribution of crystal imperfections that also yields a periodic averaged singlet density can produce a corresponding integral exhibiting a small deviation of either sign from this ideal integer value W.

The periodic nature of crystal-phase singlet densities naturally suggests a Fourier series representation:

$$\rho_1(\mathbf{r}) = \sum_{\mathbf{K}} C(\mathbf{K})\exp(i\mathbf{K}\cdot\mathbf{r}). \tag{IV.16}$$

The admissible wave vectors $\mathbf{K}$ are those consistent with the spatial periodicity of the singlet density function. This circumstance requires that for any of the vectors $\mathbf{A}$ in Eq. (IV.2) that $\mathbf{A}\cdot\mathbf{K}$ must be an integer multiple of 2π, including zero. This implies that $\mathbf{K}$ must be an integer-coefficient linear combination of generating vectors $\mathbf{b}_1, \mathbf{b}_2, \mathbf{b}_3$:

$$\mathbf{K} = \tilde{n}_1\mathbf{b}_1 + \tilde{n}_2\mathbf{b}_2 + \tilde{n}_3\mathbf{b}_3, \tag{IV.17}$$

$$\tilde{n}_1, \tilde{n}_2, \tilde{n}_3 = 0, \pm 1, \pm 2,\ldots,$$

and that this new triad of generating vectors must satisfy, and is determined by, the conditions [Brillouin, 1953]:

$$\mathbf{b}_j \cdot \mathbf{a}_l = 2\pi\delta_{jl}, \tag{IV.18}$$

where δ_{jl} is the Kronecker delta. This set of $\mathbf{K}$'s defines the "reciprocal lattice" for the original Bravais lattice, and it is itself a Bravais lattice [Ashcroft and Mermin, 1976, Chapter 5]. The primitive cell of the reciprocal lattice has content

$$\tilde{v}_0 = (2\pi)^3/v_0. \tag{IV.19}$$

The reciprocal lattice corresponding to the simple cubic Bravais lattice is itself a simple cubic lattice. That is, the simple cubic lattice is self-reciprocal. The reciprocal of the body-centered cubic lattice is face-centered cubic, and vice versa: the reciprocal of the face-centered cubic lattice is body-centered cubic.

B. Phonons

One of the most basic attributes of the basins composing the potential energy landscape of an N-body system is the set of curvatures at the enclosed inherent structures. These curvatures determine the independent normal modes of harmonic vibration about those mechanically stable configurations, as indicated in Section III.B. In the most general circumstances, determining the harmonic normal modes entails diagonalization of the full $3N \times 3N$ Hessian matrix for an N-body system. However, the translational symmetry presented by spatially periodic crystals creates drastic simplification that permits relatively facile determination of the full set of normal modes. The term "phonon" for any harmonic normal mode with frequency ω often refers conventionally to its fundamental quantum of excitation energy $\hbar\omega$ [Ashcroft and Mermin, 1976, p. 453].

Consider first the case of a perfect crystal whose inherent structure is a Bravais lattice, i.e., whose unit cell contains a basis of just one particle. This crystal is assumed to be isotopically pure, that is, all particles have identically the same mass. Suppose furthermore that this macroscopic crystal occupies a parallelepiped volume V that consists of Bravais unit cells numbering N_1, N_2, N_3 along the respective basis vector directions $\mathbf{a}_1, \mathbf{a}_2, \mathbf{a}_3$. In addition, let periodic boundary conditions apply along each of the three directions. Then it can easily be verified by substitution in the Newton equations of motion that all $3N$ harmonic normal modes for the N particles possess the form

$$\Delta\mathbf{r}_j(t) = \mathbf{B}(\mathbf{k})\exp[i\mathbf{k}\cdot\mathbf{r}_j^{(q)} - i\omega(\mathbf{k})t]. \tag{IV.20}$$

In this generic expression $\mathbf{r}_j^{(q)}$ is the inherent-structure position of particle j, and $\mathbf{k}$ is a vector of the form

$$\mathbf{k} = (n_1/N_1)\mathbf{b}_1 + (n_2/N_2)\mathbf{b}_2 + (n_3/N_3)\mathbf{b}_3, \tag{IV.21}$$

$$n_1, n_2, n_3 = 0, \pm 1, \pm 2, \ldots,$$

where $\mathbf{b}_1, \mathbf{b}_2, \mathbf{b}_3$ are the reciprocal lattice generating vectors, Eq. (IV.18). $\mathbf{B}(\mathbf{k})$ in Eq. (IV.20) depends on the normal mode under consideration but is the same vector for all N particles, and

$\omega(\mathbf{k})$ is the angular frequency for that normal mode. The set of vectors $\mathbf{k}$ is dense because N_1, N_2, N_3 must each be large to describe a macroscopic system, but this dense set contains the reciprocal lattice vectors $\mathbf{K}$ themselves as a discrete subset.

The use of form (IV.20) reduces the computation of $\omega^2(\mathbf{k})$ to diagonalization of a 3×3 matrix for each $\mathbf{k}$ [Wannier, 1959, Chapter 3]. Thus for each $\mathbf{k}$ there are three independent solutions specified by $\mathbf{B}_\zeta(\mathbf{k})$ and $\pm\omega_\zeta(\mathbf{k})$, indexed by $\zeta = 1,2,3$ (the sign of the angular frequency is irrelevant but is conventionally chosen to be positive). For each $\mathbf{k}$, the three motion-direction vectors $\mathbf{B}_\zeta(\mathbf{k})$ can be taken as mutually perpendicular because the normal modes are independent orthogonal excitations.

Because there are just $N = N_1 N_2 N_3$ particles in the crystal under consideration, and thus $3N$ independent normal modes, whereas the $\mathbf{k}$ vectors in Eq. (IV.21) form an infinite set, the $\mathbf{B}_\zeta(\mathbf{k})$ and $\omega_\zeta(\mathbf{k})$ are redundant functions throughout $\mathbf{k}$ space. Adding any reciprocal lattice vector $\mathbf{K}$ to $\mathbf{k}$ does not change the three solutions, so in fact they are periodic functions of $\mathbf{k}$ with the same periodicity as the reciprocal lattice. Each reciprocal lattice vector is at the center of its own Voronoi (nearest neighbor) polyhedron that contains N $\mathbf{k}$'s, so it is natural to take the Voronoi polyhedron surrounding the origin as the primitive period for these phonon functions. This primitive period is conventionally called the "first Brillouin zone" [Ashcroft and Mermin, 1976, Chapter 5]. Figure IV.2 shows the shapes of the first Brillouin zones for the simple cubic, body-centered cubic, and face-centered cubic Bravais lattices. These Brillouin zones exhibit 6 square faces, 12 rhombic faces, and 14 faces (6 squares, 8 hexagons), respectively.

(a) sc:

(b) bcc:

(c) fcc:

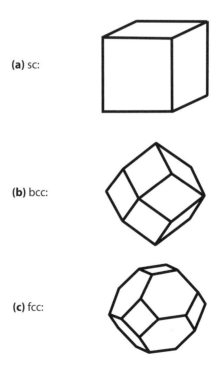

FIGURE IV.2. First Brillouin zones for (a) simple cubic (sc), (b) body-centered cubic (bcc), and (c) face-centered cubic (fcc) Bravais lattices.

By construction, the faces of the Voronoi polyhedron for the first Brillouin zone lie in planes whose points are equidistant from the origin $\mathbf{k} = 0$ and its first neighbors in the reciprocal lattice. Other analogous planes are defined as those whose points are equidistant from the origin and the second, third, fourth, … neighbors, respectively, in the reciprocal lattice. These planes are often referred to as Bragg planes. It has been conventional to define second, third, fourth, … Brillouin zones in terms of these Bragg planes. Specifically, the nth order Brillouin zone is that set of $\mathbf{k}$'s in Eq. (IV.21) that can only be reached from the origin by crossing $n - 1$ Bragg planes. Unlike the first Brillouin zone, the second and higher order Brillouin zones consist of disconnected portions, but those portions may geometrically be assembled into the shape of the first Brillouin zone by appropriate translations [Brillouin, 1953; Ashcroft and Mermin, 1976].

When $|\mathbf{k}|$ is near the center of the first Brillouin zone, the spatial wavelength of the phonon normal mode given by Eq. (IV.20) is large compared to the crystal lattice spacing. As a result, the three phonon branches can then be identified as macroscopic sound waves. In an isotropic solid, these sound waves include a longitudinal mode with $\mathbf{B}_\zeta(\mathbf{k})$'s parallel to $\mathbf{k}$, and (for cubic crystals) a degenerate pair of transverse modes with $\mathbf{B}_\zeta(\mathbf{k})$'s perpendicular to $\mathbf{k}$. In this long-wavelength regime, one then has the usual dispersion relations connecting frequencies with the longitudinal and transverse sound speeds c_l and c_t:

$$\omega_l(\mathbf{k}) = c_l |\mathbf{k}|, \tag{IV.22}$$

$$\omega_t(\mathbf{k}) = c_t |\mathbf{k}|.$$

However, this simple situation is generally not the case for crystals, and in particular need not be true for those with cubic symmetry when the phonon wavelength is comparable to the crystal lattice spacing. In general, one has three distinct sound modes with distinct speeds that are not purely longitudinal or transverse in polarization. Furthermore, they become degenerate only along lines or at points of special symmetry in the first Brillouin zone. Figure IV.3 schematically indicates the way that the three phonon frequency branches typically behave along a linear path through the center of the first Brillouin zone ($\mathbf{k} = 0$).

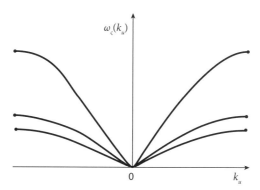

FIGURE IV.3. Typical dispersion curves in the first Brillouin zone for the three phonon branches exhibited by a Bravais lattice. The frequencies are plotted along a straight line path in $\mathbf{k}$ space passing through the origin, along some direction defined by unit vector $\mathbf{u}$.

BOX IV.1. Basin Localization

Phonon properties for crystals are traditionally analyzed (as here) on the basis of system motion within a single basin, specifically, one surrounding an inherent structure for a selected permutation of particles among the crystal sites. However, for single-component crystals, the overall potential energy landscape picture presents $N!$ equivalent basins [$W(N-1)!$ for basis-W crystals with periodic boundary conditions], differing only by particle permutation. In view of the fact that wave functions for the N-body system span the entire configuration space and so include all equivalent basins, this apparent discrepancy requires explanation and justification.

With the exception of the helium isotopes discussed in Chapter VIII, phonon calculations are directed toward substances whose constituent particles are either sufficiently massive or are positionally constrained by strong chemical bonds. Consequently, their low-lying quantum states have wave functions strongly localized near the inherent structures for the permutation-equivalent basins. Elsewhere in the multidimensional configuration space, the wave functions become extremely small. Each single-basin quantum state in principle is actually an extremely narrow band of $N!$ states [or $W(N-1)!$ for periodic boundary conditions]. But upon formally including all such basins, the corresponding orthonormal wave functions have very small weight in each basin, proportional to $(N!)^{-1/2}$ [or $(W(N-1)!)^{-1/2}$ for periodic boundary conditions]. The large number of states in each band essentially exactly compensates for the weight in each basin. Consequently, it suffices to carry out calculations just as though only a single permutational basin member were relevant.

The venerable Debye approximation to the phonon spectrum assumes that the linear small-$|\mathbf{k}|$ expressions (IV.22) for the frequencies can be extended to include all $3N$ phonons [Debye, 1912]. Furthermore, this approximation replaces the polyhedral surface of the first Brillouin zone by a sphere. The radius k_D of the Debye sphere is chosen so as to include exactly $3N$ modes, specifically

$$k_D = (6\pi^2 N/V)^{1/3}. \tag{IV.23}$$

Phonon spectra become more elaborate when the crystal structure involves a basis of $W \geq 2$ particles per unit cell, some technical aspects of which are considered in Box IV.1. Simple examples are sodium chloride (NaCl), with a face-centered cubic Bravais lattice and a basis of two, and cesium chloride (CsCl), with a simple cubic Bravais lattice and also a basis of two [Kittel, 1996, Chapter 1]. Crystalline forms of several elements provide other examples, including silicon (Si) in the diamond structure [Donohue, 1982, Chapter 7] and magnesium (Mg) in the hexagonal close-packed structure [Donohue, 1982, Chapter 4], both of which involve a basis of two atoms. In these and other cases with two or more particles in the unit cell, the three acoustic modes already described are joined by high-frequency "optical modes" that do not approach zero frequency at $|\mathbf{k}| = 0$. Figure IV.4 qualitatively indicates how the augmented phonon spectrum appears for a basis of two particles per unit cell.

In general, if W is the basis number, $3W$ phonon modes exist for each wave vector $\mathbf{k}$ in the first Brillouin zone, i.e., there are $3W$ separate phonon branches, each spanning that zone, three of which are acoustic branches. The total number of independent phonon modes, of course, is $3N$,

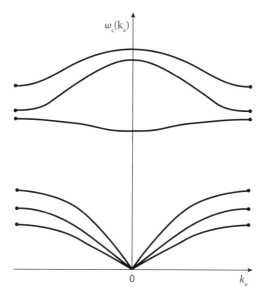

FIGURE IV.4. Schematic representation of the phonon spectrum in the first Brillouin zone for lattices with a basis of two particles. The frequencies are plotted along a straight line in **k** space passing through the origin, along some direction defined by unit vector **u**.

where N is the number of particles in the crystalline system. The number in each branch is N_u, where N_u is the number of unit cells comprised in the system. The qualitative distinction between acoustic phonons and optical phonons is that within each unit cell, the W basis particles move approximately in phase (in concert) for the former, but move out of phase (in opposition) for the latter.

In view of the fact that each of the independent phonon modes (for $\mathbf{k} \neq 0$) has a simple spectrum of quantized energy eigenstates in the harmonic approximation [Landau and Lifshitz, 1958a, p. 66]:

$$[n + (1/2)]\hbar\omega_\zeta(\mathbf{k}) \qquad (n = 0,1,2,\ldots), \qquad (IV.24)$$

the canonical partition function for the harmonic crystal possesses a rather simple form. Because of the presence of periodic boundary conditions and the free translation of the system that it permits, the center of mass motion that corresponds to the three $\mathbf{k} = 0$ modes in principle needs special attention. However, because interest normally focuses on thermodynamic properties on a per-particle basis for the large system limit, this free translation becomes substantially irrelevant. Evaluation of the free energy per particle thus can legitimately proceed with neglect of the three acoustic $\mathbf{k} = 0$ modes. Let $\varphi_0 = \Phi_0/N$ denote the potential energy per particle at one of the many equivalent crystal inherent structures. The Helmholtz free energy per particle, multiplied by $\beta = 1/k_B T$, has the following form [cf. Eq. (III.29), with $\sigma = 0$ for this structurally unique ideal crystal]:

$$\beta F/N = \beta\varphi_0 + N^{-1}\sum_{\zeta=1}^{3W}\sum_{\mathbf{k}}\{\ln[\exp(\beta\hbar\omega_\zeta(\mathbf{k})) - 1] - [\beta\hbar\omega_\zeta(\mathbf{k})/2]\}, \qquad (IV.25)$$

where ζ indexes the separate phonon branches, and the **k** sum excludes the origin but otherwise spans the first Brillouin zone. The mean energy per particle follows by differentiation:

$$E/N = \frac{\partial(\beta F/N)}{\partial \beta}$$

$$= \varphi_0 + N^{-1}\sum_{\zeta=1}^{3W}\sum_{\mathbf{k}}\left\{-\frac{\hbar\omega_\zeta(\mathbf{k})}{2} + \frac{\hbar\omega_\zeta(\mathbf{k})}{1 - \exp[-\beta\hbar\omega_\zeta(\mathbf{k})]}\right\}. \tag{IV.26}$$

The corresponding constant-volume heat capacity is

$$C_V/Nk_B = -\beta^2 \frac{\partial(E/N)}{\partial\beta}$$

$$= \frac{1}{N}\sum_{\zeta=1}^{3W}\sum_{\mathbf{k}}\frac{[\beta\hbar\omega_\zeta(\mathbf{k})]^2\exp[-\beta\hbar\omega_\zeta(\mathbf{k})]}{\{1 - \exp[-\beta\hbar\omega_\zeta(\mathbf{k})]\}^2}. \tag{IV.27}$$

The contribution of each phonon to the heat capacity is a monotonically increasing function of temperature. Consequently, the same is true for C_V/Nk_B itself. If the temperature is sufficiently high that for each phonon mode one has

$$\beta\hbar\omega_\zeta(\mathbf{k}) \ll 1, \tag{IV.28}$$

then the exponential functions in the heat capacity expression in Eq. (IV.27) can be expanded, and terms can be collected by increasing order in β. This change produces the following result:

$$C_V/Nk_B = 3 - (1/12N)\sum_{\zeta=1}^{3W}\sum_{\mathbf{k}}[\beta\hbar\omega_\zeta(\mathbf{k})]^2 + O(\beta^3). \tag{IV.29}$$

The leading term 3 for the high-temperature specific heat of crystals has long been identified as the Dulong and Petit law [Seitz, 1940, Chapter I]. Result (IV.29) shows that this limit is approached from below, subject to the assumption of harmonic motion.

In order to derive the low-temperature form of the crystal specific heat, it is useful to introduce a density of states for the phonon frequencies, applicable to the large-system limit. For this purpose, define $g(\hbar\omega)$ so that the number of values in the incremental frequency interval $\hbar\omega \pm \delta(\hbar\omega)/2$ is $Ng(\hbar\omega)\delta(\hbar\omega)$. This definition incorporates all $3W$ phonon branches and for the large-system limit transforms expression (IV.27) to the following ($x = \hbar\omega$):

$$C_V/Nk_B = \beta^2 \int_0^{x_{\max}} \frac{x^2\exp(-\beta x)}{[1 - \exp(-\beta x)]^2} g(x)dx. \tag{IV.30}$$

For the low-temperature regime, where $\beta = 1/k_BT$ is large, the integral is dominated by the integrand at small x. This domination focuses on the small-frequency portion of the phonon spectrum, namely, the acoustic phonons. In that circumstance, it is legitimate to take the density-of-states function to have the form

$$g(x) = \frac{3Vx^2}{2\pi^2(\hbar c)^3 N}. \tag{IV.31}$$

Here c stands for sound speed averaged over the three acoustic branches and propagation directions. Because the integrand in Eq. (IV.30) at low temperature declines so rapidly to zero with increasing x, the integration may be extended to $+\infty$ to yield

$$C_V/Nk_B \cong \frac{3\beta^2 V}{2\pi^2(\hbar c)^3 N} \int_0^\infty \frac{x^4 \exp(-\beta x)}{[1 - \exp(-\beta x)]^2}\, dx$$

$$= \frac{3V}{2\pi^2(\beta\hbar c)^3 N} \int_0^\infty \frac{y^4 \exp(-y)}{[1 - \exp(-y)]^2}\, dy$$

$$= \left(\frac{2\pi^2 V}{5(\hbar c)^3 N}\right)(k_B T)^3. \tag{IV.32}$$

This low-temperature result is applicable to crystalline insulators. Metals possess a similar phonon contribution, but their heat capacities are dominated at low temperatures by conduction electron excitations, which generate a term proportional to T [Ashcroft and Mermin, 1976, Chapter 2; Kittel, 1996, Chapter 6].

C. Anharmonic Effects

Although the harmonic approximation for structurally perfect crystal basins and the phonon motions it implies constitute an important level of description, the properties of real crystalline substances reveal the necessity for anharmonic corrections. The potential energy function for the N-particle system generally has a multivariable Taylor expansion about any basin's inherent structure with leading terms of the following symbolic form:

$$\Phi(\mathbf{r}_1 \ldots \mathbf{r}_N) = \Phi(\Delta\mathbf{R} = 0) + (1/2!)\mathbf{D}^{(2)}(\cdot)^2 \Delta\mathbf{R}\Delta\mathbf{R} + (1/3!)\mathbf{D}^{(3)}(\cdot)^3 \Delta\mathbf{R}\Delta\mathbf{R}\Delta\mathbf{R}$$

$$+ (1/4!)\mathbf{D}^{(4)}(\cdot)^4 \Delta\mathbf{R}\Delta\mathbf{R}\Delta\mathbf{R}\Delta\mathbf{R} + O[(\Delta\mathbf{R})^5]. \tag{IV.33}$$

Here $\Delta\mathbf{R}$ stands for the collection of all single-particle displacements $\Delta\mathbf{r}_i$ from an inherent-structure configuration, and the $\mathbf{D}^{(n)}$ are matrices of potential energy derivatives of order n evaluated at $\Delta\mathbf{R} = 0$. If periodic boundary conditions apply, the $\mathbf{D}^{(n)}$ must individually have the property that each corresponding term in Eq. (IV.33) vanishes identically under translation of the system as a whole, i.e., when all $\Delta\mathbf{r}_i$ are equal to one another.

Although harmonic oscillator degrees of freedom wander from their equilibrium positions, by symmetry their long-time average locations remain at that equilibrium position. Consequently, crystals experiencing only the harmonic portion of the basin potential energy would always exhibit the same volume (at constant external pressure) regardless of the temperature. While this constant-volume behavior may be nearly true at very low temperatures, anharmonic effects come increasingly into play as temperature rises, typically causing nonvanishing values of the isobaric thermal expansion coefficient $\alpha_p = (\partial \ln V/\partial T)_p$ to appear well below the crystal melting temperature. The majority of crystalline substances exhibit positive thermal expansion, but there are notable exceptions. Zirconium tungstate (ZrW_2O_8) supplies a vivid example, presenting at ambient pressure a nearly constant negative α_p over the wide temperature range $0.3 \leq T \leq 1050$ K [Mary et al., 1996]. Because anharmonicity can produce either sign for α_p, one cannot interpret a vanishing α_p as unambiguous evidence that the harmonic approximation is automatically applicable to other crystal properties.

Crystal heat capacities also reveal the effects of intrabasin anharmonicity. The constant-volume and constant-pressure heat capacities C_V and C_p for any equilibrium system are related by the following thermodynamic identity [Guggenheim, 1950, Chapter IV]:

$$C_p = C_V + \frac{\alpha_p^2 TV}{\kappa_T}, \tag{IV.34}$$

involving the isobaric thermal expansion α_p and the isothermal compressibility κ_T. For homogeneous single phases, crystals in particular, the second term in the right side of this identity is never negative. Thus, $C_p \geq C_V$, with any difference between the two attributable to anharmonicity. Of course, C_V itself can be influenced by anharmonic effects, one possible symptom of which could be that the maximum value 3 stated by the Dulong and Petit law for C_V/Nk_B, Eq. (IV.29), was exceeded.

Suppose external forces were to be applied to a perfect crystal so as to change its volume V (and if necessary its shape as well to retain isotropic stress) while maintaining the identity of all unit cells. This change would alter the configuration of that crystal's inherent structure and the shape of its encompassing basin, and so would alter the terms in the Taylor's expansion of Φ, Eq. (IV.33). In particular, the quadratic terms in that expansion can be expected to change in magnitude, thus causing shifts in the harmonic phonon frequencies determined by those quadratic terms. For each of the phonons specified by those V-dependent quadratic terms, define the corresponding "mode Grüneisen constant" to be

$$\lambda_\zeta(\mathbf{k}) = -\frac{V}{\omega_\zeta(\mathbf{k})}\left(\frac{\partial \omega_\zeta(\mathbf{k})}{\partial V}\right). \tag{IV.35}$$

Under the assumption that the Helmholtz free energy is adequately given by Eq. (IV.25) for any value of this changeable volume V, the pressure is determined as follows:

$$p = -\left(\frac{\partial F}{\partial V}\right)_T$$

$$= p_0 + \sum_{\zeta=1}^{3W} \sum_{\mathbf{k}} \left\{ -\frac{\hbar\omega_\zeta(\mathbf{k})}{2} + \frac{\hbar\omega_\zeta(\mathbf{k})}{1 - \exp[-\beta\hbar\omega_\zeta(\mathbf{k})]} \right\} \gamma_\zeta(\mathbf{k}). \tag{IV.36}$$

Here p_0 is the temperature-independent pressure contributed by the volume dependence of the crystal's inherent-structure energy:

$$p_0 = -\left(\frac{\partial \varphi_0}{\partial (V/N)}\right). \tag{IV.37}$$

The remaining terms in the right side of Eq. (IV.36) implicitly provide additional anharmonic contributions to the total pressure through the mode Grüneisen constants.

By applying a constant-V temperature derivative to the pressure expression (IV.36), one obtains

$$\left(\frac{\partial p}{\partial T}\right)_V = k_B \sum_{\zeta=1}^{3W} \sum_{\mathbf{k}} \frac{[\beta\hbar\omega_\zeta(\mathbf{k})]^2 \exp[-\beta\hbar\omega_\zeta(\mathbf{k})]}{\{[1 - \exp[-\beta\hbar\omega_\zeta(\mathbf{k})]]\}^2} \gamma_\zeta(\mathbf{k}). \tag{IV.38}$$

In view of the thermodynamic identity:

$$-1 = \left(\frac{\partial p}{\partial T}\right)_V \left(\frac{\partial T}{\partial V}\right)_p \left(\frac{\partial V}{\partial p}\right)_T, \tag{IV.39}$$

the anharmonic property shown in Eq. (IV.38) is also equal to α_p/κ_T.

Although the various quantities and their relations shown in this section and Section IV.B have referred specifically to structurally perfect crystals, they can in principle be extended to other cases in which the many-body system remains confined to the basin of some other single inherent structure. These structures can include defective crystals and amorphous (glassy) solids at sufficiently low temperature. For each of these, $3N$ independent harmonic normal modes exist for N-particle systems, determined by the appropriate quadratic approximation to the potential energy function in the neighborhood of the inherent structure. Those modes generally do not have the simple form of running waves, and some may be spatially localized. However, the perfect-crystal mode summations shown above simply become replaced by the frequency set $\{\omega_i\}$ summation for the structurally irregular solid of interest. This extension includes the various anharmonic measures, when the necessary mode Grüneisen constants $\{\gamma_i\}$ for the irregular solid are known.

D. Crystal Structural Defects

Although the permutation-equivalent inherent structures corresponding to the absolute Φ or $\hat{\Phi}$ minima, and their surrounding basins, suffice to describe the properties of perfect crystals, a variety of physical processes can disrupt the periodic structure and deposit these systems in several types of somewhat higher lying basins. The resulting mechanically stable structures can be identified as structurally defective crystals. The defect-producing processes can include heating an initially perfect crystal to the vicinity of its melting temperature, mechanical deformation, and radiation damage. Rapid cooling of the liquid melt, followed by nucleation and crystal growth under nonequilibrium conditions also typically generates defective crystals. This section considers some of the elementary features of crystal structural defects within the context of the inherent-structure representation. It should be mentioned that nonstructural mass defects arising from isotopic substitution also exist and are considered in Section IV.G in connection with their effect on thermal conductivity.

Unambiguous identification and classification of defects is possible only if the extent of structural disruption of the parent perfect crystal is modest. In general, it is not possible to describe amorphous particle arrangements (e.g., liquid-phase inherent structures) precisely and uniquely in terms of superpositions of basic defect types present in an underlying perfect crystal structure.

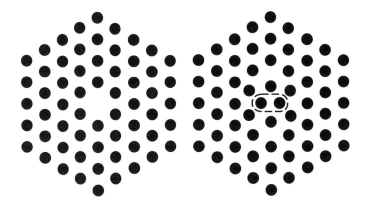

(a) Monovacancy   **(a) Split interstitial**

FIGURE IV.5. Point defects in a two-dimensional triangular crystal: (a) a monovacancy and (b) an oriented split interstitial pair outlined with a dashed contour.

But when the defect extent or concentration is low, knowing the properties of those identifiable basic defect types leads to fundamental insights into the properties of the materials involved.

The simplest defects are localized point defects. In the case of monatomic crystals, these defects include vacancies (missing particles), interstitials (inserted extra particles), and impurity-atom replacements. Crystals of molecular substances may contain orientational or conformational defects, where a molecule has been incorporated into the otherwise perfect crystal in an "incorrect" orientation or conformation. Multicomponent crystals can exhibit point defects in the form of an "incorrect" particle species present at one of the crystal positions in an otherwise periodic pattern of species arrangement.

Mechanically stable defects that are spatially more extended include both line defects (dislocations) and planar defects (stacking faults, grain boundaries, twinning structures). In comparison with point defects, these tend to have much higher excitation energies (and enthalpies) above those of the perfect crystals. Consequently, their contribution to thermal equilibrium properties of the crystal phase tends to be negligible. For brevity, this section focuses primarily on the properties of point defects. For a detailed presentation of the larger subject of crystal defects and their characteristics, readers are directed to the extensive published literature [e.g., Ashcroft and Mermin, 1976, Chapter 30; Kittel, 1996, Chapters 18, 20; Kelly and Knowles, 2012].

In the interests of visual clarity, Figure IV.5 schematically shows isolated prototype vacancy and interstitial geometries that can be present as mechanically stable local structures embedded in a single-component two-dimensional triangular crystal of structureless particles. The former (a) amounts to a single particle missing from the periodic structure and so is called a "monovacancy." The latter (b) is formed by insertion of a single extra particle, but the mechanically stable structure that results exhibits a pair of equivalent particles displaced by equal amounts in opposite directions from the nominal lattice site that would otherwise accommodate a single particle. For obvious reasons, this is called a "split interstitial," with neither member of the pair identifiable as the "intruder." Both of these types of point defects could occur at any one of the nominal lattice sites of the crystal. However, the split interstitial displays a characteristic not shared by the mon-ovacancy, namely, that it has several possible orientations. For the two-dimensional example shown in Fig. IV.5(b), three equivalent but distinguishable orientations of the split interstitial pair exist that are 60° apart, each of which is perpendicular to a row of particles in the host triangular lattice.

The existence of the geometric patterns represented in Figure IV.5 for the two kinds of point defects are amenable to confirmation by specific model calculations in two dimensions. An ex-ample would be to suppose that the particles forming the two-dimensional crystal interact via the Lennard-Jones 12,6 pair potential [Eq. (I.37)]:

$$v(r) = 4\varepsilon[(\sigma/r)^{12} - (\sigma/r)^6]. \tag{IV.40}$$

The absolute minimum of the total potential energy Φ for a large system of such particles, occur-ring at zero two-dimensional pressure (for which $\Phi = \hat{\Phi}$), is indeed the perfect triangular lattice [Theil, 2006]. Its nearest neighbor spacing r_1 and potential energy per particle $\varphi(r_1)$ in this zero-pressure condition have the following numerical values [Stillinger and Stillinger, 2006]:

$$(r_1)_{min}/\sigma \cong 1.11142, \tag{IV.41}$$

$$\varphi_{min}/\varepsilon \cong -3.38342.$$

These results, and their extensions to positive pressure (for which $\Phi < \hat{\Phi}$), provide the perfect triangular crystal baseline against which to measure the point defect structures and excitation energies (or enthalpies) for this simple two-dimensional crystal.

Defect excitation energies (or enthalpies) are defined within the inherent-structure represen-tation in the obvious way by comparing Φ (or $\hat{\Phi}$) for appropriately selected pairs of crystals that respectively are structurally perfect, and are defect-containing, but that are composed of exactly the same number N of particles. Consider first the case of monovacancies (to be denoted by a superscript "0"). Suppose that N is sufficiently large that some number $n_{mv} \ll N$ of these monova-cancies could be located in the crystal interior, sufficiently far from one another that structurally they do not interact significantly with one another. Then at any pressure, the excitation energy $\Delta^{(0)}$ for a single monovacancy can be written

$$\Delta^{(0)} = n_{mv}^{-1}[\Phi(N, n_{mv}) - \Phi(N, 0)], \tag{IV.42}$$

provided that boundary conditions have been appropriately chosen. An analogous expression involving $\hat{\Phi}$'s can also be written for the monovacancy excitation enthalpy $\hat{\Delta}^{(0)}$. It is formally pos-sible to resolve $\Delta^{(0)}$ (analogously $\hat{\Delta}^{(0)}$) into three contributions:

$$\Delta^{(0)} = \Delta_a^{(0)} + \Delta_b^{(0)} + \Delta_c^{(0)} \tag{IV.43}$$

that correspond respectively to a three-step creation process for each of the n_{mv} monovacancies. The first step "a" involves removing each of the n_{mv} particles from widely separated locations within the perfect crystal to isolation at "infinity" while the remaining $N - n_{mv}$ particles maintain their perfect-crystal positions. Assuming that only particle pair interactions are present, this step entails a loss of pair interactions for each of the n_{mv} equal to twice the binding energy per particle for the perfect crystal. In the case of the two-dimensional triangular crystal with pairwise inter-actions, such as Eq. (IV.40), this is

$$\Delta_a^{(0)} = -\sum_{v=1}^{\infty} Z_v v(r_v)$$

$$= -2\varphi(r_1) \tag{IV.44}$$

$$\xrightarrow[p=0]{} -2\varphi_{min}.$$

Here, following earlier notation, the r_v and Z_v are coordination shell distances and occupancies for the perfect triangular crystal.

The next step "b" reassembles the n_{mv} removed and isolated particles at the surface of the crystal, restoring it to an N-particle structure. In particular, doing this with periodic boundary conditions and with n_{mv} chosen so as just to complete another crystal layer eliminates irrelevant and undesirable surface corrections in the large-system limit. Consequently, one has for each of the displaced particles

$$\Delta_b^{(0)} = \varphi(r_1). \tag{IV.45}$$

This just cancels half of $\Delta_a^{(0)}$, so that

$$\Delta^{(0)} = -\varphi(r_1) + \Delta_c^{(0)}. \tag{IV.46}$$

After completion of this second stage, the area or volume of the crystal has increased over that of the starting perfect crystal by a factor $1 + (n_{mv}/N)$.

Finally, step "c" frees the entire set of N particles in the defective crystal, allowing them to relax from their perfect-crystal sites into the appropriate mechanically stable inherent-structure geometry. Such relaxation cannot cause energy to rise, so necessarily

$$\Delta_c^{(0)} \le 0. \tag{IV.47}$$

Although this third stage is primarily a local structural readjustment around each monovacancy, it generally produces an elastic strain field diminishing in distance from the monovacancy. The net result of these strains can be an overall area or volume change of the defective crystal [Kelly and Knowles, 2012, Section 10.5]. Consequently, the area or volume change factor caused by the monovacancies is modified to $1 + (\xi n_{mv}/N)$, where the positive scaling factor ξ differs from unity.

Numerical calculation for the zero-pressure triangular crystal with Lennard-Jones pair inter-actions reveals that the six immediate neighbors of an isolated monovacancy relax inward equally. Under zero-pressure conditions, their resulting distance from one another and from the monova-cancy site is estimated to be $r_1^{(0)}/\sigma \cong 1.10682$, which may be compared with the initial neighbor distance in the perfect triangular crystal shown in Eq. (IV.41). The relaxation energy and total monovacancy creation energy for this zero-pressure example are found to be the following:

$$\Delta_c^{(0)}/\varepsilon \cong -0.00773, \tag{IV.48}$$

$$\Delta^{(0)}/\varepsilon \cong 3.37569.$$

The excitation energy $\Delta^{(2)}$ for a single split interstitial defect (denoted here by superscript 2) cannot be so conveniently resolved into results from a sequence of three simple steps because of the ambiguity of which of the two particles is the "extra" one. But by analogy with the monovacancy case, one can suppose that a number n_{si} of widely separated split interstitials can be formed by inserting all of the particles from an outer layer of an initially perfect crystal into its interior. This insertion inevitably involves a substantial positive interaction energy contribution from local particle crowding, an effect largely absent for monovacancies. Such crowding would be most disruptive for close-packed crystals (triangular in two dimensions, face-centered cubic and hexagonal close-packed in three dimensions). Although precise numerical results are not yet available for split interstitials in the Lennard-Jones triangular crystal, one can safely conclude that particle crowding requires

$$\Delta^{(2)} > \Delta^{(0)} > 0. \tag{IV.49}$$

For monovacancies in three dimensions, the excitation energy $\Delta^{(0)}$ can again be resolved into the same three kinds of contributions as indicated in Eq. (IV.43). The classical Lennard-Jones pair interaction model, Eq. (IV.40), can again serve for numerical illustration. With this interaction, the stable crystal form at zero external pressure is the hexagonal close-packed structure [Kihara and Koba, 1952], with a very small axial elongation [Stillinger, 2001]. The potential energy per particle in the structurally perfect crystal at vanishing T and p is

$$\varphi_{\min}/\varepsilon \cong -8.61107. \tag{IV.50}$$

Under zero external pressure, one should expect the local relaxation contribution $\Delta_c^{(0)}$ to be relatively small, as has been observed for the two-dimensional triangular crystal, Eq. (IV.48). This expectation would be reasonable in view of the substantial number (12) of nearest neighbors that would simultaneously have to crowd inward toward the vacant site. Consequently, one estimates

$$\Delta^{(0)}(p=0)/\varepsilon \cong -\varphi_{\min}(p=0)/\varepsilon \tag{IV.51}$$

$$= 8.61107.$$

In analogy to the orientational multiplicity of the split interstitial for the two-dimensional triangular crystal, split interstitials in three-dimensional crystals can also possess several mechanically stable orientations. In all cases, those orientations locally minimize repulsive overlaps with surrounding particles, and these neighbors again would exhibit significant static displacements from their nominal crystal positions. The expected orientations for a split interstitial are those pointing to the face centers of the polyhedron whose vertices are the particles of the first neighbor shell. This simple rule suggests that three orientations are available in the body-centered cubic crystal where the polyhedron is a simple cube, and four orientations for the simple cubic crystal where the polyhedron is an octahedron. The nearest neighbor polyhedron for the face-centered cubic crystal has fourteen faces, eight of which are equilateral triangles, and the remaining six are squares. These each oppose their own kind across the center of the polyhedron, implying the possibility of two inequivalent species of split interstitials, four facing triangles and three facing squares, with unequal excitation energies.

Many numerical calculations have been devoted to the structures and energies of point defects in crystals of real substances in three dimensions [Kelly and Knowles, 2012, Chapter 10]. These calculations have not been limited to models with just pairwise interactions but have included detailed quantum mechanical approaches to provide electronic ground states for specific metallic and semiconducting elements in both face-centered cubic and body-centered cubic crystals. Both monovacancies and split interstitial structures have been obtained as mechanically stable defects, with positive excitation energies $\Delta^{(2)}$ typically several times larger than $\Delta^{(0)}$.

The next objective is to illustrate the equilibrium thermodynamics of point defects in the crystal phase. Specifically, this illustration is carried out for isobaric (constant-pressure) conditions, typical for experimental circumstances. It is assumed that the system size, and therefore the number of particles N, is very large and that the total concentration of defects on a per-crystal-site basis is small. When n_{mv} monovacancies and n_{si} split interstitials are present, the rise in inherent-structure enthalpy over that of the N-particle perfect crystal is

$$\Delta\hat{\Phi}(p) = n_{mv}\hat{\Delta}^{(0)}(p) + n_{si}\hat{\Delta}^{(2)}(p), \tag{IV.52}$$

neglecting interactions beween defects. The presence of the point defects causes the number of crystal locations (whether occupied by a defect or not) to deviate from N, becoming instead

$$N + n_{mv} - n_{si}.$$

Consequently, the total number of distinguishable ways that the dilute point defects can be arranged throughout the crystal (assuming that they are independent) is the following:

$$\frac{(N + n_{mv} - n_{si})!\, w^{n_{si}}}{(N - 2n_{si})!\, n_{mv}!\, n_{si}!} \sim \exp\left[N\hat{\sigma}\left(\frac{n_{mv}}{N}, \frac{n_{si}}{N}\right)\right]. \tag{IV.53}$$

Here w is the number of distinct orientations available to each split interstitial, and $\hat{\sigma}$ is the previously discussed enumeration function for potential-enthalpy inherent structures [Eq. (III.20)], defined for the large-system asymptotic limit, for fixed defect numbers.

Take logarithms in Eq. (IV.53) and apply Stirling's approximation to the factorials to obtain $\hat{\sigma}$ as the following expression:

$$\begin{aligned}
\hat{\sigma}(n_{mv}/N, n_{si}/N) &\approx \left(1 + \frac{n_{mv}}{N} - \frac{n_{si}}{N}\right)\ln\left(1 + \frac{n_{mv}}{N} - \frac{n_{si}}{N}\right) - \left(1 - \frac{2n_{si}}{N}\right)\ln\left(1 - \frac{2n_{si}}{N}\right) \\
&\quad - \left(\frac{n_{mv}}{N}\right)\ln\left(\frac{n_{mv}}{N}\right) - \left(\frac{n_{si}}{N}\right)\ln\left(\frac{n_{si}}{Nw}\right) \\
&\approx \left(\frac{n_{mv}}{N}\right)\left[1 - \ln\left(\frac{n_{mv}}{N}\right)\right] + \left(\frac{n_{si}}{N}\right)\left[1 - \ln\left(\frac{n_{si}}{Nw}\right)\right]. \tag{IV.54}
\end{aligned}$$

Here the latter form shown is justified by the low-concentration assumption. In order to conform to the usual convention, $\hat{\sigma}$ needs to be expressed as a function of the inherent-structure enthalpy per particle, which for convenience can be expressed as the excitation above that of the perfect crystal:

$$\Delta\hat{\varphi} = \Delta\hat{\Phi}/N \tag{IV.55}$$

$$= (n_{mv}\hat{\Delta}^{(0)} + n_{si}\hat{\Delta}^{(2)})/N.$$

This requires maximizing the latter form in Eq. (IV.54) with respect to both n_{mv} and n_{si}, subject to fixed $\Delta\hat{\varphi}$ as a constraint. Consequently, one needs to introduce a Lagrange multiplier $-\Lambda$ to enforce that fixed $\Delta\hat{\varphi}$ constraint [Korn and Korn, 1968, Section 11.3–4]. The resulting expressions for the $\hat{\sigma}$-maximizing defect concentrations are

$$n_{mv}/N = \exp(-\Lambda\hat{\Delta}^{(0)}), \tag{IV.56}$$

$$n_{si}/N = w\exp(-\Lambda\hat{\Delta}^{(2)}),$$

where Λ must be chosen so that

$$\Delta\hat{\varphi} = \hat{\Delta}^{(0)}\exp(-\Lambda\hat{\Delta}^{(0)}) + w\hat{\Delta}^{(2)}\exp(-\Lambda\hat{\Delta}^{(2)}). \tag{IV.57}$$

In general, Λ cannot be explicitly eliminated from Eqs. (IV.54)–(IV.57) so as to express $\hat{\sigma}$ directly as a function of $\Delta\hat{\varphi}$, although these implicit relations are amenable to straightforward numerical analysis. However, Eq. (IV.57) shows that in the limiting small-$\Delta\hat{\varphi}$ regime, Λ must be large and positive. Consequently, the right side of Eq. (IV.57) would then be dominated by the term possessing the smaller $\hat{\Delta}^{(i)}$ magnitude, namely the monovacancy term with $\hat{\Delta}^{(0)}$. As a result, for this limiting regime one can write

$$\Delta\hat{\varphi} = \hat{\Delta}^{(0)}\exp(-\Lambda\hat{\Delta}^{(0)}), \tag{IV.58}$$

and so the $\hat{\sigma}$ maximum occurs at

$$n_{mv}/N \approx \Delta\hat{\varphi}/\hat{\Delta}^{(0)}. \tag{IV.59}$$

Therefore, the second form of Eq. (IV.54) leads to the expression

$$\hat{\sigma}(\Delta\hat{\varphi}) \approx (\Delta\hat{\varphi}/\hat{\Delta}^{(0)})[1 - \ln(\Delta\hat{\varphi}/\hat{\Delta}^{(0)})]. \tag{IV.60}$$

While this vanishes at $\Delta\hat{\varphi} = 0$, it initially rises with infinite slope as $\Delta\hat{\varphi}$ increases above zero.

The dominance of vacancies over interstitials in crystal thermal equilibrium at low pressure has been established experimentally for several metallic elements [Simmons and Balluffi, 1960a, 1960b, 1962, 1963; Feder and Charbnau, 1966; Feder and Nowick, 1967]. This dominance involves comparison of the temperature dependence of macroscopic system volume with what would formally be implied by the temperature dependence of the crystal lattice spacing. An anomalous rise in volume upon raising the temperature toward the melting point arises from spontaneous insertion of vacancies that were evidently generated initially at the crystal surface, then subject to inward diffusion. Typical vacancy concentrations at the crystal melting temperature fall in the range 10^{-4} to 10^{-3} per lattice site [Kelly and Knowles, 2012, Table 10.3]. These experiments on metallic elements do not reveal any comparable appearance of interstitial defects at thermal equilibrium, consistent with their substantially larger excitation energy and enthalpy.

Outside of the regime of thermal equilibrium, energetic electron radiation damage to monatomic crystals can cause substantial displacement of individual atoms from their initially occupied sites, creating monovacancy-interstitial defect pairs [Makin, 1968]. These defect pairs, separated by only a few lattice spacings when initially created, are called "Frenkel defects." Unlike monatomic crystals, some ionic crystals (e.g., AgCl and AgBr) can exhibit detectable concentrations of Frenkel defects even at thermal equilibrium because of ease of displacement of the small cations within the relatively immobile framework of the larger and jammed array of anions [Weber and Friauf, 1969]; however, the resulting interstitial defects geometrically may not be simple split interstitials.

A proper descriptive formalism for point defect concentrations at thermal equilibrium of course requires inclusion of the effects of intrabasin vibrational free energies. For generality, suppose that the nondefective crystal could contain several particle species in appropriate stoichiometric ratios and that any of the possible crystallographic symmetries might apply. This assumption can produce several structurally distinct types of point defects to be indexed by a running index "i." Let B_α be a basin in the isobaric ensemble corresponding to inherent structure α, and suppose that α is a configuration that exhibits n_i point defects of species i that are sufficiently well separated so as to be unambiguously identifiable and to act independently. The isobaric partition function for B_α at $\beta = 1/k_B T$ and pressure p has the following Boltzmann factor form:

$$\exp[-\beta\hat{\Phi}_\alpha - \beta\hat{F}_{\alpha,vib}(\beta, p, \{n_i\})],$$

where $\hat{\Phi}_\alpha$ and $\hat{F}_{\alpha,vib}$ are the inherent-structure enthalpy and isobaric vibrational free energy, respectively, for B_α. The assumed independence of the point defects requires that $\hat{F}_{\alpha,vib}$ be a linear function of the n_i. If $\hat{F}_{0,vib}(\beta, p)$ denotes the vibrational free energy for an absolute minimum basin B_0 (no defects), then one has

$$\hat{F}_{\alpha,vib}(\beta, p, \{n_i\}) = \hat{F}_{0,vib}(\beta, p) + \sum_i \hat{\delta}_i(\beta, p) n_i. \tag{IV.61}$$

Here the $\hat{\delta}_i(\beta, p)$ are the isobaric vibrational free-energy increments (or decrements) resulting from insertion of a single species-i point defect.

The aggregate vibrational free energy for the distribution of numbers n_i that produce a fixed enthalpy per particle $\hat{\varphi}$ follows upon averaging the Boltzmann factors over isobaric basins $\hat{B}_\alpha$ whose depths per particle are identified as falling in the immediate neighborhood of that $\hat{\varphi}$ value:

$$\exp[-N\beta\hat{f}_{vib}(\hat{\varphi}, \beta, p, \{n_i\})] = \langle\exp[-\beta\hat{F}_{\alpha,vib}(\beta, p, \{n_i\})]\rangle_{\hat{\varphi}\pm\delta\hat{\varphi}/2}. \tag{IV.62}$$

It needs to be understood that in the large-system limit, inequalities among the $\hat{\delta}_i(\beta, p)$ have the effect in Eq. (IV.62) of causing their own dominating set of defect concentrations at equilibrium generally to differ from those that determine $\hat{\sigma}$, shown in Eqs. (IV.56) and (IV.57). It is the temperature-dependent competition between the enumerative and the vibrational distributions that establishes the physically observable defect concentrations. In particular, those observable equilibrium concentrations at T, p follow by minimizing with respect to all n_i the full per-particle free-energy expression:

$$-\beta^{-1}\hat{\sigma}(\{n_i/N\}) + N^{-1}\sum_i[\hat{\Delta}^{(i)}(p) + \hat{\delta}_i(\beta, p)]n_i, \tag{IV.63}$$

where in the low-defect-concentration regime the isobaric basin enumeration function $\hat{\sigma}$ is a straightforward extension to arbitrary numbers of independent point defect species of the second form shown in Eq. (IV.54) for two species:

$$\hat{\sigma}(\{n_i/N\}) \approx \sum_i \left(\frac{n_i}{N}\right)\left[1 - \ln\left(\frac{n_i}{Nw_i}\right)\right].$$

(IV.64)

Here, following earlier notation, w_i represents the number of distinct orientations possible for defect species i (equal to unity for simple monovacancies). Because of the assumed independence of the defects, it is possible to circumvent introduction of a Lagrange multiplier $-\Lambda$ to convert $\hat{\sigma}$ to a function of $\hat{\varphi}$. The direct minimization of free-energy expression (IV.63) with respect to defect concentrations leads to the following result:

$$\langle n_i/N\rangle_{\beta,p} = w_i \exp\{-\beta[\hat{\Delta}^{(i)}(p) + \hat{\delta}_i(\beta, p)]\}.$$

(IV.65)

As expected, Eq. (IV.65) implies a strong vanishing of all defect concentrations as temperature approaches absolute zero.

At positive temperatures below the crystal melting point, the presence of point defects and their thermally activated mobility can be a principal contributor to the typically small but non-vanishing self-diffusion rate for particles. This contribution occurs because the transition states (saddle points) permitting interbasin transitions between neighboring defect arrangements usually involve only a modest rise in potential energy or enthalpy above that of the inherent structures for those contiguous basins. The observation that temperature dependence of crystal-phase self-diffusion constants D can usually be well fitted empirically by an Arrhenius form [Kittel, 1996, p. 544]:

$$D(T) \approx D_0 \exp(-\Delta E/k_B T)$$

(IV.66)

suggests that the observed ΔE is a good approximation to the activation energy or enthalpy for the dominant defect motion process. In this respect, the crystal-phase self-diffusion process basically differs from another transport process, specifically thermal conductivity, which can proceed via intrabasin vibrational motions without any interbasin transitions [see Section IV.G].

Although this section has focused on the low-defect-concentration situation that is relevant to low temperatures, in some cases that may become inappropriate as the temperature climbs toward the crystal's thermodynamic melting point. The crystalline state of silver bromide (AgBr, known as the mineral "bromyrite") appears to offer a contrasting example. Its structurally ideal form is cubic and is isomorphous with sodium chloride (NaCl) [Huggins, 1951]. Upon heating at ambient pressure toward its melting point ($T_m \approx 700K$), the heat capacity rises continuously but rapidly to more than twice the Dulong-Petit value [Christy and Lawson, 1951]. That rise is attributed primarily to the enthalpy expenditure involved in creating high concentrations of crystal defects, predominantly vacancies. This interpretation is supported by the corresponding anomalous thermal expansion of the substance as it approaches T_m [Lawson, 1950].

E. Interchange Order–Disorder Transitions

Several binary alloys exhibit an important class of phase transitions occurring within the crystal state that involves thermally activated positional exchanges. These phase transitions have traditionally been called "order–disorder" transitions, though they represent only one category in that broader family of phase transitions. The equimolar CuZn alloy, "β-brass", is a particularly

(a)

(b)

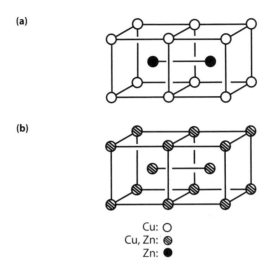

Cu: ○
Cu, Zn: ◐
Zn: ●

FIGURE IV.6. Sublattice occupancies in β-brass. As shown in (a), the ground state exhibits complete segregation of the Cu and Zn atoms on the two interpenetrating simple cubic lattices. The high-temperature situation indicated in (b) involves overall desegregation.

well-known example [Kittel, 1996, p. 615]. The atom-occupied sites in this alloy present a body-centered cubic periodic array. In the zero-temperature limit, all copper atoms reside on one of the two interpenetrating simple cubic lattices, and all zinc atoms reside on the other one. This geometric arrangement is indicated in Figure IV.6(a).

However, as temperature rises, atom exchanges begin to occur so as to reduce the extent of species segregation between the sublattices, and at a critical temperature $T_c = 741$ K, the intermixing becomes complete, as shown in Figure IV.6(b). This disordering temperature is still well below the alloy's melting temperature, $T_m = 1107$ K. Associated with the species disordering between the sublattices is a rise in heat capacity, apparently diverging at T_c. Figure IV.7 shows a plot of the experimentally measured C_p through this order–disorder phase transition. The divergence is frequently identified as a "lambda transition" because of its rough resemblance to that Greek character. The experimental observation for the β-brass alloy system suggests that C_p diverges upon approach to T_c both from below and from above. However, the divergence is not symmetrical. For a small temperature change $\Delta T > 0$, $C_p(T_c - \Delta T)$ substantially exceeds $C_p(T_c + \Delta T)$.

The dominant kinetic mechanism for atom interchanges in β-brass appears not to have been determined experimentally. An obvious candidate relies on the presence of a small concentration of monovacancies whose individual hops between nearest neighbor sites have the effect of moving atoms from one simple cubic sublattice to the other. Any arrangement of Cu and Zn atoms over the body-centered cubic crystal in principle can be reached from any other one by a suitable (possibly very long) sequence of monovacancy hops. Another conceivable kinetic mechanism not requiring vacancies would instead involve cyclic rotation of atoms around a closed loop of nearest neighbors.

Suppose that the β-brass crystal consists of an even number N of sites, with $N/2$ belonging to each of the simple-cubic sublattices. The relevant inherent structures for this phase are distinguishable by the atomic species inhabiting each of the N sites. For purposes of accounting, attach

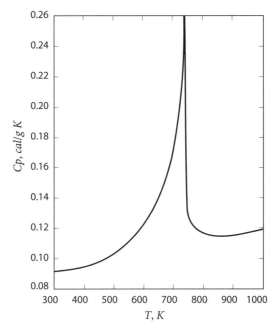

FIGURE IV.7. Constant-pressure heat capacity C_p at 1 atmosphere for β-brass. Redrawn from Nix and Shockley, 1938, Figure 4.

an occupation variable $\mu_j = \pm 1$ to site j, where -1 represents Cu and $+1$ represents Zn. Consequently, the alloy's equimolar composition implies

$$\sum_{j=1}^{N} \mu_j = 0. \tag{IV.67}$$

For notational simplicity, require that site numbering be assigned such that $1 \leq j \leq N/2$ corresponds to one sublattice "A" while $(N/2) + 1 \leq j \leq N$ corresponds to the other sublattice "B". Using this convention, it is possible to define in the following way an intensive long-range order parameter $-1 \leq x \leq +1$ that measures sublattice segregation:

$$x = (2/N) \sum_{j=1}^{N/2} \mu_j \tag{IV.68}$$
$$= -(2/N) \sum_{j=(N/2)+1}^{N} \mu_j.$$

Consequently, the numbers of Cu and Zn atoms on sublattice A are $(1 + x)N/4$ and $(1 - x)N/4$, respectively. The roles are reversed on sublattice B, where the respective numbers of Cu and Zn atoms are $(1 - x)N/4$ and $(1 + x)N/4$. In a state of thermal equilibrium at any positive temperature, x fluctuates about zero as time passes. But below temperature T_c, the alloy macroscopically exhibits its long-range order, so in the large-system limit, x would exhibit a persistent average magnitude $\langle |x| \rangle > 0$. Above T_c, the average magnitude collapses to $\langle |x| \rangle = 0$ in the large-N limit.

For a fixed value of x, the number of distinguishable atom arrangements Ω_d over the full set of N sites follows from elementary combinatorial considerations:

$$\Omega_d(x) = \left\{ \frac{(N/2)!}{[(1 - x)N/4]![(1 + x)N/4]!} \right\}^2. \tag{IV.69}$$

Using Stirling's approximation for the large-system limit, this result implies

$$\ln \Omega_d(x) \sim N\{\ln 2 - (1/2)[(1-x)\ln(1-x) + (1+x)\ln(1+x)]\}. \tag{IV.70}$$

This symmetrical function of x passes through a maximum at $x = 0$ and vanishes as $x \to \pm 1$.

Because β-brass is a metal, with arbitrarily small electronic excitations possible above the ground state in the large-system limit, the atoms of this alloy in principle do not reside on a single Born-Oppenheimer potential energy surface when $T > 0$. However, it has been pointed out that the electronic heat capacity caused by those excitations is far outweighed by the interchange order–disorder contribution to the heat capacity [Nix and Shockley, 1938]. Therefore, in the absence of compelling evidence to the contrary, it is reasonable to suppose that electronic-excitation and atomic exchange degrees of freedom can be regarded as acting independently. Consequently, it is presumed that a single potential energy function Φ of atomic positions, or a single potential enthalpy function $\hat{\Phi}$ of atomic positions and container volume, can be utilized for analyzing the order–disorder transition. This situation implies that the general "landscape" and inherent structure formalism developed in previous chapters continues to be applicable for the nonelectronic portion of the crystalline alloy system's properties.

In view of the fact that experimental measurements on alloys generally, and β-brass specifically, tend to be performed at constant pressure, an isobaric version of the canonical partition function is the natural choice to represent thermal equilibrium properties. The corresponding Gibbs free energy expression appeared as Eq. (III.30) in the preceding chapter and is now repeated here:

$$\beta G_N/N = \beta\{\hat{\varphi}^*(\beta, p) + \hat{f}_{vib}[\hat{\varphi}^*(\beta, p), \beta, p]\} - \hat{\sigma}[\hat{\varphi}^*(\beta, p)]. \tag{IV.71}$$

In this expression, which is formally exact in the large-system limit, $\hat{\varphi}^*$ locates the enthalpy depth (per particle) of the dominant basins at the given β, p, whereas $\hat{f}_{vib}$ is the corresponding intrabasin vibrational free energy per particle. The enthalpy depth quantity $\hat{\varphi}^*$ is determined by the requirement that the $\beta G_N/N$ expression be at a minimum with respect to $\hat{\varphi}$ at constant β, p, which then leads to the equation

$$\left[\frac{d\hat{\sigma}(\hat{\varphi})}{d\hat{\varphi}}\right]_{\hat{\varphi}=\hat{\varphi}*} = \beta\left[1 + \frac{\partial \hat{f}_{vib}(\hat{\varphi}, \beta, p)}{\partial \hat{\varphi}}\right]_{\hat{\varphi}=\hat{\varphi}*}. \tag{IV.72}$$

As indicated in Eq. (IV.71), $\hat{\varphi}^*(\beta, p)$ is required to evaluate the relevant vibrational free energy $\hat{f}_{vib}$ and the depth-dependent basin enumeration function $\hat{\sigma}$.

The right side of Eq. (IV.71) can be straightforwardly estimated by invoking a "mean field approximation" (MFA). This is an adaptation to the present isobaric "landscape" representation of the historically prominent "Bragg-Williams" approximation for order–disorder phenomena [Bragg and Williams, 1934]. Simply put, the mean field approximation in the present context presumes that $\hat{f}_{vib}$, $\hat{\sigma}$, and $\hat{\varphi}$ depend configurationally only on the long-range order parameter x. This is an appropriate presumption for the case of long-range interactions that effectively sample a large local population of particles. In that circumstance, the collection of neighbors with which each particle interacts tends to be closely representative of the overall average composition on each of the sublattices. In the event that the effective interaction for the system at its inherent structures is dominated by pair potentials, the mean field approximation leads formally to an assignment of the intensive depth parameter $\hat{\varphi}$ for the enthalpy basins, which is simply a symmetric quadratic function of x:

$$\hat{\varphi}(x) = \hat{\varphi}_0 - \hat{\varphi}_2 x^2, \tag{IV.73}$$

$$x(\hat{\varphi}) = \pm [\hat{\varphi}_0 - \hat{\varphi})/\hat{\varphi}_2]^{1/2}.$$

Here the coefficients $\hat{\varphi}_0$ and $\hat{\varphi}_2$ generally depend on the external pressure p. To be consistent with the fact that low temperature causes the system to order spontaneously in a state with $x = \pm 1$, Eq. (IV.73) must have $\hat{\varphi}_2 > 0$, at least within the pressure range for which the β-brass phase is stable. Furthermore, $\hat{\varphi}_0$ can be identified as the enthalpy per particle of the highest lying isobaric inherent structures for the crystalline alloy. Equation (IV.73) leads in turn, via Eq. (IV.70), to an expression for the enumeration function:

$$\hat{\sigma}(\hat{\varphi}) = \lim_{N \to \infty} N^{-1} \ln \Omega_d [x(\hat{\varphi})]$$

$$= \ln 2 - (1/2)\{[1 - x(\hat{\varphi})] \ln[1 - x(\hat{\varphi})] + [1 + x(\hat{\varphi})] \ln[1 + x(\hat{\varphi})]\}$$

$$= \ln 2 - \frac{1}{2} \left[1 - \left(\frac{\hat{\varphi}_0 - \hat{\varphi}}{\hat{\varphi}_2} \right)^{1/2} \right] \ln \left[1 - \left(\frac{\hat{\varphi}_0 - \hat{\varphi}}{\hat{\varphi}_2} \right)^{1/2} \right] \tag{IV.74}$$

$$- \frac{1}{2} \left[1 + \left(\frac{\hat{\varphi}_0 - \hat{\varphi}}{\hat{\varphi}_2} \right)^{1/2} \right] \ln \left[1 + \left(\frac{\hat{\varphi}_0 - \hat{\varphi}}{\hat{\varphi}_2} \right)^{1/2} \right]$$

This mean field estimate of the inherent structure enumeration function for β-brass has been plotted in Figure IV.8. The most notable features of this function are its positive slope at its upper end ($\hat{\varphi} = \hat{\varphi}_0$) and its diverging slope upon approach to its lower end ($\hat{\varphi} = \hat{\varphi}_0 - \hat{\varphi}_2$). The latter

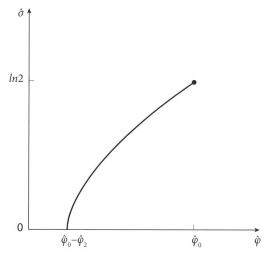

FIGURE IV.8. Plot of the mean field approximation for the inherent structure enumeration function $\hat{\sigma}(\hat{\varphi})$ versus the potential enthalpy per particle $\hat{\varphi}$.

characteristic stems from a few misplaced particles residing on the "wrong" sublattice, acting therefore as point defects, as described by Eq. (IV.60). It should be kept in mind that this plot only refers to inherent structures relevant to the unmelted crystal, and therefore it excludes those that underlie the molten state.

In a fashion similar to that of the mean field assignment for $\hat{\varphi}$, the vibrational free energy per particle is also assigned as a symmetric quadratic function of x:

$$\hat{f}_{vib}(\hat{\varphi}, \beta, p) = \hat{f}_0(\beta, p) - \hat{f}_2(\beta, p)x^2. \tag{IV.75}$$

This is consistent with the underlying assumption of very long-range interactions, where the only relevant feature would be how many particle pairs of each type are on the same or on different sublattices. As explicitly indicated in this last equation, the two coefficients $\hat{f}_0$ and $\hat{f}_2$ may be pressure dependent, but the latter (unlike $\hat{\varphi}_2$) is not necessarily positive.

Eq. (IV.72) for determining the free-energy-minimizing quantity $\hat{\varphi}^*$ can, for the mean field approximation, be expressed equivalently as an equation to determine the optimal value of the long-range order parameter

$$\beta[(\partial\hat{\varphi}/\partial x)_{\beta,p} + (\partial\hat{f}_{vib}/\partial x)_{\beta,p}] - (\partial\hat{\sigma}/\partial x)_p = 0. \tag{IV.76}$$

Upon inserting the mean field expressions Eqs. (IV.73) and (IV.75) as well as the enumeration function in Eq. (IV.74) into this last relation, the resulting equation for determining $x^* \equiv x(\hat{\varphi}^*)$ has the following simple form:

$$B(\beta, p)x^* = \ln(1 + x^*) - \ln(1 - x^*), \tag{IV.77}$$

$$B(\beta, p) = 4\beta(\hat{\varphi}_2 + \hat{f}_2).$$

Figure IV.9 presents in graphical form the two members of this transcendental equation for the case relevant to β-brass, namely $\hat{\varphi}_2 + \hat{f}_2 < 0$. For high temperature (small $\beta = 1/k_BT$), the only real solution is $x^* = 0$, stating that the alloy is disordered and devoid of long-range order. At low temperature, a symmetric pair of nonzero solutions also appears, which represents the mean field prediction for long-range order in the alloy, i.e., species segregation between the sublattices. The value $x^* = 0$ continues to be a solution to transcendental Eq. (IV.77) at low temperature, but in the presence of the flanking symmetric pair, it no longer locates the dominant contributions to free-energy expression Eq. (IV.71).

The crossover point between the two types of solution, which is the predicted critical point for the order–disorder transition, occurs when the slopes at $x = 0$ of the linear and logarithmic members of Eq. (IV.77) are equal. This leads to the following criterion for location of the order–disorder critical temperature in the mean field approximation:

$$k_BT_c \equiv \beta_c^{-1} \tag{IV.78}$$

$$= 2[\hat{\varphi}_2(p) + \hat{f}_2(\beta_c, p)].$$

Figure IV.10 presents the solutions to transcendental Eq. (IV.77) in graphical form, with x^* plotted versus the positive quantity $B(\beta, p)$, which under constant-pressure conditions can reasonably be expected to be a monotonically decreasing function of temperature T.

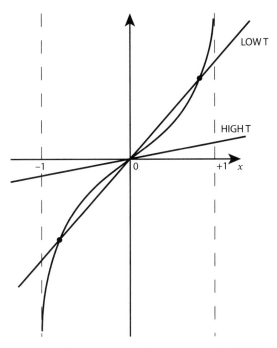

FIGURE IV.9. Graphical representation of the two members of transcendental Eq. (IV.77) to determine the β-brass long-range order parameter x as a function of temperature. The intersections between the linear and the logarithmic curves locate the mean field approximation to the long-range order, $x*(\beta, p)$.

Thermodynamics allows the constant-pressure heat capacity C_p to be obtained in general and without approximation from the Gibbs free energy by differentiation:

$$\frac{C_p}{Nk_B} = -\beta^2 \left(\frac{\partial^2 (\beta G_N / N)}{\partial \beta^2} \right)_p . \tag{IV.79}$$

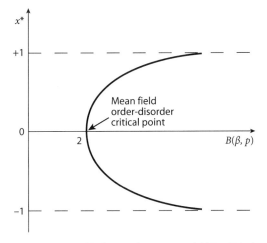

FIGURE IV.10. Plot of the solutions $x*$ versus $B(\beta, p)$, according to mean field Eq. (IV.77).

By using expression (IV.71) for G_N, and carrying out the differentiation subject to the determining condition Eq. (IV.72) for $\hat{\varphi}^*$, one obtains the following:

$$\frac{C_p}{Nk_B} = \frac{C_p^{(0)}}{Nk_B} + \frac{C_p^{(1)}}{Nk_B}, \tag{IV.80}$$

where

$$\frac{C_p^{(0)}}{Nk_B} = -\beta^2 \left(\frac{\partial^2 \beta \hat{f}_{vib}}{\partial \beta^2} \right)_{\hat{\varphi}^*, p}, \tag{IV.81}$$

$$\frac{C_p^{(1)}}{Nk_B} = -\beta^2 \left(\frac{\partial \hat{\varphi}^*}{\partial \beta} \right)_p.$$

The first term $C_p^{(0)}/Nk_B$ is the contribution to heat capacity due only to intrabasin thermal excitation, holding $\hat{\varphi}^*$ fixed at its prevailing value for the given β, p. The second term $C_p^{(1)}/Nk_B$ represents the heat capacity contribution due to the temperature-induced changes in basin occupancy probabilities, including those attributable to the particle population shifts between the sublattices.

Within the mean field approximation, only $C_p^{(0)}/Nk_B$ survives above the order–disorder critical temperature T_c because $x^* \equiv 0$ and so $\hat{\varphi}^* \equiv \hat{\varphi}_0$ in this regime. Consequently,

$$\frac{C_p}{Nk_B} = -\beta^2 \left(\frac{\partial^2 \beta \hat{f}_0}{\partial \beta^2} \right)_p \qquad (\text{MFA}, \ T > T_c). \tag{IV.82}$$

In the low-temperature regime, by contrast, the long-range order parameter x is virtually pinned to the values ± 1. In this limit, the MFA once again attributes the entire heat capacity to the first term in Eq. (IV.80), but now with the shifted value

$$\frac{C_p}{Nk_B} = -\beta^2 \left(\frac{\partial^2 \beta (\hat{f}_0 - \hat{f}_2)}{\partial \beta^2} \right)_p \qquad (\text{MFA}, \ T \ll T_c). \tag{IV.83}$$

In the intermediate sub-T_c regime, the temperature dependence of the long-range order provides an additional contribution to the heat capacity. This causes the term $C_p^{(1)}/Nk_B$ in Eq. (IV.80) to be nonvanishing. Just below the critical temperature where $T_c - T \geq 0$ is small, a Taylor series expansion of the logarithmic terms in transcendental Eq. (IV.77) leads to the conclusion that x has the values

$$x^* = \pm A(T_c - T)^{1/2} + O(T_c - T), \tag{IV.84}$$

$$A = \left\{ \left[\frac{3}{2} k_B \beta^2 \left(\frac{\partial B}{\partial \beta} \right)_p \right]_{\beta = \beta_c} \right\}^{1/2}.$$

Using this leading-order result, one finds that the mean field prediction for the heat capacity just below the order–disorder critical temperature is the following:

$$\frac{C_p}{Nk_B} = -\beta^2 \left(\frac{\partial^2 \beta \hat{f}_0}{\partial \beta^2} \right)_p + A\hat{\varphi}_2 + O(T_c - T). \qquad (\text{MFA}) \tag{IV.85}$$

Thus the mean field prediction is that C_p/Nk_B undergoes a discontinuous increase by $A\hat{\varphi}_2 > 0$ as T declines through T_c, hence predicting an asymmetry about T_c that nominally agrees with the experimental observation, Figure IV.7. However, the mean field heat capacity remains finite through the order–disorder transition, in apparent disagreement with experiment.

The failure of the MFA to reproduce the diverging "lambda" behavior of C_p at the order–disorder critical point is an inadequacy of its $C_p^{(1)}/Nk_B$ assignment since the roughly harmonic intrabasin contribution to heat capacity $C_p^{(0)}/Nk_B$ is expected to remain finite. This inadequacy stems from the fact that particle interactions in β-brass are relatively short-ranged spatially. As a consequence, the basic presumption of the MFA is misleading, namely, that each particle interacts with a population of others locally identical in sublattice composition to that of the entire macroscopic system. Instead, a more accurate statistical description of the order–disorder phenomenon must account for deviations of short-range particle order from that described simply by the long-range order parameter x. To illustrate the implications of this distinction, it is useful to examine the opposite extreme, namely that only nearest neighbor particles interact. This amounts to describing the collection of inherent structures for the order–disorder phenomenon in terms of the three-dimensional Ising model (with vanishing external field) on the body-centered cubic lattice [Domb, 1960a, 1960b; Huang, 1963, Chapter 16].

The site occupancy variables $\mu_j = \pm 1$ introduced earlier serve as Ising model "spins." The potential enthalpy of the inherent structures for the β-brass system then can be written in terms of a sum over all nearest neighbor (n.n.) pairs of sites in the alloy, subject to the fixed composition constraint Eq. (IV.67):

$$\hat{\Phi}(\mu_1 \ldots \mu_N) = \hat{\Phi}_{BL}(p) + J(p) \sum_{(n.n.)} \mu_i \mu_j. \tag{IV.86}$$

Here $\hat{\Phi}_{BL}(p)$ is a baseline enthalpy for the crystal that is independent of the distribution of particles among the sites, and $J(p)$ is the assumed nearest neighbor interaction. In the large-system limit that is of interest here, the formal constraint Eq. (IV.67) on the N spins carries no weight (it is satisfied on average) and so can be neglected in the following. Assuming that periodic boundary conditions apply to the alloy system, where each particle is surrounded by eight nearest neighbors, the number of terms in the sum in Eq. (IV.86) must be $4N$. Because the ground-state structure of the β-brass system has the two species segregated fully between the two sublattices, this requires that

$$J(p) > 0, \tag{IV.87}$$

which in conventional Ising model magnetic terminology is an "antiferromagnetic" interaction [ter Haar, 1954, Chapter XII]. Therefore, for all inherent structures, one has

$$\hat{\Phi}(\mu_1 \ldots \mu_N) \geq \hat{\Phi}_{BL}(p) - 4J(p)N. \tag{IV.88}$$

An upper limit for $\hat{\Phi}$ corresponds to surrounding each particle with eight nearest neighbors of its own species, i.e., segregating the species in their own macroscopic domains. With neglect of the interface between such domains, one concludes that

$$\hat{\Phi}(\mu_1 \ldots \mu_N) \leq \hat{\Phi}_{BL}(p) + 4J(p)N. \tag{IV.89}$$

Both of these limiting situations (segregation between sublattices, or in pure-species domains) are configurationally special, and so they correspond to small values of the enumeration function $\hat{\sigma}(\hat{\varphi})$. However, a maximum of this enumeration function can be expected to occur when $N\hat{\varphi} \equiv \hat{\Phi}$ lies between these lower and upper limits, with the particles irregularly distributed over the N sites so that the species composition of the nearest neighbors of any particle is random. This indicates that the $\hat{\sigma}(\hat{\varphi})$ maximum should occur at $\hat{\varphi} = \hat{\Phi}_{BL}(p)/N$. Consequently, it represents a qualitative departure from the mean field result presented in Figure IV.8, which displays a maximum located at the upper $\hat{\varphi}$ limit.

The conventional context for discussion of Ising models avoids consideration of intrabasin vibrational contributions to thermodynamic properties [Huang, 1963, Chapter 16]. Returning to the general Gibbs free energy expression Eq. (IV.71), this amounts to neglecting the $\hat{\varphi}$ dependence of $\hat{f}_{vib}$. For the moment, this simplification is accepted in order to establish a "baseline" behavior. However, the proper influence of $\hat{f}_{vib}$ is subsequently fully restored. With this simplification, Eq. (IV.72) for the determination of $\hat{\varphi}^{*}(\beta)$ reduces to

$$[\partial\hat{\sigma}(\hat{\varphi})/\partial\hat{\varphi}]_{\hat{\varphi}=\hat{\varphi}*} = \beta, \tag{IV.90}$$

i.e., the slope of $\hat{\sigma}(\hat{\varphi})$ at $\hat{\varphi}^{*}(\beta)$ is equal precisely to the inverse temperature quantity β. The temperature dependence of $\hat{\varphi}^{*}(\beta)$ in the neighborhood of the critical point, which is directly connected to the diverging heat capacity contribution $C_p^{(1)}/Nk_B$ in Eq. (IV.81), determines the shape of the enumeration function $\hat{\sigma}(\hat{\varphi})$. This shape thus can be determined by using known properties of the field-free Ising model [Stillinger, 1988b].

Although three-dimensional Ising models currently do not possess exact solutions analogous to that discovered by Onsager for the two-dimensional version [Onsager, 1944; Kaufmann, 1949], reliable analyses are available that suffice for present purposes. Whether the three-dimensional lattice involved is the 8-coordinate body-centered cubic structure or an alternative periodic structure (simple cubic, face-centered cubic, …), the corresponding nearest neighbor Ising model critical points are characterized by a common set of critical exponents [Stanley, 1971, Chapter 3]. In particular, the diverging heat capacity is described by a pair of positive Ising exponents α_{I-} and α_{I+}, both approximately equal to 0.1. For the body-centered cubic case relevant to β-brass, these results imply the following divergence behaviors at T_c:

$$C_p^{(1)}/Nk_B \cong c_-(T_c - T)^{-\alpha_{I-}} + O(1) \qquad (T < T_c), \tag{IV.91}$$

$$\cong c_+(T - T_c)^{-\alpha_{I+}} + O(1) \qquad (T > T_c),$$

where c_- and c_+ are positive constants. The value of the critical temperature T_c for the Ising model on this lattice has been estimated to be [Domb, 1960b, p. 288]:

$$k_B T_c(p) \cong 6.35J(p). \tag{IV.92}$$

The inverse-power divergences shown in Eq. (IV.91) must be the same kind of divergences exhibited by $\partial\hat{\varphi}^{*}/\partial\beta$, except for numerical coefficients. In order for Eq. (IV.90) to produce those inverse-power divergences, $\hat{\sigma}(\hat{\varphi})$ must have just the right shape in the neighborhood of the point at which its slope is $\beta_c = 1/k_B T_c$. In particular, $d^2\hat{\sigma}/d\hat{\varphi}^2$ must vanish there to produce a sufficiently rapid change of $\hat{\varphi}^*$ with changing temperature. Therefore, the leading-order shape must conform to the following expression:

$$\hat{\sigma}(\hat{\varphi}) = \hat{\sigma}(\hat{\varphi}_c) + \beta_c(\hat{\varphi} - \hat{\varphi}_c) - a_\pm |\hat{\varphi} - \hat{\varphi}_c|^{q_\pm} + \ldots, \tag{IV.93}$$

where the +, – subscripts refer respectively to $T > T_c$ and $T < T_c$. In order to produce the particular singularities exhibited in Eq. (IV.91) via the determining condition in Eq. (IV.90), the coefficients $a_\pm$ and exponents $q_\pm$ must be the following:

$$a_\pm = \frac{(1 - \alpha_{I\pm})k_B\beta_c^2}{(2 - \alpha_{I\pm})}\left[\frac{k_B c_\pm}{1 - \alpha_{I\pm}}\right]^{1/(\alpha_{I\pm} - 1)}, \tag{IV.94}$$

and

$$q_\pm = \frac{2 - \alpha_{I\pm}}{1 - \alpha_{I\pm}} > 2. \tag{IV.95}$$

It was noted earlier that the mean field approximation, though relatively crude, manages to produce a heat capacity asymmetry about the order–disorder critical point, yielding a discontinuously larger value upon cooling through T_c. Asymmetry also emerges from the physically more accurate Ising model solution with nearest neighbor interactions [Domb, 1960b, Section 4.7.3]. This is due at least in part to the comparative magnitudes of the "$O(1)$" terms indicated in Eqs. (IV.91).

The Ising model description also deviates from that of the mean field approximation with respect to the behavior of the long-range order quantity x in the neighborhood of the critical point. The former reveals another critical exponent $\beta_I \cong 0.31$ [Stanley, 1971, Chapter 3] such that

$$x^*(T) = \pm x_0(T_c - T)^{\beta_I} + \ldots \qquad (T < T_c), \tag{IV.96}$$

where x_0 is a positive constant, and the terms following the one explicitly shown vanish more rapidly as $T \to T_c$ from below. Eq. (IV.84) indicated that the mean field analog of β_I had the larger value 1/2, implying a slower development of long-range order in the alloy as T declines below T_c. It should be noted that the set of Ising critical exponents is thought to remain unchanged as the interaction range increases to encompass a finite number of lattice coordination shells, and only in the infinite-range limit would the mean field set become relevant [Stanley, 1971, Chapter 6].

It is now necessary to return to examine the effect of $\hat{f}_{vib}(\hat{\varphi}, \beta, p)$ on the order–disorder transition. This examination requires considering its role in Eq. (IV.72) for the determination of $\hat{\varphi}^*$ as a function of temperature. If $\partial\hat{f}_{vib}(\hat{\varphi}, \beta, p)/\partial\hat{\varphi}$ were independent of $\hat{\varphi}$ and could thus be represented merely by $\gamma(\beta, p)$, then the only vibrational influence on $\hat{\varphi}^*$ would be to introduce an effective value β_{eff} for the inverse temperature:

$$\beta_{eff} = \beta[1 + \gamma(\beta, p)], \tag{IV.97}$$

which could then formally be inserted in place of β in the right side of Eq. (IV.90). Although this would shift the critical temperature (up if $\gamma > 0$, down if $\gamma < 0$), the critical exponents would be unchanged.

More generally, one must expect $\hat{f}_{vib}$ to depend upon the enthalpic depth parameter $\hat{\varphi}$. However, this vibrational free-energy quantity as a function of $\hat{\varphi}$ must be subject to a shape constraint so that its effect on the thermodynamics is consistent with the observed heat capacity divergence at the critical point. In particular, this demands that

$$[\partial^2 \hat{f}_{vib}(\hat{\varphi}, \beta, p)/\partial \hat{\varphi}^2]_{\hat{\varphi}_c, \beta_c} = 0. \tag{IV.98}$$

That is, the $\hat{\varphi}$-direction curvature of $\hat{f}_{vib}$ must vanish identically for the values of its variables at the critical point. Otherwise, the condition in Eq. (IV.72) would eliminate the heat capacity divergence, in spite of the singular character of $\hat{\sigma}(\hat{\varphi})$ exhibited in Eqs. (IV.93) and (IV.95). Although at first sight this might seem to be an ad hoc requirement for $\hat{f}_{vib}$, it should be kept in mind that this function depends directly on the statistical nature of the relevant basins within which the vibrations occur. The short-range interactions that in the Ising model cause like spins (same-species particles) to cluster together on the same sublattice for β-brass and to produce a heat capacity divergence, also supply the force constants that underlie the intrabasin vibrational frequencies. This clustering generates the singular behavior in $\hat{\sigma}(\hat{\varphi})$ at $\hat{\varphi}_c$, and it may be implicated in producing an analogous curvature-eliminating singularity in $\hat{f}_{vib}$. It needs to be remarked in passing that this aspect of the interchange order–disorder transition is relatively poorly understood at the time of this writing and so deserves more and careful research attention in the future.

Figure IV.11 graphically illustrates the terms in Eq. (IV.72) that determine $\hat{\varphi}^*$. This is an isobaric version of Figure III.1 for isochoric (constant-volume) conditions, now representing only those inherent structures and basins relevant to the β-brass crystalline alloy. The slope matching position that locates $\hat{\varphi}^*$ has been indicated for several temperatures, including the critical temperature at which the curvatures vanish.

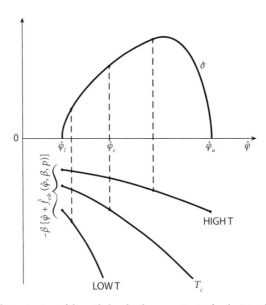

FIGURE IV.11. Graphical determination of the enthalpy depth parameter $\hat{\varphi}*$ for the interchange order–disorder transition. The $\hat{\sigma}(\hat{\varphi})$ curve possesses a nonanalytic singular point at $\hat{\varphi}_c$ connected to system critical behavior.

If the fixed composition of the β-brass alloy deviates from strict equimolar stoichiometry ($N_{Cu} = N_{Zn}$), the continuous transition just described becomes replaced by a first-order transition with a latent heat. Correspondingly, the heat capacity divergence becomes replaced by a simple discontinuity and a latent heat. Also, the long-range order parameter undergoes a discontinuous change at this first-order transition.

The first-order-transition scenario appears in another type of binary alloy (symbolically, AB_3) that exhibits an interchange order–disorder transition. A specific example is $AuCu_3$, each pure-element component of which has the face-centered cubic structure as its stable crystal form, as indeed does the alloy. The sizes of the atoms are sufficiently close to one another that exchange substitution within the face-centered structure is possible. The fully ordered arrangement that is the equilibrium form at low temperature has one of the four simple-cubic sublattices fully occupied by all of the Au atoms, while the Cu atoms occupy the other three sublattices. The first-order transition (i.e., with a latent heat) to a disordered high-temperature alloy phase occurs at approximately 650 K [Nix and Shockley, 1938].

The most straightforward application of an Ising model to represent an AB_3 alloy of the $AuCu_3$ type defines spin μ_j on any site j of the lattice to be -1 if an A occupies that site and to be $+1$ if a B occupies that site. With this protocol, one has for an N-site face-centered cubic lattice

$$\sum_{j=1}^{N} \mu_j = N/2. \tag{IV.99}$$

To produce the sublattice-segregated pattern of the ground state, it suffices to have the Ising spins interact only with their six second neighbors because these are on the same simple cubic sublattice, and those pair interactions would be negative for a pair of spins with the same sign. Under the constraint Eq. (IV.99), the relevant Ising model's overall interaction potential is simply the following analog to Eq. (IV.86):

$$\hat{\Phi}(\mu_1 \ldots \mu_N) = \hat{\Phi}_{BL}(p) + J_2(p) \sum_{n.n.n.} \mu_i \mu_j, \tag{IV.100}$$

with $J_2(p) < 0$ representing the interaction between all $3N$ second, or next-nearest neighbor (*n.n.n.*) site pairs.

It should be noted in passing that the binary alloy $AlFe_3$ presents a distinct transition scenario. Its lattice of occupied sites is body-centered cubic. At low temperature, one of the simple-cubic sublattices is occupied solely by Fe atoms, while the other simple-cubic sublattice displays a strict alternation of Al and Fe atoms from one position to any nearest neighbor on that same sublattice. Raising the temperature produces a second-order transition, analogous to that of β-brass, with Al–Fe interchange disordering occurring only on the mixed-composition sublattice [Seitz, 1940, p. 37]. At least nominally, a nearest neighbor Ising model description for that sublattice seems appropriate.

F. Orientational Order–Disorder Transitions

Beyond particle position exchanges that produce an order–disorder transition below the crystal melting point (as in β-brass), there are other forms of partial structural disorder that can be thermally excited so as to generate a continuous phase transition within the crystalline state, with a

TABLE IV.5. Observed order–disorder transition temperatures for ammonium chloride crystals at 1 atm pressure

Substance	T_c, K	T_{subl}, K
NH$_4$Cl	242	613
ND$_4$Cl	250	—

Source: Shumake and Garland, 1970.

lambda-type heat capacity anomaly. Considering the wide range of molecular shapes and flexibilities that can be present in the crystalline state, many different molecular-scale interactions can be involved in producing such continuous phase changes. For simplicity, only a single class of these is considered in detail in this section, but this serves to illustrate several basic features of the broader family of conformational disorder transitions. The class considered involves orientational transitions in crystals and is illustrated by the relatively simple examples of ammonium chloride (NH$_4$Cl) and its deuterated analog (ND$_4$Cl). The transition temperatures T_c experimentally observed for these substances at 1 atm pressure are listed in Table IV.5. Simon appears to have been the first to report the heat capacity lambda anomaly in ammonium chloride crystals [Simon, 1922]. No corresponding melting points appear in the table. The crystal–liquid–vapor triple points for both of these substances occur at pressures above 1 atm, so heating the crystals at ambient pressure bypasses the melt and causes sublimation directly into the vapor phase at T_{subl}. A noticeable feature of these measured temperatures is the effect of replacing the light H atoms with the heavier D isotope in the ammonium cations, causing T_c to rise.

The crystal structures of these ammonium chlorides have the CsCl form: The N atoms of the ammonium cations (NH$_4^+$) occupy one of two interpenetrating simple cubic lattices, and the chloride anions (Cl$^-$) occupy the other simple cubic lattice. That is, the heavy atoms N and Cl reside, respectively, on the sites of the sublattices of a body-centered cubic lattice. In the ground-state structure applicable as $T \to 0$, all of the tetrahedral ammonium cations possess the same nominal orientation, with the four attached hydrogens or deuteriums pointing toward four of the eight neighboring halide anions. This arrangement is illustrated in Figure IV.12, which is an orientational preference that arises from hydrogen-bonding interactions. That kind of energetically preferred geometry with parallel ammonium orientations has each chloride acting as an acceptor for exactly four hydrogen bonds. Reversing the orientation of a single ammonium anion within the crystal would not alter the total number of hydrogen bonds present, but the result would be a higher energy arrangement in which four neighboring chlorides were reduced to three hydrogen bonds, while the other four were forced to accept five. Of course, reversing the orientations of all ammoniums simultaneously in the ordered structure illustrated in Figure IV.12 would lead to another equivalent ordered structure. Because simple on-site rotation of ammonium ions is involved in the disordering process, no further kinetic mechanism has to be invoked to produce the transition, such as vacancy-hopping or cyclic interchange processes that were considered for the preceding β-brass transition.

The analysis of these phase transitions is initially constrained to constant-volume conditions. In adapting the general inherent structure formalism to this class of order–disorder phenomena, it suffices to restrict attention to a relevant subset of the potential energy basins. In particular, one

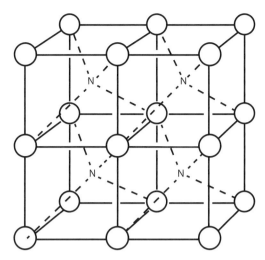

FIGURE IV.12. Atomic arrangement in the ground-state structures of NH$_4$Cl and ND$_4$Cl. The chloride anions have been indicated by open circles, and the directions of N–H or N–D covalent bonds in the ammonium cations are shown simply as dashed lines toward four of eight neighboring chlorides.

is interested only in those inherent structures and their basins for which (a) all ammonium ions remain chemically intact five-atom units with near-tetrahedral symmetry, (b) the system remains in a cubic crystal form, and (c) the nitrogens and chlorines strictly remain on their own simple-cubic sublattices of the CsCl structure. In principle, point defects might occur as equilibrium features of the system in the temperature range of interest, but their concentrations are likely to be very low, and no compelling evidence exists to suggest that they play an important role in the order–disorder thermodynamics. As a consequence of these constraints, the remaining inherent structures can be distinguished simply by which of the two orientations each of the ammonium ions has adopted. The obvious description of the system's configurational state (i.e., basin and inherent structure identification) thus utilizes Ising spins $\mu_i = \pm1 (1 \leq i \leq N)$ for each of the ammonium cations. [Presumably, the reader is not misled by the coincidence of the standard choices of N for "nitrogen" and for "particle number"!] Under constant-volume conditions, the potential energy of the 2^N hydrogen-bonded inherent structures is a function of these Ising spins, $\Phi(\mu_1 \ldots \mu_N)$, with the property that it is invariant to simultaneous sign change of all spins:

$$\Phi(\mu_1 \ldots \mu_N) = \Phi(-\mu_1 \ldots -\mu_N). \tag{IV.101}$$

Because each chloride anion is surrounded by eight ammonium cations, which either can or cannot hydrogen-bond to it, the total number of local configuration possibilities for a chloride formally is $2^8 = 256$. Many pairs of these are equivalent, differing only by symmetry operations. Nevertheless, this still leaves a substantial number of distinct hydrogen bonding patterns that are possible at each chloride. Figure IV.13 indicates these schematically, where the eight vertices of the cube surrounding a central chloride are either directions along which a hydrogen bond occurs (identified by a dark circle), or a direction with no hydrogen bond (no dark circle). Only those cases with zero to four hydrogen bonds are listed because the remaining cases simply involve exchanging vertex types (bare to dark circle, dark circle to bare).

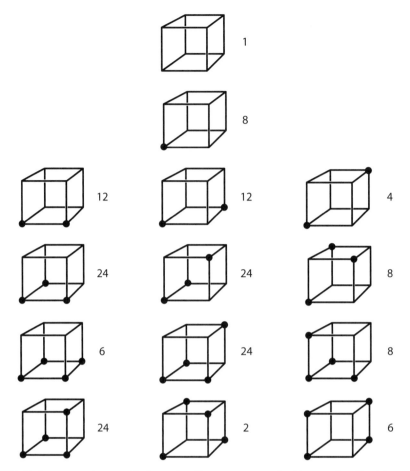

FIGURE IV.13. Distinct geometric types of hydrogen bonding patterns around a chloride anion. The eight vertices represent directions along which a hydrogen bond from a neighboring ammonium cation could be present, and if it is, that vertex has been embellished with a dark circle. Integers to the right of each cube indicate how many distinct occurrences of each type are possible as a result of symmetry. Cases with more than four hydrogen bonds correspond simply to exchange of bare and dark circle vertex types.

Figure IV.13 reveals that 22 geometrically distinct patterns of hydrogen bonding exist to a chloride from its neighboring ammonium cations, spanning the range from 0 to 8 hydrogen bonds. The pairs of ammoniums contributing to each chloride's neighbor cube are first, second, and third neighbors of one another on the ammonium simple-cubic sublattice. Assuming that interactions are sufficiently short ranged to be confined to these neighbor cubes, the inherent structure potential function Φ then resolves into a sum of terms, one for each chloride, and involving the eight Ising spins for its neighbor cube. This phenomenon may be indicated in the following manner:

$$\Phi(\mu_1 \ldots \mu_N) = \Phi_0 + \sum_{j=1}^{N} \varepsilon_8(\mu_{j1} \ldots \mu_{j8}), \tag{IV.102}$$

where Φ_0 establishes the mean energy for all inherent structures under consideration, j is a running index for the chloride anions, the $\mu_{j1}, \mu_{j2}, \ldots, \mu_{j8}$ are the eight Ising spins describing the ammonium orientations immediately surrounding chloride j, and ε_8 is the eight-spin energy function for each chloride. Of course, these spin octuplets are not independent of one another because each Ising spin serves simultaneously as a member of eight such octuplets. The eight-spin interaction functions ε_8 in principle can take on 22 numerically different values, one for each of the distinct hydrogen bond arrangements illustrated in Figure IV.13.

Eight-spin interaction potentials are not present in the usual format for which the statistical mechanics of Ising models is conventionally developed. Nevertheless, it is straightforward in principle at least to produce a rough assessment of the orientational order–disorder transition using the mean field approximation. One starts with the applicable definition of a long-range order parameter for this system:

$$x = N^{-1}\sum_{i=1}^{N}\mu_i. \tag{IV.103}$$

The spin probability $P(\mu)$ for any one of the ammonium ions therefore is

$$P(-1) = (1 - x)/2, \tag{IV.104}$$

$$P(+1) = (1 + x)/2.$$

The mean field approximation assigns a common value $N\bar{\varepsilon}_8(x)$ to all of the inherent structures that have the same long-range order x, and this value is an average of the actual interaction, assuming that all spins are independently distributed, subject to the constraint in Eq. (IV.103). The precise statement of this average is

$$\bar{\varepsilon}_8(x) = \sum_{\mu_{j1}}\ldots\sum_{\mu_{j8}}P(\mu_{j1})\ldots P(\mu_{j8})\varepsilon_8(\mu_{j1}\ldots\mu_{j8}), \tag{IV.105}$$

which is an eighth-order polynomial in x containing only even powers of that parameter. This function is expected to exhibit a simple maximum at $x = 0$, falling off to its minimum value at the fully ordered states $x = \pm 1$. The number of distinguishable inherent structures sharing this x and assigned average value $\bar{\varepsilon}_8(x)$ is the elementary combinatorial expression

$$\Omega_d(x) = \frac{N!}{[(1 - x)N/2]![(1 + x)N/2]!}, \tag{IV.106}$$

to be compared with the corresponding β-brass version, Eq. (IV.69). Under the reasonable assumption that isochoric vibrational free energies f_{vib} for the basins surrounding inherent structures have only weak dependence on x, then the total free energy emerging from the last two equations has the usual characteristics of the mean field approximation, specifically that the heat capacity remains bounded at the order–disorder transition, with no lambda-type divergence but only a jump discontinuity.

In order eventually to develop a quantitative Ising model theory or numerical simulation for this transition, feasibility would seem to require some kind of simplification of the potential energy expression Eq. (IV.102). Too little information is currently available to evaluate separately each of the 22 values that the eight-spin interaction ε_8 can display. One possibility rests on the observation mentioned above that only first, second, and third neighbor spins (ammonium ions) enter each ε_8, and so it is attractive and reasonable to suppose that the collection of inherent structure energies Φ could be represented approximately by a sum of first, second, and third neighbor spin-pair interactions. In other words, replace Eq. (IV.102) with the much simpler expression

$$\Phi(\mu_1 \ldots \mu_N) \cong \Phi_0 - J_1 \sum_{(1st)} \mu_i \mu_k - J_2 \sum_{(2nd)} \mu_i \mu_l - J_3 \sum_{(3rd)} \mu_i \mu_m, \tag{IV.107}$$

in which the previously required 22 distinct values collapse to just 3 in this approximation. These three spin-pair interaction parameters J_1, J_2, and J_3 are intended to represent the best fit with form (IV.107) to the more complete expression in Eq. (IV.102). In order for the lowest energy structure to have the orientational order shown in Figure IV.12, it is necessary that

$$3J_1 + 6J_2 + 4J_3 > 0, \tag{IV.108}$$

where the integer coefficients are determined by the relative numbers of first, second, and third neighbors in the simple cubic lattice (Table IV.3). It is worth remarking at this juncture that orientation-dependent interactions between pairs of ammoniums more widely separated than third neighbors should be very weak. Because of its high symmetry, the lowest order electrostatic multipole moment that an ammonium cation has (beyond the monopole net charge) is an octupole, and octupole–octupole interactions die away rapidly with separation.

The Ising model on the simple cubic lattice, with pair interactions limited just to the first three neighbor shells, has the same critical exponents as for other three-dimensional Ising models with just nearest neighbor spin-pair interactions [Stanley, 1971, Chapter 3]. These include the same three exponents cited in Section IV.E for β-brass, namely, the low- and high-temperature heat capacity divergence exponents α_{1-} and α_{1+}, respectively, and the critical exponent β_I, describing the increase of mean long-range order as temperature declines below the critical temperature. Some limited infrared spectroscopic measurements at low pressure and thus at nearly constant volume [Shumake and Garland, 1970] suggest that vibrational motions in the crystal are only moderately sensitive to long-range order, i.e., f_{vib} is not strongly dependent on the Ising spin assignments. Consequently, the pure Ising model critical exponents should survive the inclusion of vibrational degrees of freedom in the system free energy, under constant-volume conditions, without having their values modified.

However, vibrational degrees of freedom do play a role in measurable properties. In particular, the isotope effect on the order–disorder transition temperature T_c noted in Table IV.5 is an obvious example. Assuming that the Born-Oppenheimer approximation is valid for separation of electronic and nuclear degrees of freedom, the inherent-structure energies and basin shapes for the system must be independent of whether the light or the heavy hydrogen isotope is present. However, the amplitudes and energies of zero-point vibrational and librational motions for NH_4^+ exceed those of ND_4^+, and both can be expected to depend somewhat on the specific patterns of ammonium ion orientations present in the crystal. In other words, they should depend on the patterns of hydrogen bonds connecting ammonium cations to chloride anions. The corresponding quantum vibrational and librational effects that are larger for H than for D are still present as temperature rises above absolute zero, with excited quantum states becoming more and more populated. The observation at 1 atm that T_c is lower for NH_4Cl than for ND_4Cl indicates that the quantum effects on the basin vibrational free energies act to diminish the effect of the range of inherent structure energies (from most ordered to least ordered) and that this diminishing is greater for the light isotope H than for the heavier isotope D.

Experimental observations of the order–disorder transition in ammonium chloride are performed conventionally under constant-pressure (isobaric), not constant-volume (isochoric), conditions. In this connection, it is important to note that the critical temperature T_c increases with rising pressure, reaching approximately 280 K at 5 kbar [Garland and Renard, 1966]. Furthermore, varying the temperature through T_c under constant-pressure conditions apparently produces continuous but singular changes in sample volume as well as in its elastic constants [Garland and Jones, 1963; Weiner and Garland, 1972]. These observations indicate that both the relevant Ising model interactions (ε_g, or the triad J_1, J_2, J_3) and the basin vibrational free energies must be treated as functions of pressure or of volume. Lawson's measurements of elastic moduli and of thermal expansion through the transition lead to the conclusion that C_V has a substantially weaker divergence at T_c than does C_p [Lawson, 1940]. It is also noteworthy that a measurement of the constant-pressure heat capacity C_p at 1 atm has formed the basis of a report that its divergence exponents deviate significantly from the traditional Ising (C_V) values α_{I-} and α_{I+}. In particular, these are replaced by $\alpha_- \approx 0.7$ and $\alpha_+ \approx 0.8$ [Schwartz, 1971]. These various laboratory observations indicate that coupling of ammonium ion orientational degrees of freedom to the isobaric vibrational free-energy contribution substantially increases the strength (if not the thermodynamic order) of the order–disorder transition.

G. Thermal Conductivity of Insulators

In the case of crystalline metals, conduction electrons tend to dominate both electrical and thermal conductivity. But crystalline insulators are substantially free of those electronic effects, and in particular their thermal conductivity is due principally to intrabasin vibrational degrees of freedom. The temperature dependence of insulator thermal conductivity consequently provides an independent statistical probe of crystal basin shapes. This justifies examining the specific dynamical details of intrabasin vibrations that are the mechanism for thermal energy transport in insulators and which would underlie the corresponding current autocorrelation function expression, Eq. (II.102).

The macroscopic property thermal conductivity λ expresses the magnitude of heat flow rate $\mathbf{q}$ in the normal direction across area A, within the interior of a bulk material sample, produced by a temperature gradient ∇T. In isotropic media and in crystals with cubic symmetry, λ is a scalar quantity. The Fourier heat conduction law formalizes the definition as follows [Chapman, 1981]:

$$\frac{\mathbf{q}}{A} = -\lambda \nabla T. \tag{IV.109}$$

Because the spatial divergence of the heat flow causes a local temperature change, one is led to the following partial differential equation for the position and time dependence of temperature $T(\mathbf{r}, t)$:

$$\nabla \cdot (\lambda \nabla T) = \rho C_x \frac{\partial T}{\partial t}, \tag{IV.110}$$

where ρ is the number density, C_x is the heat capacity appropriate for the boundary conditions, and the possible temperature dependence of λ has been taken into account. This heat flow equation (IV.110) is substantially isomorphous with a diffusion equation for time-dependent matter redistribution by Brownian motion. That is, heat flow and the temperature change it produces is macroscopically a diffusion process, specifically of energy, wherein λ plays a role analogous to that of the self-diffusion constant D [Section II.G]. But in contrast to the phenomenon of self-diffusion, which intrinsically entails transitions among distinct configuration space basins, thermal conduction can proceed while the many-particle system remains confined to a single basin. Obviously, this feature is especially relevant to the crystalline state at low temperature.

The elementary Einstein model of crystal-phase vibrations [Einstein, 1907, 1911] assumes that each particle dynamically acts independently as an isotropic harmonic oscillator. Consequently, for a single-component system this amounts to assuming that the perfect crystal basin in configuration space possesses full hyperspherical symmetry. Furthermore, the independence of those individual particle vibrations makes thermal conduction impossible. Any initial-condition "hot spot" at some particle would remain localized there forever. One should realize that such a $\lambda = 0$ scenario is not confined to the assumption of full hyperspherical basin symmetry. A more general class of quadratic basin shapes for which every harmonic normal mode was localized at its own small set of vibrating particles would also prevent macroscopic diffusion of thermal energy, as required by Eq. (IV.110).

Structurally (and isotopically) perfect crystals exhibit harmonic normal modes that are fully delocalized as running waves spanning the entire periodic structure. This situation places its own distinct constraint on the class of quadratic forms that determine the configuration space basin shape. However, one must recognize that this purely harmonic situation provides the other extreme in which heat transfer occurs ballistically, not via diffusive Brownian motion, but essentially as undamped sound waves. In other words, $\lambda \rightarrow +\infty$ for periodic media possessing unperturbed harmonic normal modes. In such a situation, the formal expression for λ as a current autocorrelation function time integral, Eq. (II.102), must diverge in the large-system limit.

Experimentally measured thermal conductivities $0 < \lambda(T) < +\infty$ owe their finite magnitudes to phenomena that interfere with the free propagation of running-wave harmonic normal modes, i.e., to phenomena that provide attenuation. Structural defects in the crystal are one source of attenuation by acting as scattering centers for propagating sound waves. These scattering centers can be point defects, such as vacancies or interstitials in a chemically pure crystal, or impurity substitutional defects. Local mass defects in the form of isotopic substitution also provide attenuation. With respect to the last of these, the shape of a perfect crystal's basin is unaltered by isotopic substitution, but the internal vibrational dynamics, whether quantized or classical, depend upon the distribution of masses among the atoms present.

Even in a structurally perfect and isotopically pure crystal, anharmonic corrections to the basin shape contribute to damping of vibrational normal-mode propagation. Let $\mathbf{R}_\alpha$ be the configuration vector for the inherent structure at the bottom of one of the permutation-equivalent perfect-crystal basins. The potential energy function near that inherent structure at energy Φ_α has a Taylor expansion [presented in Eq. (IV.33) in slightly different format] with the following leading-order terms:

$$\Phi(\mathbf{R}_\alpha + \Delta\mathbf{R}) = \Phi_\alpha + \frac{1}{2}[\nabla_\mathbf{R}\nabla_\mathbf{R}\Phi]_\alpha : \Delta\mathbf{R}\Delta\mathbf{R} + \frac{1}{6}[\nabla_\mathbf{R}\nabla_\mathbf{R}\nabla_\mathbf{R}\Phi]_\alpha(\cdot)^3\Delta\mathbf{R}\Delta\mathbf{R}\Delta\mathbf{R}$$

$$+ \frac{1}{24}[\nabla_\mathbf{R}\nabla_\mathbf{R}\nabla_\mathbf{R}\nabla_\mathbf{R}\Phi]_\alpha(\cdot)^4\Delta\mathbf{R}\Delta\mathbf{R}\Delta\mathbf{R}\Delta\mathbf{R} + \dots. \tag{IV.111}$$

A similar expansion exists for the enthalpy function $\hat{\Phi}$ about its pressure-dependent inherent structure, in which the volume appears as a "configurational" variable. The generic expression (IV.111) is valid whether the crystal under consideration is a single-species Bravais lattice with just three phonon branches or a more complicated periodic structure with several basis vectors for the unit cell and correspondingly more phonon branches. Anharmonic contributions beyond the cubic and quartic terms explicitly shown in this expansion should have a relatively minor effect well below the melting temperature for the crystal of interest, where vibrational amplitudes measured by $\Delta\mathbf{R}$ are still relatively small. One should note in passing that the independence of purely harmonic modes by itself would not permit attainment of thermal equilibrium; the influence of anharmonicity and the energy sharing that it induces between normal modes is an absolute necessity for that to occur.

A complete theory of thermal conduction in crystals is a sufficiently complex subject to fall beyond the scope of this presentation. However, a variety of specialized references dealing in depth with this subject is available for interested readers to consult [e.g., Ashcroft and Mermin, 1976, Chapter 25; Ziman, 2001; Tritt, 2004]. For present purposes, it will suffice to highlight a few key aspects. Because a comprehensive view of thermal conductivity in crystals includes low temperatures, a quantum mechanical description is mandatory.

Phonon scattering processes produced by the lowest order anharmonicity, the cubic terms in expansion (IV.111), involve (a) the creation of two phonons from one, or inversely (b) the merging of two phonons to produce just one. These transitions change the quantum numbers of the normal-mode oscillators by ± 1 and are subject to conservation laws governing the phonon crystal momenta and their energies. For (a), one requires the constraints

$$\mathbf{k}_{1,\zeta_1} = \mathbf{k}_{2,\zeta_2} + \mathbf{k}_{3,\zeta_3} + \mathbf{K}, \tag{IV.112}$$

$$\hbar\omega_{\zeta_1}(\mathbf{k}_{1,\zeta_1}) = \hbar\omega_{\zeta_2}(\mathbf{k}_{2,\zeta_2}) + \hbar\omega_{\zeta_3}(\mathbf{k}_{3,\zeta_3}).$$

Here the wave vectors $\mathbf{k}_{i,\zeta_i}$ are to be understood as belonging to the first Brillouin zone, with ζ_i denoting the phonon branch involved. $\mathbf{K}$ stands for a reciprocal lattice vector (which can include $\mathbf{K} = 0$). Processes of type (b) must obey analogous constraints:

$$\mathbf{k}_{1,\zeta_1} + \mathbf{k}_{2,\zeta_2} = \mathbf{k}_{3,\zeta_3} + \mathbf{K}, \tag{IV.113}$$

$$\hbar\omega_{\zeta_1}(\mathbf{k}_{1,\zeta_1}) + \hbar\omega_{\zeta_2}(\mathbf{k}_{2,\zeta_2}) = \hbar\omega_{\zeta_3}(\mathbf{k}_{3,\zeta_3}).$$

Phonon scattering events with $\mathbf{K} = 0$ are conventionally called "normal" processes, while those with $\mathbf{K} \neq 0$ are called "umklapp" processes. The energy conservation requirement rules out processes in which three phonons spontaneously appear or disappear.

The normal and the umklapp scattering processes, respectively, possess a basic distinction. The former are able to redistribute phonon energy among the available harmonic normal modes, but they cannot attenuate the directed energy flow. By contrast, the umklapp scattering processes indeed are the ones that create the attenuation that thermal conductivity $\lambda(T)$ measures. At very low temperature, the few phonons present tend to be those of low energy, i.e., they are acoustic modes close to $\mathbf{k} = 0$ and thus unable to engage in umklapp scattering. As a result, the intrinsic material property $\lambda(T)$ for a structurally and isotopically pure crystal diverges to infinity in the

$T \rightarrow 0$ limit. A simple argument based on this observation concludes that the divergence would have the leading-order behavior of the following type [Ashcroft and Mermin, 1976, Chapter 25]:

$$\lambda(T) \sim \exp(const./T) \qquad (T \rightarrow 0). \tag{IV.114}$$

The crystal momentum and energy conservation conditions (IV.112) and (IV.113) enforce stringent constraints on the cubic anharmonicity scattering possibilities. In particular, in any given temperature range they limit the attenuation magnitude that umklapp processes can produce. Quartic anharmonicity is subject to the same crystal momentum and energy conservation conditions, but their net effect is less stringent. Although the quartic perturbations formally have higher order than the cubic perturbations, their relative contribution to $\lambda(T)$ can be comparable in magnitude over much of the temperature range up to the thermodynamic melting point. Consequently, both orders of perturbation are usually considered together in analysis of crystal thermal conduction. As in cubic order, the quartic phonon scattering processes involve both normal ($\mathbf{K} = 0$) and umklapp ($\mathbf{K} \neq 0$) events, with only the latter contributing to attenuation of energy flow along a temperature gradient. Three types of quartic-order scattering events arise: (a) one input phonon creates three output phonons, (b) two input phonons scatter off one another to yield two output phonons, and (c) three input phonons combine to produce a single output phonon. The conservation conditions applicable in quartic order eliminate the spontaneous appearance or disappearance of four phonons.

The presence of phonon attenuation can alternatively be described in terms of a phonon mean free path. The divergence exhibited in Eq. (IV.114) amounts to stating that the mean free path becomes arbitrarily large as temperature declines toward absolute zero. Therefore, in order to measure the intrinsic material property $\lambda(T)$ in that limit, it becomes necessary to use larger and larger crystal samples to ensure that the observations reflect bulk properties of that material. However, laboratory reality restricts attention to limited and modest sample sizes. For these samples, lowering the temperature produces a crossover from bulk scattering caused by umklapp processes as just described to a regime in which scattering from the sample surfaces dominates the measurements. Consequently, the diverging behavior indicated by Eq. (IV.114) gives way to a turnover in the apparent thermal conductivity and a decline toward zero as the population of acoustic phonons drops to zero. The position of this turnover depends on the sample shape and size [Thacher, 1967]. Figure IV.14 schematically illustrates this observed behavior.

Mass variations resulting from isotopic substitution within a structurally perfect crystal also affect thermal conductivity. This phenomenon does not require anharmonicity but is present even for the harmonic normal modes that describe vibrational motions close to the crystal's inherent structure. With mass variation across particle locations, the harmonic normal modes are no longer simply running waves described merely by a wave vector $\mathbf{k}$ and a polarization direction $\mathbf{B}_\zeta(\mathbf{k})$ for phonon branch ζ. Instead, their determination in principle requires a far more demanding diagonalization of the $3N$-dimensional matrix whose elements are those of the matrix of second Φ derivatives with appropriate mass weighting. In contrast to the perturbation theory for anharmonicity, mass deviation from the average mass of the isotopically variable species serves as the relevant perturbation [Carruthers, 1961, Section V]. The resulting scattering processes reduce thermal conductivity and are not required to be umklapp processes. Measurements showing this effect have been performed on crystals of lithium fluoride (LiF) in which the composition of the

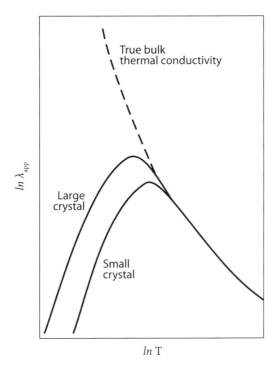

FIGURE IV.14. Qualitative behavior of the apparent thermal conductivity λ_{app} in a structurally perfect and isotopically pure single crystal in the low-temperature regime. Surface scattering processes for phonons limit the rise of the intrinsic $\lambda(T)$ toward infinity, causing a turnover and decline toward zero as $T \to 0$.

lithium species was varied from nearly pure ^{7}Li to approximately equal amounts of this isotope and ^{6}Li (the only stable isotope of fluorine is ^{19}F); the observations show a modest but clear reduction in $\lambda(T)$ as the isotopic purity is degraded [Thacher, 1967].

Structural disorder embedded in the interior of a crystalline material switches the system to the interior of a potential energy basin whose internal geometry fundamentally differs from that of the perfect crystal state. The various structural disorder elements (vacancies, interstitials, dislocations, …) act as scattering agents for the phonons, and thus reduce their mean free paths. Thermal conductivity $\lambda(T)$ is reduced accordingly. To illustrate that this can have a strong effect, one can cite the measured thermal conductivity for quartz crystals (SiO_2) that have been subjected to varying amounts of radiation damage caused by exposure to fast neutrons [Berman et al., 1950]. In the temperature range $10 \leq T \leq 100$ K, the resulting structural defects reduce $\lambda(T)$ from its perfect-crystal value by one to two orders of magnitude, though it still remains about an order of magnitude above that of the amorphous SiO_2 glass.

H. Static Dielectric Constant of Insulators

The macroscopic description of the static linear response of an insulating material to an applied electric field $\mathbf{E}(\mathbf{r})$ within that material can be expressed as an induced polarization density $\mathbf{P}(\mathbf{r})$. If the applied field is spatially uniform and the material is isotropic or possesses cubic symmetry, these two vector fields are parallel, and their ratio defines the static dielectric constant ε:

$$\mathbf{P} = \left(\frac{\varepsilon - 1}{4\pi}\right)\mathbf{E}. \tag{IV.115}$$

In the case of a crystalline solid well below its melting temperature, the polarization density results from a local redistribution of charge around the $\mathbf{E} = 0$ equilibrium position of each atom in the medium because of the net electric field at that atom. In the static (zero frequency) case of initial interest here, this redistribution can come from two sources whose effects are additive: (a) displacement of the electron cloud relative to its binding nucleus and (b) a field-induced static displacement of the nucleus relative to its position before application of the external electric field. The induced dipole moment $\mathbf{m}_i$ at atom (or ion) i then is proportional to the induced local electric field $\mathbf{E}_{loc,i}$ at that site:

$$\mathbf{m}_i = (\alpha_{a,i} + \alpha_{b,i})\mathbf{E}_{loc,i}, \tag{IV.116}$$

involving the two types of polarizabilities (a) and (b) for that particle. The local field consists of the field $\mathbf{E}_0$ that would be externally applied before putting the dielectric substance in place, and the net field at i caused by all dipole moments induced in the other particles because of displacements (a) and (b). The macroscopic polarization field results from the local density of induced dipole moments:

$$\mathbf{P} = \sum_j \rho_j \mathbf{m}_j, \tag{IV.117}$$

where ρ_j is the number density of particles of species j.

In the event that the insulating cubic crystal of interest is composed of identical (neutral) atoms all in equivalent sites, then symmetry requires $\alpha_{b,j}$ to vanish. This situation is illustrated by the ambient pressure crystalline phases of the noble gases neon (Ne), argon (Ar), krypton (Kr), and xenon (Xe) [Donohue, 1982, Chapter 2]. Therefore, the application of an external electric field to such materials has no effect in linear order on the configurational distribution within the inhabited potential energy basin.

When each particle i resides at a site of cubic symmetry (true for the noble gases mentioned, and also for the alkali halides), and the sample is exposed to a spatially uniform external electric field, one can show that the local field within the sample has the following simple form [Ashcroft and Mermin, 1976, p. 541]:

$$\mathbf{E}_{loc,i} = \mathbf{E} + \left(\frac{4\pi}{3}\right)\mathbf{P}. \tag{IV.118}$$

Substituting expression (IV.115) for $\mathbf{P}$ leads to the following:

$$\mathbf{E}_{loc,i} = \left(\frac{\varepsilon + 2}{3}\right)\mathbf{E}. \tag{IV.119}$$

Finally, by equating expressions (IV.115) and (IV.117), followed by use of Eqs. (IV.116) and (IV.119) to eliminate dipole moment vectors and electric fields, one is led to the Clausius-Mossotti equation [Kittel, 1956, p. 163]:

$$\frac{\varepsilon - 1}{\varepsilon + 2} = \left(\frac{4\pi}{3}\right) \sum_j \rho_j (\alpha_{a,j} + \alpha_{b,j}). \tag{IV.120}$$

Local particle displacements that significantly disrupt the assumed cubic symmetry, whether by intrabasin vibrations or by structural defects, would require corrections to this simple result.

The right side of the Clausius-Mossotti equation is positive and for dielectric insulators is always smaller than unity. Compression increases the number densities ρ_j and would act to increase the magnitude of the right member. As a formal property of the Clausius-Mossotti equation, one recognizes that if the right side were to approach unity from below, the static dielectric constant would diverge to $+\infty$. This change would amount to the dielectric behaving as a metal, thus contradicting the assumption of insulating behavior and of well-defined individual particle polarizability on which the derivation of Eq. (IV.119) is based.

The last point can be illustrated with a specific example. The electronic polarizability of an isolated Xe atom can be extracted from the properties of its dilute vapor. The result is the following [Pauling, 1988, p. 396]:

$$\alpha_{a,Xe} = 4.16 \,\text{Å}^3. \tag{IV.121}$$

The number density of this element in its face-centered cubic crystal at absolute zero and ambient pressure is [Donohue, 1982, p. 27]:

$$\rho = 1.735 \times 10^{-2} \text{Å}^{-3} \qquad (1\ \text{bar}). \tag{IV.122}$$

A combination of electronic structure calculations [Ross and McMahan, 1980] and various high-pressure experiments [Nellis, van Thiel, and Mitchell, 1982] indicate that Xe transforms from an insulator to a metal as the pressure is raised to 1.3 Mbar, at which point the number density has increased to

$$\rho \cong 6.7 \times 10^{-2} \text{Å}^{-3} \qquad (1.3\ \text{Mbar}). \tag{IV.123}$$

Presuming that indeed this is the state for which the right side of the Clausius-Mossotti equation (IV.120) (with $\alpha_{b,Xe} = 0$) has just risen to unity, the implied value of the Xe polarizability is

$$\alpha_{a,Xe} = 3.56 \text{Å}^3. \tag{IV.124}$$

The fact that this polarizability result is somewhat lower than that shown in Eq. (IV.121) for an isolated Xe atom may be attributed to interference between the electron clouds of strongly crowded atoms in the compressed crystal.

Alkali halide crystals M^+X^- exhibit either the sodium chloride (NaCl) or the cesium chloride (CsCl) structure, both of which possess the cubic symmetry upon which the Clausius-Mossotti equation (IV.120) is based. The presence of two distinct particle species leads to nonvanishing nuclear displacement polarizabilities $\alpha_{b,+}$ and $\alpha_{b,-}$. The opposite net electrostatic charges on the ions cause opposite directions of displacement in the presence of the applied electric field **E**. This displacement pattern is similar to that involved in a $\mathbf{k} \to 0$ optical-phonon longitudinal mode and thus involves the curvature of the multidimensional Φ basin in the same direction as that mode. In spite of the nominal elementary charge $\pm\, e$ informally attributed to the alkali halide ions, the overlap and distortion of their electron clouds as they reside in the crystal can produce somewhat different effective charges $\pm\, q_{eff}$. The forces on the ions initiated by the external field would then be $\pm\, q_{eff}\mathbf{E}_{loc}$. The effective charges depend on the state of compression of the crystal under consideration, formally approaching some integer multiple $(0, \pm 1)$ of the fundamental charge e as the lattice spacing of the alkali halide formally approaches infinity.

The wide discrepancy in response times to an applied electric field for electronic vs. nuclear degrees of freedom offers the opportunity for separate determinations of the $\alpha_{a,j}$ and $\alpha_{b,j}$ polarizabilities. In the frequency range of visible light, the electronic response in ionic crystals is close to that expected at zero frequency, whereas the nuclear response is so sluggish as to be essentially absent. Recognizing that the visible-light frequency dielectric constant is n^2, the square of the refractive index, the effective form taken by the Clausius-Mossotti equation is

$$\frac{n^2 - 1}{n^2 + 2} = \left(\frac{4\pi}{3}\right)\sum_j \rho_j\, \alpha_{a,j}, \qquad (\text{IV.125})$$

which is identified as the Lorentz-Lorenz equation [Price, 1969]. Using this last relation, tables of electronic polarizabilities have been prepared containing numerically estimated values for a large number of monatomic ions [Tessman et al., 1953].

I. Melting

The equilibrium isobaric melting process for crystals normally is a first-order phase transition to an isotropic liquid accompanied by discontinuities in density and enthalpy. The latter thermodynamically implies a discontinuous rise in entropy. The melting transition thus entails a sudden shift in configuration-space occupancy from a relatively small population of basins that have crystalline inherent structures, possibly containing a low concentration of defects, to a substantially larger and higher basin set with amorphous inherent structures. As a result of this sudden shift, discontinuities also appear in heat capacity, isothermal compressibility, and thermal expansion. At the atomic or molecular level, distribution functions of all orders exhibit sudden changes as well. The general features of the inherent structure description at phase transitions appear in Section III.E.

It is natural to begin a discussion of melting by recalling the thermodynamic Clausius-Clapeyron equation, which relates the pressure dependence of the equilibrium melting temperature $T_m(p)$ to the changes in molar volume $\Delta V = V_{liq} - V_{crys}$ and to the molar latent heat $\Delta H = H_{liq} - H_{crys}$ at the transition [Guggenheim, 1950, p. 122; Landau and Lifshitz, 1958b, p. 257]:

$$\frac{dT_m}{dp} = \frac{T_m \Delta V}{\Delta H} \qquad (\text{IV.126})$$

$$\equiv \frac{\Delta V}{\Delta S}.$$

Here $\Delta S = S_{liq} - S_{crys}$ is the corresponding entropy change across the phase transition. There are rare exceptions to the conventional expectation that $\Delta S > 0$, as defined (specifically "inverse melting" [Stillinger and Debenedetti, 2003; Feeney et al., 2003]); however, these exceptions are put aside for the moment. Consequently, the sign of the melting curve slope dT_m/dp usually is the same as that of ΔV.

The present discussion of melting phenomena restricts attention initially to single-component systems composed of structureless particles. An especially transparent example is provided by the three-dimensional classical model with inverse-power pair interactions, so that the system's potential energy function is

$$\Phi(\mathbf{r}_1 \ldots \mathbf{r}_N) = \varepsilon \sum_{i<j} (\sigma/r_{ij})^n. \qquad (IV.127)$$

Here the energy and length parameters ε and σ are positive, and the exponent $n > 3$. This family of models is introduced in Chapter I, Section I.D (where n was written as $\bar{n}$). The lowest potential energy for this interaction in the large-system limit at fixed number density is attained with the face-centered cubic structure for all allowed values of n. However, for $3 < n < 6$, heating causes a polymorphic transition to the body-centered cubic structure before melting [Dubin and Dewitt, 1994]. It is a straightforward exercise to show that in the classical limit the excess (nonideal) part of the Helmholtz free energy per particle for these models is a function of a single reduced variable ξ that combines number density ρ and temperature T (see Box IV.2):

$$\xi = \varepsilon(\rho\sigma^3)^{n/3}/k_B T. \qquad (IV.128)$$

Consequently, if any T, ρ pair corresponds to a melting point with $\xi = \xi_m$, then any other T, ρ with the same ξ value ξ_m must also lie on the melting curve. Analogously, a smaller ξ value ξ_f is constant along the equilibrium freezing curve. The virial equation of state for this family of models has the following form:

$$p = \rho k_B T[1 + F_n(\xi)], \qquad (IV.129)$$

with F_n an n-dependent continuous function. Using definition (IV.128) to express ρ in terms of ξ and T, one obtains a simple relation between T and p along the melting curve:

$$p = C_n T_m^{1+3/n}, \qquad (IV.130)$$

where C_n is a suitable temperature-independent positive coefficient for the exponent-n case. This leads in turn to

$$\frac{dT_m}{dp} = \left(\frac{n}{n+3}\right) C_n^{-n/(n+3)} p^{-3/(n+3)}. \qquad (IV.131)$$

Inclusion of quantum corrections to the melting behavior of these inverse-power-potential models, however, would upset the simple results displayed in Eqs. (IV.128)–(IV.131).

Equation (IV.130) is a special case of an empirical melting curve relation proposed long ago and known as the "Simon equation" [Simon and Glatzel, 1929]. This equation has the following simple form:

$$p = A_1 T_m^a + A_2. \qquad (IV.132)$$

BOX IV.2. Scaling Behavior for Inverse-Power Pair Potentials

Consider a model N-body system at number density $\rho = N/V$ for which the interaction potential Φ is given by Eq. (IV.127). This potential is positive and a homogeneous function of degree $-n$:

$$\Phi(\lambda\mathbf{r}_1\dots\lambda\mathbf{r}_N) = \lambda^{-n}\Phi(\mathbf{r}_1\dots\mathbf{r}_N).$$

The excess Helmholtz free energy F_{ex} (in the classical limit), in volume V and at $\beta = 1/k_BT$ is given by

$$\exp(-\beta F_{ex}) = V^{-N}\!\int\! d\mathbf{r}_1\dots\!\int\! d\mathbf{r}_N \exp[-\beta\Phi(\mathbf{r}_1\dots\mathbf{r}_N)].$$

The integration limits for each of the $3N$ Cartesian configuration coordinates can be taken as 0 to $V^{1/3}$. For all particles, set $\mathbf{r}_i = \rho^{-1/3}\mathbf{s}_i$ so that this last equation becomes

$$\exp(-\beta F_{ex}) = N^{-N}\!\int\! d\mathbf{s}_1\dots\!\int\! d\mathbf{s}_N \exp[-\xi\sum_{i<j} s_{ij}^{-n}],$$

$$\xi = \beta\varepsilon\sigma^n\rho^{n/3}.$$

The integration limits on the transformed configuration variables are now 0 and $N^{1/3}$, so that in this representation the geometric size of the system scales only with the number of particles present, i.e., its apparent density is fixed. The single parameter ξ combines the effects of the two intensive properties temperature and density. It is the only control parameter whose value locates the system's melting and freezing points, and its respective values at those points can be denoted as ξ_m and ξ_f, both of which depend on the interaction exponent n. In particular, this simplifying scaling behavior implies that the neighbor spacing in the crystal at its melting temperature T_m is proportional to $T_m^{-1/n}$.

The same transformation of configurational coordinates can be applied to the configurational integrals that determine the particle correlation functions of various orders for this family of models. In particular, this change demonstrates that the pair distribution function ρ_2 has the following scaling behavior:

$$\rho_2(\mathbf{r}_1, \mathbf{r}_2 \mid T, \rho) = \rho^2\gamma_n[\rho^{1/3}\mathbf{r}_1, \rho^{1/3}\mathbf{r}_2 \mid \xi(T, \rho)],$$

thus reducing its configurational behavior to a single function γ_n of density-scaled positions and of ξ. The reduced correlation function γ_n, of course, varies with exponent n.

As mentioned in Chapter I, Section I.D, the multidimensional Φ landscape for these inverse-power-potential models retains a constant number of inherent structures and basins, as well as constant numbers of saddle points of all orders, as volume V varies at fixed N.

The exponent a and the constants A_1 and A_2 are to be fitted to measured melting data for any substance of interest. Any non-negligible A_2 value so obtained implies that interactions between the particles of that substance deviate substantially from simple inverse-power form, and its sign and magnitude can serve to suggest the nature of that deviation.

An important attribute of thermal motion within configuration-space basins is the distribution of displacements it produces from the bottom of the occupied basins, i.e., the separation from the relevant inherent structure in the multidimensional configuration space. Equivalently, this is the displacement distribution that an ensemble of system initial configurations experiences under a constant-volume (isochoric) steepest descent mapping to that basin's minimum. Let $D_\alpha(\Delta\mathbf{R}, \beta)$ stand for the normalized distribution of configuration-space displacements from the inherent structure in basin α at $\beta = 1/k_B T$. Presuming that the classical statistical-mechanical limit is applicable, this is a normalized Boltzmann factor for the system interaction potential if the given displacement lies within the interior region B_α for basin α but is zero outside that region:

$$D_\alpha(\Delta\mathbf{R}, \beta) = C_\alpha(\beta)\exp[-\beta\Phi(\mathbf{R}_\alpha + \Delta\mathbf{R})] \qquad (\Delta\mathbf{R} \text{ inside } B_\alpha) \qquad \text{(IV.133)}$$

$$= 0 \qquad (\Delta\mathbf{R} \text{ outside } B_\alpha).$$

If $P_\alpha(\beta)$ represents the occupancy probability of distinguishable basin type α at thermal equilibrium, then a thermally averaged displacement probability $D(\Delta\mathbf{R}, \beta)$ is the correspondingly weighted average over all such basin types:

$$D(\Delta\mathbf{R}, \beta) = \sum_\alpha P_\alpha(\beta)D_\alpha(\Delta\mathbf{R}, \beta), \qquad \text{(IV.134)}$$

where

$$\sum_\alpha P_\alpha(\beta) = 1. \qquad \text{(IV.135)}$$

The α sums in these last two equations are intended to include only single examples of each distinguishable basin type.

Thermally activated displacements of single particles from their nominal crystal sites have been a long-standing topic of interest in connection with the melting transition. The normalized distribution function $D^{(i)}$, $i = 1\ldots N$, for these single particle displacements can formally be extracted from $D(\Delta\mathbf{R}, \beta)$ by integrating over $N - 1$ particle positions, e.g.,

$$D^{(1)}(\Delta\mathbf{r}_1, \beta) = \int d(\Delta\mathbf{r}_2)\ldots\int d(\Delta\mathbf{r}_N)D(\Delta\mathbf{R}, \beta), \qquad \text{(IV.136)}$$

recognizing that $\Delta\mathbf{R}$ is just the direct sum of individual particle displacements:

$$\Delta\mathbf{R} = \Delta\mathbf{r}_1 \oplus \Delta\mathbf{r}_2 \oplus \ldots \oplus \Delta\mathbf{r}_N. \qquad \text{(IV.137)}$$

In the basic case of a single-species crystal phase, the mean-square particle displacement from its inherent-structure location is the following:

$$\langle(\Delta\mathbf{r})^2\rangle_\beta = N^{-1}\sum_{i=1}^{N}\int(\Delta\mathbf{r})^2 D^{(i)}(\Delta\mathbf{r}, \beta)d(\Delta\mathbf{r}). \qquad \text{(IV.138)}$$

While the system remains in the same crystal phase, this intrabasin mean-square displacement can be expected to increase monotonically as temperature rises toward the melting point. This should be the case whether the constant-volume mappings to inherent structures begin with system configurations all at the same volume or whether those initial configurations corresponded to fixed system pressure.

A technical point needs to be emphasized in connection with this last expression. For simplicity, assume that the perfect crystal structure is such that all particles reside on geometrically equivalent sites (the unit cell may have a basis of several particles). Specifically, the definitions (IV.133) to (IV.136) have incorporated only single examples chosen from the $N!$ members in each of the permutation-equivalence classes of inherent structures. If only the structurally perfect crystal inherent structure is a significant contributor, all N particles would produce identical contributions to the i sum in Eq. (IV.138). In that case, the average over the N particles is redundant. However, if inherent structures for defective configurations are thermally relevant and need to be included in the α summation, Eq. (IV.134), the specific inherent structures chosen generally do not have a chosen particle i situated at all possible locations relative to the defects. That would influence its displacement distribution. However, the average over particles in Eq. (IV.138) completely disposes of that distribution-biasing distinction.

The elementary Einstein model of the crystal state offers an exceptionally simple representation of these displacement distributions [Einstein, 1907, 1911]. As mentioned in Section IV.G, this approximation assumes that each of the N particles vibrates independently, harmonically, and isotropically about its nominal lattice position (defects are not explicitly included). The common Einstein frequency ω_E can then be expressed as

$$\omega_E = (K_E/m)^{1/2}. \tag{IV.139}$$

The assumed restoring force constant for each of the mass-m particles has been denoted by K_E, which can vary with number density or equivalently with external pressure p. Within this picture, one easily verifies that the single-particle mean-square displacement has the following form:

$$\langle (\Delta \mathbf{r})^2 \rangle_\beta = \frac{3}{\beta K_E}. \tag{IV.140}$$

Needless to say, it is necessary to go beyond the Einstein approximation to obtain a physically more relevant description of particle displacement statistics in the harmonic regime. A proper analysis in terms of independent normal modes is required. Recall that a coordinate rotation in the multidimensional configuration space diagonalizes the potential energy function Φ in the neighborhood of an inherent structure at a fixed system volume V. This property is displayed in Eq. (III.8) in slightly different notation, which we now write as follows:

$$\Phi = \frac{1}{2} \sum_q K_q u_q^2. \tag{IV.141}$$

Here the u_q are the normal mode displacement coordinates resulting from the rotation. Assuming that confining walls are present to prevent the system from executing free translation (as opposed to the case with periodic boundary conditions), there are $3N$ positive-frequency normal modes present, where the qth normal mode angular frequency is

$$\omega_q = (K_q/m)^{1/2}. \tag{IV.142}$$

The independence of these normal modes implies that they each provide separate additive contributions to the mean-square displacement of all particles. The contribution to the entire system's mean-square displacement from mode q is

$$\langle u_q^2 \rangle_\beta = \frac{1}{\beta K_q}, \tag{IV.143}$$

so that summing over normal modes yields

$$\left\langle (\Delta \mathbf{R})^2 \right\rangle_\beta = \frac{1}{\beta} \sum_q \frac{1}{K_q}. \tag{IV.144}$$

Keeping in mind the multidimensional Pythagorean relation,

$$(\Delta \mathbf{R})^2 = \sum_{i=1}^{N} (\Delta \mathbf{r}_i)^2, \tag{IV.145}$$

the corresponding single-particle result in a one-component crystal phase is

$$\left\langle (\Delta \mathbf{r}_i)^2 \right\rangle_\beta = \frac{1}{\beta N} \sum_q \frac{1}{K_q}. \tag{IV.146}$$

$$= \frac{1}{\beta m N} \sum_q \frac{1}{\omega_q^2}.$$

In the simple circumstance of a structurally nondefective crystal, the latter form in Eq. (IV.146) can be interpreted as a sum over phonon modes in the first Brillouin zone, and the lowest frequency modes contribute a disproportionately large amount to that sum. Recall the earlier Eqs. (IV.22), which state that the longitudinal and transverse phonons near the center of the Brillouin zone ($\mathbf{k} = 0$) are essentially macroscopic sound waves whose frequencies ω_q obey dispersion relations $c_l|\mathbf{k}|$ and $c_t|\mathbf{k}|$ involving positive longitudinal and transverse sound velocities. The total of their individually diverging contributions to the latter sum in Eq. (IV.146) as $|\mathbf{k}| \to 0$ remains finite in the large-system limit, where the sum passes to a $\mathbf{k}$-space integral by virtue of the differential volume element $4\pi k^2 dk$ in three dimensions. But note that this situation does not apply for two-dimensional crystals, where the "weaker" differential area element $2\pi k dk$ fails to produce cancellation, allowing instead a logarithmic divergence. Thus, one sees that in the large-system limit $\left\langle (\Delta \mathbf{r})^2 \right\rangle_\beta$ diverges in two dimensions for crystals at positive temperature. The consequence is elimination of the possibility of long-range periodic order for the crystal's density distribution. This is a result that extends beyond the assumption of basin harmonicity, assumed in Eqs. (IV.141) to (IV.146) [Mermin, 1968].

It is important to note that particle displacement analysis using isobaric mapping in the configuration space augmented by system volume V would have produced qualitatively different and less useful results. This difference arises from the thermal expansion effect caused by intrabasin anharmonicity. Inclusion of V as a variable in steepest descent mapping applied to $\hat{\Phi}$ then causes a macroscopic change in volume as that thermal expansion contribution is eliminated. Individual particles thus would be displaced by amounts proportional to the linear dimensions of the system, i.e., by $O(N^{1/3})$. This is not a measure of local particle displacement within its immediate crystal neighborhood but rather a measure of macroscopic crystal shrinkage or expansion.

The thermodynamic criteria for a phase transition require strict equality of temperature, pressure, and chemical potential for the two phases at coexistence. However, approximate phase transition conditions have sometimes been proposed that involve properties only of one phase. In the case of crystal melting to an isotropic liquid, an especially prominent example of this one-phase view is the Lindemann melting criterion [Lindemann, 1910]. This approximation for the melting of single-component crystals focuses on the ratio $\theta(\beta)$ of the root-mean-square particle

TABLE IV.6. Lindemann melting ratios for single-component inverse-power-potential models[a]

Exponent n	Crystal Structure at Melting	Lindemann Ratio $\theta_{mp}(n)$
4	fcc[b]	0.18
6	fcc	0.17
9	fcc	0.16
12	fcc	0.15
$+\infty$	fcc	0.13

[a]Hansen and McDonald, 1976, p. 361.
[b]Stable form at low temperature but metastable with respect to bcc near the melting temperature.

displacement from its nominal crystal location, to the nearest neighbor lattice spacing a, at temperature $T = (k_B\beta)^{-1}$:

$$a^{-1}\left[\langle(\Delta\mathbf{r})^2\rangle_\beta\right]^{1/2} = \theta(\beta). \tag{IV.147}$$

The melting point is then associated with this dimensionless ratio rising to a threshold value θ_{mp} as temperature increases. The numerical value of θ_{mp} at melting has been found to depend to some modest extent on the crystal structure involved. The Lindemann melting criterion is traditionally viewed as a statement just about intrabasin vibrational amplitudes for the perfect-crystal inherent structure alone, but it allows for the presence of anharmonicity in those vibrations.

The classical single-component models with inverse-power pair potentials provide a significant illustration of the Lindemann melting criterion. The scaling properties of these models indicated in Box IV.2 imply that any ratio of characteristic lengths at the equilibrium melting curve must be constant along that curve. The Lindemann threshold quantity θ_{mp} is one of these ratios. Its value has been determined by numerical simulations for several values of the exponent n, including the $n \to +\infty$ hard-sphere limit. Results are listed in Table IV.6. They show a modest downward drift as n increases.

For some other model potentials that do not have the scaling behavior of the inverse-power family, the Lindemann ratio continues to supply a useful tool for understanding the melting phenomenon. One example is the Yukawa pair potential, which has the form (Section I.D)

$$v_2(r) = \varepsilon(\sigma/r)\exp[-(r/\sigma)], \tag{IV.148}$$

containing the usual energy (ε) and length (σ) scale parameters. Molecular dynamics simulations at variable number density ρ indicate that for $2 < (\sigma\rho^{1/3})^{-1} < 4.6$ the stable crystal form is body-centered cubic, with $\theta_{mp} \cong 0.18$; for $4.6 < (\sigma\rho^{1/3})^{-1} < 13.5$ the stable crystal form is face-centered cubic, with $\theta_{mp} \cong 0.17$ [Stevens and Robbins, 1993, Table I]. These results lend support to the credibility of the Lindemann melting criterion as a useful approximation, at least for a broad range of model systems composed of structureless particles.

The Debye-Waller factor representing vibrational reduction of Bragg diffraction peaks offers a direct experimental means to test the Lindemann melting criterion because it is determined by atomic mean-square displacements [Kittel, 1996, p. 632]. Using Mössbauer γ rays for this purpose, the criterion has been tested both for metallic elements and for alkali halide crystals [Martin and

TABLE IV.7. Test of the Lindemann melting criterion using Mössbauer radiation diffraction[a,b]

Substance	T_m, K	Crystal Structure	θ_{mp}
Al	934	fcc	0.075
Cu	1,357	fcc	0.084
LiF	1,118	NaCl	0.114
NaCl	1,074	NaCl	0.112
KCl	1,043	NaCl	0.110
KBr	1,007	NaCl	0.114

[a]Martin and O'Connor, 1977.
[b]Measurements carried out at 1 atm.

O'Connor, 1977]. For the latter, the mean-square displacement was taken to be the average over both anions and cations. The results are listed in Table IV.7. It is noteworthy that the measured θ_{mp} values are substantially smaller than those listed in Table IV.6 for model system simulations. This is especially the case for metals. However, within each class (metals, alkali halides), there appears to be at least rough consistency of θ_{mp} values.

Quantum effects also significantly influence θ_{mp}. Solid helium, a hexagonal close-packed crystal, offers an extreme example. Theoretical calculations suggest that θ_{mp} at the 0 K melting point of ^{4}He (occurring as the pressure is reduced to ≈ 25 bar) lies in the range 0.25–0.27 [Hansen and Pollock, 1972; Moroni and Senatore, 1991]. In this case, the crystal phase vibrational displacements are caused entirely by zero-point motions of the helium atoms. These zero-point motions have amplitudes sufficiently large that they carry the crystallized helium into a large number of basins with defect-containing inherent structures. This situation is discussed in detail in Chapter VIII.

It is not clear at this point whether a useful generalization of the Lindemann melting criterion can be identified for a wide range of polyatomic molecules. Needless to say, intrabasin vibrational distribution functions can be defined for various locations fixed within each molecule, such as any one of the nuclei of its constituent atoms, the geometric centroid, or the center of mass. But whether these distribution functions and the moments they imply suffice to characterize the melting process is currently not well known. In addition, it remains to be determined whether the Lindemann approximation can provide useful descriptions for the melting of crystalline substances that under heating have first undergone order–disorder transitions. These issues remain challenging possibilities for future research.

V.

Liquids at Thermal Equilibrium

With the exception of those substances capable of forming liquid crystals, passage through the thermodynamic melting temperature upon heating a crystal of a pure substance normally creates an isotropic liquid phase devoid of translational or orientational order. The number and geometric diversity of distinguishable inherent structures and their basins that underlie the liquid are far greater than the corresponding crystal set. Furthermore, the detailed dynamical sequences encountered in the equilibrium liquid state (interbasin transitions), and how they are connected to linear transport properties, are also much more complex issues than in the case of crystals. This chapter is devoted to this equilibrium liquid state and to its linear transport phenomena within the regime of classical statistical mechanics. Chapter VI adapts the energy landscape/inherent-structure representation to description of metastable states created by liquid supercooling and glass formation.

A. Some Experimental Observations

For single-component systems, an informal but conventional identification of the equilibrium homogeneous liquid state presumes that it is bounded below in temperature by the triple point at T_{tp} and above by the critical point at T_c. This condition identifies a subregion of the entire continuously connected fluid phase region in the temperature–pressure (or temperature–density) plane that surmounts the critical point and includes the dilute vapor region. In this chapter, it is convenient to expand the working definition of "liquid" to include a neighborhood that encloses the critical point. For purposes of orientation with respect to experiment, Table V.1 presents measured triple-point and critical-point temperatures and pressures for some selected substances. The entries in Table V.1 illustrate that the dimensionless ratio T_c/T_p is not constant but varies modestly from case to case. By comparison, the corresponding dimensionless pressure ratio p_c/p_{tp} can exhibit a much wider variation.

The capacity for a substance to exhibit liquid–vapor coexistence in its equilibrium phase diagram hinges upon the presence of attractive forces with adequate range acting between the constituent particles. Model substances whose interactions are purely repulsive, such as those with hard-sphere or inverse-power pair potentials of course cannot display such coexistence. Other model substances are known whose attractive interactions are short-ranged compared to the size of their repulsive core diameters, with the result that they fail to possess equilibrium liquid–vapor coexistence [ten Wolde and Frenkel, 1997]. Some real substances may also fail to exhibit equilib-

TABLE V.1. Liquid-phase temperature and pressure ranges for selected materials.[a] The triple-point and critical-point values are indicated by subscripts *tp* and *c*, respectively

Substance	T_{tp}, K	T_c, K	p_{tp}, bar	p_c, bar
Argon (Ar)	83.81	150.87	0.688	48.98
Xenon (Xe)	161.4	289.77	0.817	58.41
Chlorine (Cl_2)	170	416.9	0.01054	79.91
Methane (CH_4)	90.69	190.56	0.1170	45.99
Carbon dioxide (CO_2)	216.58	304.13	5.180	73.75
Hydrogen cyanide (HCN)	259.83	456.7	0.1862	53.9
Water (H_2O)	273.16	647.14	0.00612	220.64
Ammonia (NH_3)	195.4	405.5	0.0612	113.5
Silicon tetrafluoride (SiF_4)	186.3	259.0	2.208	37.2
Sulfur hexafluoride (SF_6)	223.1	318.69	2.327	37.7
Neopentane (C_5H_{12})	256.58	433.8	0.358	31.96

[a]Lide, 2003, Chapter 6.

rium liquid–vapor coexistence for the same reason. It has been suggested that the carbon allotrope "buckminsterfullerene" (C_{60}) may be such an example [Hagen et al., 1993].

Table V.2 offers some illustrative examples of relative volume changes and of molar entropy changes for the melting ($s \rightarrow l$) of single-component systems at atmospheric pressure. These properties are related to the slope of the melting curve in the temperature–pressure plane by the Clausius-Clapeyron thermodynamic identity [Guggenheim, 1950, p. 122; Landau and Lifshitz, 1958b, p. 257]:

$$\frac{dp}{dT} = \frac{S_l - S_s}{V_l - V_s}. \tag{V.1}$$

TABLE V.2. Melting transition characteristics for some selected single-component substances at atmospheric pressure[a]

Substance	T_m, K	$(V_l - V_s)/V_s$	$S_l - S_s$, e.u.[b]
Argon (Ar)	83.8	0.144	3.35
Xenon (Xe)	161.4	0.151	3.40
Methane (CH_4)	89.2[c]	0.0869	2.47
Water (H_2O)	273.2	−0.083	5.25
Carbon disulfide (CS_2)	161.1	0.067	6.51
Carbon tetrachloride (CCl_4)	250.5	0.0530	2.31
Silicon tetrachloride ($SiCl_4$)	205.5	0.1242	9.08
Sodium chloride (NaCl)	1074	0.250	6.7
Benzene (C_6H_6)	278.6	0.1332	8.44
Chlorobenzene (C_6H_5Cl)	228.2	0.071	7.89

[a]Ubbelohde, 1978.
[b]e.u. (entropy unit) = cal/mol deg K.
[c]Solid-phase rotational transition at 20.5 K.

TABLE V.3. Phase-separating binary liquid mixtures at 1 atm (1.01325 bar). Upper and lower critical (consolute) temperatures are indicated by T_{uc} and T_{lc}, respectively

Binary System	Critical Temperature, K	Reference
Oxygen (O_2) + ozone (O_3)	$T_{uc} = 93.3$	[a]
Methane (CH_4) + carbon tetrafluoride (CF_4)	$T_{uc} = 94.5$	[b]
Iso-octane (C_8H_{18}) + perfluoro-n-heptane (C_7F_{16})	$T_{uc} = 296.4$	[c]
Water (H_2O) + triethylamine ($C_6H_{15}N$)	$T_{lc} = 291.9$	[c]
Water (H_2O) + 2,6-dimethyl pyridine (C_7H_9N)	$T_{lc} = 307.2$	[d]
Aniline (C_6H_7N) + cyclohexane (C_6H_{12})	$T_{uc} = 302.8$	[e]

[a]Jenkins et al., 1955.
[b]Croll and Scott, 1958.
[c]Jura et al., 1953.
[d]Cox and Herington, 1956.
[e]Rowden and Rice, 1951.

The well-known decrease in molar volume upon melting exhibited by water is unusual but hardly unique; the elements silicon (Si), germanium (Ge), bismuth (Bi), and gallium (Ga) all share this attribute, and the Clausius-Clapeyron equation declares that their melting temperatures therefore decrease with increasing pressure.

Liquid mixtures provide examples of limited mutual solubility and separation into distinct coexisting liquid phases. Table V.3 presents a few examples of binary systems at atmospheric pressure that show either upper or lower critical (consolute) points. The respective temperatures are denoted by T_{uc} and T_{lc}, with phase separation below T_{uc} and above T_{lc}. These examples illustrate the chemical diversity of the components that can exhibit critical solution behavior.

The traditional expectation has been that the regime of thermodynamic equilibrium single-component systems can only exhibit a single isotropic liquid phase that connects smoothly to the dilute fluid. However, it appears that some exceptions can exist. An experimental investigation of elemental phosphorus (P) has established that a first-order phase transition line exists in the temperature–pressure plane (though well above the critical temperature and pressure), separating two distinct but interconvertible dense fluids [Katayama et al., 2004]. Furthermore, the molecular liquids triphenyl phosphite ($P(OC_6H_5)_3$) and n-butanol (C_4H_9OH) also show evidence of phase changes separating two distinguishable isotropic liquid phases, though these occur in the super-cooled regime [Kurita and Tanaka, 2004, 2005].

B. Enumeration of Liquid-Phase Basins

In the case of those substances whose crystal phases are "simple" (i.e., exhibit no rotational or order–disorder transitions upon heating from $T = 0$ to the melting point), the set of distinct inherent structures and their basins determining properties of those crystals is manageably small. If thermally excited point defects in the crystals can be neglected, only the basins corresponding to the absolute minimum of the potential energy Φ or potential enthalpy $\hat{\Phi}$ need to be considered. The entropic contributions $\sigma(\varphi^*)$ [Eq. (III.29)] and $\hat{\sigma}(\hat{\varphi}^*)$ [Eq. (III.30)] to crystal Helmholtz and Gibbs free energies, respectively, then are both zero. As emphasized in Section IV.C, anharmonic vibrational effects within the perfect-crystal basins nevertheless can have significant thermodynamic consequences.

The first-order melting transition profoundly increases the number of contributing inherent structures and basins. One of the major challenges of the statistical-mechanical theory is how to enumerate, or at least to estimate, the number of distinguishable inherent structures/basins (i.e., those unrelated by permutations of identical particles) as a function of depth that suddenly come into play upon melting. Because of primary interest in the large-system limit, this amounts to evaluation of the temperature-dependent intensive quantities $\sigma[\varphi^*(\beta, \rho)]$ and/or $\hat{\sigma}[\hat{\varphi}^*(\beta, p)]$ for the liquid.

In order to provide a simple context for assessing more detailed and accurate approaches, it is worth starting with a rough order-of-magnitude estimate of the entire number Ω_d of distinguishable isochoric-mapping inherent structures for the case of structureless particles at typical liquid densities. This would be relevant specifically to those elements that do not engage in covalent bonding, such as the heavier noble gases, and most metals. One such estimate starts with the observation that the particles essentially possess hard repulsive cores that limit close approaches. This effectively reduces the configuration space for N particles in volume V, from V^N to a content diminished by an amount determined by the negative hard-sphere "excess" entropy S_{ex}/k_B (compared to an ideal gas) in the following manner:

$$V^N \rightarrow V^N \exp[S_{ex}/k_B]. \tag{V.2}$$

This remaining "available" portion of the $3N$-dimensional configuration space is then divided among basins for each of the inherent structures. If the content of each of these basins were to be rearranged (if necessary) into a convex region, its linear dimensions would reasonably be expected to be comparable to some modest fraction, f, of the mean neighbor spacing $(V/N)^{1/3} \equiv \rho^{-1/3}$. This rests on the observation that most interbasin transitions involve only localized displacements of $O(1)$ particles. The total number of basins then would equal the ratio of the available configuration space content to the mean content of one basin $(f\rho^{-1/3})^{3N}$. After accounting for the number $N!$ of permutation-related images of any basin type (assuming that rigid confining walls are present), these considerations suggest that

$$\Omega_d(N, V) \cong \frac{V^N \exp[S_{ex}/k_B]}{N![f\rho^{-1/3}]^{3N}}$$

$$\cong \exp\{N[1 - 3\ln f + (S_{ex}/Nk_B)]\}. \tag{V.3}$$

Let the effective repulsive core diameter be denoted by σ_0. A simple liquid near its triple point has a dimensionless number density substantially below the hard-sphere close-packed density $(\rho\sigma_0^3 = 2^{1/2})$, so one might reasonably choose a range in the vicinity of

$$\rho\sigma_0^3 = 0.5. \tag{V.4}$$

The corresponding excess entropy can be adequately represented by the closed-form expression that emerges from the scaled particle theory (SPT) for hard spheres [Reiss et al., 1959]:

$$S_{ex}(\rho\sigma_0^3)/Nk_B = \ln(1 - (\pi\rho\sigma_0^3/6)) - (3/2)\{[1 - (\pi\rho\sigma_0^3)]^{-2} - 1\} \quad \text{(SPT)}. \tag{V.5}$$

Along with estimate (V.4), this equation leads to

$$\Omega_d \cong \exp(N(-3\ln f - 1.556)). \tag{V.6}$$

To make any sense, the exponent in this last expression must be positive, which requires $0 < f < 0.595$. The alternative values $f = 0.5$ and 0.4 imply, respectively

$$\Omega_d \cong \exp(0.523N) \qquad (f = 0.5), \tag{V.7}$$

$$\cong \exp(1.193N) \qquad (f = 0.4).$$

The implication of this crude estimate is that the exponential rise rate with N of Ω_d involves a coefficient in the exponent that is of order unity for simple dense liquids. This conclusion is qualitatively consistent with an analysis by Wallace, based on experimental data for the melting of metals, which led to a coefficient 0.8 [Wallace, 1997].

In principle, inherent-structure enumerations can be accomplished within the context of computer simulations for many-body systems at equilibrium. The basic representation formally effects an exact separation between the distribution of inherent-structure energies (or enthalpies), and intrabasin vibrational contributions. It follows from free-energy expression (III.29) [or (III.30)] that the system entropy S is likewise separable into a "configurational" part identified with $\sigma(\varphi^*)$ [or $\hat{\sigma}(\hat{\varphi}^*)$] and a vibrational part generated by intrabasin vibrational displacements. When expressed on a per-particle basis, this separation leads to

$$S/Nk_B = \sigma[\varphi^*(\beta, \rho)] + S_{vib}/Nk_B \qquad \text{(constant } V), \tag{V.8}$$

$$= \hat{\sigma}[\hat{\varphi}^*(\beta, p)] + \hat{S}_{vib}/Nk_B \qquad \text{(constant } p).$$

In these expressions, one has

$$S_{vib}/Nk_B = \beta(\partial \beta f_{vib}/\partial \beta)_{\varphi^*,\rho} - \beta f_{vib}, \tag{V.9}$$

$$\hat{S}_{vib}/Nk_B = \beta(\partial \beta \hat{f}_{vib}/\partial \beta)_{\hat{\varphi}^*,p} - \beta \hat{f}_{vib},$$

the forms of which rely upon the fact that the Helmholtz and Gibbs free energy expressions (III.29) and (III.30) are extrema with respect to φ and $\hat{\varphi}$, respectively. The total entropy per particle can numerically be evaluated by thermodynamic integration of the equilibrium pressure equation of state from the low-density, high-temperature regime where the entropy is that of an ideal gas, and thus known. Vibrational entropy can usually be approximated adequately by assuming that harmonic normal modes describe intrabasin motions, although in principle anharmonic corrections could also be included. This permits evaluation of the configurational entropies (inherent-structure enumeration functions) $\sigma(\varphi)$ and $\hat{\sigma}(\hat{\varphi})$.

Sheng and Ma have applied this technique to a model monatomic liquid whose particles interact in pairs with a truncated and shifted Lennard-Jones potential [Sheng and Ma, 2004]. Their study involved systems of $N = 864$ and $N = 10,976$ particles, both confined to reduced density 0.9. Their results indicate that over that portion of the φ range probed, the enumeration function $\sigma(\varphi)$ appeared to have a single maximum, at which

$$\max_{\varphi} \sigma(\varphi) = 1.5. \tag{V.10}$$

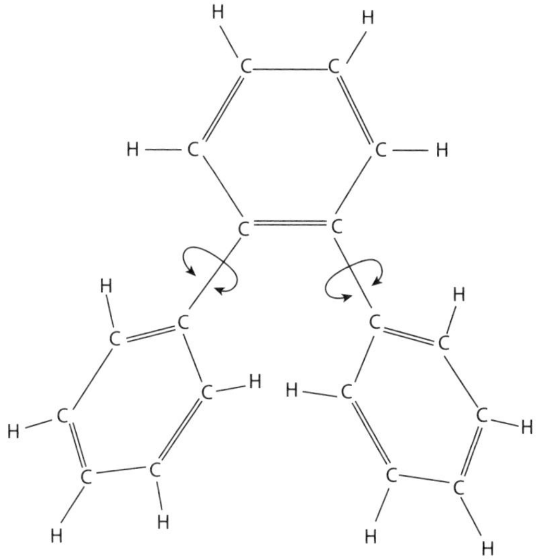

FIGURE V.1. Chemical structure of the *ortho*-terphenyl molecule ($C_{18}H_{14}$).

This notion is consistent with the suggestion of the crude estimates (V.7) that the exponential rise rate with N of the total number of distinguishable inherent structures should be of order unity for simple liquids.

One of the materials frequently examined experimentally in studies of liquid supercooling and glass formation is the organic substance *ortho*-terphenyl ($C_{18}H_{14}$). Discussion of its behavior in these low-temperature metastable states is deferred to Chapter VI. At atmospheric pressure, this substance exhibits a thermodynamically stable liquid phase over a significant temperature range. Its melting and boiling points occur at $T_m = 329$ K and $T_b = 605$ K. The constituent molecules have the chemical structure shown in Figure V.1, with three linked benzene rings that are capable of somewhat hindered relative rotational motion, as indicated by the curved arrows in the figure. This is a good example of a moderately "nonsimple" liquid because of its nonspherical symmetry and its internal flexibility.

A simplified model of *ortho*-terphenyl has been devised by Lewis and Wahnström for use in computer simulations of that substance [Lewis and Wahnström, 1993, 1994; Wahnström and Lewis, 1993]. This model represents each of the three benzene units as a single Lennard-Jones interaction center (thus suppressing the intramolecular rotations), and it treats the bond lengths and bond angles between those centers as rigidly fixed. Thus each *ortho*-terphenyl molecule becomes viewed as a rigid rotor possessing C_{2v} symmetry, with angle 75° between two bonds with lengths 0.483 nm. This rigid three-site model has been examined by molecular dynamics simulation, with emphasis on its liquid-state inherent structures and vibrational properties within the basins surrounding those inherent structures [Mossa et al., 2002; La Nave et al., 2002]. That group of investigators examined systems of $N = 343$ model molecules over a range of temperatures and densities covering the stable and supercooled liquid region. Using the entropy separation mentioned above for constant-volume mapping to inherent structures, it was found that the values for the enumeration function at the temperatures examined fell in the range

$$2.7 < \sigma(\varphi^*) < 3.9. \tag{V.11}$$

That this numerical range involves larger values than what has been indicated above for "simple" liquids is not surprising, given the additional degrees of freedom. Furthermore, the observed temperature trends at all densities investigated indicated that these $\sigma(\varphi^*)$ values are well below the maximum that in principle would be attained at infinite temperature. It is also worth noting that isothermal compression of the liquid had the effect of reducing the number of distinct inherent structures accessed.

Experimental constant-pressure heat capacity (C_p) measurements at 1 atm for *ortho*-terphenyl permit an independent estimate of its isobaric inherent-structure enumeration function [Stillinger, 1998]. These measurements reveal that the heat capacity of the stable crystal phase and that of the amorphous solid (glassy) phase below the experimental glass transition temperature $T_g \cong$ 240 K are nearly identical [Chang and Bestul, 1972]. Under the assumption that the crystal phase is dominated by a single perfect crystal inherent structure and its basin (the measurements show no evidence for a crystal-phase disordering transition), it is possible to interpret the data so as to yield at least a portion of the enumeration function $\hat{\sigma}_m(\hat{\varphi})$ for the disordered structures that underlie the stable, supercooled, and superheated liquids (see Section III.G). [Recall that subscript m on enumeration function $\hat{\sigma}_m$ refers to the configuration space projection operation that underlies "metastable" state extension.] Details of this calculation are deferred until Chapter VI, Section VI.E. Figure V.2 displays graphically the result of the analysis. Although not exactly the case, the shape of the curve is close to that of an inverted parabola with its axis slightly off vertical. Because the pressure is low and because the volumes of the *ortho*-terphenyl condensed-phase crystal, glass, and liquid are also small, the quantitative distinction between isobaric and isochoric enumerations is relatively small in this case.

One sees from Figure V.2 that as temperature increases from T_m to T_b, $\hat{\sigma}_m(\hat{\varphi})$ rises monotonically from approximately 6.3 to 10.6. It is not surprising that these values considerably exceed those shown in Eq. (V.11) for the Lewis-Wahnström model because of the additional molecular degrees of freedom possessed by the real substance. The maximum value of the $\hat{\sigma}_m(\hat{\varphi})$ curve is approximately 13.1, occurring at about

$$\hat{\varphi} - \hat{\varphi}_{crys} \cong 6.0 \times 10^4 \, \text{J/mol}, \tag{V.12}$$

which would hypothetically be attained in thermal equilibrium for the constrained superheated liquid as temperature diverged to infinity. It should be remarked in passing that the two points at which Figure V.2 shows $\hat{\sigma}_m = 0$ are artifacts of the underlying extrapolation (dashed portions of curve) away from the temperature range of experimental data used. In particular, the simple finite-slope intersections with the horizontal axis violate basic arguments presented in Chapter VI.

Models for binary liquid mixtures have been the focus of a large number of computer simulation studies, a substantial portion of which have analyzed the collections of inherent structures, and how they and their surrounding basins have determined both thermodynamic and kinetic properties. One of the more popular model types involves two species A and B interacting through

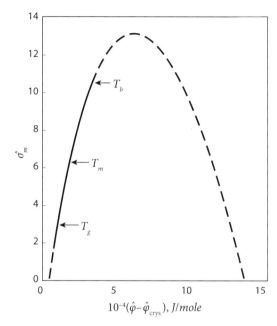

FIGURE V.2. Isobaric enumeration function $\hat{\sigma}_m$ for stable and supercooled liquid phases of *ortho*-terphenyl inferred from $C_p(T)$ measurements at one atmosphere. The horizontal axis measures enthalpy relative to that of the crystal. The portions of the enumeration function accessed at the experimentally observed glass, melting, and boiling temperatures are designated by T_g, T_m, and T_b, respectively.

pair potentials that are qualitatively like the full Lennard-Jones 12,6 pair function but with species dependence affecting only energy and distance scaling:

$$v_{\mu\nu}(r) = \varepsilon_{\mu\nu} v(r/\sigma_{\mu\nu}) \qquad (\mu, \nu = A, B). \qquad (\text{V.13})$$

The original version of this family of binary mixture models was devised to emulate the deep-eutectic metal–metalloid system composed of nickel (Ni) and phosphorus (P) in approximate number ratio 4:1 [Weber and Stillinger, 1985a, 1985b; Büchner and Heuer, 1999]. With this motivation (and identifying A with Ni, B, with P), the relative values of the scaling parameters were initially chosen to be [Weber and Stillinger, 1985a]:

$$\sigma_{AB}/\sigma_{AA} = 2.00/2.49 \cong 0.8032,$$

$$\sigma_{BB}/\sigma_{AA} = 2.20/2.49 \cong 0.8835, \qquad (\text{V.14})$$

$$\varepsilon_{BB}/\varepsilon_{AA} = 0.5,$$

$$\varepsilon_{AB}/\varepsilon_{AA} = 1.5.$$

A slightly later variant [Weber and Stillinger, 1985b], presumably yielding a better description of the 4:1 Ni–P system, changed the first of these ratios to

$$\sigma_{AB}/\sigma_{AA} = 2.20/2.49 \cong 0.8835. \qquad (\text{V.15})$$

The generic interaction function v(x) was selected in that early version to vanish beyond a cutoff distance, at which the function and all its derivatives were continuous. More recent versions have chosen v(x) to have a cutoff and shifted version of the Lennard-Jones 12,6 pair potential [Kob and Andersen, 1995; Sastry et al., 1997; Gleim et al., 1998]. These latter versions have tended to utilize longer ranged interactions than in the original version but have accepted relatively low-order derivative discontinuities at their cutoffs.

Enumeration functions deduced for the binary mixture models at the 4:1 particle–species concentration ratio have uniformly been consistent with the qualitative pattern presented in Figure V.2 for the single component liquid *ortho*-terphenyl. Specifically, $\sigma(\varphi)$ appears to have the qualitative shape of an inverted parabola over a significant density range:

$$\sigma_m(\varphi) \cong \sigma_{max} - A(\varphi - \varphi_0)^2. \tag{V.16}$$

Using the first parameter set (V.14) in a molecular dynamics simulation, Sastry has found that as the particle density increases by a factor of about 1.23, the inferred values of σ_{max} decline from 1.010 to 0.849, while the width-determining parameter A decreases by a factor of about 8.3 over the same density range [Sastry, 2001]. That is, for this binary mixture model over the liquid density range examined, increasing the density reduces the total number of distinct inherent structures but spreads them out in energy.

In addition to the references just cited, several additional published papers have also examined aspects of the inherent-structure distributions for the 4:1 binary liquid mixture models [Sciortino et al., 1999, 2000; Heuer and Büchner, 2000; La Nave, Mossa, and Sciortino, 2002; Doliwa and Heuer, 2003]. Results reported in these papers conform to the type of distribution shown in Eq. (V.16).

After establishing at least roughly the enumeration range for distinct inherent structures and their encompassing basins in liquids, one is in a position to discuss the basic concept of thermal equilibrium for macroscopic systems. This leads in a direction that is unconventional, but certainly revealing, if not disturbing. Consider a single-component material sample containing a mole (Avogadro's number $N_{Av} \cong 6.022 \times 10^{23}$) of particles. For purposes of qualitative illustration, one can take $\sigma_{max} = 1$. The number of distinct inherent structures under isochoric (constant-density) conditions is essentially given by

$$\Omega_d = \exp(N_{Av}\sigma_{max})$$

$$\cong 10^{2.6 \times 10^{23}}. \tag{V.17}$$

In order for any single dynamical system of this sort to exhibit true equilibration, it would have to be able to visit the basin for every distinct inherent structure at least once. This naturally leads to the question about how long a time t_{visit} it would take on average to complete that basin circuit.

For simplicity, suppose that this collection of distinguishable basins were to be visited one by one via classical mechanics, at a fixed rate, with no returns until the entire set had been visited. The interbasin transition rate is an extensive quantity, thus proportional to N_{Av} in this example. For the purposes of an order of magnitude estimate, this transition rate R_{trans} could reasonably be chosen as

$$R_{trans} = 10^{10}N_{Av}s^{-1}, \tag{V.18}$$

which amounts to assuming that each particle is at or near a localized system rearrangement 10^{10} times per second. The estimated time for completion of the visitation circuit therefore becomes

$$t_{visit} \approx \Omega_d / R_{trans}$$

$$\approx 10^{2.6 \times 10^{23}} \text{ s} \tag{V.19}$$

$$\approx 10^{2.6 \times 10^{23}} \text{ yr.}$$

This result is so dominated by the enormous quantity Ω_d for the macroscopic system that it becomes substantially a typographical invariant, regardless of the conventional time unit used. Furthermore, even this absurdly large result is probably a gross underestimate, in that it has assumed no repeated visits to any distinguishable basin type.

In the light of this rough but indicative estimate, it is clear that thermal equilibrium in a macroscopic particle system only requires that the dynamics provide a *representative* sampling of the full set of inherent-structure and basin types. That representative sampling can occur on an experimentally accessible time scale, presuming that temperature and density conditions are appropriate. Of course, for some metastable phase circumstances, the conditions may indeed not be appropriate for representative sampling, i.e., may be non-ergodic. In any event, this brings to mind the analogy of voter polling before a national election, in which a proper selection of a tiny fraction of an electorate can have powerful predictive capacity, or if it is improperly implemented, can be grossly misleading.

C. Separation of Contributions

The multidimensional configuration space for many-particle systems has been described in Chapter III in terms of its tiling by potential energy landscape basins. This representation leads naturally to resolution of thermodynamic properties into separate contributions that are associated, respectively, with the distribution of static inherent structures and with intrabasin vibrational displacements. Such a formal separation has explicitly appeared in Helmholtz and Gibbs free energy expressions, Eqs. (III.29) and (III.30). This section provides some illustrations of that separation for other properties of single-component isotropic liquids at equilibrium.

(1) ISOTHERMAL COMPRESSIBILITY

The pair distribution function in an isotropic fluid phase depends spatially only on relative coordinates of the participating particle pair. If the constituent particles are structureless, this distribution function then depends geometrically only on the scalar distance between the particles and may be written as

$$\rho_2(r) \equiv \rho^2 g_2(r). \tag{V.20}$$

Here ρ is the number density, and $g_2(r)$ is the "pair correlation function," normalized as shown to approach unity at large particle separations r. To be precise, $g_2(r)$ in this section is the infinite-system limit function. The isothermal compressibility $\kappa_T = (\partial \ln \rho / \partial p)_T$ for the isotropic liquid [from Eq. (II.87)] is the following:

$$\rho k_B T \kappa_T = 1 + \rho \int [g_2(r) - 1] d\mathbf{r}. \tag{V.21}$$

In liquids, the pair correlation function deviates from its asymptotic value unity only at small distances, so it is valid formally to extend the integration in this last expression to infinity.

The structure factor $S(k)$ for the liquid phase under consideration is defined by the expression

$$S(k) = 1 + \rho \int \exp(i\mathbf{k} \cdot \mathbf{r}) [g_2(r) - 1] d\mathbf{r}, \tag{V.22}$$

which immediately implies that

$$\rho k_B T \kappa_T = S(0). \tag{V.23}$$

The definition of the collective density variables $\rho(\mathbf{k})$, Eq. (I.60), leads to an alternative expression for the structure factor as a non-negative ensemble average:

$$S(k) = N^{-1} \langle \rho^*(\mathbf{k}) \rho(\mathbf{k}) \rangle \tag{V.24}$$

$$\geq 0.$$

At $k = 0$, this confirms the fact that κ_T cannot be negative for a liquid system at equilibrium.

The equilibrium pair correlation function $g_2(r)$ can be associated with a corresponding function $g_2^{(q)}(r)$ that results from "quenching" the vibrational displacements of configurations contributing to the former function. In other words, $g_2^{(q)}(r)$ is the pair correlation function for the collection of inherent structures that results from the steepest descent mapping, applied to the thermal equilibrium configurations for the liquid of interest. This mapping is carried out at constant volume, so that the particle number density ρ remains unchanged. Because of the configurational "smearing" effect of intrabasin vibrational displacements, the inherent pair correlation function $g_2^{(q)}(r)$ typically exhibits more pronounced local order in comparison with its premapped companion $g_2(r)$ [LaViolette and Stillinger, 1986; Stillinger and Weber, 1988]. However, for a liquid phase, $g_2^{(q)}(r)$ possesses the same unity limit at large-r that $g_2(r)$ does. The distinction between these two correlation functions leads to a natural separation of the isothermal compressibility into two components [Stillinger et al., 1998]:

$$\kappa_T = \kappa_T^{(q)} + \kappa_T^{(vib)}, \tag{V.25}$$

where

$$\kappa_T^{(q)} = (\rho k_B T)^{-1} \{ 1 + \rho \int [g_2^{(q)}(r) - 1] d\mathbf{r} \}, \tag{V.26}$$

$$\kappa_T^{(vib)} = (\rho k_B T)^{-1} \int [g_2(r) - g_2^{(q)}(r)] d\mathbf{r}.$$

Because of its description in terms of collective density variables analogous to that shown in Eqs. (V.23) and (V.24), $\kappa_T^{(q)} \geq 0$. However, $\kappa_T^{(vib)}$ has no similar intrinsic sign constraint.

Under constant-volume conditions for simple monatomic liquids, the dominant contribution to the temperature dependence of the pair correlation function arises from intrabasin vibrational motions. That is, the mapped pair correlation function $g_2^{(q)}(r)$ is rather insensitive to temperature. So too would $\kappa_T^{(q)}$ be rather insensitive to temperature, thus leaving $\kappa_T^{(vib)}$ to exhibit most of the temperature dependence of the isothermal compressibility at constant density.

(2) VIRIAL PRESSURE

The virial pressure expression was displayed in Eq. (II.84) for classical-limit systems subject only to additive pair interactions v_2. For classical isotropic liquids composed of structureless particles, that expression reduces to the following:

$$p = \rho k_B T - (2\pi\rho^2/3)\int_0^\infty r^3 v_2{}'(r) g_2(r) dr. \qquad (V.27)$$

Once again, the separation of $g_2(r)$ into an isochorically mapped inherent-structure part $g_2^{(q)}(r)$ plus a correction to pair correlation due to intrabasin vibrational deformations is immediately applicable. As a result one formally has

$$p = p^{(q)} + p^{(vib)}, \qquad (V.28)$$

where

$$p^{(q)} = -(2\pi\rho^2/3)\int_0^\infty r^3 v_2{}'(r) g_2^{(q)}(r) dr, \qquad (V.29)$$

$$p^{(vib)} = \rho k_B T - (2\pi\rho^2/3)\int_0^\infty r^3 v_2{}'(r)[g_2(r) - g^{(q)}{}_2(r)] dr.$$

Unlike the inherent-structural contribution to isothermal compressibility, $p^{(q)}$ can be negative, i.e., it can indicate that on average the relevant inherent structures exist in a state of tension. Indeed, simulations coupled to inherent-structure mapping have revealed just this property for a diverse group of model substances and over a suitable density range for each of those substances. This characteristic is discussed at greater length later in Section VII.B and is illustrated in Figure VII.8.

Although the separation just exhibited refers to simple systems with only spherically symmetric pair interactions, generalization to a wider family of systems is formally straightforward. When three-body, four-body, … interactions are present, the correspondingly extended form of the virial pressure involves integrals over higher order correlation functions $g_n \equiv \rho_n/\rho^n$. These g_n can also be separated into their isochorically quenched versions $g_n^{(q)}$ plus a correction for intrabasin displacements, and so the pressure again can be separated into an inherent-structure part plus a vibrational contribution.

(3) ENERGY AND HEAT CAPACITY

The energy per particle is obtainable from the basic Helmholtz free energy expression Eq. (III.29) by temperature differentiation:

$$\bar{E}/N = \left(\frac{\partial \beta F_N/N}{\partial \beta}\right)_{N,V} \qquad (V.30)$$

$$= \varphi^*(\beta, \rho) + \left(\frac{\partial \beta f_{vib}(\varphi^*, \beta, \rho)}{\partial \beta}\right)_{\varphi^*, \rho}.$$

The two terms shown are just the mean inherent-structure potential energy for the given temperature and density, and the mean intrabasin excitation energy (including kinetic energy) above the inherent-structure energy. The fact that free energy is at a minimum with respect to variation in φ at $\varphi = \varphi^*$ has been used to simplify Eq. (V.30).

Constant-volume heat capacity follows by application of a further temperature derivative:

$$C_V/Nk_B = -\beta^2\left(\frac{\partial \bar{E}/N}{\partial \beta}\right)_{N,V}$$

$$\equiv C_V^{(q)}/Nk_B + C_V^{(vib)}/Nk_B, \tag{V.31}$$

in which

$$C_V^{(q)}/Nk_B = -\beta^2\left(\frac{\partial\varphi^*}{\partial\beta}\right)_\rho, \tag{V.32}$$

and

$$\frac{C_V^{(vib)}}{Nk_B} = -\beta^2\left[2\left(\frac{\partial f_{vib}}{\partial\beta}\right)_{\varphi^*,\rho} + \left(\frac{\partial f_{vib}}{\partial\varphi^*}\right)_{\beta,\rho}\left(\frac{\partial\varphi^*}{\partial\beta}\right)_\rho\right] - \beta^3\left[\left(\frac{\partial^2 f_{vib}}{\partial\beta^2}\right)_{\varphi^*,\rho} + \left(\frac{\partial^2 f_{vib}}{\partial\beta\partial\varphi^*}\right)_\rho\left(\frac{\partial\varphi^*}{\partial\beta}\right)_\rho\right]. \tag{V.33}$$

The first of these contributions $C_V^{(q)}/Nk_B$ arises simply from the temperature dependence of the dominating basin depth φ^*. The second $C_V^{(vib)}/Nk_B$ includes both the intrinsic heat capacity attributable to temperature rise in a fixed group of basins as well as the effect on vibrations of change in average basin shape caused by shift in φ^*.

At least at the notational level, there is a strong correspondence between the energy and constant-volume heat capacity on one side and the enthalpy and constant-pressure heat capacity on the other. The isobaric-mapping analogs of Eqs. (V.31) to (V.33) are easily seen to be the following:

$$C_p/Nk_B \equiv C_p^{(q)}/Nk_B + C_p^{(vib)}/Nk_B; \tag{V.34}$$

$$C_p^{(q)}/Nk_B = -\beta^2\left(\frac{\partial\hat{\varphi}^*}{\partial\beta}\right)_p; \tag{V.35}$$

$$\frac{C_p^{(vib)}}{Nk_B} = -\beta^2\left[2\left(\frac{\partial\hat{f}_{vib}}{\partial\beta}\right)_{\hat{\varphi}^*,p} + \left(\frac{\partial\hat{f}_{vib}}{\partial\hat{\varphi}^*}\right)_{\beta,p}\left(\frac{\partial\hat{\varphi}^*}{\partial\beta}\right)_p\right] - \beta^3\left[\left(\frac{\partial^2\hat{f}_{vib}}{\partial\beta^2}\right)_{\hat{\varphi}^*,p} + \left(\frac{\partial^2\hat{f}_{vib}}{\partial\beta\partial\hat{\varphi}^*}\right)_p\left(\frac{\partial\hat{\varphi}^*}{\partial\beta}\right)_p\right]. \tag{V.36}$$

(4) THERMAL EXPANSION

Isobaric thermal expansion α_p is defined by the thermodynamic expression

$$\alpha_p = (\partial \ln V/\partial T)_{N,p} \tag{V.37}$$

$$= -(\partial \ln\rho/\partial T)_p.$$

The usually observed behavior in liquids is that $\alpha_p > 0$; however, there are important exceptions. Perhaps the most prominent real-substance case of $\alpha_p < 0$ is that of liquid water just above its melting point and at low to moderate pressures [Eisenberg and Kauzmann, 1969]. Negative thermal expansion has also been verified in simulation studies of specialized models, such as the Gaussian core model [Stillinger and Stillinger, 1997].

For the case of classical fluids that involve only structureless particles with just additive pair interactions, one can obtain the thermal expansion from the virial equation of state, Eq. (V.27). After applying the differential operator $(\partial/\partial T)_p$ to both members of that equation and rearranging the result, one obtains

$$\alpha_p = (2p - \rho k_B T)^{-1}\{\rho k_B - (2\pi\rho^2/3)\int_0^\infty r^3 v_2'(r)[\partial g_2(r)/\partial T]_p dr\}. \tag{V.38}$$

The isobaric temperature derivative $[\partial g_2(r)/\partial T]_p$ appearing in this expression resolves naturally into two separate temperature-dependence contributions from distinct basins within the multidimensional enthalpy ($\hat{\Phi}$) landscape [Stillinger and Debenedetti, 1999]. These contributions arise, respectively, from (a) temperature variation in $g_2(r)$ caused by the isobaric temperature shift in the quenched ("q") basin population identifier $\hat{\varphi}^*(\beta, p)$ while holding the intrabasin vibrational amplitudes fixed, and (b) changes in $g_2(r)$ from increases in intrabasin vibrational ("vib") amplitudes as T increases, occurring at fixed $\hat{\varphi}^*(\beta, p)$. Thus, Eq. (V.38) formally produces the separation

$$\alpha_p = \alpha_p^{(q)} + \alpha_p^{(vib)}. \tag{V.39}$$

A more general approach to resolution of α_p can be followed that includes particles with more complicated molecular structures, as well as involving possible nonpairwise intermolecular interactions. Let $\hat{P}_\gamma(\beta, p)$ be the isochoric ensemble probability that distinguishable basin type γ is occupied at the given temperature and pressure, with normalization

$$\sum_\gamma \hat{P}_\gamma(\beta, p) = 1. \tag{V.40}$$

Of course, these probabilities are strongly concentrated around the value specified by enthalpic basin depth $\hat{\varphi}^*(\beta, p)$. For each basin type γ, one can resolve the average of its fluctuating volume into two parts:

$$\langle V_\gamma \rangle_{\beta, p} = V_\gamma^{(q)}(p) + V_\gamma^{(vib)}(\beta, p), \tag{V.41}$$

where $V_\gamma^{(q)}$ is the nonfluctuating volume at the isobaric inherent structure γ, and $V_\gamma^{(vib)}$ is the fluctuating volume change of the system arising from anharmonic intrabasin vibrational effects. By averaging the last relation with the distribution probabilities $\hat{P}_\gamma$, one obtains the following resolution of the equilibrium average system volume:

$$\langle V \rangle_{\beta, p} = \hat{V}^{(q)}(\beta, p) + \hat{V}^{(vib)}(\beta, p), \tag{V.42}$$

where

$$\hat{V}^{(q)}(\beta, p) = \sum_\gamma \hat{P}_\gamma(\beta, p) V_\gamma^{(q)}(p), \tag{V.43}$$

$$\hat{V}^{(vib)}(\beta, p) = \sum_\gamma \hat{P}_\gamma(\beta, p) V_\gamma^{(vib)}(\beta, p).$$

Because $\langle V \rangle_{\beta,p}$ is identified as the thermodynamic extensive variable V, one can apply the differential operator $(\partial / \partial T)_p$ to the logarithm of both sides of Eq. (V.42) to reach the following expressions:

$$\alpha_p^{(q)} = \frac{1}{\langle V \rangle_{\beta,p}} \left(\frac{\partial \hat{V}^{(q)}}{\partial T} \right)_p, \tag{V.44}$$

$$\alpha_p^{(vib)} = \frac{1}{\langle V \rangle_{\beta,p}} \left(\frac{\partial \hat{V}^{(vib)}}{\partial T} \right)_p.$$

These quantities reduce to the forms generated from the virial equation of state, as indicated in Eqs. (V.38) and (V.39), for structureless particles with just pairwise additive interactions.

It is appropriate now to recall the following thermodynamic identity for macroscopic systems [Guggenheim, 1950, p. 87]:

$$C_p - C_V = \frac{(\alpha_p)^2 VT}{\kappa_T}. \tag{V.45}$$

This expression involves isothermal compressibility, volume, constant-volume and constant-pressure heat capacities, and isobaric thermal expansion, all of which have been individually separated into "q" and "vib" contributions above, some isochorically and others isobarically. It needs to be recognized that Eq. (V.45) entangles the magnitudes of all of these contributions in a nontrivial way so as to satisfy this identity.

D. Hard-Sphere Limit

Because of its prominence in the history and published literature of statistical mechanics, the hard-sphere model for many-body phenomena deserves special attention. The continuing importance of the hard-sphere idealization stems in large part from its well-established fluid-crystal phase transition [Alder and Wainwright, 1957] and from its role in rationalizing the short-range order present in liquids of simple particles [Bernal, 1964]. It has been similarly related to jammed amorphous structures that arise in a wide variety of real materials [Donev et al., 2005; Torquato and Stillinger, 2010]. The hard-sphere model has also served as the starting point for many-body theories of solids and fluids in which attractive interactions serve as a perturbation [Zwanzig, 1954; Longuet-Higgins and Widom, 1964; Tang and Lu, 1993]. In the present context, it is important to establish how the hard-sphere model connects formally to the potential energy and potential enthalpy landscape descriptions, with their emphasis on identifiable inherent structures and surrounding basins.

Because the hard-sphere model possesses a singular potential, a useful approach to its analysis within the inherent-structure formalism treats its pair potential as the limiting case for a sequence of continuous potentials. This can be accomplished in a variety of ways, but perhaps the simplest [mentioned in Chapter I, Section I.D] utilizes the family of inverse-power pair potentials. The hard-sphere pair interaction $v_{hs}(r)$ then emerges as the power n diverges to infinity $(\varepsilon_0, \sigma_0 > 0)$:

$$\mathrm{v}_{hs}(r) = \lim_{n \to +\infty} \varepsilon_0 (\sigma_0/r)^n$$

$$= +\infty \qquad (r < \sigma_0) \tag{V.46}$$

$$= 0 \qquad (\sigma_0 < r).$$

Placing subscript "0" on the energy and length parameters here should avoid confusion with the isochoric enumeration function $\sigma(\varphi)$.

The classical mechanics of hard-sphere systems involves the basic simplification that between each pair of subsequent instantaneous elastic collisions, all particles traverse linear trajectories. This feature implies that the (classical) thermal energy of the hard-sphere system at temperature T is identical to that of an ideal gas comprising the same number of particles:

$$\bar{E}_{hs} = (3/2)Nk_B T. \tag{V.47}$$

However, the collisions contribute to the virial pressure, augmenting the ideal gas contribution with a term involving the contact value of the pair correlation function [Hill, 1956]:

$$p = \rho k_B T + [2\pi \rho^2 \sigma_0^3 k_B T/3] g_2(\sigma_0). \tag{V.48}$$

The equilibrium first-order transition from fluid to face-centered cubic crystal occurs as the hard-sphere system is compressed through the reduced density range [Hoover and Ree, 1968]:

$$0.943 \leq \rho\sigma_0^3 \leq 1.041, \tag{V.49}$$

and over this coexistence density range under isothermal conditions, the pressure p remains at a constant value given by [Fernández et al., 2012]:

$$p\sigma_0^3/k_B T = 11.57. \tag{V.50}$$

While exponent n is still finite, the many-particle potential energy function is finite and well defined for all particle configurations that avoid vanishing pair distances:

$$\Phi(\mathbf{r}_1 \ldots \mathbf{r}_N) = \varepsilon_0 \sum_{i<j} (\sigma_0/r_{ij})^n. \tag{V.51}$$

The corresponding potential energy landscape at constant volume displays the usual features: Distinct minima and saddle points of various orders distributed over nonzero ranges in φ. Furthermore, any pair of minima (inherent structures) can be continuously connected by a path on the landscape along which Φ remains finite. But as exponent $n \to +\infty$, these features change drastically: Basins surrounding inherent structures develop flat bottoms tending toward a common depth $\varphi = 0$, and saddles of all positive orders either tend to vanishing height ($\Phi \to 0$) or to infinite height ($\Phi \to +\infty$). If the number density is such that most pairs of inherent structures continue to have connecting paths in the configuration space, those paths can be chosen to have arbitrarily small elevation change by choosing n to be sufficiently large. However, if the reduced density $\rho\sigma_0^3$ is high enough, there is a small fraction of basins that become isolated in the sense that any connection of their enclosed inherent structure to those of all other basins require surmounting Φ barriers whose heights diverge as n diverges. In the hard-sphere limit, these are absolutely trapped configurations from which the system cannot escape without violating the hard-sphere nonoverlap condition.

While n is finite, the force exerted on particle i by particle j in the N-particle system has the purely repulsive form

$$\mathbf{F}_{i,j} = (n\varepsilon_0/\sigma_0)(\sigma_0/r_{ij})^{n+1}[(\mathbf{r}_i - \mathbf{r}_j)/r_{ij}]. \tag{V.52}$$

As n increases, with ε_0 and σ_0 held fixed, the magnitude of this pair force increases without bound for $r_{ij} < \sigma_0$ but declines monotonically to zero for $\sigma_0 < r_{ij}$. The isochoric (constant-volume) version of steepest descent mapping [Eq. (III.2)] identifies inherent structures as N-particle configurations for which the vector sum of all forces on each particle i vanishes identically, and that are local Φ minima. Although for any finite n, the number, scaled geometry, and relative ordering of inherent-structure Φ values are invariant under changes in the system volume V, those invariance properties do not apply to changes in n.

The maximum number density ρ_{max} for a large system of hard spheres with diameter σ_0 is given by [Hales, 2005]:

$$\rho_{max}\sigma_0^3 = 2^{1/2}. \tag{V.53}$$

This maximum is attained (though not uniquely) by placing the spheres in a face-centered cubic lattice, with each sphere in contact with 12 nearest neighbors. A consequence of that fact is that if

$$\sigma_0 > 2^{1/6}\rho^{-1/3}, \tag{V.54}$$

then Φ in Eq. (V.51) diverges for all particle configurations $\mathbf{r}_1 \ldots \mathbf{r}_N$ in the $n \to +\infty$ limit (ρ held fixed). Obviously, no inherent structures can be identified in that circumstance, regardless of initial conditions, so one limits the analysis to cases for which

$$\sigma_0 < 2^{1/6}\rho^{-1/3}. \tag{V.55}$$

Under this restriction, one identifies isochoric inherent structures whose potential energy approaches zero in the $n \to +\infty$ limit, and these are necessarily configurations for which no pair of particles is closer than σ_0, i.e., nonoverlapping configurations for identical hard spheres with diameter σ_0.

In the asymptotic large-n regime, the particle rearrangements produced by the isochoric steepest descent mapping have a distinctive hierarchical character [Stillinger and Weber, 1985b]. This stems from the increasingly steep drop-off of the pair force magnitude with increasing n, indicated in Eq. (V.52). The implication is that during the initial phase of steepest descent a few close-distance pairs are forced apart, while particles at larger distance from all neighbors remain virtually undisplaced. But as the steepest descent operation proceeds, the minimum distance between all pairs in the system increases, so more and more pairs participate in the displacement. Those pairs moving apart begin to contact other particles and thus begin to create larger and larger expanding contact clusters. With very large n, this reconfiguring behaves as though the individual particles were hard spheres with diameters that increase with the passage of the virtual "time" s governing the steepest descent, Eq. (III.2). Such a process must eventually cease when the nearly-hard-sphere particles "jam up," with contacts to neighbors that percolate throughout the entire system and that ultimately show force balance throughout the system. At this final stage, the geometric arrangement of particles is substantially equal to one of the possible "collectively jammed" arrangements of hard spheres [Torquato and Stillinger, 2001] but with an effective diameter that can be denoted by $\sigma_1 \geq \sigma_0$. In the $n \to +\infty$ limit, the corresponding potential energy of the inherent structures converges to $\Phi = 0$, and the actual-diameter-σ_0 hard spheres remain uniformly spaced apart. The effective diameter σ_1 varies from case to case, depending on the initial condition to which the steepest descent is applied.

In view of the fact that all isochoric inherent structures in the hard-sphere limit collapse to the same value of the depth parameter, $\varphi = 0$, this intensive parameter is not a useful way to classify those inherent structures. The obvious alternative is the effective diameter σ_1, or equivalently the fraction of the system volume covered by those σ_1-diameter spheres, $\zeta = \pi\rho\sigma_1{}^3/6$. In principle, this fraction can range up to the close-packed value $\zeta_{max} = \pi/(18)^{1/2}$. A strict lower ζ limit for collectively jammed spheres is not currently known; however, the existence of "tunneled crystals" of hard spheres that are at least collectively jammed [Torquato and Stillinger, 2007] demonstrates that such a lower limit cannot be larger than $2\zeta_{max}/3$.

A collectively jammed state has the property that no subset of the jammed hard spheres can be continuously displaced, while observing the diameter-σ_1 effective nonoverlap conditions, so as to move that subset out of contact with one another and with all other spheres [Torquato and Stillinger, 2001]. But though the majority of the spheres are thus collectively trapped in place, a small fraction of the spheres, typically about 2–3%, remain as "rattlers"; these are spheres that in the final state are out of contact with all others but are imprisoned within a surrounding cage of jammed spheres [Lubachevsky et al., 1991]. Because of the force balance condition that applies to all particles in an inherent structure for finite n, the $n \to +\infty$ limit for the strict hard-sphere model has all rattlers located away from caging particles, i.e., near the cage "center."

Figure V.3 indicates qualitatively how the hard-sphere pair correlation function transforms under the $n \to \infty$ limit steepest descent mapping. The initial configuration for which g_2 is the pair correlation function shown at the top of the figure is from the dense equilibrium fluid, and beyond the contact distance σ_0, it is bounded, continuous, and at least once differentiable with damped oscillations of modest amplitude. The corresponding mapped configuration possesses pair correlation function $g_2{}^{(q)}$, shown at the bottom of the figure. This latter function is infinite at its effective contact distance σ_1, is unbounded just beyond contact, and is only piecewise continuous. Furthermore, as r increases well beyond contact, the oscillations displayed by $g_2{}^{(q)}$ involve greater amplitude excursions about unity than those of its precursor function g_2. Figure V.3 shows that distinctive cusps appear in $g_2{}^{(q)}$ at $r/\sigma_1 = 3^{1/2}$ and 2, with a substantial jump discontinuity associated with the latter [Donev et al., 2005].

E. Critical Points

Table V.1 presents temperature and pressure values at liquid–vapor critical points for several single-component systems. Table V.3 indicates analogous critical temperatures for some representative binary liquid mixtures at 1 atm, including cases of both upper and lower consolute points (solution demixing critical points). These latter binary mixture critical temperatures depend on pressure, and the resulting critical loci $T_c(p, x)$ in the temperature–pressure–composition

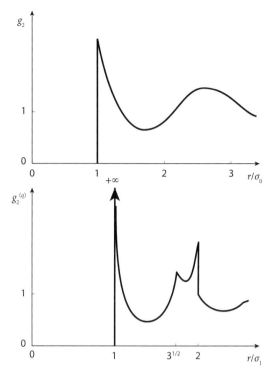

FIGURE V.3. Schematic representation of the pair correlation functions for a dense equilibrated hard-sphere fluid config-uration (g_2) and that for the corresponding inherent structure after isochoric mapping ($g_2^{(q)}$). The latter exhibits a discon-tinuity at $r/\sigma_1 = 2$.

(x) space can exhibit a substantial range of shapes, but always with finite lengths [Rowlinson, 1959, Chapter 6].

As indicated earlier, the equilibrium phase behavior of single-component systems shows non-negligible variation in the ratio of critical to triple-point temperatures T_c/T_{tp}. Among those few substances appearing in Table V.1, this ratio ranges from 1.390 for silicon tetrafluoride (SiF_4) to 2.452 for chlorine (Cl_2). The corresponding discrepancy in pressure ratios p_c/p_{tp} for these same two substances is far more dramatic, specifically 16.8 (SiF_4) vs. 7,582 (Cl_2). This difference empha-sizes the fact that the liquid–vapor portions of equilibrium phase diagrams for single-component systems are not quantitatively related to one another by simple scaling (i.e., by "corresponding states").

But the nonscalable behavior for the entire fluid regions of phase diagrams does not eliminate the possibility of identifying characteristic behaviors displayed in common by many substances in the much more localized neighborhood of their critical points. In particular, it is widely recog-nized that both for single-component systems, as well as for mixtures, there exist sets of positive critical exponents that describe singular thermodynamic behavior upon approach to the critical (or consolute) points [Fisher, 1967a]. Some of these kinds of critical exponents were introduced in Section IV.E in connection with the analogous critical points for interchange order–disorder transitions in crystals.

One of the critical exponents, to be denoted here by β' [conventionally denoted by β, inviting confusion with the same symbol used elsewhere in this text for $(k_B T)^{-1}$], describes the leading-order behavior of the coexistence curve shape. For a pure substance, this exponent involves the liquid–vapor number density difference at coexistence in the immediate vicinity of the critical point as the temperature is lowered along the vapor-pressure curve:

$$(\rho_l - \rho_v)/(2\rho_c) \approx K_1(T_c - T)^{\beta'}. \tag{V.56}$$

Here K_1 is a substance-dependent positive constant, and ρ_l, ρ_v, and ρ_c stand for liquid, vapor, and critical-point number densities, respectively. For binary liquid mixtures that are typically examined at constant pressure, the corresponding definitional expression can be written in terms of the mole fraction x_c of either one of the components at the critical (consolute) point, as well as its values x_A and x_B in the coexisting liquid solutions slightly displaced in temperature from that of the critical (consolute) point:

$$|x_A - x_B|/(2x_c) \approx K_2 |T_c - T|^{\beta'}; \tag{V.57}$$

here $K_2 > 0$, and the use of absolute values accommodates either choice of phases for the mole fractions, as well as both upper and lower consolute point cases.

Many experimental studies of critical regions both for single-component liquid–vapor transitions and for binary mixtures have been focused on determining the critical exponents [Heller, 1967]. In most cases, the inferred coexistence curve exponent β' lies near 1/3. In view of the inevitable imprecision for such challenging experiments, this value may well be indistinguishable from the number β_I suggested by careful studies of the three-dimensional Ising model [Essam and Fisher, 1963], specifically

$$\beta' \cong \beta_I = 5/16 = 0.3125. \tag{V.58}$$

The present state of the theory cannot absolutely rule out the possibility that β' possesses a small dependence on the type of interparticle interactions that are present or whether binary mixture β' values differ slightly for isochoric versus isobaric conditions.

It is pointed out in Section III.E that when the number densities and temperature are fixed at values corresponding to phase coexistence in the equilibrium phase diagram, those coexisting phases with overwhelming probability are macroscopically separated. Indeed, the macroscopic requirement that interfacial free energy be minimized can cause each of those phases to occupy single macroscopic regions within the encompassing system volume. When the constant-volume steepest descent mapping is applied to identify inherent structures, these regions can change size and shape somewhat as a result of differing anharmonic vibrational characteristics for the two phases. However, the macroscopic fractions of the particles contained in each of those subvolumes is not expected to change. This fact should be especially clear in the case of binary liquid mixtures, where particle packing is everywhere quite dense, thus substantially limiting particle excursions relative to their neighbors in the steepest descent mapping operation. But as discussed in Chapter VII, an analogous conclusion for constant-volume mapping applies to single-component liquid–vapor critical points, where substantial open space is present at the molecular length scale.

Consequently, one can consider together both the cases for pure substances and for binary solutions where the overall particle densities are fixed at their critical-point values. Then upon applying the constant-volume steepest descent mapping, the differences in densities or mole fractions of the phases that appear in the inherent structures versus temperature difference from the critical point should be described by the same coexistence curve exponent β' as for the thermodynamic equilibrium states before mapping. Specifically, Eqs. (V.56) and (V.57) modify to the following postmapping forms, where superscript q denotes averaging over the collections of isochoric inherent structures:

$$(\rho_l^{(q)} - \rho_v^{(q)}/(2\rho_c) \approx K_1^{(q)}(T_c - T)^{\beta'}; \tag{V.59}$$

$$|x_A^{(q)} - x_B^{(q)}|/(2x_c) \approx K_2^{(q)}|T_c - T|^{\beta'}. \tag{V.60}$$

Note that these expressions leave open the possibility that the positive coefficients $K_1^{(q)}$ and $K_2^{(q)}$ may be different from those in Eqs. (V.56) and (V.57) because of mapping-induced changes in the relative sizes of the subvolumes occupied by the coexisting phases.

Two other pairs of critical-point exponents describe additional basic features of the critical region. These are denoted by α, α', and γ, γ'. They describe, respectively, the behaviors of heat capacity and of isothermal compressibility. The notation with the prime (') refers to the coexistence situation, while the absence of the prime refers to the single-phase situation.

In the single-species case with the overall number density fixed at ρ_c, the constant-volume heat capacity per particle diverges both above and below T_c, and these divergences are described in leading order in $|T - T_c|$ by the following forms (K_3, $K_4 > 0$):

$$C_V(T) \approx K_3(T - T_c)^{-\alpha} \qquad (T > T_c) \tag{V.61}$$

$$\approx K_4(T_c - T)^{-\alpha'} \qquad (T < T_c).$$

The same kinds of representations apply to constant-volume heat capacities for the cases of binary mixtures with upper consolute points, where the composition is fixed at that of the consolute point and the temperature change occurs while the system volume remains fixed. For lower consolute point mixtures α and α', refer, respectively, to $T < T_c$ and $T > T_c$, that is, to the single-phase and to the coexisting two-phase regimes. It should be acknowledged that binary solution experimental measurements are often performed at fixed pressure (e.g., 1 atm), under which the system volume changes slightly as temperature changes. However, the leading-order divergence exponents for C_V's measured along that altered thermodynamic path should be little if any different from those for fixed system volume.

Measurements on both single-component fluid critical points and binary mixture consolute points conclude that both α and α' are indeed positive but relatively small, usually lying in the range [Heller, 1967, Section 4]:

$$0 < \alpha, \alpha' \leq 0.3. \tag{V.62}$$

The width of this range likely reflects the difficulty of performing decisive heat capacity measurements in the critical region. Nevertheless, this range is consistent with what is mentioned in Section IV.E in connection with interchange order–disorder transitions and three-dimensional Ising models. Based on detailed analysis for those three-dimensional Ising models with nearest neighbor interactions, the following exponents have been inferred for the C_V critical-point divergences [Fisher, 1967a]:

$$\alpha \cong 1/8 = 0.125, \tag{V.63}$$

$$\alpha' \cong 1/16 = 0.0625.$$

At present, it is not possible to decide whether the heat capacity divergence exponents for fluids with short-range interactions theoretically should agree precisely with these Ising model values. However, it can be argued on the basis of "renormalization group" analysis [Wilson, 1971] that short-range interaction details are largely irrelevant provided simply that they do indeed produce a critical point; thus, one would conclude that both kinds of three-dimensional systems should belong to a single "universality class" with common critical exponents.

In accord with the conclusion reached for interchange order–disorder critical points in crystals (Section IV.E), it appears reasonable to suppose that the C_V singularities arise principally from the temperature dependence of contributing inherent structures, not from the vibrational degrees of freedom within the basins of those inherent structures. In the case of a single-component fluid of structureless particles with simple pair interactions $v_2(r)$, this assumption implies a relationship between the temperature dependences (at fixed density) of the premapped (g_2) and postmapped ($g_2^{(q)}$) pair correlation functions. The corresponding average interaction energies per particle (whose T derivatives give C_V) should exhibit the same kinds of leading-order singularities at the critical point. For $T > T_c$, this can be expressed thus:

$$(\rho_c/2)\int v_2(r)g_2(r, T, \rho_c)d\mathbf{r} = [K_3/(1 - \alpha)](T - T_c)^{1-\alpha} + \dots, \tag{V.64}$$

$$(\rho_c/2)\int v_2(r)g_2^{(q)}(r, T, \rho_c)d\mathbf{r} = [K_3^{(q)}/(1 - \alpha)](T - T_c)^{1-\alpha} + \dots,$$

where background contributions possessing weaker singularities have been typographically suppressed, and one admits the possibility that the positive coefficients K_3 and $K_3^{(q)}$ might be unequal. Likewise, below T_c one has

$$(\rho_c/2)\int v_2(r)g_2(r, T, \rho_c)d\mathbf{r} = [K_4/(1 - \alpha')](T_c - T)^{1-\alpha'} + \dots, \tag{V.65}$$

$$(\rho_c/2)\int v_2(r)g_2^{(q)}(r, T, \rho_c)d\mathbf{r} = [K_4^{(q)}/(1 - \alpha')](T_c - T)^{1-\alpha'} + \dots,$$

where again K_4 and $K_4^{(q)}$ might differ, and the pair correlation functions are those appropriate for macroscopically separated coexisting phases [Section II.H; Section III.E].

Exponents γ, γ' describe singular behavior for the rate of change of critical region thermodynamics as the system density or composition varies infinitesimally from that of the critical point. For the single-component fluid case, γ specifies the divergence of isothermal compressibility $\kappa_T = (\partial \ln \rho/\partial p)_T$ in the one-phase region at ρ_c for $T > T_c$. Specializing Eq. (V.21) to this circumstance, one has

$$\rho_c k_B T \kappa_T(T, \rho_c) = 1 + \rho_c \int [g_2(r, T, \rho_c) - 1]d\mathbf{r} \tag{V.66}$$

$$= K_5(T - T_c)^{-\gamma} + \dots, \qquad (T > T_c),$$

where both γ and K_5 are positive. For the same single-component systems, exponent γ' specifies the divergence of κ_T for $T < T_c$, evaluated for either the liquid or the vapor phase at coexistence. For the liquid, the analog of Eq. (V.66) is

$$\rho_l(T)k_BT\kappa_T[T, \rho_l(T)] = 1 + \rho_l(T)\int\{g_2[r, T, \rho_l(T)] - 1\}d\mathbf{r} \qquad (V.67)$$

$$= K_6(T_c - T)^{-\gamma'} + \ldots\ldots$$

These last two equations (V.66) and (V.67) make it clear that the critical-point divergence of κ_T is intrinsically related to the development of long-range density correlations that produce divergences of the spatial integrals in those equations. Within the single-phase portion of the critical region, a plausible generic form for the long-range behavior of g_2 is the following [Fisher, 1966]:

$$g_2(r, T, \rho) - 1 \sim Ar^{-1-\eta}\exp[-r/\xi(T, \rho)], \qquad (V.68)$$

where coefficient $A > 0$, and ξ is a coherence length for equilibrium density fluctuations that diverges to infinity at the critical point. The algebraic exponent appearing here lies in the range $1 \leq 1 + \eta < 3$ so as to produce the κ_T critical divergence as the exponential damping term reduces to unity at all r.

In the cases of binary solution consolute points, both upper and lower, the exponents γ and γ' describe divergence of an "osmotic condition" compressibility for either component, in the single-phase or in the phase coexistence circumstance, respectively. These osmotic condition compressibilities need to be distinguished from the more conventional overall isothermal compressibility for a fixed-composition binary mixture, a property that also can be displayed in terms of pair correlation function integrals [Kirkwood and Buff, 1951]. The osmotic pressure, usually denoted by Π, refers to the difference in pressure between the binary solution of interest at a given T and the pressure of the pure "solvent" component at the same temperature and chemical potential that it has in the solution. An equivalent picture imagines the binary solution of interest to be surrounded by a semipermeable membrane through which only the solvent species can pass. Osmotic pressure Π is then the equilibrium pressure inside the enclosing membrane minus that outside, after thermal equilibrium has been established between the two. This conceptual situation is symmetric in the sense that either one of the two components could have been the designated solute trapped by the membrane, and the other one would then be the membrane-permeating solvent. If ρ_1 denotes the number density of the designated solute, and μ_0 is the chemical potential of the solvent that is equal on both sides of the membrane, then the osmotic condition compressibility has the following representation:

$$\kappa_{T,1}{}^{(osm)} = (\partial \ln \rho_1/\partial \Pi)_{T, \mu_0}. \qquad (V.69)$$

Pair correlation function expressions (V.66) and (V.67) adapt straightforwardly to the osmotic condition isothermal compressibilities, where of course the included number densities and g_2 functions in those expressions refer just to the solute species. In view of the expectation that local density fluctuations for the two components of a dense binary liquid should be strongly anticorrelated, it is a reasonable assumption that osmotic-condition exponents γ and γ' remain unchanged upon exchanging the roles of solute and solvent.

As pointed out earlier, Eqs. (V.25) and (V.26), the formal separation of κ_T into inherent-structure and intrabasin vibrational contributions is formally straightforward, depending only on the difference between the isochorically premapped and postmapped pair correlation functions g_2 and $g_2^{(q)}$. In accord with the discussions above regarding coexistence curve and constant-volume heat capacity singularities, one would propose that the same γ and γ' describe the inherent-structure portions of the observable compressibility divergences, both for single-component fluids and for osmotic behavior of binary mixtures. Furthermore, this would include the presumption that the asymptotic expression (V.68) exhibited for $g_2 - 1$ would also apply to $g_2^{(q)} - 1$ with the same exponents, though possibly with shifted coefficients.

Determining the κ_T critical singularities experimentally for single-component fluids appears to be quite challenging. No very precise results are currently available. However, the rough consensus appears to suggest that their exponents lie in the range $1.0 < \gamma, \gamma' < 1.5$ [Heller, 1967, Table 3]. The corresponding experimental determinations that would be required to characterize osmotic compressibilities in binary liquids apparently are even more difficult, so no reliable exponent values can be cited. Consequently, it may be reasonable for the γ and γ' values that have emerged from careful analyses of three-dimensional Ising models to serve as targets for future experimental investigations. This possible connection assumes that real-substance critical phenomena belong precisely, or nearly, to the Ising model universality class. The Ising model estimates are the following [Fisher, 1967a]:

$$\gamma \cong 5/4 = 1.25, \tag{V.70}$$

$$\gamma' \cong 21/16 = 1.3125.$$

Critical exponents are subject to rigorous inequalities. Rushbrooke derived one such inequality that has direct relevance in the present context [Rushbrooke, 1963]. It relates the three exponents that describe behavior of systems at phase coexistence:

$$\alpha' + 2\beta' + \gamma' \geq 2. \tag{V.71}$$

Notice that the values indicated above for three-dimensional Ising models ($\alpha', \beta', \gamma' = 1/16, 5/16, 21/16$) satisfy this Rushbrooke relation as an equality. It might be worth noting in passing that the so-called "classical" or "mean field" exponents ($\alpha', \beta', \gamma' = 0, 1/2, 1$), which are predicted by the venerable van der Waals equation of state, also satisfy the Rushbrooke relation as an equality.

F. Static Dielectric Constant

At the macroscopic level of description, the linear response of an isotropic liquid to an externally imposed uniform electric field is described by a scalar dielectric "constant" ε. For any given substance, this intensive macroscopic property varies with temperature and pressure. Although other characteristics of liquids are also relevant, the magnitude of ε is important for the solvent properties of liquids, particularly when ionic solutes are involved.

Consider the case of a large pair of parallel capacitor plates, respectively bearing equal but opposite charges. If the space between these plates initially is a vacuum (dielectric constant unity), let $\mathbf{E}_0$ denote the uniform electric field in that empty space caused by the charged capacitor plates.

Without changing those charges, suppose next that an isotropic liquid dielectric is placed in that space. As a result, the electric field changes to a reduced uniform field value $\mathbf{E}$, where

$$\mathbf{E} = \varepsilon^{-1}\mathbf{E}_0. \tag{V.72}$$

In the geometry specified, the state of the dielectric liquid can also be described by a macroscopic displacement field $\mathbf{D}$, and a polarization field (induced electric moment per unit volume) $\mathbf{P}$, as follows [Landau and Lifshitz, 1960, Chapter II]:

$$\mathbf{D} = \varepsilon\mathbf{E} \equiv \mathbf{E}_0, \tag{V.73}$$

$$\mathbf{P} = \left(\frac{\varepsilon - 1}{4\pi}\right)\mathbf{E} \equiv \left(\frac{\varepsilon - 1}{4\pi\varepsilon}\right)\mathbf{E}_0.$$

This section is concerned with the static (i.e., low frequency limit) electrical response of insulating liquids to which ε refers, with primary emphasis on how that response is determined by the properties of the constituent particles and their interactions with neighbors. This situation naturally involves connecting dielectric response to the characteristics of the multidimensional potential energy landscape for the substance of interest.

To provide an empirical context for the statistical mechanical formalism concerning static dielectric constants ε, Table V.4 presents some representative measured values for liquids at 1 atm. The listing is restricted to pure substances but includes representatives of a chemically diverse range of liquid insulators. The entries include both "nonpolar" and "polar" cases, according to whether the symmetry of the isolated constituent particles (atoms or molecules) at their stable electronic ground state structures, respectively, prohibits or allows the presence of a nonvanishing electric dipole moment. The isolated particle (i.e., gas-phase) dipole moments that are also listed in Table V.4 show considerable variation even for those cases classified as polar. Of course, a molecule that has high symmetry in isolation and is classified as nonpolar (e.g., CCl_4) can instantaneously exhibit a small dipole moment in the liquid phase as a result of asymmetric intramolecular vibrations or from shape-distorting interactions with neighbors.

The ε values exhibited in Table V.4 illustrate a marked quantitative difference between most (but not all) examples of nonpolar and polar substances, respectively. Evidently, the presence of a permanent dipole moment on molecules comprising the liquid can boost the dielectric constant by well over an order of magnitude. However, this effect is strongly substance dependent and, as is stressed below, this dependence is due only in part to the magnitude of the permanent dipole moment of the isolated molecule involved.

For the sake of clarity in the following, the case of nonpolar substances is examined first, followed by extension to polar substances. It is natural to start by recalling the Clausius-Mossotti Eq. (IV.120), which is discussed in Section IV.H in connection with dielectric constants of crystalline insulators. For systems of uncharged particles, that equation adopts the form

$$\frac{\varepsilon - 1}{\varepsilon + 2} = \left(\frac{4\pi}{3}\right)\sum_j \rho_j \alpha_j, \tag{V.74}$$

in which ρ_j is the number density and α_j is the isotropic electronic polarizability of the species-j particles present in the phase. The validity of this relation hinges upon the local symmetry of

TABLE V.4. Static dielectric constants e for some representative liquids at 1 atm

Substance	Dipole moment, D[a]	Temperature, K	ε^b
Acetic acid (CH_3COOH)	1.70	293.2	6.20
Acetylene (C_2H_2)	0	195.0	2.4841
Argon (Ar)	0	140.00	1.3247
Benzene (C_6H_6)	0	293.2	2.2825
Bromine (Br_2)	0	297.9	3.1484
Carbon disulfide (CS_2)	0	293.2	2.6320
Difluoromethane (CH_2F_2)	1.99	152.2	53.74
Dimethyl ether (($CH_3)_2O$)	1.30	258.0	6.18
Ethanol (C_2H_5OH)	1.69	293.2	25.3
Formamide (CH_3NO)	3.73	293.2	111.0
Hydrogen cyanide (HCN)	2.99	293.2	114.9
Hydrogen fluoride (HF)	1.83	273.2	83.6
Hydrogen peroxide (H_2O_2)	1.57	290.2	74.6
Methanol (CH_3OH)	1.70	293.2	33.0
Nitrogen (N_2)	0	63.15	1.4680
Sulfur (S)[c]	0	407.2	3.4991
Sulfur hexafluoride (SF_6)	0	223.2	1.81
Tetrachloromethane (CCl_4)	0	293.2	2.2379
Water (H_2O)	1.85	293.2	80.100
Xenon (Xe)	0	161.35	1.880

[a]Values selected from Lide, 2003, pp. **9**-45 to **9**-51, refer to isolated molecules in their electronic ground states. Units denoted by D are Debye (10^{-18} e.s.u. cm).
[b]Values selected from Lide, 2003, pp. **6**-155 to **6**-177.
[c]Sulfur atoms can chemically bond to one another to form cyclic aggregates and linear polymers [Meyer, 1965; Steudel, 1984].

neighbors surrounding each particle, specifically cubic symmetry. At least at low temperature, vibrational motions in a crystal which nominally break that symmetry have small amplitudes and thus have little net effect. The same negligible effect can be claimed for crystal defects, which normally have insignificant concentrations at low-temperature thermal equilibrium. As a result, the dielectric constant specified by the Clausius-Mossotti equation is substantially the same for cubic crystals as that after constant-volume steepest descent mapping of the system configuration onto the mechanically stable crystal inherent structure. However, the situation for the equilibrium liquid phase is basically different. Typical configurations encountered after melting have particles surrounded by a distribution of asymmetric arrangements of neighbors, a situation that persists in the inherent structures underlying those thermally occupied configurations. Unlike the situation for crystals, the particles in liquids are able to diffuse, thereby changing both premapped and postmapped configurations. Furthermore, it is by no means clear a priori that the premapped and postmapped dielectric constants for the liquid would be the same or nearly so. An especially vivid case is provided by ionic crystals such as those of the alkali halides with finite ε for which the Clausius-Mossotti equation provides a good description. Upon melting those crystalline substances, the resulting liquid is a conducting electrolyte for which the formal static dielectric constant ε has diverged to $+\infty$; see Section V.G. However, the localization of ions at amorphous inherent structures underlying the liquid prevents conductivity and restores ε to finite values.

Consider now the elementary case of a single-component liquid composed of N structureless but polarizable particles, e.g., the noble gas liquids. Let the instantaneous positions of those particles be denoted by $\mathbf{r}_1 \ldots \mathbf{r}_N$. If α is the polarizability of each particle, then the set of induced dipole moments $\mathbf{m}_1 \ldots \mathbf{m}_N$ when those particles are subject to a uniform (empty capacitor) external electric field $\mathbf{E}_0$ is determined by the following set of $3N$ linear equations [Kirkwood, 1936; Brown, 1950a]:

$$\mathbf{m}_i = \alpha \left[\mathbf{E}_0 - \sum_{j(\neq i)} \frac{1}{r_{ij}^3} \left(1 - \frac{3 \mathbf{r}_{ij} \mathbf{r}_{ij}}{r_{ij}^2} \right) \cdot \mathbf{m}_j \right], \qquad (1 \leq i \leq N). \tag{V.75}$$

The solution to this set of equations is proportional to $|\mathbf{E}_0| = E_0$, provided that those solutions exist. The corresponding change in potential energy for the collection of N polarizable particles, to be added to the field-free potential energy function $\Phi(\mathbf{r}_1 \ldots \mathbf{r}_N)$, is the following:

$$\Delta\Phi(\mathbf{r}_1 \ldots \mathbf{r}_N, \mathbf{E}_0) = -\frac{1}{2} \mathbf{E}_0 \cdot \sum_{i=1}^{N} \mathbf{m}_i. \tag{V.76}$$

This is a quantity proportional to E_0^2, the implication of which is that for the linear response defining ε for the nonpolar liquid, it suffices to use the field-free potential energy landscape Φ.

It is pointed out in Section IV.H that the Clausius-Mossotti relation (V.74) would formally predict a divergent ε if its right side containing density-polarizability products were to approach unity from below. This would arise from the interactions between the polarizable particles that reinforce one another so as to cause the individual induced dipoles all to become formally unbounded. Normally this is not the case because the density-polarizability products tend to be rather small. It needs to be recognized that a qualitatively similar situation applies to the configurationally more general situation described by Eq. (V.75). If the determinant of coefficients for those coupled linear equations approaches zero, all components of the solution set $\mathbf{m}_1 \ldots \mathbf{m}_N$ would diverge. But once again, this does not occur while the liquid remains in an insulating state. The repulsive forces acting between particles at small separations normally prevent such divergent configurations from occurring. If high pressures were to be applied to the system of particles forcing them into very close neighbor separation, the simple description of interactions between induced point dipoles with fixed scalar polarizabilities would require substantial modification, and extreme compression would convert the insulator to a metallic state exhibiting electronic conductivity ($\varepsilon \to +\infty$).

The full set of dipole–dipole interactions included in the coupled linear equations (V.75) represents a many-body phenomenon that cannot be resolved into independent pair interactions. Similarly, $\Delta\Phi$ in Eq. (V.76) cannot be resolved exactly into just pair potentials but instead depends in principle on the full details of the N-particle configuration. This implies that an exact accounting of the liquid-phase dielectric constant ε would at least require accessing configurational probability information carried in the full set of correlation functions $g_2, g_3, g_4, \ldots$, which according to the the remark following Eq. (V.76), should be those for the field-free liquid. However, specifying the occurrence probability for a set of n particles in the liquid does not directly determine the corresponding set of n induced dipoles $\mathbf{m}_i$ in the presence of the applied field because of the many-body character of the linear equation set (V.75).

In the case of a single-component nonpolar substance, it is possible to account for local fluctuation effects in dielectric fluids starting with a trivially transformed version of the Clausius-Mossotti equation that is only an approximation for liquids:

$$\left(\frac{\varepsilon + 2}{\varepsilon - 1}\right)\left(\frac{4\pi\rho\alpha}{3}\right) \approx 1, \tag{V.77}$$

the right side of which is then formally extended to a density expansion:

$$\left(\frac{\varepsilon + 2}{\varepsilon - 1}\right)\left(\frac{4\pi\rho\alpha}{3}\right) = 1 + b_1\rho\alpha + b_2(\rho\alpha)^2 + b_3(\rho\alpha)^3 + \dots. \tag{V.78}$$

The fact that a density series has been invoked does not imply that this form is appropriate only for dilute gases because all b_i for $i \geq 1$ vanish for dense cubic crystals. A systematic analysis [Brown, 1950a], extending earlier work [Yvon, 1937], shows that each b_n depends on the mean moment of a particle in the external field when $n - 1$ others are at specified locations in its neighborhood. Dielectric data for gas and liquid phases of several nonpolar dielectrics have been fitted well in the form shown in Eq. (V.78) but with its right side truncated into a quadratic or cubic polynomial in density [Brown, 1950b].

Steepest descent mapping at constant volume of a nonpolar liquid onto its relevant family of inherent structures would be expected to change its linear response to an external field. This would be determined by the corresponding configurational changes for the particle arrangements that affect the solutions of the coupled linear equations (V.75). However, the limitation to present knowledge does not permit one to draw general conclusions about how the quenched system dielectric constant $\varepsilon^{(q)}$ for nonpolar substances compares with the prequenched value ε.

Polar dielectrics present an even greater challenge for theoretical analysis. First, even in the absence of an externally applied field, the polarizing electrical interactions between molecules in the liquid cause the individual dipole moments to change from their isolated-molecule values. Second, the orientational correlations that exist between neighboring molecules can statistically produce either a reinforcement or a partial neutralization between those molecular dipole moments. Third, in contrast to the situation for nonpolar liquids described by Eq. (V.76), the change in potential energy for the system due to interaction with the external field $\mathbf{E}_0$ contains a contribution linear in that field:

$$\Delta\Phi(\mathbf{x}_1\dots\mathbf{x}_N, \mathbf{E}_0) = \mathbf{M}^{(0)}(\mathbf{x}_1\dots\mathbf{x}_N)\cdot\mathbf{E}_0 + O(E_0^2). \tag{V.79}$$

Here $\mathbf{x}_i$ stands for the set of configurational coordinates for molecule i, and $\mathbf{M}^{(0)}$ is the instantaneous net moment for the system in the absence of the external field, due to the vector sum of molecular moments:

$$\mathbf{M}^{(0)}(\mathbf{x}_1\dots\mathbf{x}_N) = \sum_{i=1}^{N}\mathbf{m}_i^{(0)}(\mathbf{x}_i). \tag{V.80}$$

An important descriptor of local order in polar dielectrics relevant to the second point here is the Kirkwood orientational correlation factor [Kirkwood, 1939], to be denoted here by its conventional symbol g_K (but not to be confused with a K-particle configurational correlation function). This refers to the correlation between the permanent moment $\mathbf{m}_i^{(0)}$ of a typical interior molecule and the net moment $\tilde{\mathbf{m}}_i^{(0)}$ in a local sphere of molecular size surrounding i and its locally coordinated neighbors. Then one has

$$\langle \mathbf{m}_i^{(0)} \cdot \tilde{\mathbf{m}}_i^{(0)} \rangle = g_K \langle \mathbf{m}_i^{(0)} \cdot \mathbf{m}_i^{(0)} \rangle \tag{V.81}$$

$$\equiv g_K \mu^2,$$

where, following frequently used convention, μ represents the root-mean-square value of the molecular moment in the field-free liquid:

$$\mu = \langle \mathbf{m}_i^{(0)} \cdot \mathbf{m}_i^{(0)} \rangle^{1/2}. \tag{V.82}$$

Although g_K must be non-negative, it can represent either positive correlation between local moments ($g_K > 1$) or anticorrelation ($1 > g_K \geq 0$). After constant-volume steepest descent mapping, it is natural to expect that the resulting quantity $g_K^{(q)}$ will reflect an amplified extent of local correlation or anticorrelation, implying

$$| g_K^{(q)} - 1 | > | g_K - 1 |. \tag{V.83}$$

A general theory of polar dielectrics would have to account at least for local orientational correlations between molecules, the change in magnitude and direction of molecular dipoles as a result of molecular interactions, and the generally nonisotropic polarizabilities of those molecules. Such a general exact theory does not exist at present. However, a qualitatively useful and often-quoted approximation due to Kirkwood leads to the following expression [Kirkwood, 1939]:

$$\frac{(\varepsilon - 1)(2\varepsilon + 1)}{3\varepsilon} = 4\pi\rho\left(\alpha + \frac{g_K \mu^2}{3k_B T} \right). \tag{V.84}$$

Here α stands for a directionally averaged molecular polarizability. The reader should consult available detailed reviews of dielectric theory for more elaborate and precise formulations [Brown, 1956; Hill, 1969].

Evidently, the local orientational order expressed by the Kirkwood factor g_K can exert a strong influence on the dielectric constant ε. The entry in Table V.4 for pure ("glacial") acetic acid at 293.3 K provides a good example. In spite of the fact that its isolated-molecule dipole moment is in the same range as for other polar substances that display considerably higher ε's, the acetic acid ε has the anomalously low value 6.20. In part at least, the explanation lies in the known tendency for pairs of acetic acid molecules in vapor and solution phases to form doubly hydrogen-bonded dimers [Togeas, 2005]. In their undeformed geometry, these dimers possess a twofold rotational symmetry that requires the two molecule dipole moments to cancel one another. This unperturbed dimer structure is illustrated in Figure V.4. Presuming that a substantial fraction of the molecules in liquid acetic acid is involved in dimers of this kind, though subject to some inevitable deformation, the effect would be to depress g_K below unity and, as the Kirkwood approximation Eq. (V.84) suggests, the measured ε in turn would be depressed. The presence of other hydrogen-bonded structures in the acetic acid liquid, such as larger cyclic clusters, might be significant [Briggs et al., 1991], but their electrostatic characteristics would have to be consistent with the low measured dielectric constant and low g_K. It is worth noting that the structure of crystalline acetic acid also shows each molecule engaged in two hydrogen bonds, but with two neighbors rather

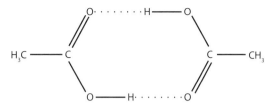

FIGURE V.4. Minimum energy structure of the acetic acid dimer. The two intermolecular hydrogen bonds are signified by dotted lines.

than just one in such a way as to form long chains [Jones and Templeton, 1958]. Evidently this alternative is chosen to permit low potential energy packing of molecules into the higher density of the crystal compared to that of the liquid.

The inclusion in the total potential energy function of a term linear in external field $\mathbf{E}_0$, Eq. (V.79), implies that the local minima that define the inherent structures under isochoric steepest descent shift configurationally by an amount proportional to $\mathbf{E}_0$. The shift direction is always to align the net system dipole along that field. The shift amount depends directly on the curvatures of the inhabited basins: The higher the curvature in the relevant direction, the lower the shift. Consequently, the electrical response via $\varepsilon^{(q)}$ for polar liquids if it were to be available would provide a measure of the multidimensional potential energy topography.

G. Electrolytic Liquids

Although the distinction is empirical rather than absolute, two classes of electrolytic liquids can be identified. In terms of common experience at ambient conditions, the more familiar class includes solutions of ionic substances in a substantially nonconducting solvent, often water. These solutions can span a wide range of ionic concentrations, depending on the choice of solute and solvent. The other class comprises molten salts, including pure ionic substances such as the alkali halides, heated above their usually high-temperature thermodynamic melting points, as well as liquid mixtures of those ionic substances. This section considers some basic properties of all electrolytic liquids, including the behavior of those properties under steepest descent mapping to inherent structures.

All cases to be covered can be described as consisting of $2 \leq v$ particulate species, including anionic, cationic, and neutral solvent particles. For $1 \leq \mu \leq v$, the respective number densities and electrostatic charges are denoted by ρ_μ and $Z_\mu e$. Overall electroneutrality can be assumed for a macroscopic sample, which obviously requires that

$$\sum_{\mu=1}^{v} Z_\mu \rho_\mu = 0. \tag{V.85}$$

Chemical tradition assigns charges to the species that are integer multiples of the fundamental proton charge e. However, there can be circumstances for which a detailed quantum mechanical description of the electron distribution surrounding the nuclei present in the system might require a more general description with Z_μ's deviating from standard integer values. For example, this might be the case with fused salts at high pressure, where electron clouds surrounding the ions are forced to overlap strongly, leading to possible transfer of some electronic charge among the nuclei. The analysis to be presented here includes that possibility, but in any case, electroneutrality condition (V.85) continues to be satisfied.

Just as with neutral molecules, ionic species can occur with chemically large and complex structures. However, for the purposes of the present analysis, the ionic species occurring in the electrolytes are assumed to have rigid shapes that include a center of symmetry. This restriction is trivially satisfied by monatomic ions, such as the alkali metal cations and the halide anions. But it also includes highly symmetric polyatomic ions that are structurally rigid so as to adhere closely to a symmetrical shape, such as the tetrahedral cases of the ammonium cation (NH_4^+), tetrafluoroborate anion (BF_4^-), and the sulfate anion (SO_4^{2-}). It also includes examples with octahedral symmetry such as the hexafluorophosphate anion (PF_6^-), the ferrocyanide anion ($Fe(CN)_6^{4-}$), and the ferricyanide anion ($Fe(CN)_6^{3-}$). Because of this restriction, it is natural to identify the center of symmetry as the formal location of the ion's charge. Position assignment for net charge is not straightforward in the cases of ions with low, or no, symmetry, and those subject to substantial structural flexibility. Consequently, these will not be considered explicitly here.

Because they are electrical conductors, liquid electrolytes at equilibrium macroscopically shield their interiors fully from any static electrical fields that might be applied by external sources. This is equivalent to a static macroscopic dielectric constant equal to $+\infty$. For present purposes, a more general description of the response is adopted. Specifically, it is useful to examine the linear response of the electrolytic liquid to a single Fourier component of an applied static electrical field with nonzero wave vector $\mathbf{k}$ and to obtain an expression for the corresponding dielectric response function $\varepsilon(k)$ [Stillinger and Lovett, 1968b]. Subsequently, it is of interest to examine the molecular-scale consequences of the long-wavelength limiting response requirement:

$$\lim_{k \to 0} [1/\varepsilon(k)] = 0. \tag{V.86}$$

Suppose that the macroscopic liquid electrolyte is contained in a large rectangular-solid volume V to which periodic boundary conditions apply, and suppose that $\mathbf{k} \neq 0$ is one of the natural wave vectors corresponding to that containment geometry. In the absence of the electrolyte, the Poisson equation verifies that an applied electrostatic potential of the form

$$\psi_{ap}(\mathbf{r}) = \psi_0 \sin(\mathbf{k} \cdot \mathbf{r}) \tag{V.87}$$

would be established in the empty volume V by the following hypothetical continuous charge distribution:

$$q_{ap}(\mathbf{r}) = (\psi_0 k^2 / 4\pi) \sin(\mathbf{k} \cdot \mathbf{r}). \tag{V.88}$$

With this continuous charge distribution already in place, if volume V were then to be filled with a liquid nonelectrolyte, the resulting mean electrostatic potential could be written thus:

$$\bar{\psi}(\mathbf{r}) = [\psi_0 / \varepsilon_0(k)] \sin(\mathbf{k} \cdot \mathbf{r}). \tag{V.89}$$

Here the presumption is that multiplier ψ_0 is sufficiently small that the response of the insulating dielectric liquid is a linear function of that parameter, with a k-dependent dielectric function $\varepsilon_0(k)$ that approaches a well-defined finite limit for that nonconducting liquid as $k \to 0$. If instead the volume V were filled with an electrolytic liquid, the corresponding expression would be written thus:

$$\bar{\psi}(\mathbf{r}) = [\psi_0/\varepsilon(k)] \sin(\mathbf{k} \cdot \mathbf{r}), \tag{V.90}$$

where now Eq. (V.86) is relevant.

It is important to recognize that if the ionic species in the liquid were to have their centers constrained to remain at their instantaneous positions, and then the sinusoidal electric field component were to be imposed, the resulting response would qualitatively be that of an insulating liquid. The average electrostatic charge density due to the ions would remain zero under this positional constraint. The corresponding dielectric response $\varepsilon_0(k)$ would involve the effects of electronic polarization of the ions. This is the appropriate description of the "dielectric medium" in which the ionic charges $Z_\alpha e$ find themselves, and to which they contribute. It is worth noting that experimental measurements of frequency-dependent dielectric response for various concentrated aqueous electrolytes have been interpreted as providing $\varepsilon_0(0)$ values for those solutions [Hasted et al., 1948].

After removing the constraint on their positions, the ions respond to the prevailing field so as to establish an average ionic charge density that is spatially sinusoidal. The amplitude of this average ionic charge density is such that its own electric field tends to oppose the applied field within the electrolyte. The average induced charge is a sum of charges for the individual species present and may be denoted by

$$q_{in}(\mathbf{r}, \mathbf{k}) = \sum_{\mu=1}^{v} q_{in,\mu}(\mathbf{r}, \mathbf{k}). \tag{V.91}$$

In combination with the externally applied charge density, this quantity by definition must obey the following identity that expresses the additional shielding caused by the ions:

$$q_{ap}(\mathbf{r}) + q_{in}(\mathbf{r}, \mathbf{k}) = q_{ap}(\mathbf{r})\varepsilon_0(k)/\varepsilon(k), \tag{V.92}$$

or equivalently

$$1/\varepsilon(k) = \{1 + [q_{in}(\mathbf{r}, \mathbf{k})/q_{ap}(\mathbf{r})]\}/\varepsilon_0(k). \tag{V.93}$$

In principle, the polarization process that determines $\varepsilon_0(k)$ for the ion-constrained situation depends upon the fine-grained structural details of the particulate medium and in its minute details is thus locally a very complicated phenomenon for liquid media. But fortunately, the small-k (large-wavelength) regime of interest allows use of a locally coarse-grained description of the applied electric field without incurring substantial error. This implies that the linear perturbations of the species densities ρ_μ can be calculated as arising from an effective single-particle potential:

$$\varphi_{eff,\mu}(\mathbf{r}) = [\psi_0 Z_\mu e/\varepsilon_0(k)] \sin(\mathbf{k} \cdot \mathbf{r}). \tag{V.94}$$

As a consequence, the change in number densities for species can be expressed in terms of the equilibrium pair correlation fuctions $g_{\mu\varsigma}(r)$ in the absence of the applied electric field. This leads to the following expressions for the induced charge densities:

$$q_{in,\mu}(\mathbf{r}, \mathbf{k}) = [\psi_0 e^2/k_B T \varepsilon_0(k)] \sin(\mathbf{k} \cdot \mathbf{r})[Z_\mu^2 \rho_\mu + Z_\mu \rho_\mu \sum_{\varsigma=1}^{v} Z_\varsigma \rho_\varsigma \int_V g_{\mu\varsigma}(s) \cos(\mathbf{k} \cdot \mathbf{s}) d\mathbf{s}]. \tag{V.95}$$

As a result, it becomes possible to transform Eq. (V.93), via Eqs. (V.91) and (V.92), to

$$1/\varepsilon(k) = [1/\varepsilon_0(k)]\left\{1 - \frac{4\pi e^2}{k_B T \varepsilon_0(k)k^2}\left[\sum_{\mu=1}^{v} Z_\mu^2 \rho_\mu + \sum_{\mu,\zeta=1}^{v} Z_\mu Z_\zeta \rho_\mu \rho_\zeta \int g_{\mu\zeta}(s)\,\frac{\sin(ks)}{ks}\,d\mathbf{s}\right]\right\}. \qquad (V.96)$$

Subsequently, Eq. (V.86) leads to the requirement

$$0 = \lim_{k\to 0}\left\{1 - \frac{4\pi e^2}{k_B T \varepsilon_0(k)k^2}\left[\sum_{\mu=1}^{v} Z_\mu^2 \rho_\mu + \sum_{\mu,\zeta=1}^{v} Z_\mu Z_\zeta \rho_\mu \rho_\zeta \int g_{\mu\zeta}(s)\,\frac{\sin(ks)}{ks}\,d\mathbf{s}\right]\right\}. \qquad (V.97)$$

The quantity enclosed in square brackets in Eq. (V.97), [...], formally possesses an expansion in even non-negative powers of k. In order for Eq. (V.97) to be satisfied, the $O(k^0)$ contribution must vanish and the $O(k^2)$ contribution must have a value to cause cancellation of the leading unity term in the full expression {...}. The first of these requirements follows from the full electrostatic shielding that any ion experiences because of the diffuse charge distribution that its neighbors provide at equilibrium:

$$-Z_\mu e = \int\left[\sum_{\zeta=1}^{v} Z_\zeta e \rho_\zeta g_{\mu\zeta}(s)\right]d\mathbf{s} \qquad (V.98)$$

$$= \sum_{\zeta=1}^{v} Z_\zeta e \rho_\zeta \int [g_{\mu\zeta}(s) - 1]d\mathbf{s},$$

where the second form follows from the overall electroneutrality condition Eq. (V.85) in the infinite system limit. This local electroneutrality condition Eq. (V.98) is another consequence of the electrical conductivity of the electrolyte, and has been recognized and used in electrolyte theory since its early pioneering work [Debye and Hückel, 1923].

The second requirement hinges upon the second term in the elementary expansion:

$$\frac{\sin(ks)}{ks} = 1 - \frac{(ks)^2}{6} + \frac{(ks)^4}{120} - \dots . \qquad (V.99)$$

It leads to the following condition that must be satisfied by the equilibrium pair correlation functions in the electrolyte [Stillinger and Lovett, 1968b]:

$$-\frac{6}{\kappa^2} = \left(\sum_{\mu=1}^{v} Z_\mu^2 \rho_\mu\right)^{-1}\int\left[\sum_{\mu,\zeta=1}^{v} Z_\mu Z_\zeta \rho_\mu \rho_\zeta g_{\mu\zeta}(s)\right]s^2 d\mathbf{s} \qquad (V.100)$$

$$= \left(\sum_{\mu=1}^{v} Z_\mu^2 \rho_\mu\right)^{-1}\sum_{\mu,\zeta=1}^{v} Z_\mu Z_\zeta \rho_\mu \rho_\zeta \int [g_{\mu\zeta}(s) - 1]s^2 d\mathbf{s},$$

where the second form again has been obtained using the overall electroneutrality condition Eq. (V.85) in the infinite system limit. Here κ is the Debye inverse length defined by

$$\kappa^2 = \frac{4\pi e^2}{k_B T \varepsilon_0(0)}\sum_{\mu=1}^{v} Z_\mu^2 \rho_\mu. \qquad (V.101)$$

Equation (V.100) is a single constraint involving a second spatial moment of a linear combination of all pair correlation functions for pairs of ionic species. In this respect, it differs from the set of zeroth spatial moments in Eq. (V.98), one for each ionic species.

In its linearized version for point ions, the Debye-Hückel theory for electrolytes assigns the following simple form to the ion pair correlation functions ($r > 0$):

$$g_{\alpha\gamma}(r) \cong 1 - \left[\frac{Z_\alpha Z_\gamma e^2}{k_B T \varepsilon_0(0)} \right] \frac{\exp(-\kappa r)}{r}. \qquad (V.102)$$

By substitution, one easily verifies that this approximation satisfies the electroneutrality conditions (V.98) as well as the single second moment condition (V.100). However, it is valid only in the low ion-concentration limit, where the average charge densities surrounding each ion are spatially widely extended and monotonically declining with increasing distance. At higher concentrations, these charge densities are more compact but must reflect short-range ion–ion repulsions, and in order to satisfy the second moment condition may be forced to become nonmonotonic functions of r [Stillinger, 1968]. Such nonmonotonic behavior (i.e., charge oscillation with increasing radial distance) should be especially relevant for fused salts at and just above their freezing points. Box V.1 presents a simple model electrolyte consisting of electrostatically charged hard spheres to which these moment conditions apply.

The constant-volume steepest descent mapping process applied to electrolytic liquids replaces the equilibrium pair correlation functions $g_{\mu\zeta}(r)$ with corresponding "quenched" functions $g_{\mu\zeta}^{(q)}(r)$, representing averages over the relevant distribution of inherent structures. As usual, one expects that the removal of intrabasin vibrational displacements by steepest descent mapping should produce an enhanced image of short-range order in each of the pair correlation functions. But additional questions can be asked about the fate under steepest descent mapping of the zeroth and second moment conditions (V.98) and (V.100) that have no analogs for nonelectrolytes. Each of the zeroth moment conditions (V.98) simply specifies the magnitude of the total charge in the diffuse ion atmosphere surrounding an ion of species μ, without constraining the shape of its spatial distribution. That spatial distribution is modified by the steepest descent mapping, but the net charge it contains is conserved at the stated value $-Z_\mu e$. However, the single second moment condition (V.100) is indeed sensitive to the spatial distribution and is not a statement of a conserved quantity. By contrast with the situation for the zeroth moment conditions, the second moment condition does not remain valid under the mapping to inherent structures. An alternative statement is that the left side of Eq. (V.100), $-6/\kappa^2$, does not have a straightforward replacement after the mapping to inherent structures is applied.

The analysis above leading to the zeroth and second moment conditions has assumed that the electrolytic fluid was present as a single phase. However, if the thermodynamic state involved coexistence of a concentrated liquid and a dilute vapor, both of which are electrolytes, the conclusions need modification. On one side, the zeroth moment conditions in Eq. (V.98) would continue to hold, provided that the correct coexistence-state pair correlation functions were utilized. This follows from the fact that those are merely charge conservation conditions insensitive to details of the local ion atmospheres surrounding each ion species, those conditions are valid in both phases, and the required average over the entire system with coexisting phases does not violate those conservation conditions. On the other hand, the single second moment condition in Eq. (V.100) is an explicit statement concerning the spatial distributions of electrostatic charge within ion atmospheres, and those distributions differ between the phases. Even if proper pair correlation functions for coexistence were to be inserted into Eq. (V.100), there is no reason to

BOX V.1. Restricted Primitive Model Electrolyte

In order to focus on the statistical mechanical effects of the long-ranged Coulomb interactions in ionic systems, theoretical and computational effort has frequently been directed to the "restricted primitive model" electrolyte. This classical model possesses basic anion–cation symmetry in that both ion types are uniformly charged hard spheres, all with the same diameter σ_0, present in equal numbers N in a volume V. The anions and cations respectively bear charges $-Ze$ and $+Ze$. It is furthermore postulated that the ions are suspended in a structureless medium with fixed (k-independent) dielectric constant ε_0 and that each spherical ion also behaves dielectrically as though it were composed of the same uniform dielectric constant ε_0. Given these assumptions, the potential energy for all configurations of nonoverlapping ions is simply a sum of coulombic pair interactions renormalized by the dielectric constant

$$\Phi(\mathbf{r}_1 \ldots \mathbf{r}_{2N}) = [(Ze)^2/\varepsilon_0] \left\{ \sum_{i=1}^{N-1} \sum_{j=i+1}^{N} r_{ij}^{-1} - \sum_{i=1}^{N} \sum_{j=N+1}^{2N} r_{ij}^{-1} + \sum_{i=N+1}^{2N-1} \sum_{j=i+1}^{2N} r_{ij}^{-1} \right\}$$

$$(\text{all } r_{ij} \geq \sigma_0)$$

Here the numbering scheme covers all N ions of one charge first, followed by the same number of those of opposite charge.

This model's intrinsic symmetry reduces the number of distinct ion–ion pair correlation functions from three to two as a result of the obvious identity

$$g_{++}(r) \equiv g_{--}(r).$$

As a consequence, the second-moment condition reduces to the following form:

$$-\frac{6}{\kappa^2} = \frac{4\pi N}{V} \int_0^\infty r^4 [g_{++}(r) - g_{+-}(r)] dr.$$

For this model, the Debye inverse-length parameter κ is determined by the expression

$$\kappa^2 = \frac{8\pi (Ze)^2 N}{\varepsilon_0 k_B T V}.$$

At low pressure, the crystal state exhibited by the restricted primitive model has the cesium chloride (CsCl) structure, in which the spherical cores of the ions are arranged in alternating fashion on the sites of a body-centered cubic lattice [Stillinger and Lovett, 1968a, Figure 1]. Upon melting, the model produces a concentrated electrolytic fluid that possesses a vapor pressure, i.e., it displays a phase coexistence with a dilute vapor. Monte Carlo computer simulation shows that upon raising the temperature these coexisting electrolytic fluids approach one another in density and merge at a critical point, with numerical evidence strongly suggesting that the critical exponents fall into the Ising model universality class [Luijten et al., 2002].

A nominal application of the steepest descent mapping to the potential function Φ would be strongly dominated by the net cohesive property of the Coulomb interactions. The end result would be inherent structures with the charged hard spheres in contact with one another. This would be inconsistent with harmonic vibrational motions near the basin bottoms, in contrast to more realistic potentials. This simple interpretation assumes that the continuum dielectric medium housing the ions is an entirely passive parameter, thus not acting as a realistic particulate solvent would under steepest descent mapping. However, it may be worth noting in passing that the continuum dielectric medium could be treated as an incompressible fluid whose viscosity increases to infinity at an intermediate stage of that mapping. The result would be that many ions would experience descent arrest before contacts with other ions occurred, much as would happen with a more realistic particulate solvent.

believe that the result would equal $-6/\kappa^2$, where κ was computed using the systemwide average number densities for all species as well as a spatially averaged $\varepsilon_0(0)$.

This leads to the question of whether raising the temperature within the coexistence region causes the second moment condition Eq. (V.100) to return to validity at the electrolyte's critical point. A Monte Carlo numerical study of the restricted primitive model electrolyte concluded that at that model's critical point the second moment condition remained violated, though its validity appeared to be intact elsewhere within the single-phase region, perhaps even arbitrarily close to the critical point [Das et al., 2011, 2012]. This would be consistent with the fact that in the infinite system limit a system precisely at its critical point displays divergent responses to external perturbations. That is, it does not have well-defined finite linear responses to all perturbations as has been presumed to be the case in the derivation of the second moment condition.

H. Liquid Interfaces

Two classes of liquid interfaces are considered in this section. One involves the surface of a liquid in equilibrium with its own vapor. The other concerns the surface of contact between two immiscible liquids. Both are amenable to description in terms of inherent structures and their multidimensional basins [Stillinger, 2008]. The principal objectives are description of the matter distribution across the interface, the surface tension γ and related thermodynamic properties, and how these attributes vary with temperature and external fields.

The earliest statistical thermodynamic description of liquid interfaces originated with van der Waals [Rowlinson and Widom, 1982, Chapter 3]. The van der Waals approach introduced a molecular interaction length scale into the mathematical description, which had the effect of predicting a smooth monotonic density (or concentration) profile in the direction normal to the interface. The width of that intrinsic profile diverged as the temperature was raised to the critical point of a pure liquid, or in the case of immiscible liquids as it was adjusted to the upper or lower consolute point. The predicted surface tensions vanished continuously upon approach to the critical or consolute point, in qualitative agreement with experiments. These principal characteristics have survived the construction of more modern variants of the van der Waals interface theory [Rowlinson and Widom, 1982].

It has been suggested that liquid interfaces should exhibit effects from the presence of thermally excited "capillary waves" and that these collective surface density fluctuations would affect light scattering from those interfaces [Mandelstam, 1913; Frenkel, 1955, Chapter VI]. Presuming that this phenomenological point of view has physical validity, capillary waves should be included in a complete description of the statistical thermodynamics of liquid interfaces. They would act to broaden the equilibrium interface density profile, and in particular they would be sensitive to the gravitational field strength g such that the density profile should diverge (weakly) as that gravitational field strength goes to zero [Buff et al., 1965]. Assuming that capillary waves on the surface with two-dimensional wave vectors lesser in magnitude that an upper cutoff k_u can be treated as independent classical degrees of freedom, and assuming that each of these has an associated energy resulting from work against both surface tension and gravity, the theory deduces that the mean-square vertical displacement of an otherwise unexcited macroscopic liquid–vapor interface would be

$$\langle (\delta z)^2 \rangle = \left(\frac{k_B T}{4\pi\gamma_0} \right) \ln\left[1 + \frac{k_u{}^2\gamma_0}{mg(\rho^{(l)} - \rho^{(v)})} \right]. \tag{V.103}$$

This result explicitly displays a dependence on a "bare" surface tension γ_0 (i.e., that in the absence of capillary waves), as well as on the number densities $\rho^{(l)}$ and $\rho^{(v)}$ of the coexisting liquid and vapor phases, respectively. The main significance of this result is that it diverges logarithmically as $g \to 0$, principally because of the contribution of long-wavelength surface waves, and these are just those for which the independence assumption is most justified. The result at fixed $g > 0$ also diverges upon approach to the critical point at least in part because of the vanishing density difference between the liquid and vapor (perhaps modified somewhat by a weaker vanishing of γ_0), thus qualitatively reinforcing the critical-point width divergence of the van der Waals formalism. It should be stressed that the existence of these divergences is not sensitive to the specific choice for the ill-defined cutoff wave vector k_u. Nevertheless, on physical grounds it might be reasonable to suppose that k_u corresponds to a wavelength approximately equal to the nearest neighbor separation in the liquid phase.

Although Eq. (V.103) treats the capillary waves as classical modes, the formalism can be adapted to incorporate quantum effects [Cole, 1970, 1980]. The logarithmic interface width divergence as $g \to 0$ persists under this extension. These quantum effects are especially relevant for liquid interfaces of the helium isotopes, Chapter VIII.

X-ray scattering experiments have been applied to the examination of planar liquid–vapor interfaces but thus far only in the terrestrial gravitational field. Water [Schwartz et al., 1990] and ethanol [Sanyal et al., 1991] are examples of insulating liquids that have been studied. In both of these cases, the researchers concluded that the capillary wave model for surface roughness provided a satisfactory description of the experimental observations. Furthermore, computer simulation suggests that capillary waves play a significant role in the interaction of nanoparticles with liquid surfaces [Cheung and Bon, 2009].

Theories in the van der Waals tradition do not explicitly incorporate capillary wave fluctuations, and so do not predict the diverging interface width as $g \to 0$. There have been previous attempts to combine the van der Waals and capillary wave aspects of the interface theory, but these have proceeded only at the expense of introducing arbitrary elements into the formalism, such as dimensions of descriptive coarse-graining cells [Weeks, 1977; Sedlmeier et al., 2009] or of a percolation process connectivity length [Stillinger, 1982; Chacón and Tarazona, 2003]. The objective of this section is to determine the extent to which the inherent-structure representation is applicable to the family of interface problems and to see how it might help to reconcile the van der Waals and the capillary wave aspects of the interface description without invoking artificial or arbitrary concepts.

The analysis to follow is restricted to planar interfaces, although extension of the strategy to curved interfaces is possible. Initially, the formalism is developed for the single-component liquid–vapor case, in which the constituent particles are structureless. A mechanism must be present in order to localize the interface, and the gravitational field plays that role. The macroscopic system resides in a rectangular box with volume $V = L_x L_y L_z$, horizontal cross-sectional area $A = L_x L_y$, and with impenetrable walls acting as "floor" and "ceiling." Periodic boundary conditions are applicable in the two lateral directions (x, y). With inclusion of the coupling to the gravitational field, the potential energy Φ_g of this closed system in a uniform gravitational field takes the form

$$\Phi_g(\mathbf{r}_1 \ldots \mathbf{r}_N) = \Phi(\mathbf{r}_1 \ldots \mathbf{r}_N) + mg \sum_{i=1}^{N} z_i. \qquad (\text{V.104})$$

Here Φ as usual includes the intermolecular interactions but now includes the top and bottom molecule-wall interactions as well. The mass-m particles are located at positions $\mathbf{r}_1 \ldots \mathbf{r}_N$. The z coordinate measures vertical position, and $g \geq 0$ is the gravitational coupling strength (or its centrifugal equivalent) for the environment of interest. The fixed number N of particles should be such that in the volume available to the system, an equilibrium liquid–vapor coexistence obtains over some nonzero temperature range.

For ordinary terrestrial applications, the gravitational acceleration g is approximately 980 cm s^{-2}. But under other circumstances, this quantity can vary widely. Contrasting but attainable examples include $g = 0$ in an orbiting space vehicle, and $g \approx 0.1$ cm s^{-2} at the surface of a small asteroid whose diameter is in the kilometer range. Furthermore, g equal (effectively) to 10^5 to 10^6 times its terrestrial-surface strength can be attained in ultracentrifuges [Svedberg and Pedersen, 1940]. Astronomy supplies the extreme case of neutron stars, at the surfaces of which the gravitational fields lie in the range of 10^{11} times that at the Earth's surface [Shapiro and Teukolsky, 1983; Baym and Lamb, 2005]. The present approach in principle can accommodate any nonzero g value and should ultimately be able to address the behavior of interfaces as g varies over the entire available range. Box V.2 provides rough estimates of the gravitational field effect on the water interface at room temperature.

A basic concept for interface theory has been that of the Gibbs equimolar dividing surface [Gibbs, 1906]. Conventionally, for the planar interface under consideration, this involves comparing the actual number density $\rho_1(z)$ as a function of height z, with the ideal density function $\rho_{1,0}(z)$ that switches discontinuously from a constant value equal to the thermodynamic coexistence liquid density $\rho^{(l)}$, to a constant value equal to the corresponding coexistence vapor density $\rho^{(v)}$ as altitude z increases:

$$\rho_{1,0}(z \,|\, z_G) = \rho^{(l)} \qquad (z \leq z_G),$$

$$= \rho^{(v)} \qquad (z_G < z). \qquad (\text{V.105})$$

Here z_G denotes the altitude of the horizontal Gibbs dividing surface. Its value is uniquely determined in this traditional approach by the condition that both $\rho_1(z)$ and $\rho_{1,0}(z \,|\, z_G)$ represent exactly the same number N of particles in the system. That is, one requires

$$\int_{z_f}^{z_c} [\rho_1(z) - \rho_{1,0}(z \,|\, z_G)] dz = 0, \qquad (\text{V.106})$$

BOX V.2. Estimate of Capillary Wave Broadening vs. Gravitational Field

---

In order to provide a rough indication of how gravitational field strength affects interface broadening as viewed from the capillary wave picture, Eq. (V.103) may be used to evaluate the root mean square average vertical displacement, inserting values specifically for water at room temperature and pressure. In particular, choose $m\rho^{(l)} \approx 1.0$ g cm^{-3}, $m\rho^{(v)} \approx 0$ g cm^{-3}, $\gamma_0 \approx 73$ dyne cm^{-1}, and $k_u \approx 20$ nm^{-1}. These values lead to the following set of results for $T = 293.2$ K that span the relevant $g > 0$ range indicated in the text:

Location	g, cm s^{-2}	$\langle(\delta z)^2\rangle^{1/2}$, nm
Small asteroid	0.1	0.445
Earth's surface	980	0.397
Ultracentrifuge	10^8	0.326
Ultracentrifuge	10^9	0.310
Neutron star	10^{14}	0.213

This illustrates the very weak but monotonically decreasing g dependence that is a characteristic of the simple capillary wave picture.

---

where the integral spans the macroscopic height of the system between the "floor" at z_f and the "ceiling" at z_c. The idealized matter distribution represented by $\rho_{1,0}(z\,|\,z_G)$, if it were locally an accurate indicator of thermodynamic properties in the system, would imply values for free energy, average energy, and entropy that are determined simply by the amounts of matter below and above z_G; in particular, this involves no surface tension (i.e., surface free energy) per se. The actual values of these thermodynamic quantities differ, of course, because of the presence of the true interface with nonzero-range particle interactions and correlations. The surface tension $\gamma(T)$ is the Helmholtz free energy excess per unit area of interface, and the excess values per unit area of energy (E_s) and entropy (S_s) can be obtained from $\gamma(T)$ by differentiation [Rowlinson and Widom, 1982, Chapter 2]:

$$E_s(T) = d(\gamma/T)/d(1/T), \tag{V.107}$$

$$S_s(T) = -d\gamma/dT.$$

While the piecewise-constant profile $\rho_{1,0}(z\,|\,z_G)$ may be appropriate for systems subject to small external field strengths g, it clearly becomes inappropriate for large field strengths. In that latter circumstance, hydrostatic pressure increases from top to bottom of the system through both the compressible vapor and liquid phases, creating height-dependent density changes in each bulk phase that are not directly related to the presence of the planar interface. Such hydrostatic effects become particularly important in the neighborhood of the critical temperature. This circumstance warrants replacing $\rho_{1,0}(z\,|\,z_G)$ with a generalization $\rho_{1,g}(z\,|\,z_G)$ that is applicable for arbitrary field strength g. This generalized form should also possess a discontinuity and should have the property

$$\lim_{g\to 0} \rho_{1,g}(z\,|\,z_G) = \rho_{1,0}(z\,|\,z_G). \tag{V.108}$$

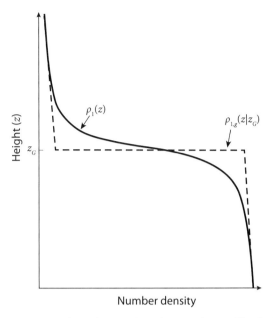

FIGURE V.5. Schematic diagram showing the qualitative relation between the actual height-dependent density function ρ_1 (solid curve) and the ideal density distribution $\rho_{1,g}$ (dashed curve) generated by the barometer formula, Eq. (V.109).

The natural definition of $\rho_{1,g}(z\,|\,z_G)$ utilizes a thermodynamic "barometer formula" [Rowlinson and Widom, 1982, p. 73]:

$$\mu[\rho_{1,g}(z\,|\,z_G),\ T] + mgz = C(N). \tag{V.109}$$

The chemical potential function for the field-free system has been denoted by $\mu(\rho,\ T)$, and the constant $C(N)$ must be chosen so that the solution to Eq. (V.109) represents the given number of particles in the closed system:

$$A\int_{z_f}^{z_c}\rho_{1,g}(z\,|\,z_G)dz = N, \tag{V.110}$$

where the surface area is

$$A = L_x L_y = V/(z_c - z_f). \tag{V.111}$$

This condition in Eq. (V.110) automatically fixes the height z_G of the generalized Gibbs dividing surface. Figure V.5 schematically indicates how the actual density profile and the ideal version defined by the barometer formula, Eq. (V.109), conform to one another, except in the immediate vicinity of the Gibbs dividing surface at z_G. The vertical position z_G generally is temperature dependent.

Once the Gibbs dividing surface has been located, one can formally define the numbers of particles "belonging" to the liquid and vapor phases. Denote these by $N^{(l)}$ and $N^{(v)} \equiv N - N^{(l)}$, respectively. They are the following:

$$N^{(l)} = A\int_{z_f}^{z_G}\rho_1(z)dz, \tag{V.112}$$

$$N^{(v)} = A\int_{z_G}^{z_c}\rho_1(z)dz.$$

In the strict mathematical sense implied by these definitions, $N^{(l)}$ and $N^{(v)}$ may not be integers because the density profile represents a canonical average for the prevailing temperature and includes kinetic excursions of particles across the Gibbs dividing surface. However, in the large-system limit that is primarily of interest here, these two numbers may be replaced by their nearest integers without incurring any significant error. Because of the continuous and smooth nature of the actual density profile compared to that generated by Eq. (V.109), the liquid-phase region just below z_G experiences a local matter deficit, whereas the vapor phase region just above z_G experiences a local surplus, as illustrated in Figure V.5. Consequently, one expects the following inequalities to be satisfied, compared with Eqs. (V.112):

$$N^{(l)} < A\int_{z_f}^{z_G} \rho_{1,g}(z\,|\,z_G)dz, \tag{V.113}$$

$$N^{(v)} > A\int_{z_G}^{z_c} \rho_{1,g}(z\,|\,z_G)dz.$$

The strategy just outlined for the arbitrary-g circumstance requires a shift in the definition of surface excess quantities γ, E_s, and S_s. The "ideal" values of thermodynamic quantities now locally follow the barometric formula profile $\rho_{1,g}(z\,|\,z_G)$ rather than the simple piecewise-constant $\rho_{1,0}(z\,|\,z_G)$. However, with that generalization implemented, the surface tension continues to be excess free energy, and the surface excess energy E_s and entropy S_s continue to be given by Eqs. (V.107).

Two technical matters deserve at least brief mention at this point. In general, the impenetrable walls that form the floor and ceiling produce their own interfaces and so can generate static local density perturbations in $\rho_1(z)$ just above z_f and just below z_c, respectively, that would not be represented in a barometric formula function $\rho_{1,g}(z\,|\,z_G)$. This distinction would have the effect of displacing the Gibbs dividing surface due to contributions that are fundamentally irrelevant to analysis of the liquid–vapor interface. We shall take the point of view that appropriately chosen wall forces at the floor and ceiling boundaries can eliminate the density perturbations at those lower and upper locations. Alternatively, one could quantitatively document those floor and ceiling location density variations and modify the z_G criterion Eq. (V.112) accordingly so as to eliminate those irrelevant contributions.

The second technical matter is the following. If the gravitational field strength g is large, and/or the container depth $z_c - z_f$ is large, the hydrostatic pressure near the bottom of the container could be sufficiently large that it might produce a crystalline layer below the liquid. That would entail a second interface and would appear as another discontinuity in $\rho_{1,g}$. It will henceforth be assumed that $z_c - z_f$ can be chosen small enough to avoid this complication, while still large enough so that the liquid–vapor interface structure is not subject to perturbation by the nearby presence of container "floor" or "ceiling."

The next step is to examine the effect of steepest descent mapping on the canonical distribution of particle configurations, holding the system volume fixed. For the current interface application, the mapping is generated by the $3N$ coupled first-order differential equations that are summarized by the equation

$$d\mathbf{R}(s)/ds = -\nabla_{\mathbf{R}}\Phi_g(\mathbf{R}) \qquad (0 \le s), \tag{V.114}$$

$$\mathbf{R} \equiv \mathbf{r}_1 \oplus \mathbf{r}_2 \oplus \ldots \oplus \mathbf{r}_N.$$

BOX V.3. Competing Forces

In order to provide a context for the steepest descent mapping in the presence of gravity, it may be instructive to compare typical magnitudes of intermolecular and gravitational forces. Specifically, one can ask: At what distance between a pair of particles does the magnitude of the intermolecular attractive force equal the magnitude of the gravitational force? Consider the example of two argon atoms, and suppose that their interaction can be approximated by the Lennard–Jones pair potential:

$$v(r) = 4\varepsilon_0[(\sigma_0/r)^{12} - (\sigma_0/r)^6].$$

Only the second term is relevant for this estimate and leads to the equal-force-magnitude condition:

$$r = \sigma_0(24\varepsilon_0/mg\sigma_0)^{1/7}.$$

For argon atoms, the various quantities in this last expression can be assigned the following values (assuming terrestrial gravity):

$$\varepsilon_0 = 1.654 \times 10^{-14} \text{ erg}$$

$$\sigma_0 = 3.405 \times 10^{-8} \text{ cm}$$

$$m = 6.6336 \times 10^{-23} \text{ g}$$

$$g = 980 \text{ cm s}^{-2}$$

Substituting these values into the formula leads to the result

$$r = 370 \times 10^{-8} \text{ cm.}$$

This distance, when interpreted as a mean separation between near-neighbor particles, corresponds to argon at roughly one-millionth of its liquid density at the triple point. For most applications, however, the vapor phase is likely to be considerably denser, so that it would be reasonable to assume that intermolecular forces dominate the steepest descent mapping operation in comparison with gravity. Of course, this qualitative conclusion requires revision for cases involving extreme gravitational (or centrifugal) forces.

As usual, inherent structures are identified as the distinct set of configurations $\mathbf{R}(s \rightarrow +\infty)$ that emerge from this operation, using configurations selected from the continuous canonical distribution as initial conditions for Eq. (V.114). The displacements of individual particles as virtual time s increases depend on the forces experienced by those particles, and in the present case those forces are a combination of intermolecular, wall, and gravitational components. As a general rule, the inherent structures continue to display a vertical separation between liquid and vapor portions. The mechanically stable particle arrangements in the mapped vapor portion may have a very porous architecture (described in detail in Section VII.B) but subject to some degree of downward crushing that depends on the magnitude of the field strength g. The denser and less compressible liquid usually exhibits the same effect but to a lesser extent. Box V.3 offers a specific comparison of competing intermolecular and gravitational effects.

Because of the periodic boundary conditions that have been applied in the lateral (x, y) directions, the inherent structures must themselves exhibit lateral translational symmetry, presuming that the floor and ceiling wall interactions do. This means that each inherent structure is a connected two-dimensional manifold in the $3N$-dimensional configuration space. Alternatively, if hard vertical walls had been imposed at the sides of the system, the inherent structures would reduce to single points in that configuration space. However, the interface then would show a perturbed structure near those vertical walls.

As a result of the steepest descent mapping, one must expect the relative volumes of the liquid and vapor phases to change. In particular, the mapping eliminates anharmonic vibrational effects that normally contribute to thermal expansion of the liquid, and so removing those effects would change the volume of the liquid by an amount on the order of $[N^{(l)}]^{1/3}$. If it were disconnected from the liquid, the vapor would be subject to the same phenomenon, but much more weakly because of its lower density. Consequently, in the mapping of the coexisting pair of phases, the liquid would tend to dominate movement of the interface. Regardless of whether that movement was to greater or to lesser height, it is possible to identify uniquely a displaced Gibbs dividing surface position $z_G^{(q)}$ after mapping, such that the average numbers of particles $N^{(l)}$, $N^{(v)}$, respectively, below and above this displaced surface are the same as previously. Denoting by $\rho_1^{(q)}(z)$ the density profile after applying the steepest descent mapping, one requires $z_G^{(q)}$ to satisfy the following equivalent conditions:

$$N^{(l)} = A \int_{z_f}^{z_G^{(q)}} \rho_1^{(q)}(z) dz, \tag{V.115}$$

$$N^{(v)} = A \int_{z_G^{(q)}}^{z_c} \rho_1^{(q)}(z) dz.$$

The postmapping density profile $\rho_1^{(q)}(z)$ in the large-system limit is expected to be continuous and smooth across the displaced Gibbs surface at $z_G^{(q)}$ and in that respect would be similar to the behavior of $\rho_1(z)$ where it crosses the undisplaced Gibbs surface at z_G, as indicated in Figure V.5. In addition to the vertical displacement shift of the interfacial zone, one should expect some interface-region shape change in passing from $\rho_1(z)$ to $\rho_1^{(q)}(z)$.

The mapped density profile $\rho_1^{(q)}(z)$ contains contributions from the entire set of inherent structures that are produced from the initial canonical distribution of particle configurations. Let α index the distinct inherent structures that are involved, and let P_α stand for the occurrence probability of inherent-structure type α (lumping together all permutation-equivalent cases) in the canonical distribution. Then one can write

$$\rho_1^{(q)}(z) = \sum_\alpha P_\alpha \rho_{1,\alpha}^{(q)}(z), \tag{V.116}$$

$$\sum_\alpha P_\alpha = 1,$$

thereby resolving the mapped density profile into its separate contributions from each of the distinct inherent-structure types.

The separate inherent-structure contributions differ from one another in the vicinity of $z_g^{(q)}$. In particular, the numbers of particles that they present below and above the displaced Gibbs

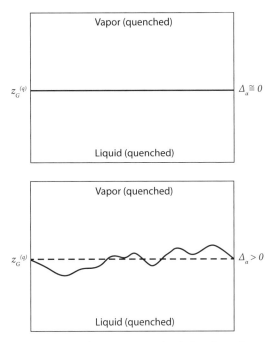

FIGURE V.6. Geometric distinction between inherent structures for the liquid interface situation. The highly simplified schematic diagram at the top represents mesoscopically an inherent structure whose density change from the lower liquid zone to the upper vapor zone is confined at all lateral positions to the immediate neighborhood of the planar dividing surface at $z_G^{(q)}$, and for which the number variance Δ_α is therefore small. By contrast, the schematic diagram at the bottom refers to a rough contour of density transition, with an expected relatively large positive Δ_α.

surface almost always deviate from the numbers $N^{(l)}$ and $N^{(v)}$. For a given α, the actual numbers can be written as $N^{(l)} - \Delta_\alpha$ and $N^{(v)} + \Delta_\alpha$. To a substantial extent, these deviations can be attributed to surface roughness between the mapped phases, lifting dense "liquid" from below to above z_G and filling the resulting void with more dilute mapped vapor. This is illustrated qualitatively in Figure V.6.

The smallest absolute value of the number-deviation parameter, $|\Delta_\alpha|$, tends to occur for those inherent structures that align their sudden density drop-off most closely with z_G across the entire planar interface. Denote this set of distinct inherent structures by Λ. It is then natural to consider the interface structure just for this restrictive set:

$$\rho_{1,\Lambda}{}^{(q)}(z) = \sum_{\alpha\in\Lambda} P_\alpha \rho_{1,\alpha}{}^{(q)}(z) \Big/ \Big[\sum_{\alpha\in\Lambda} P_\alpha\Big]. \tag{V.117}$$

This is a mapped interface profile with minimal effects of surface roughness. It is the natural precursor of the intrinsic density profile considered and approximated in the van der Waals interface theory and its subsequent revisions. That intrinsic density profile is just the basin-restricted re-excitation for the temperature of interest, i.e., the same form as shown in Eq. (V.117), but with the superscripts "(q)" removed:

$$\rho_{1,\Lambda}(z) = \sum_{\alpha\in\Lambda} P_\alpha \rho_{1,\alpha}(z) \Big/ \Big[\sum_{\alpha\in\Lambda} P_\alpha\Big]. \tag{V.118}$$

The re-excitation is expected to move the midpoint of the intrinsic density profile back to the vicinity of the prequench dividing surface at z_G.

The intrinsic profile $\rho_{1,\Lambda}(z)$ has an associated surface tension $\gamma^{(0)}$, the excess surface free energy per unit area for this "flat" interface in comparison with that implied by the local-equilibrium barometer formula, Eq. (V.109). It is tempting and perhaps natural to identify this $\gamma^{(0)}$ with the "bare" surface tension γ_0 involved in the capillary wave approximation, Eq. (V.103), but that would require independent verification. In any event, $\gamma^{(0)}$ refers to a proper subset of all inherent structures, and because of this constraint it must be a larger free-energy quantity than that for the unconstrained interface, namely, the macroscopically measurable surface tension γ:

$$\gamma^{(0)} > \gamma. \tag{V.119}$$

Now that the intrinsic density profile has been identified in Eq. (V.118), the obvious next step is to inquire about whether the actual density profile results from a broadening of the intrinsic profile as suggested by the capillary wave picture. In particular, this would require that capillary wave surface deformations inevitably must be due to interbasin transitions on the multidimensional potential energy landscape, rather than due to intrabasin vibrational displacements. This proposition receives support from an X-ray photon correlation spectroscopy study of the glycerol ($C_3H_8O_3$) surface, showing a dramatic slowing of capillary wave relaxation with declining temperature because of viscous damping [Seydel et al., 2001]. One way to assess the capillary wave picture is to ask what level of precision the actual profile $\rho_1(z)$ can be represented by the convolution of the intrinsic profile $\rho_{1,\Lambda}(z)$ with a non-negative smoothing kernel:

$$\rho_1(z) \cong \int_{z_f}^{z_c} K(z - z')\rho_{1,\Lambda}(z')dz', \tag{V.120}$$

where the kernel is normalized:

$$1 = \int_{-\infty}^{+\infty} K(t)dt. \tag{V.121}$$

The capillary wave picture implies that $K(t)$ would specifically have a Gaussian form:

$$K(t) = \left[2\pi\langle(\delta z)^2\rangle\right]^{-1/2} \exp\left[-\frac{t^2}{2\langle(\delta z)^2\rangle}\right], \tag{V.122}$$

with a width given by the $\langle(\delta z)^2\rangle$ expression shown in Eq. (V.103). Of course, passing such a convolution test would not certify the capillary waves as independent modes, nor would it unambiguously identify a cutoff wavelength k_u.

Only partial verification of the physical relevance of capillary wave excitations would result from experimental or simulational observations exhibiting the predicted logarithmic divergence of interface width measure $\langle(\delta z)^2\rangle$ with vanishing gravitational field strength, as indicated by Eq. (V.103). It has to be stressed once again that this divergence arises just from surface waves in the long-wavelength limit, so that the postulated full spectrum of capillary waves with $|\mathbf{k}| \leq k_u$ as independent normal modes would not be tested. For long-wavelength interface deformations away from the flat intrinsic profile, the steepest descent mapping to inherent structures should at least qualitatively preserve those deformations. If indeed the proposed logarithmic divergence of width as $g \to 0$ applies to $\rho_1(z)$, it should also apply to $\rho_1^{(q)}(z)$.

A more detailed and stringent test of the capillary wave picture would have to assign wave amplitudes to arbitrarily chosen inherent structures $\alpha' \notin \Lambda$, i.e., all those not involved in the intrinsic profile. Because of the lateral-direction periodic boundary conditions that have been assumed, this would mean assigning wave amplitudes to any one configuration from the two-dimensional manifold of translation-equivalent configurations for that α'. Conceivably, if this could be done for a large sample of inherent structures, the resulting database of amplitudes might be statistically analyzed for independence, Gaussian character, and limitation by an upper wave vector cutoff k_u. However, it has been argued [Stillinger, 2008] that no such procedure can be found which (as required by the definition of an intrinsic profile) is devoid of arbitrary parameters or procedures that are not contained within the system Hamiltonian itself.

Although the particles thus far have been treated as structureless, an extension to include polyatomic particles with various types of internal degrees of freedom is nominally straightforward. In that generalization, let each particle $1 \leq i \leq N$ be described by a $(3 + v)$-component vector $\mathbf{r}_i^*$, which includes both center-of-mass position $\mathbf{r}_i$ as well as v orientational and (if necessary) conformational coordinates. The extended version of the potential energy Eq. (V.104) is

$$\Phi_g(\mathbf{r}_1^* \ldots \mathbf{r}_N^*) = \Phi(\mathbf{r}_1^* \ldots \mathbf{r}_N^*) + mg\sum_i z_i, \tag{V.123}$$

where z_i is the elevation of the center of mass of polyatomic particle i. The correspondingly extended form of the steepest descent Eq. (V.114) then becomes

$$d\mathbf{R}^*(s)/ds = -\nabla_{\mathbf{R}^*}\Phi_g(\mathbf{R}^*), \tag{V.124}$$

$$\mathbf{R}^* = \mathbf{r}_1^* \oplus \mathbf{r}_2^* \oplus \ldots \oplus \mathbf{r}_N^*.$$

Just as with structureless particles, the identification of a set Λ of inherent structures with minimal values of $|\Delta_\alpha|$ is possible, and the intrinsic density profile again follows Eqs. (V.117) and (V.118). For these purposes, z_i serves as the natural vertical "position" of particle i.

The relatively straightforward description and resolution of interfacial matter distribution offered by the inherent-structure formalism applies to many kinds of material systems with liquid–vapor interfaces. However, the reader should be aware that some cases exist in which a proper presentation may merit a more elaborate treatment than has just been presented. One of these cases concerns some of the normal, or straight-chain, alkanes (C_nH_{2n+2}) just above their thermodynamic melting points. It has been established that for a small temperature interval, these substances in the size range $16 \leq n \leq 50$ spontaneously have their liquid surfaces covered with an ordered layer of molecules that are apparently arranged as a two-dimensional crystal [Wu et al., 1993; Ocko et al., 1997]. Fracture phenomena have been observed for these ordered layers [Prasad and Dhinojwala, 2005]. This deserves at least an extension of the theory to account for preferred molecular orientation through the interface. Another case deserving special attention is that of molten metals, for which the conduction-electron-mediated effective interactions tend to produce an oscillatory density stratification locally through the interface region [Harris et al., 1987; Magnussen et al., 1995; DiMasi et al., 1998]. These density oscillations would be especially prominent in the intrinsic density profile as defined above through the classification of inherent structures. But in weak external fields ($g \to 0$), the expected large capillary wave amplitudes would smooth out those oscillations, leaving a less structured monotonic density profile. It should also be mentioned that X-ray reflectivity experiments have shown that nonmetallic liquids at sufficiently low temperature can also display oscillatory interfacial density profiles [Mo et al., 2006]. Furthermore, a class of theoretical models with pairwise interactions between structureless particles, the so-called "Madrid" liquids, has been determined by Monte Carlo simulation to exhibit interfacial density oscillations [Li and Rice, 2004].

Liquid–vapor equilibrium with a planar separating interface also can occur with two or more particle species present, each with its own mass m_ζ. An important class of examples involves surfactant species as a second component added to an otherwise pure liquid in coexistence with its own vapor [Rosen, 2004]. For structureless particles involved in a multicomponent liquid–vapor scenario, the potential energy function in a gravitational field is a straightforward extension of the one-species version in Eq. (V.104), specifically

$$\Phi_g(\mathbf{r}_1\ldots\mathbf{r}_N) = \Phi(\mathbf{r}_1\ldots\mathbf{r}_N) + g\sum_{i=1}^{N}m_{\zeta(i)}z_i, \tag{V.125}$$

where $\zeta(i)$ denotes the species of particle i. When intraparticle degrees of freedom must be included, Eq. (V.123) permits a similar generalization. In either of these cases, the gravitational field serves to stratify the two bulk fluid phases according to their respective mass densities. The matter distribution in a state of thermal equilibrium can be described by the singlet density functions $\rho_{1(\zeta)}(z)$ for each species ζ. Here it has been supposed that the same kind of rectangular solid container with volume $V = L_xL_yL_z = A(z_c - z_f)$ serves to house the system, with z the coordinate measuring altitude in the gravitational field. Once again, it is possible to introduce "barometric formulas" that specify ideal matter distributions for the given temperature and gravitational field strength, with one such function for each species: $\rho_{1(\zeta),g}(z|z_G)$. These are determined by a set of generalizations of the one-species formula (V.109), for each of the species now present:

$$\mu_\zeta[\{\rho_{1(\xi),g}(z|z_G)\}, T] + m_\zeta gz = C_\zeta(\{N_\xi\}). \tag{V.126}$$

By virtue of the coexistence phase equilibrium that is present, the chosen discontinuity position z_G must be the same for all species. However, there exists a choice for which of those species is to be accorded the special role of determining precisely what value of z_G to define. But having made the species choice, say ζ, z_G then is determined by the analog of the previous Eq. (V.110):

$$N_\zeta = A\int_{z_f}^{z_c}\rho_{1(\zeta),g}(z|z_G)dz, \tag{V.127}$$

which generally causes the corresponding integrals for the other species to yield values differing from the actual numbers of particles of those species present in the system. Needless to say, a different species choice in Eq. (V.127) would be expected to identify a different location for the dividing surface.

The same kind of strategy as detailed above for the multicomponent liquid–vapor case applies to interfaces separating pairs of immiscible liquids. As before, this strategy requires selecting one of the species ζ to define the dividing surface. Configurations sampled by the canonical distribution are quenched as usual to their inherent structures α, and a displaced location for the dividing surface resulting from the mapping is identified. Then a subset Λ_ζ of those inherent structures is designated by the criterion of minimum in number deviation $|\Delta_{\alpha,\zeta}|$ for species ζ relative to that displaced dividing surface. Thermal re-excitation of the set Λ_ζ creates the intrinsic density profile. The remainder set of inherent structures supplies surface-roughening distor-

---

BOX V.4. Adjustable Mass Densities in a Two-Liquid, Three-Component System

---

Water (H_2O), carbon tetrachloride (CCl_4), and diethyl ether [$(C_2H_5)_2O$] are all liquids at temperature 20°C and 1 atm pressure. In pure form under these conditions, their mass densities are [Weast, 1978]:

> Water: 0.9982 g/cm^3
> Carbon tetrachloride: 1.5940 g/cm^3
> Diethyl ether: 0.7138 g/cm^3

The first two of these are virtually insoluble in one another, so placing comparable volumes of them together in the same vessel results in the nearly pure water phase floating atop the nearly pure carbon tetrachloride phase.

Low-mass-density diethyl ether is soluble in all proportions in carbon tetrachloride but only very slightly soluble in water. Consequently, isothermally adding some of this third liquid (with adequate mixing) to the system with water floating on carbon tetrachloride causes the mass density of the lower phase to decrease toward that of the upper phase. Adding more diethyl ether continues to decrease the difference in mass densities. Upon including a volume of diethyl ether roughly equal to 2.1 times the original carbon tetrachloride volume, the mass densities of the two liquids become equal. Further addition would reverse the relative mass densities of the two immiscible phases and so would cause inversion of their vertical positions.

---

tions, and these would be available for analysis in the light of capillary wave approximation predictions.

The diversity of possible immiscible liquid pairs offers a kind of parameter control not directly available in the single-species liquid–vapor case. Certain ranges in the choice of relative amounts for some three-species systems can cause the relative mass densities of the two phases to become arbitrarily small and even to change sign. When the mass densities are equal, the gravitational work of exciting capillary waves vanishes, which would produce a logarithmic divergence of the type anticipated in Eq. (V.103). In a real system under experimental observation, this equal-mass-density situation would have an interface shape and position largely controlled by boundary conditions, and in most cases would not remain planar. Box V.4 describes a specific example of this kind of system involving three chemical species.

I. Liquid-Phase Transport Properties

(1) SELF-DIFFUSION

As discussed in Section II.G, the self-diffusion constant D in a macroscopic single-component system measures the mean-square displacement of any one particle i over long time intervals:

$$f(t) = \langle [\mathbf{r}_i(t) - \mathbf{r}_i(0)]^2 \rangle \sim 6Dt \quad (t \to \infty). \tag{V.128}$$

The self-diffusion constant can be evaluated directly as a time integral of the single-particle velocity autocorrelation function, presuming that the system is in a state of thermal equilibrium:

$$D = (1/3) \int_0^\infty \langle \mathbf{v}_i(0) \cdot \mathbf{v}_i(t) \rangle dt. \tag{V.129}$$

These expressions are independent of the choice of time origin. Because of time-reversal symmetry in the equilibrium state, $f(t)$ formally is an even function of t, i.e., it is a function of $|t|$. Furthermore, if the interactions in the system that determine the equations of motion are continuous and differentiable along all possible particle paths, then this function has an expansion in even powers of t. The leading terms in this time series are straightforward to evaluate:

$$f(t) = (3k_BT/m)t^2 + \left[(k_BT/3m^2)\langle\nabla_i\cdot\mathbf{F}_i(0)\rangle + (1/4m^2)\langle\mathbf{F}_i^2(0)\rangle\right]t^4 + O(t^6) \tag{V.130}$$

$$\equiv (3k_BT/m)t^2 + \left[-(k_BT/3m^2)\langle\nabla_i^2\Phi\rangle + (1/4m^2)\langle(\nabla_i\Phi)^2\rangle\right]t^4 + O(t^6).$$

Here as usual m is the particle mass, and $\mathbf{F}_i = -\nabla_i\Phi$ is the force exerted on particle i by its neighbors. Terms subsequent to those explicitly shown involve higher order spatial derivatives of the force. That the leading term in the right side is independent of interactions simply reflects the fact that particle trajectories are linear at arbitrarily short times with the Maxwell-Boltzmann velocity distribution. The coefficient of t^4 in this series is presumably negative because it represents the effect of perturbations on the initial linear trajectories, which if undeflected, or not decelerated on average, would be a nondiffusive means for particle dispersal with the initial velocity distribution.

If the interparticle interactions consisted entirely of spherically symmetric pair potentials $v_2(r_{ij})$, then the $O(t^4)$ terms in Eq. (V.130) could be expressed as integrals over pair and triplet correlation functions for the ρ, T state under consideration. Specifically, one has the following identities:

$$\langle\nabla_i\cdot\mathbf{F}_i(0)\rangle = -\rho\int d\mathbf{r}[\nabla^2 v_2(r)]g_2(r), \tag{V.131}$$

$$\langle\mathbf{F}_i^2(0)\rangle = \rho\int d\mathbf{r}[v_2'(r)]^2 g_2(r)$$

$$+ (\rho^2/4)\int d\mathbf{r}_2\int d\mathbf{r}_3 v_2'(r_{12})v_2'(r_{13})[(r_{12}^2 + r_{13}^2 - r_{23}^2)/r_{12}r_{13}]g_3(\mathbf{r}_1, \mathbf{r}_2, \mathbf{r}_3).$$

Analogously, the coefficients of higher powers t^{2n} in Eq. (V.130) could also be expressed as integrals involving equilibrium correlation functions for larger and larger numbers of particles.

The mean-square displacement function $f(t)$ is well defined for the singular model of diameter-σ_0 hard spheres, but the series expansion indicated in Eq. (V.130) is not applicable in that case. The force functions are not defined for hard spheres. Instead of continuous and differentiable particle trajectories, the hard spheres execute piecewise constant-velocity linear trajectories, with energy-conserving collisions of infinitesimal duration that send the participants instantaneously in different directions. Time reversal symmetry continues to be exhibited by the hard-sphere version $f_{hs}(t)$, but instead of an expansion in even powers t^{2n}, the short-time behavior requires inclusion of odd powers of $|t|$. The hard-sphere analog of Eq. (V.130) has the following format:

$$f_{hs}(t) = (3k_BT/m)|t|^2 - A(\rho\sigma_0^3)(k_BT)^{3/2}|t|^3 + O(|t|^4), \tag{V.132}$$

where both the density-dependent function $A(\rho\sigma_0^3)$ and its derivative $A'(\rho\sigma_0^3)$ are positive, the latter reflecting the diffusion-retarding effect of the increased collision rate produced by isothermal compression. Of course, the mean-square path length between subsequent collisions for any particle diverges as density decreases, so $A(0) = 0$. Note that a mean-square displacement function

$f(t)$ requiring a short-time expansion in both even and odd powers of $|t|$ would also be required whenever a hard-core portion of a model's interparticle interaction is present, even if continuous and differentiable longer range portions were also present.

Isochoric steepest descent mapping, or quenching, configurations along dynamical trajectories that contribute to $f(t)$ leads to an alternative view of diffusion [Shell et al., 2004]. This mapping connects each particle position $\mathbf{r}_i(t)$ to its corresponding location $\mathbf{r}_i^{(q)}(t)$ within the inherent structure that is relevant at time t. The time-dependent mean-square displacement function for mapped positions is then defined as the obvious analog of Eq. (V.128):

$$f^{(q)}(t) = \langle [\mathbf{r}_i^{(q)}(t) - \mathbf{r}_i^{(q)}(0)]^2 \rangle. \tag{V.133}$$

In agreement with its predecessor $f(t)$, this is likewise an even function of the time difference t. But unlike the continuous dynamical trajectory positions $\mathbf{r}_i(t)$, the $\mathbf{r}_i^{(q)}(t)$ are piecewise constant, displaying discontinuous jumps only when the system executes an interbasin transition. In view of the fact that those transitions between successive inherent structures primarily affect only an order-unity subset of the N particles present, and only those few shift by order-unity displacements, $f^{(q)}(t)$ must exhibit the same leading-order asymptotic linear rise with t as does $f(t)$:

$$f^{(q)}(t) \sim 6Dt \qquad (t \to \infty). \tag{V.134}$$

At least from this one viewpoint, the self-diffusion constant does not permit separation into distinct contributions attributable to inherent structures and to intrabasin displacements, respectively.

In contrast to the inherent-structure analysis of thermodynamic properties, which allows all permutation-equivalent basins to be lumped together (Chapter III), the analysis required for the self-diffusion constant necessarily treats all basins and their inherent structures individually. As time proceeds in a diffusive many-particle system, the multidimensional configuration point can enter, or come close to, permutation variants of the $t = 0$ basin. At very large times t, the closest permutation variant can indeed have all N particles far displaced from their $t = 0$ locations. The $f^{(q)}(t)$ expressions (V.133) and (V.134) are based on this individual-basin interpretation.

The discontinuities between the temporarily constant values of the quenched positions $\mathbf{r}_i^{(q)}(t)$ influence the behavior of $f^{(q)}(t)$ near the origin. It needs to be kept in mind that interbasin transition rates for the N-body system as a whole are extensive, i.e., proportional to N or to volume V. But as emphasized above, so far as any single particle i is concerned, the vast majority of these localized particle shifts from one inherent structure to the next in a macroscopic system are remote and thus produce virtually no displacement for i. It is only those less frequent interbasin transitions localized at or close to i that matter and that dominate the small-time behavior of $f^{(q)}(t)$. These close-by transitions have an order-unity rate, and the first of them to occur at $t > 0$ confers a positive contribution to $f^{(q)}(t)$. After averaging over all possible time origins, and thus over the distribution of first times for a transition localized around i, one obtains an initial form linear in time:

$$f^{(q)}(t) = B(\rho, T)|t| + o(t), \tag{V.135}$$

where there is no obvious relation between the positive coefficient $B(\rho, T)$ appearing here for small t and the asymptotic large-t slope $6D(\rho, T)$ for the same mean-square displacement function.

Whereas both $f(t)$ and $f^{(q)}(t)$ share the same asymptotic slope at large $|t|$, linear fits carried to the next order in that asymptotic regime produce distinct constant offsets:

$$f(t) \sim 6D(\rho, T)|t| + C(\rho, T), \qquad\qquad (V.136)$$

$$f^{(q)}(t) \sim 6D(\rho, T)|t| + C^{(q)}(\rho, T).$$

These two additive constants have been examined in the course of a molecular dynamics simulation with the spherically symmetric and pairwise-additive Dzugutov potential [Shell et al., 2004]. For that model examined at constant liquid density, $C^{(q)} > C$ over the stable liquid range. However, the relative difference between the two declines as temperature declines, and a reversal was observed within the supercooling regime. The same qualitative trend has been observed for a simulated Lennard-Jones liquid [Keyes and Chowdhary, 2002].

In the long-time asymptotic regime for liquids, $f(t)$ possesses a correction to its linear form in Eq. (V.136) because of the presence of a "hydrodynamic tail" [Alder and Wainwright, 1970; Hansen and McDonald, 1986, Section 8.7]. This cooperative phenomenon involves the average response of surrounding liquid to the impulse it receives from the initial velocity possessed by particle i. The motion of the surroundings amounts to a growing vortex with i at its approximate center, tending at intermediate to long times to drag i along in the direction of its initial velocity. As a result, the velocity autocorrelation function displays only an algebraic decay toward zero:

$$\langle \mathbf{v}_i(0) \cdot \mathbf{v}_i(t) \rangle \sim J(\rho, T)|t|^{-3/2}, \qquad\qquad (V.137)$$

where $J > 0$. The velocity persistence, of course, has an effect on the mean-square displacement function:

$$f(t) - 6Dt - C(\rho, T) \sim -(1/2)J(\rho, T)|t|^{-1/2}. \qquad\qquad (V.138)$$

At present, it has not been directly established whether the same hydrodynamic tail phenomenon appears in the quench version $f^{(q)}(t)$.

Setting aside the hard-sphere model for the moment and focusing attention on models with continuous and differentiable potentials, it is important to recognize the relevance for diffusion of Hessian matrix eigenvalues (potential energy landscape curvatures, Section III.B). As diffusion occurs, and as the system's multidimensional configuration point passes through a sequence of inherent-structure basins, that point must traverse the neighborhood of transition states that are located in shared boundaries of basin pairs. Each simple transition state provides a single negative Hessian eigenvalue, and as stressed above, that transition involves just a local rearrangement of a small number [$O(1)$] of particles. But in a macroscopic N-particle system throughout which diffusion is everywhere underway, the local configurational approach to local transitions, and subsequent passage to the rearranged geometry, is occurring throughout the system at an extensive [$O(N)$] rate. This implies that on average an $O(1)$ fraction of the $3N$ Hessian eigenvalues would have to be negative. It should be recalled that in principle some basins might contain interior saddles that are not associated with interbasin transitions, a possibility that would slightly increase the average fraction of negative Hessian eigenvalues.

These concepts have generated suggestions to link the fraction of negative Hessian eigenvalues, as a function of temperature and density, quantitatively to the diffusion constant [Keyes,

1994, 1997]. The distribution of both positive and negative eigenvalues can be formally interpreted in terms of the presence of instantaneous effective normal modes with both real and
imaginary frequencies, the latter group of which might involve "reaction trajectories" that cross
saddles to neighboring basins [Seeley and Keyes, 1989]. However, the appropriateness and accuracy of using even some subsets of instantaneous normal modes for predicting kinetic properties
have been called into question because of the presence of "false barriers" that do not produce
diffusive structural changes [Gezelter et al., 1997]. It remains to be determined whether Hessian
eigenvalues and their corresponding effective normal modes can provide a definitive characterization of the self-diffusion phenomenon.

Experimental measurements of self-diffusion constants in pure liquids require isotope or spin
labelling techniques. Results for thermodynamically stable liquid ranges are scattered throughout
the published literature, though a few collections are available [e.g., Tyrrell and Harris, 1984]. By
contrast, continuing theoretical efforts to understand diffusion in simple classical liquids have
produced a large number of publications reporting results of computer simulations. Many of
these results have focused on the venerable Lennard-Jones 12,6 pair potential model. A recent
tabulation of results for that model is now available [Meier et al., 2004b].

An analogy can be drawn between self-diffusion kinetics and chemical reaction kinetics,
whose rates are traditionally described by the Arrhenius law or Eyring's transition state theory
[Connors, 1990]. Both are controlled quantitatively by details of the underlying potential energy
landscape. Such an analogy suggests that D might usefully be represented as follows for variable
temperature T and pressure p:

$$D(T, p) = D_0 T^{1/2} \exp[-\Delta G_D^{\ddagger}(T, p)/k_B T]. \tag{V.139}$$

Here $\Delta G_D^{\ddagger}$ has dimension energy. The constant D_0 is chosen to be independent of T, p and has
dimensions $(\text{length})^2(\text{time})^{-1}(\text{deg})^{-1/2}$. The reason for explicitly including pre-exponential factor
$T^{1/2}$ is to account directly for the mean particle speed dictated by the Maxwell-Boltzmann velocity
distribution; this is the exact temperature scaling at constant density for D in the case of a flat
available potential landscape (i.e., for hard spheres). All of the dynamical diversity involved in
traversing a rugged landscape involving sequences of transition states between neighboring basins during the diffusion process is by definition lumped into the quantity $\Delta G_D^{\ddagger}$. This quantity can
informally be identified as a "Gibbs free energy of activation" for self-diffusion. With this convention, one can use standard thermodynamic formulas to identify the corresponding activation
volume $\Delta V_D^{\ddagger}$, activation entropy $\Delta S_D^{\ddagger}$, and activation energy $\Delta E_D^{\ddagger}$:

$$\Delta V_D^{\ddagger} = \left(\frac{\partial \Delta G_D^{\ddagger}}{\partial p} \right)_T,$$

$$\Delta S_D^{\ddagger} = -\left(\frac{\partial \Delta G_D^{\ddagger}}{\partial T} \right)_p, \tag{V.140}$$

$$\Delta E_D^{\ddagger} = \Delta G_D^{\ddagger} + T\Delta S_D^{\ddagger} - p\Delta V_D^{\ddagger}.$$

The first of these three derived quantities can be interpreted as a measure of the local expansion
of the surrounding medium that on average occurs to allow the diffusing particle to escape from

the cage of neighbors that may enclose it. The last of the three is a measure of the average potential energy landscape rise that must be surmounted for the self-diffusion to proceed stepwise. Finally, the second of the three indicates on average how wide or constricted the pathways are that lead to permanent particle displacement during the diffusion process.

It needs to be emphasized that the descriptions provided by Eqs. (V.139) and (V.140) do not refer necessarily to the average dynamics just of single-barrier traversals. They generally incorporate details about intrabasin lifetimes, relative importance of low versus high potential energy barriers, and barrier transition probabilities once the configuration point approaches a transition state. The subtle effects of landscape topographic diversity are implicit in the description. Nevertheless, fitting experimental, simulational, or theoretical results for D to such a simple fitting form can emphasize the extent to which the actual potential energy landscape and its dynamics differ, say, from those of the flat landscape for hard-particle models.

Self-diffusion constants D_ξ for each species ξ in liquid mixtures at thermal equilibrium are defined by the obvious generalization of Eqs. (V.128) and (V.129), measuring the long-time mean-square displacement of labeled individual particles. Likewise, the inherent-structure function $f^{(q)}(t)$ generalizes straightforwardly to a set for individual-species functions $f_\xi^{(q)}(t) \sim 6D_\xi |t|$. Also, the functional form Eq. (V.139) can be individually adapted to each species in mixtures and the corresponding activation volumes, entropies, and energies extracted from the multicomponent versions of Eqs. (V.140).

Liquid mixtures present an important distinction that needs to be recognized between the self-diffusion constants D_ξ for mean-square displacement just discussed and a different set of diffusion constants $\widetilde{D}_\xi$ that describe transport under concentration gradients. This latter set describes macroscopic matter currents $\mathbf{J}_\xi$ in the presence of number density (concentration) gradients according to Fick's "first law" [Tyrrell and Harris, 1984]:

$$\mathbf{J}_\xi = -\widetilde{D}_\xi \nabla \rho_\xi. \tag{V.141}$$

In particular, this equation can refer to a steady-state situation where particles of the various species are added or removed at the boundaries of the system to maintain a fixed set of density gradients. Under such circumstances, the time-dependent displacements of individual particles on average do not have a spherically symmetric distribution but are biased in the direction of steady-state current flow for that species. These currents generally interact with one another, either reinforcing or inhibiting the flow rates. In extreme cases, it is possible for the current of a species to flow up its concentration gradient, i.e., for $\widetilde{D}_\xi$ to be negative. The basic driving forces for the matter flow in liquid mixtures are not the number density gradients but the gradients of thermodynamic activity coefficients α_ξ (see Box V.5).

The relation between the two sets of diffusion coefficients involves an isobaric–isothermal derivative of activity coefficient logarithm with respect to number density (or concentration) logarithm [Hirschfelder et al., 1954, p. 631]:

$$\widetilde{D}_\xi = \left(\frac{\partial \ln a_\xi}{\partial \ln c_\xi}\right)_{T,\,p,\,\{c_\xi\}} D_\xi. \tag{V.142}$$

BOX V.5. Activity Coefficients

In an equilibrated system that may contain a mixture of distinct species, the chemical potentials thermodynamically are defined by equivalent isochoric derivatives of the Helmholtz free energy F or isobaric derivatives of the Gibbs free energy G:

$$\mu_\xi = \left(\frac{\partial F}{\partial N_\xi}\right)_{T,V,\{N_\zeta\}} = \left(\frac{\partial G}{\partial N_\xi}\right)_{T,p,\{N_\zeta\}}.$$

In a mixture of classical ideal gases (i.e., no interactions), the chemical potentials adopt the simple density-dependent form

$$\mu_\xi = \mu_{\xi,0}(T) + k_B T \ln \rho_\xi,$$

where $\mu_{\xi,0}$ collects density-independent contributions such as mean thermal de Broglie wavelengths. For the cases of liquid solutions, it is often convenient to restate the number densities in terms of "concentrations" c_ξ involving a scale factor K (which may have dimensions); the noninteracting form above trivially becomes

$$\mu_\xi = \mu'_{\xi,0}(T) + k_B T \ln c_\xi,$$

$$c_\xi = K\rho_\xi,$$

$$\mu'_{\xi,0}(T) = \mu_{\xi,0}(T) - k_B T \ln K.$$

Particle interactions produce changes in the chemical potentials, which can formally be incorporated by replacing the concentrations c_ξ with "activities" a_ξ, each of which in general depends on temperature and the full set of densities (concentrations):

$$\mu_\xi = \mu'_{\xi,0}(T) + k_B T \ln a_\xi(T, \{c_\xi\}).$$

(2) SHEAR VISCOSITY

In contrast to the situation for experimental determination of the self-diffusion constant D in liquids, it is much easier to measure the shear viscosity η. This leads to higher precision results, as well as permitting determinations over a much larger numerical range. This last feature is especially important for probing the kinetics in the supercooled liquid regime as is discussed in Chapter VI. Not surprisingly, tables listing shear viscosities for many liquids, including temperature variations, are widely available [e.g., Lide, 2003, pp. **6**-186 to **6**-194]. Also, numerical simulation results for the Lennard-Jones model of structureless particles have been published [Meier et al., 2004a].

Considering the fact that the self-diffusion constant D is a statistical property of single particle motions in the liquid phase, whereas shear viscosity η intrinsically measures collective relaxation and dissipation on a macroscopic scale, it would be natural to expect that these two properties should be quantitatively uncorrelated. Yet for the majority of single-component systems composed of relatively compact particles, there is an empirical connection between D and η in dense liquid states that typically includes both the equilibrium liquid regime as well as that of modest supercooling. This empirical connection is supplied by the Stokes-Einstein relation [Einstein, 1905, 1906; Milne-Thomson, 1960, p. 584]. It has the following form:

$$D\eta = k_B T / C R_{eff}. \tag{V.143}$$

The quantity R_{eff} is an effective hydrodynamic radius for the particle species forming the liquid, and C is a dimensionless numerical constant whose value depends on the applicable hydrodynamic boundary condition at the surface of the particle. The simple model picture that underlies the Stokes-Einstein relation is that of a rigid sphere executing Brownian motion in a structureless viscous medium, and consequently

$$4\pi \le C \le 6\pi, \tag{V.144}$$

where the lower and upper limits correspond to hydrodynamic "slipping" and "sticking" boundary conditions at the surface of the diffusing sphere, respectively. What is surprising about the Stokes-Einstein relation is that while D and η individually tend to display substantial temperature variations, the combination $D\eta/T$ tends to be much more nearly temperature independent. Numerical study of the simple hard-sphere and Lennard-Jones models, where R_{eff} is well-defined or nearly so, show that indeed C tends to fall in the range (V.144) when the particle packing density corresponds to the liquid state [Cappelezzo, et al., 2007].

It is useful to recall from Section II.G the formal current autocorrelation function expression in an equilibrium ensemble for η. The expressions are repeated here for convenience:

$$\eta = (1/k_B T V) \int_0^\infty \langle J^{xy}(0) J^{xy}(t) \rangle dt, \tag{V.145}$$

$$J^{xy}(t) = \sum_{j=1}^N \{ p_{jx}(t) p_{jy}(t)/m + (1/2)[x_j(t)F_{jy}(t) + y_j(t)F_{jx}(t)] \}.$$

In the second of these expressions, p_{jx}, p_{jy} and F_{jx}, F_{jy} are the x and y components, respectively, of the momentum of particle j, and of the force on particle j. It should also be noted that an alternative current autocorrelation function exists, but for $1/\eta$, based on the persistence of spontaneous shear flow fluctuations at equilibrium [Stillinger and Debenedetti, 2005].

When the many-particle system resides statically at one of its inherent structures, J^{xy} vanishes identically. Consequently, Eq. (V.145) reflects the fact that η strictly emerges from the time dependence of particle motions, including interbasin transitions, during the course of which both particle momenta and forces vary. Spontaneous fluctuations that produce shear stress distributions across the system tend to increase the potential energy (including inherent-structure energy) of the basins into which the system has traveled. The subsequent configurational relaxation then would be "downhill," although this typically entails traversing over a rough landscape. The average indicated in the autocorrelation function expression above expresses the time required for that statistical traversal.

In the hard-sphere limit, under constant density conditions, one can show that η scales with temperature as $T^{1/2}$ [Wannier, 1966, p. 422], a feature that can be extracted by careful analysis from Eqs. (V.145). This is the behavior that obtains when the topography of the available potential energy landscape for system dynamics is precisely flat and is similar to the analogous scaling behavior mentioned above for D. Consequently, one is led as for D to collect the effects of landscape topographical ruggedness on the exact autocorrelation function expression Eq. (V.145) by an analog of Eq. (V.139), specifically,

$$\eta(T, p) = \eta_0 T^{1/2} \exp[\Delta G_\eta^{\ddagger}(T, p)/k_B T]. \tag{V.146}$$

In this formula, $\Delta G_\eta^{\ddagger}$ is identified as an activation Gibbs free energy for shear flow, and it appears in the exponent with a positive sign because any increase in this quantity makes shear flow more difficult, and η increases. The quantity η_0 is chosen to be independent of T and p and has units (mass)(length)$^{-1}$(time)$^{-1}$(deg)$^{-1/2}$. The activation volume, entropy, and energy for viscous shear flow are then the following:

$$\Delta V_\eta^{\ddagger} = \left(\frac{\partial \Delta G_\eta^{\ddagger}}{\partial p} \right)_T,$$

$$\Delta S_\eta^{\ddagger} = -\left(\frac{\partial \Delta G_\eta^{\ddagger}}{\partial T} \right)_P, \tag{V.147}$$

$$\Delta E_\eta^{\ddagger} = \Delta G_\eta^{\ddagger} + T\Delta S_\eta^{\ddagger} - p\Delta V_\eta^{\ddagger}.$$

Upon inserting the two phenomenological expressions (V.139) and (V.146) into the combination $D\eta/T$ suggested by the Stokes-Einstein relation Eq. (V.143), one obtains

$$\frac{k_B}{CR_{eff}} = D_0\eta_0 \exp\left[\frac{\Delta G_\eta^{\ddagger}(T, p) - \Delta G_D^{\ddagger}(T, p)}{k_B T} \right]. \tag{V.148}$$

Liquids obeying the Stokes-Einstein relation over a nontrivial temperature range involve either cancellation of the activation Gibbs free energies, or more generally a difference in those activation Gibbs free energies that is proportional to absolute temperature T. As is discussed in Chapter VI, such simplification typically fails in the regime of strongly supercooled liquids.

(3) THERMAL CONDUCTIVITY

Tables of thermal conductivities $\lambda(T)$ for insulating liquids measured at ambient pressure reveal that in most cases raising the temperature causes this quantity to decrease [Lide, 2003, pp. **6**-197 to **6**-198]. However, there are a few notable exceptions, specifically, water (H_2O) and glycerol ($C_3H_8O_3$). Both of these anomalous cases have liquid short-range structure dominated by hydrogen bonds. Furthermore, ethylene glycol ($C_2H_6O_2$), also a hydrogen-bonded liquid, provides a borderline case with a $\lambda(T)$ that is virtually temperature independent over its ambient-pressure stable liquid range. These exceptional behaviors can be overturned by extending the temperature range by including pressure variation. For example, following λ for liquid water along its liquid–vapor coexistence curve in the p,T plane, one finds that the thermal conductivity rises with T from the triple point 273 K to $T \cong 410$ K, at which it attains a maximum, but then declines monotonically to the critical temperature $T_c \cong 647$ K [Schmidt, 1969, p. 167].

The measurement protocols that in principle are used for determination of self-diffusion constants and shear viscosities involve liquid states that are macroscopically uniform, i.e., contain no density or temperature gradients. But measurement of thermal conductivity λ necessitates a temperature gradient, whether heat flow is determined under steady-state conditions between thermal reservoirs at different fixed temperatures, or whether it is determined by rate of relaxation in a thermally isolated sample with an initial temperature nonuniformity toward a final uniform equilibrium state. However, the presence of a thermal gradient implies a density gradient as well, the magnitude and sign of which depend on the isobaric thermal expansion α_p of the substance under consideration. Of course, this effect requires orienting a measured sample with downward density gradient to avoid convective flow. The only rare exceptions would be those substances that happen to have equilibrium states with $\alpha_p = 0$. Barring these exceptions, the inherent structures underlying the nonuniform gradient-containing states would themselves be spatially nonuniform and thus would not exhibit a probability distribution identical to that of the thermal equilibrium state. The magnitude of this gradient effect on the inherent structures would depend on the results of separation of measured thermal expansion and of heat capacity between intrabasin vibrational contributions on one side and the distribution of inherent structures on the other, as discussed in Section V.C.

By contrast with low-temperature heat flow in structurally ordered crystals of electrical insulators, the corresponding processes in insulating liquids are more diverse and complicated. In the former, the rate of heat flow depends on the presence of well-defined phonons as well as the presence of phonon scattering mechanisms (anharmonicity, isotopic substitution, crystal structural defects) that interrupt their propagation [Section IV.G]. In liquids, the absence of an underlying periodic order eliminates phonons as relevant collective motions. Furthermore, the dynamic coupling between neighboring particles at the higher temperatures involved in the liquid state is considerably more anharmonic than in the crystalline state of the same substance and can often be described simply as sequences of collisions. In addition, particle mobility in liquids contributes to energy transport and thus thermal conductivity, as "hot" particles move toward the colder region and "cold" molecules move toward the warmer region. This last effect can involve energy transport not only as translational and rotational kinetic energy but also as intramolecular vibrational and conformational excitation and de-excitation. Thermal conduction in low-temperature crystals basically involves dynamics confined to a single inherent-structure basin. But due largely to the particle mobility that is present, liquid-phase thermal conduction proceeds with interbasin transitions.

The heat current vector $\mathbf{J}_H$ that appears in the general autocorrelation function expression for λ [Eqs. (II.101)–(II.102)] encompasses all of these liquid-state processes; it is repeated here from Section II.G:

$$\lambda = (3Vk_BT^2)^{-1}\int_0^\infty \langle \mathbf{J}_H(0)\cdot\mathbf{J}_H(t)\rangle\,dt, \tag{V.149}$$

$$\mathbf{J}_H(t) = (d/dt)\sum_{i=1}^{N}[E_i(t) - h_0]\mathbf{r}_i(t).$$

Here $E_i(t)$ stands for the energy that is to be assigned to particle i at time t, including its proportional share of all n-particle potential functions in which it participates at that moment. The quantity h_0 is the equilibrium average enthalpy per particle in the system at the state of interest. The obvious implication of Eqs. (V.149) is that liquids possessing a high λ experience relatively long-lasting fluctuations of energy flow, whereas in poor thermal conductors, those energy-current fluctuations simply die out quickly.

The observed increases for $\lambda(T)$ as temperature rises in water and glycerol around ambient conditions evidently arise from energy transport dominated by the dynamic behavior of the hydrogen bond networks that pervade those liquids. A reasonable assumption is that with rising temperature the resulting increase in concentration of bonding defects in those networks in the form of strained or broken hydrogen bonds are the source of this anomalous behavior. Such network defects possess structural excitation energy, and those defects can jump along the network while it is above its percolation threshold, giving rise to directionally correlated jumps and thus enhancement of energy-current persistence. However, at substantially elevated temperature, the hydrogen bond network would become so pervaded with seriously strained and broken hydrogen bonds that this energy propagation mode would no longer be possible, consistent with the nonmonotonic $\lambda(T)$ measurements for water heated along its vapor pressure curve to the critical point.

J. Freezing Transitions

The equilibrium first-order phase transition between a liquid and a crystalline solid thermodynamically requires equality of temperature, pressure, and chemical potentials of all particle species that reside in both phases. Consequently, it can be argued that any criterion claiming to locate this transition that is based solely on properties of just one of the two coexisting phases must necessarily be inaccurate. Yet that is just what the well-known Lindemann melting criterion advocates; see Section IV.I. The Lindemann criterion proposes that at least for single-component systems composed of spherically symmetric particles, the equilibrium melting transition occurs when the mean-square particle displacement caused by vibrations rises to a critical fraction of the nearest neighbor distance in that crystal. Computer simulations provide at least semiquantitative support for this proposition [Hansen and McDonald, 1976, Chapter 10].

There is a counterargument that can be formulated in support of at least some single-phase transition criteria, though its detailed application might be impractical in most cases. In particular, suppose that the thermodynamic and structural details of a liquid phase were known to very high accuracy over a wide range of temperatures and pressures, including the properties of the liquid's inherent structures and their basins. Suppose furthermore for simplicity that classical statistical mechanics applies. It can then be argued that in order to replicate those detailed and accurately determined properties, only the actual Hamiltonian or an extremely close approximation to that Hamiltonian would suffice. This implies that the two-body, three-body, ... interactions would have been determined, with minor error. But if that were the case, then use of those interaction potentials in an accurate statistical mechanical theory or numerical simulation would be able to identify that system's crystal structure and to locate its equilibrium melting and freezing behavior. Put succinctly, the liquid phase has hidden within it encoded information about that substance's other phases and the thermodynamic transitions that yield them.

Such a line of reasoning at least partially justifies formulating freezing criteria for liquids just in terms of their own properties. In particular, a natural "partner" for the venerable Lindemann melting criterion applicable to single-species spherically symmetric particles would be an "inverse Lindemann" freezing criterion for those same particles. The inherent-structure formalism provides a context within which both criteria can be viewed on a common basis. For both crystal and liquid phases, attention focuses on a dimensionless ratio:

$$\theta = a^{-1}\langle u_i^2\rangle^{1/2},\tag{V.150}$$

in which u_i represents the scalar displacement of particle i from its isochoric inherent-structure location, and a stands for a nearest neighbor distance. For a structurally perfect crystal, only a single type of inherent structure and its basin are involved in the vibrational average $\langle....\rangle$ and in the definition of a. However, the liquid phase at equilibrium samples a large family of basins and inherent structures, so the vibrational average required for θ intrinsically requires averaging over those basins as well. The nearest neighbor distance a in the liquid can be identified with the position of the first peak of the pair correlation function $g^{(2)}(r)$ for the state of interest, or perhaps more precisely and consistently with that of the postmapped pair correlation function $g_q^{(2)}(r)$ with its enhanced structural detail. It should be noted in passing that crystal phases containing significant defect concentrations would also require similar averaging over inherent-structure basins.

As the temperature of the crystal phase rises to its melting point under constant pressure conditions, its Lindemann ratio θ rises monotonically to a value θ_{mp}. Analogously, the liquid phase θ declines monotonically to θ_{fp} as temperature is reduced at the same fixed pressure to the freezing point. Although only limited evidence is currently available, it appears that θ_{fp} is significantly larger than θ_{mp} at coexistence. Whereas the conventional θ_{mp} for simple nonmetallic monatomic structureless particles lies in the neighborhood of 0.10–0.20, the ratio apparently jumps discontinuously by at least a factor of 2 to θ_{fp} for a model utilizing a foreshortened version of the Lennard-Jones pair potential [LaViolette and Stillinger, 1985, Figure 7]. Evidently, this substantial jump can be attributed to the increase in anharmonicity, including an enhanced fraction of negative Hessian eigenvalues (unstable displacement directions) of the sampled portions of the liquid-phase basins compared to those of the crystal.

Another liquid-phase freezing criterion has been proposed for systems containing a single species of spherically symmetric particles [Hansen and Verlet, 1969]. It is based on the structure factor

$$S(k) = 1 + \rho\int \exp(i\mathbf{k}\cdot\mathbf{r})[g_2(r) - 1]d\mathbf{r}.\tag{V.151}$$

In particular, the Hansen-Verlet freezing criterion utilizes the fact that the maximum of the structure factor (at k_0) increases as temperature decreases or density increases. Based upon simulations for the Lennard-Jones and hard-sphere pair potential models, the thermodynamic freezing transition is identified as occurring when

$$S(k_0) \approx 2.85.\tag{V.152}$$

Although it remains yet to be explored, there may be merit in re-examining the Hansen-Verlet freezing criterion in terms of the postmapping structure factor $S^{(q)}(k)$ corresponding to the postmapping pair correlation function $g_2^{(q)}(r)$, leading presumably to a numerically modified version of Eq. (V.152).

BOX V.6. Elemental Sulfur: A Melting Anomaly

Under atmospheric pressure and in the vicinity of room temperature, elemental sulfur exists in crystalline form with all sulfur atoms chemically bound as puckered octahedral ring molecules (S_8) [Donohue, 1982, pp. 328–338]. Several allotropic versions of this crystalline molecular form have been identified, the most stable of which is "β-monoclinic sulfur." However, sulfur can also exist as ring molecules of other sizes, as well as linear polymers, and many of these alternatives have been observed forming their own metastable crystal phases [Steudel, 1984]. The chemical reaction rates that control interconversion of these various molecular forms are very slow unless temperature is relatively high. This sluggishness of chemical kinetics produces an unusual melting scenario. On one side, rapid melting of β-monoclinic sulfur on the time scale of seconds is observed to occur at 120.14°C [Thackray, 1970], with the resulting liquid consisting essentially of just S_8 molecules. But maintaining the liquid isobarically at this temperature for hours causes the apparent melting/freezing temperature to decline by up to 5°C [Wiewiorowski et al., 1968] as the S_8 molecules convert slowly to a mixture of other molecular forms. In other words, the slow chemical kinetics cause the liquid sulfur phase to create its own antifreeze.

Extension of the inverse Lindemann or the Hansen-Verlet freezing criteria to multicomponent systems is not straightforward. The Lindemann measure for melting has been applied with some success to the cubic-symmetry alkali halide crystals, where the nearest neighbor distance is well defined (between ions of opposite charge), and the root-mean-square displacement is defined to be the simple average over all ions present [Martin and O'Connor, 1977]. Similarly, it is possible to identify the appropriate nearest neighbor distance a for the molten alkali halide salts from the position of the first peak in the unlike-ion pair correlation function $g_{+-}(r)$, and the root-mean-square intrabasin displacement as an average over all ions. However, the proper definitions to use for a Lindemann melting criterion become somewhat obscure for crystals with low symmetry, where the distance to near neighbors can depend on direction to those neighbors and where the distributions of particle displacements are not isotropic. It still needs to be recognized, though, that melting to an isotropic liquid phase eliminates such low-symmetry problems, so in principle an inverse Lindemann freezing criterion could still be formulated. It is less clear how to formulate a multicomponent Hansen-Verlet freezing criterion. Polyatomic molecules pose yet another class of challenges for both melting and freezing, in that several choices exist as to what nuclei or other positions in those molecules should serve as the "centers" for evaluation of distance a and the root-mean-square displacement quantity. Box V.6 presents details of the unusual melting and freezing behavior of a common element.

Table V.2 shows that melting entropies at atmospheric pressure for low-molecular-weight insulating substances exhibit a substantial variability. Such observations raise a basic question, currently unanswered, about whether a lower bound could be established for melting entropies. In particular, it would be important to know if there are circumstances for which melting entropy could be arbitrarily close to zero. It is not possible for a fluid-crystal critical point to exist at finite temperature and pressure because of the different point-group symmetries of the crystal and fluid phases [Alexander and McTague, 1978]. However, this situation leaves open the possibility that along an unbounded melting/freezing curve in the temperature–pressure plane that $\Delta S[T_m(p)]$ might continuously approach zero as $p \to +\infty$. The Gaussian core model presents one such possibility (it has no vapor–liquid critical point). In three dimensions, this model's stable crystal form at high density is body-centered cubic (bcc), and as pressure rises in this regime, its melting curve monotonically declines toward zero [Prestipino et al., 2005; Ikeda and Miyazaki, 2011]. One of the fundamental characteristics of this model's potential energy landscape is that increasing density is equivalent (with an appropriate energy rescaling) to a convolution of that landscape with another Gaussian [Stillinger and Stillinger, 1997]. The effect of this convolution is to smooth the landscape, eliminating many basins but increasing the extent of those that remain. This modification disproportionately eliminates those basins whose inherent structures underlying the fluid phase are most unlike the crystal but always preserves those basins supporting the bcc crystal. Consequently, the isobaric enumeration function $\hat{\sigma}(\hat{\varphi}^*)$ for the fluid at coexistence declines with increasing density, and the fluid locally approaches the crystal in both structural order and vibrational character. Among other expectations, the Lindemann and inverse Lindemann ratios for the melting crystal (θ_{mp}) and freezing fluid (θ_{fp}) at coexistence should approach one another ($\theta_{fp}/\theta_{mp} \to 1$) in this high-density asymptotic limit.

In order for the freezing process to initiate and eventually carry to completion in a reasonable time of observation, a liquid must be supercooled by more than an infinitesimal amount. Detailed description of the kinetics for the crystallization phase transition involves specifying the required nucleation process within the supercooled fluid medium. This phenomenon therefore falls naturally within the scope of Chapter VI and is considered in detail in Section VI.F.

VI.

Supercooled Liquids and Glasses

Given the proper conditions of purity, containment characteristics, and isolation from mechanical disturbance, most liquids can be cooled substantially below their thermodynamic freezing temperature without reverting immediately to the crystalline state. Such undercooling can be continued until either a crystal nucleation event finally intervenes, or else the system encounters kinetic arrest at a glass transition [Debenedetti, 1996; Debenedetti and Stillinger, 2001]. Metastable supercooled liquid states display experimentally reproducible properties that are smooth extrapolations of those in the higher temperature equilibrium liquid regime, which therefore become a legitimate object for theoretical attention. Furthermore, glasses (amorphous solids) that form as a result of deep supercooling enjoy considerable scientific and technological interest in their own right. The following sections adapt the energy landscape/inherent structure approach to description of supercooled liquid and glassy states. The development presented in this chapter is largely restricted to substances with low to moderate molecular weights. Chapter X examines supercooling and glass formation in polymeric media.

A. Basic Phenomena

Table VI.1 lists experimentally observed ambient-pressure melting/freezing temperatures T_m and glass transition temperatures T_g for a few pure nonmetallic substances that are frequently cited in the scientific literature about liquid supercooling and glass formation. Many further examples could have been listed, including mixtures and metallic alloys. Thermodynamic melting temperatures $T_m(p)$ are equilibrium attributes that are precisely defined at any given external pressure p. By contrast, glass transition temperatures $T_g(p)$ do not share such unambiguous precision but can appear to depend weakly upon cooling rate. The entries in the table display a nontrivial variation in the T_g/T_m ratio, which is outside this weak cooling-rate variability, and that diversity of the ratios is thus an indicator of intrinsic differences among substances.

During an experimentally imposed cooling-rate schedule, a liquid in a moderately supercooled state with $T > T_g$ is able to adjust its configuration-space sampling rapidly enough to continue reversibly on the smooth extension of the equilibrium liquid above T_m. This can be the case even though the rate of adjustment (i.e., relaxation rate) declines substantially with cooling. But below T_g, that adjustment can no longer occur, as the experimental cooling rate outruns the ability of the system to relax properly. The crossover from relaxed to unrelaxed status at a given moderate-to-slow cooling rate is typically rather sudden but nevertheless continuous. The isobaric heat capacity $C_p^{(liq)}$ of the supercooling liquid is affected by passage through the glass transition, as illustrated

TABLE VI.1. Characteristic temperatures of single-component nonmetallic glass formers at $p = 1$ atm

Substance	Chemical Formula	T_m, K	T_g, K	T_g/T_m
Boron trioxide	B_2O_3	723	539[a]	0.75
Silica (cristobalite)	SiO_2	1996	1446[b]	0.72
Arsenic trisulfide (orpiment)	As_2S_3	573	454[c]	0.79
Zinc chloride	$ZnCl_2$	556	371[b]	0.67
Ethanol	C_2H_5OH	160	95[d]	0.59
Toluene	$C_6H_5(CH_3)$	178	113[d]	0.63
Ortho-terphenyl	$C_6H_4(C_6H_5)_2$	329	240[e]	0.73
Salol (phenyl salicylate)	$C_{13}H_{10}O_3$	403	220[f]	0.55
Diethyl phthalate	$C_6H_4(COOC_2H_5)_2$	270	180[g]	0.67

[a]Angell and Rao, 1972, Figure 1.
[b]Angell et al., 2000, Figure 2.
[c]Angell, 1985, Figure 1.
[d]Debenedetti, 1996, Table 4.5.
[e]Chang and Bestul, 1972.
[f]Wagner and Richert, 1999.
[g]Chang et al., 1967.

qualitatively in Figure VI.1 for a constant slow cooling rate. At the thermodynamic melting/freezing temperature T_m, it is typically the case that $C_p^{(liq)} > C_p^{(crys)}$, and furthermore, this discrepancy tends to increase upon penetrating further and further into the supercooled liquid regime. This difference in heat capacities can be attributed to the fact that liquid-phase configurational relaxation (absent in the ordered crystal) causes the system in that phase to sink into lower and lower basins on the potential enthalpy landscape as cooling proceeds. However, this trend is largely suspended upon passage through the glass transition at T_g, causing C_p to drop precipitously as indicated in the figure. A dashed extension of the supercooled liquid curve to $T < T_g$, as shown in the figure, is a simple extrapolation that might seem reasonable if the kinetic arrest at the glass transition could somehow be avoided while still avoiding crystallization. The specific observed supercooling and glass transition behavior of the experimentally well-studied substance *ortho*-terphenyl illustrates these concepts. The fact that the increment $\Delta C_p = C_p^{(liq)} - C_p^{(crys)}$ of its constant-pressure heat capacity rises significantly as T decreases through the supercooled liquid interval toward T_g [Greet and Turnbull, 1967; Chang and Bestul, 1972] strongly implies that the relevant altitude range of the preferentially occupied portion of the potential enthalpy multidimensional landscape for the supercooled liquid is significantly temperature dependent.

The corresponding behavior of the system enthalpy H as a function of temperature is illustrated schematically in Figure VI.2, where the curves shown result from temperature integration of the heat capacities $C_p^{(crys)}$ and $C_p^{(liq)}$, with accounting for the latent heat of melting ΔH_m at T_m. Results for several cooling rates are indicated for the resulting glass phase after passing through the temperature range of kinetic arrest. Laboratory cooling rates can usually be varied over several orders of magnitude, but they cannot be arbitrarily slow because of limited time available for experimentation, and usually because of a nonzero rate of crystal nucleation. The curves shown in Figure VI.2 are intended to represent systems for which crystal nucleation did not detectably occur.

The temperature dependence of the volume V of a glass-forming substance, under isobaric conditions, superficially resembles the curves shown in Figure VI.2 for enthalpy, at least for liquids with positive thermal expansion. Upon cooling through T_g, the rate of decline of V suddenly

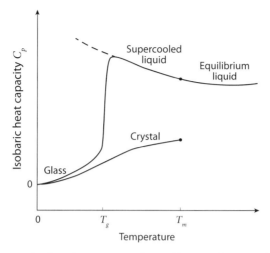

FIGURE VI.1. Typical behavior of isobaric heat capacities for equilibrium and supercooled liquid, glass, and crystal phases.

diminishes, as relaxation processes involving large-scale exploration of the enthalpy landscape become kinetically bypassed. Presuming that crystallization does not interfere, slower cooling rates lead to lower final glass volumes in the $T \to 0$ limit. Figure VI.3 illustrates this behavior.

At least for some glass formers, it is possible temporarily to circumvent the usual kinetic arrest that frustrates enthalpy and volume relaxations upon passing downward through T_g in bulk

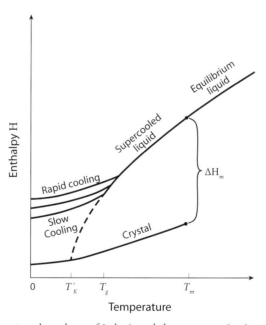

FIGURE VI.2. Typical temperature dependence of isobaric enthalpy upon entering into the supercooled and glass-forming regimes, for a liquid subject to a range of cooling rates. Assignment of the glass transition temperature T_g has been identified with the slow limit of attainable cooling rates. The thermally equilibrated crystal enthalpy is shown as well, and the latent heat of melting has been denoted by ΔH_m.

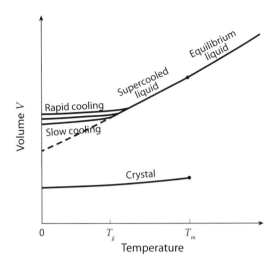

FIGURE VI.3. Typical temperature dependence of the volume of a glass-forming supercooled liquid, for various cooling rates under isobaric conditions. The behavior illustrated refers to the common case of liquids with positive thermal expansion.

samples, even for their slowest attainable cooling rates. This alternative involves slow vapor deposition of the glass-forming substance on a surface maintained at a temperature substantially below the usual bulk T_g range [Swallen et al., 2007; Kearns et al., 2008, 2009]. Evidently, the outermost layers of the growing amorphous film permit substantially higher molecular mobility than in the bulk, thereby permitting those molecules to relax into lower enthalpy and lower volume configurations that would otherwise be missed by relaxation processes within the interior of a bulk sample. But in order for this to be a viable alternative to bulk-sample cooling, the film geometry must be relatively unfavorable for crystal nucleation.

The measurable shear viscosity $\eta(T)$ of a supercooled liquid rises rapidly upon approach to the glass transition region. In order to eliminate uncertainty due to cooling rate differences, a frequently used convention for supplying a precise (but somewhat arbitrary) definition of T_g relies upon this characteristic. It utilizes the following empirical criterion [Angell, 1988]:

$$\eta(T_g) = 10^{13} \text{ poise.} \tag{VI.1}$$

To put this into context, the shear viscosity of liquid water at its normal freezing point 0°C is 0.018 poise. Actually attaining this criterion, Eq. (VI.1), in a fully relaxed sample of a glass-forming substance experimentally requires very slow cooling of the liquid medium.

The dashed extension for $T < T_g$ of the supercooled liquid enthalpy curve shown in Figure VI.2 corresponds to the dashed $C_p^{(liq)}$ presumed extension in Figure VI.1. Its slope is larger than that of the crystal enthalpy curve because of the large heat capacity difference between crystal and supercooled liquid phases. Consequently, the simple enthalpy extrapolation for the supercooled liquid seems to require an intersection with the crystal enthalpy at a temperature $T_K' > 0$, as indicated in Figure VI.2. Kauzmann was apparently the first to point this out as a general situation [Kauzmann, 1948]. In view of the fact that the crystal phase is the configurational ground state (aside

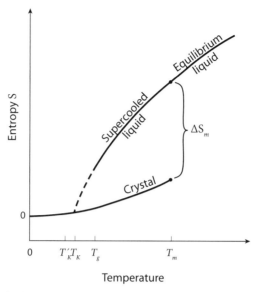

FIGURE VI.4. Entropies vs. absolute temperature for crystal and supercooled liquid under isobaric conditions. The dashed extension relies on an extrapolated heat capacity indicated in Figure VI.1. The intersection of the two curves defines the "Kauzmann temperature" $T_K > 0$.

from very small defect concentrations that it might contain at positive temperatures), this naive extrapolation scenario surprisingly implies that the disordered supercooled liquid enthalpy would continuously approach equality with that of the ordered crystal state as T declines toward T_K'. Furthermore, formal continuation of the extrapolation beyond the intersection would lead to the unacceptable conclusion that disordered particle configurations would lie lower in enthalpy or energy than the ordered crystalline configuration.

A more frequently invoked extrapolation procedure, analogous to that for enthalpy, involves the entropy difference between the crystal and the supercooled liquid phases. This was examined early on specifically for glycerol by Simon [Simon, 1930], but as a more general observation it is usually attributed to Kauzmann [Kauzmann, 1948]. This has been illustrated in Figure VI.4. The implication of the plot appears to be that the entropies of the two phases become equal at another temperature $T_K > 0$. It is this second intersection temperature that is conventionally called the "Kauzmann temperature." Note that the temperature dependence of the difference in entropy between the supercooled liquid (with its extension below T_g) and the crystal leads to an inequality when $T < T_m$:

$$\Delta S(T) = S^{(liq)}(T) - S^{(crys)}(T)$$

$$= \frac{\Delta H_m}{T_m} - \int_T^{T_m} [C_p^{(liq)}(T') - C_p^{(crys)}(T')(dT'/T')$$

$$< (1/T_m)\left\{\Delta H_m - \int_T^{T_m} [C_p^{(liq)}(T') - C_p^{(crys)}(T')]dT'\right\}$$

$$= \Delta H(T)/T_m. \tag{VI.2}$$

Consequently, the vanishing of $\Delta S(T_K)$ requires that $\Delta H(T_K) > 0$. This in turn requires that $T_K > T_K'$, as indicated in Figure VI.4.

The so-called "Kauzmann paradox" arises from any attempt to continue the dashed entropy extrapolation for $S^{(liq)}(T)$ to temperatures lower than T_K. In view of the general understanding that the thermal equilibrium low-temperature crystal is a spatially ordered arrangement of its constituent particles and that the extrapolated liquid would necessarily be configurationally disordered, Kauzmann argued that extrapolating the entropy below T_K would constitute a claim that "disorder" had a lower entropy than "order." Consequently, he maintained that something must intervene to prevent the possibility of the crossing. One suggestion is that an "ideal glass transition" occurs at T_K, thus terminating the extrapolation and yielding a zero-entropy amorphous state (i.e., exhibiting an inherent structure that is substantially unique). Inequality (VI.2) then would imply that this ideal glass state has an enthalpy exceeding that of the crystal at T_K, and so it would remove the earlier implication, based on enthalpy curves in Figure VI.2, that the extrapolated liquid would continuously approach and attain the crystal-state enthalpy upon cooling to T_K'. A critical analysis of the ideal glass concept and hypothesis from the statistical mechanical viewpoint appears in Section VI.D.

It is necessary to acknowledge at this point that the argument just presented implicitly supposes that the vibrational contributions to heat capacities, enthalpies, and entropies of the two phases are substantially equal at the intersection point. A more detailed version of the argument could accommodate modest differences in these vibrational contributions and would result in small shifts of T_K' and T_K. But in particular, the tentative inference that an ideal glass transition in principle might exist at a positive Kauzmann temperature would survive that elaboration.

The reader should be cautioned that equality of liquid and crystal entropies is not merely confined to the behavior of supercooling and glass-forming substances. Indeed, it even appears occasionally in a strict thermal equilibrium regime. This is illustrated by the phase behavior of the helium isotopes, where it arises from extreme quantum effects to be discussed in Chapter VIII. Furthermore, polymers possessing large numbers of intramolecular degrees of freedom that are coupled to translational degrees of freedom in a special way can also exhibit equal-entropy crystal-fluid transitions at equilibrium within the context of classical statistical mechanics [Feeney et al., 2003; Stillinger and Debenedetti, 2003]. These exceptional polymeric cases are put aside for later consideration in Chapter X. The existence of both of these kinds of equal-entropy occurrences has no direct implications for the present analysis of supercooling and glass formation.

Well below the Kauzmann temperatures T_K and T_K', and upon approach to $T = 0$, the heat capacities of glasses formed from supercooled liquids and of other amorphous solids display a behavior that sets them apart qualitatively from crystalline solids. As pointed out in Chapter IV [Eq. (IV.32)], low-frequency phonons dominate the low-temperature heat capacity of insulating crystals, producing a result proportional to T^3. However, calorimetric measurements on nonmetallic glasses indicate that the ratio C_p/Nk_BT^3 (or the nearly equivalent C_V/Nk_BT^3) diverges as $T \to 0$. In fact, for these noncrystalline solids, the heat capacities typically in the temperature range around 1 Kelvin and below empirically exhibit a combination of the usual T^3 type of phonon contribution, supplemented by a contribution that displays a substantially smaller positive power of T, i.e.,

$$C_V/Nk_B \cong K_u T^u + K_3 T^3, \qquad \text{(VI.3)}$$

where K_u and K_3 are positive constants. An early experimental study of several inorganic glasses indicated that $u \approx 1$ [Zeller and Pohl, 1971]; later work on amorphous silica found that $u \approx 1.3$ [Hunklinger, 1997; Trachenko et al., 2002]. In any case, this T^u contribution becomes dominant at sufficiently low temperature. It is a characteristic attribute that indicates a basic difference in the potential landscape topographies sampled by the many-particle system, respectively, in the crystalline and in the amorphous portions of the configuration space. This is a distinguishing property to be discussed in Section VI.C. It should be noted that the magnitude of this contribution has been shown experimentally to depend on the preparation method for the amorphous solid involved [Ediger, 2014; Pérez-Castañeda et al., 2014].

A functional form frequently used to fit the strong temperature dependence of the shear viscosity at constant pressure is the Vogel-Tammann-Fulcher (VTF) equation [Debenedetti, 1996, p. 257]:

$$\eta(T) \approx \eta_0 \exp\left[\frac{A}{T - T_0}\right], \tag{VI.4}$$

where η_0, A, and T_0 are positive constants. A simple Arrhenius form is recovered by setting $T_0 = 0$, but this seldom describes liquids well over their full equilibrium and supercooling temperature range. More generally, a positive value for this parameter is required for a satisfactory fit to viscosity measurements. But of course, that requirement formally implies a divergence of η as $T \to T_0$, a physical impossibility. If indeed the fit is accurate down to the glass transition, Eq. (VI.1) would obviously require $T_0 < T_g$. It should be noted in passing that other three-parameter functions have been proposed to fit shear viscosity measurements, but they do not involve divergence at a positive temperature [Avramov and Milchev, 1988; Mauro et al., 2009].

Unfortunately, this VTF form does not universally supply an accurate fit to shear viscosities for most substances over the entire temperature range of measurements. However, it can provide a useful empirical fit over at least limited temperature intervals in the equilibrium and supercooled regimes. For example, its three adjustable parameters can be fixed by requiring that an isobarically measured $\eta(T)$ and its first two temperature derivatives at some chosen temperature $T^* > T_g$ be reproduced. Set

$$l_0 = \ln \eta(T^*),$$

$$l_1 = [(d/dT) \ln \eta(T)]_{T=T*} = \frac{\eta'(T^*)}{\eta(T^*)}, \tag{VI.5}$$

$$l_2 = [(d/dT)^2 \ln \eta(T)]_{T=T*} = \frac{\eta''(T^*)}{\eta(T^*)} - \left[\frac{\eta'(T^*)}{\eta(T^*)}\right]^2.$$

Then the three VTF parameters are determined as follows:

$$\eta_0 = \exp\left[l_0 + \frac{2(l_1)^2}{l_2}\right],$$

$$A = -\frac{4(l_1)^3}{(l_2)^2}, \tag{VI.6}$$

$$T_0 = T^* - \frac{2l_1}{l_2}.$$

Glass-forming liquids can be classified on a continuous "strong" vs. "fragile" scale, according to the extent to which their isobaric shear viscosities do, or do not, adhere closely to Arrhenius temperature behavior [Angell, 1988]. In terms of a VTF fit, this classification depends simply on the value assigned to the dimensionless parameter ratio A/T_0. Very large values (≥ 100) signify nearly Arrhenius temperature dependence as exhibited, for example, by molten SiO_2 and GeO_2. Smaller values of the ratio (≈ 5) indicate behavior at the fragile end of the scale, observed for *ortho*-terphenyl in some portion of its supercooled temperature range. However, it must be stressed that any given substance may display a substantially variable strength/fragility measure A/T_0 when examined over its full equilibrium and supercooled liquid range. Indeed, *ortho*-terphenyl itself provides an example of this variability, for which an "Arrhenius plot" of $\log_{10}\eta(T)$ vs. $1/T$ changes from distinctly curved to nearly linear upon cooling to its T_g, and salol behaves similarly [Laughlin and Uhlmann, 1972].

As mentioned in Section V.I, the Stokes-Einstein relation provides an approximate connection between the shear viscosity $\eta(T)$ and the translational self-diffusion constant $D(T)$ for a single-component liquid:

$$D(T) = k_B T/C\eta(T)R_{eff}. \tag{VI.7}$$

The underlying elementary picture assigns an effective hydrodynamic radius R_{eff} to a diffusing particle and presumes that it moves by Brownian motion in an incompressible fluid medium with the given viscosity. The positive constant C depends on the hydrodynamic boundary condition applicable at the particle surface (assumed to be spherical), equaling 6π or 4π for "stick" or "slip," respectively. The $D(T)$ predictions supplied by the Stokes-Einstein relation, using the "stick" value of C and a reasonable choice for a temperature-independent R_{eff}, are surprisingly good for non-polymeric glass formers in their equilibrium and modestly supercooled ranges. Considering the fact that several orders of magnitude change in $\eta(T)$ can be involved, this is a nontrivial observation [Walker et al., 1988]. However, the simple Stokes-Einstein situation disappears for at least some strongly supercooled liquids, particularly those that display "fragile" behavior. In particular, self-diffusion constants do not decline with temperature reduction as steeply as the Stokes-Einstein formula would indicate, implying onset of a decoupling between the self-diffusion and the shear viscosity processes. As an example, when liquid *ortho*-terphenyl supercools toward its T_g, the measured translational diffusion constant becomes approximately two orders of magnitude larger than its viscosity and the Stokes-Einstein relation would predict [Fujara et al., 1992]. Paradoxically, the analogous Debye-Stokes-Einstein relation predicting that the rotational diffusion constant $D_{rot}(T)$ is also proportional to $T/\eta(T)$ remains substantially valid down to the *ortho*-terphenyl glass transition [Fujara et al., 1992; Cicerone et al., 1995].

An important aspect of the temperature dependence of structural relaxation in deeply supercooled liquids can be revealed by the frequency-dependent dielectric response function $\varepsilon(v) = \varepsilon_r(v) - i\varepsilon_i(v)$. Here v is the frequency, and ε_r and ε_i represent the real and imaginary parts of that function, respectively, the latter indicating the presence of dissipative processes. For supercooling temperatures somewhat above the glass transition, the imaginary part often exhibits a pair of broad maxima in the range 10^{-2}–10^6 Hz when plotted vs. frequency, indicating the presence of distinguishable types of structural relaxation modes. Traditionally, these have been designated as the "α" (or "primary") and "β" ("secondary," or alternatively, "Johari-Goldstein") relaxation pro-

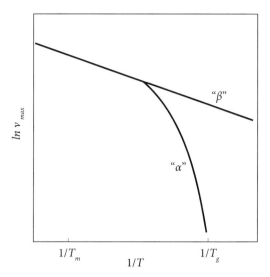

FIGURE VI.5. Typical temperature dependences of "α" and "β" relaxation process categories, as inferred from maxima exhibited by dielectric relaxation measurements. The plot indicates how the logarithms of the peak dielectric absorption frequencies v_{max} vary with inverse temperature.

cesses for the low- and high-frequency modes, respectively [Johari and Goldstein, 1970; Johari, 1973; Angell, 1990; Yardimci and Leheny, 2006]. But as the temperature rises, these distinguishable dissipation frequency maxima approach one another and become indistinguishable, leaving just a single broad maximum. Figure VI.5 illustrates these characteristics. One basic observation presented by Figure VI.5 is that the "β" relaxation tends to display an Arrhenius temperature dependence below the "α," "β" bifurcation point and that this temperature dependence is approximately an extension of that displayed by the single maximum at higher temperature. A second basic observation is that the "α" process has a distinctly non-Arrhenius temperature dependence. This latter behavior can be described empirically at least approximately by a VTF-type function, Eq. (VI.4), with an apparent divergence temperature $T_0 > 0$ that could be compared to the corresponding divergence temperature obtained by fitting shear viscosity measurements. A third basic observation is that the "β" process survives passage through the glass transition, in contrast to the "α" process, which becomes so slow at that stage as to be unobservable. Shape details of the distributions presenting the "α" and "β" maxima depend on the glass-forming material involved.

A brief notational comment is in order. Because of established but conflicting conventions, the second letter in the Greek alphabet is used both for inverse temperature ($\beta = 1/k_B T$) as well as for one of the low-temperature relaxation processes ("β"). Consistent use of surrounding quotation marks for the latter in the present discussion should forestall confusion.

When a viscous liquid, in particular one just above its T_g, is subjected to a shear stress at sufficiently high frequency, it behaves as an elastic solid with a shear modulus G_∞. Experimentally, this shear modulus is largely unaffected by passage through the glass transition. If the shear stress is suddenly applied by deforming the sample homogeneously and then subsequently the shape of the deformed volume is held fixed, the shear stress slowly relaxes to zero as the particle configuration within the system kinetically rearranges. A simple overall measure of the time required for this structural relaxation to occur (in the linear response regime) is provided by the ratio

$$\bar{\tau} = \eta(T)/G_\infty. \tag{VI.8}$$

If the stress relaxation strictly exhibited a simple exponential decay with time [i.e., was proportional to $\exp(-t/\bar{\tau})$] the process would be called a "Maxwell relaxation" [Landau and Lifshitz, 1959, pp. 130–131]. However, following an initial fast response that precedes slow structural relaxation, strongly supercooled glasses typically exhibit a more complicated "stretched exponential" relaxation profile, also known as a Kohlrausch-Williams-Watts (KWW) relaxation function:

$$f_\theta(t) = \exp[-(t/\tau)^\theta], \tag{VI.9}$$

where the stretching exponent lies in the range $0 < \theta < 1$ [Debenedetti, 1996, p. 269]. In this empirical KWW expression, both τ and θ are functions of temperature and pressure. The relaxation time-scale quantity τ increases strongly upon cooling toward T_g, as does the Maxwell time $\bar{\tau}$ in Eq. (VI.8), but detailed examinations of experimental data for many glass formers do not support the concept that these relaxation times might diverge at a positive temperature [Hecksher et al., 2008; Elmatad et al., 2009]. The stretching exponent θ is often observed to change at a modest rate from a value near unity in the thermodynamically stable liquid above T_m, toward approximately the value 1/2 at T_g [Ediger, 2000]. In the strongly supercooled liquid, approaching T_g, the time scale τ of the KWW relaxation function is roughly comparable to that of the "α" relaxation processes just described, strongly suggesting that both are controlled by the same class of kinetic processes, though possibly with different weightings.

The empirical applicability of the stretched exponential KWW relaxation function is not limited to shear stress relaxation. It has also been used to describe time-dependent dielectric response [Angell and Smith, 1982] and response to temperature jumps [Dixon and Nagel, 1988]. In most of these cases, upon cooling toward T_g the qualitative behavior of the temperature dependence for fitting parameters τ and θ is similar for distinct external perturbations, but not identical. Any quantitative distinctions presumably are due to the different roles that various specific molecular-level kinetic processes play in responding to each of the different physical perturbations applied.

Relaxation functions found experimentally for supercooled *ortho*-terphenyl are rather well fitted by stretched-exponential (KWW) form, with exponent $\theta(T)$ declining apparently monotonically as T is reduced toward T_g [Fytas et al., 1981; Fujara et al., 1992]. A similar result has been found for supercooled triphenylphosphite [Silence et al., 1992]. Both of these substances are conventionally classified as fragile glass formers.

The stretched exponential relaxation function $f_\theta(t)$ can formally be resolved into a continuous spectrum of simple exponential decay processes occurring in parallel by expressing it as a Laplace transform:

$$f_\theta(t) = \int_0^\infty F_\theta(s)\exp(-st)ds. \tag{VI.10}$$

The distribution represented by $F_\theta(s)$, where s is the decay rate, is non-negative. This distribution approaches a Dirac delta function at $s = 1/\tau$ as $\theta \to 1$ (the Maxwell relaxation limit), but it is other-

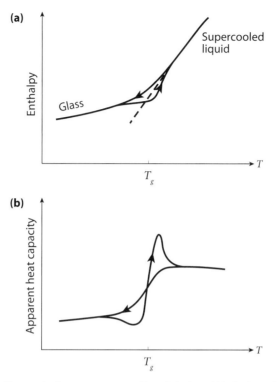

FIGURE VI.6. Illustration of hysteresis phenomena generated by relatively rapid isobaric cooling through the glass tran-sition region, followed by isobaric heating at a rate reversing that of the cooling. Part (a) indicates the behavior of system enthalpy during these sequential operations, whereas part (b) presents the corresponding apparent heat capacities.

wise finite and positive for all $s \geq 0$. For certain θ values, closed-form expressions are available for $F_\theta(s)$ [Helfand, 1983]. One specific example occurs when the stretching exponent $\theta = 1/2$:

$$F_{1/2}(s) = [2\pi^{1/2}\tau^{1/2}s^{3/2}]^{-1}\exp[-(4\tau s)^{-1}]. \tag{VI.11}$$

For other values of stretching exponent θ, the relaxation distribution $F_\theta(s)$ shares with this last expression a strong vanishing as $s \to 0$, the dominant leading term of which is [Stillinger, 1990]:

$$\ln[F_\theta(s)] = -[(1-\theta)\theta^{\theta/(1-\theta)}](\tau s)^{-\theta/(1-\theta)} + o(s^{-\theta/(1-\theta)}). \tag{VI.12}$$

It is this behavior near the origin that controls the asymptotic long-time behavior of the KWW stretched exponential relaxation function.

The sluggishness of relaxation processes in the temperature vicinity of T_g gives rise to hyster-esis phenomena. Figure VI.6(a) illustrates this feature by showing qualitatively how system en-thalpy typically behaves during a scenario that first cools the sample through T_g at some modest rate $dT/dt = -R$ well into the vitreous glass state, followed by heating back through T_g at the re-versed rate $dT/dt = +R$ [Jäckle, 1986, p. 184]. Figure VI.6(b) presents the hysteresis phenomenon in an alternative manner, simply indicating the apparent isobaric heat capacities for the cooling and heating stages, which are nothing more than the temperature derivatives of the enthalpy curves in Figure VI.6(a). The existence of this, and similar, hysteresis loops around T_g can informally be rationalized by the statement that "It is easier to fall down than to fall up." Microscopic reversibility implies that transitions between pairs of states must have the greater rate from the higher energy or enthalpy state to the lower one compared to the reverse, and the rate discrepancy increases with increasing difference in energy or enthalpy of the states involved. In the glass transition context, this indicates that the heating stage exhibits slower enthalpy increase around T_g than the corresponding enthalpy decrease during cooling. However, as temperature continues to climb at positive rate R, producing enhancement of all rates, the warming system eventually has the chance to "catch up," displaying a maximum in the apparent "postponed" heat capacity, and then retracing the enthalpy curve well above T_g.

Although downward passage through the conventional T_g to a substantially lower temperature effectively switches off most structural relaxation processes, a few remain barely active in most glasses. This is consistent with the broad distribution of relaxation rates represented by $F_\theta(s)$ for $\theta < 1$. If the cooled vitreous sample is stored isothermally and isobarically for a long annealing period, perhaps even stretching over months or years, measurable changes in physical properties result, the presence of which is unsurprisingly called "aging" [Struik, 1978]. The affected properties can include volume, elastic constants, plastic flow measures, and enthalpy. The last of these properties drifts slowly downward during the extended annealing period, a result of which is that subsequent heating displays an enhancement of the apparent heat capacity peak that was illustrated in Figure VI.6(b). It should also be noted that the structural relaxation resulting from aging can be reversed (the sample "rejuvenated") by plastic deformation [Utz et al., 2000].

B. Landscape Characteristics

A many-particle system that is initially in a well-equilibrated liquid state above its melting temperature $T_m(p)$ inhabits a portion of its potential energy (or potential enthalpy) landscape whose inherent structures are amorphous. Subsequent slow and undisturbed cooling down through that $T_m(p)$ normally avoids crystal nucleation, at least for moderate supercooling. As a result, the system continues to remain in the same characteristic portion of its potential landscape that is composed of steepest descent basins surrounding amorphous inherent structures, and it is the topographic details of this portion that underlie the phenomenology described in Section VI.A. Failure of the dynamics to discover a thermodynamically required exit path in the configuration space to a crystalline phase, but instead to get "lost" elsewhere, illustrates the complexity of landscape topography that controls the dynamics. That complexity is magnified in the case of good glass formers, where crystal nucleation rates are especially low over the supercooling temperature interval $T_g < T < T_m$. This is the kind of physical situation that led earlier to the projection approach for the configuration space, Section III.G, to facilitate generally the description of metastable states.

As applied to the liquid supercooling and glass-forming situation, the basic presumption of the projection approach is that inherent structures and their basins can be classified into those that exhibit substantial regions of crystalline order and those that do not. The qualifier "substantial" is not uniquely defined but involves identifying within the inherent structures local regions that are large enough to act as nuclei for crystal growth under some significant amount of supercooling; see Section VI.F for additional details. The projection must eliminate structural patterns not only for the crystal form that is thermodynamically stable below $T_m(p)$ but also for those of any other near-stable crystal structures. Experimental procedures for studying supercooled liquids demonstrate empirical reproducibility of measured properties, with no nucleation, at least for some temperature interval below T_m. This indicates that within that temperature interval the system dynamics manage to avoid the transition zone between obviously fully amorphous, and obviously partly crystalline, inherent structures and basins. This implicitly indicates that those reproducible properties are insensitive to the precise details of the inherent structure and basin classification rule.

For the purpose of analyzing liquid supercooling and glass formation, let subscript "a" denote those inherent structures and their basins identified as fully amorphous. This replaces the more generic subscript "m" used in Chapters III and V to stand for "metastable." The totality of these amorphous basins defines (i.e., tiles) a part of the full configuration space that may be denoted as a multidimensional region $\mathbf{A}$ for isochoric (constant-volume) conditions, or $\hat{\mathbf{A}}$ for isobaric (constant-pressure) conditions. These regions respect permutational symmetry in that inclusion of any one basin implies inclusion of all its permutation-equivalent replicas. It is reasonable to assume in all cases of interest that both $\mathbf{A}$ and $\hat{\mathbf{A}}$ individually are topologically connected, so that no subset of amorphous basins is isolated geometrically from all the others within the multidimensional configuration space. Following the usual strategy for the large-system asymptotic limit, the basins in $\mathbf{A}$ and $\hat{\mathbf{A}}$ can be described by their depth on a per-particle basis, that is, by potential energy per particle φ or by potential enthalpy per particle $\hat{\varphi}$, for isochoric or isobaric circumstances, respectively. Then in the large-system limit, the enumeration functions for distinct (unrelated by particle permutations) inherent structures are written as $\exp[N\sigma_a(\varphi)]$ and $\exp[N\hat{\sigma}_a(\hat{\varphi})]$. It should be emphasized at this point that because system volume V is one of the steepest descent variables for the isobaric case that defines $\hat{\sigma}_a(\hat{\varphi})$, this enumeration function characterizes the mapped many-particle system at a density different from that to which the isochoric enumeration function $\sigma_a(\varphi)$ refers.

There are lowest energy and enthalpy values $\varphi_{\min,a}$ and $\hat{\varphi}_{\min,a}$ for amorphous inherent structures, at which, respectively, $\sigma_a(\varphi)$ and $\hat{\sigma}_a(\hat{\varphi})$ can be taken to equal zero, and below which they are undefined. These lowest amorphous-structure energy and enthalpy values nevertheless lie well above the corresponding lowest values for the entire configuration space because of the energetic and enthalpic favorability of crystalline order:

$$\varphi_{\min} < \varphi_{\min,a}, \tag{VI.13}$$

$$\hat{\varphi}_{\min} < \hat{\varphi}_{\min,a}.$$

By contrast, it seems reasonable to expect that the highest lying inherent structures, whether isochoric or isobaric conditions apply, lie within region $\mathbf{A}$ or $\hat{\mathbf{A}}$, indicating that $\varphi_{\max,a} = \varphi_{\max}$ and $\hat{\varphi}_{\max,a} = \hat{\varphi}_{\max}$. It is not a priori clear whether the basins of the lowest lying amorphous inherent structures occur entirely within the interiors of the regions $\mathbf{A}$ and $\hat{\mathbf{A}}$, or whether they lie at the boundaries shared with the crystal-pattern-containing regions that have been projected out of consideration. It is possible that this distinction is substance dependent.

Inability of a supercooled liquid system to access its stable crystal form, at least over the time scale of experimental or simulational observation, by definition means it is non-ergodic. However, in the temperature range $T_g < T < T_m$, the empirically established reproducibility of measurable properties for the supercooled liquid indicates that the dynamics is at least quasi-ergodic over the restricted configuration-space regions **A** and **Â**. This justifies use over that range of the following metastable-state Helmholtz and Gibbs free-energy expressions, respectively (from Section III.G), for the large-system (large-N) limit:

$$\beta F_{N,a}/N = \beta\{\varphi_a{}^*(\beta,\rho) + f_{vib,a}[\varphi_a{}^*(\beta,\rho),\beta,\rho]\} - \sigma_a(\varphi_a{}^*(\beta,\rho)), \qquad \text{(VI.14)}$$

$$\beta G_{N,a}/N = \beta\{\hat{\varphi}_a{}^*(\beta,p) + \hat{f}_{vib,a}[\hat{\varphi}_a{}^*(\beta,p),\beta,p]\} - \hat{\sigma}_a[\hat{\varphi}_a{}^*(\beta,p)].$$

In these expressions, $f_{vib,a}$ and $\hat{f}_{vib,a}$ are mean vibrational free energies per particle at those portions of the regions **A** and **Â** that are identified, respectively, by the values $\varphi_a{}^*$ and $\hat{\varphi}_a{}^*$ of the intensive order parameters. The asterisk designations indicate that they are those special values which minimize the respective free energies. The metastable-state free-energy expressions (VI.14) become inapplicable as soon as lowering the temperature causes a system to enter and pass through its glass transition region, consequently causing it to fail even to be quasi-ergodic.

The projection operations that define **A** or **Â** are chosen specifically to suppress identifiable localized "clusters" or "droplets" of the crystal phase that would occupy the entire system below T_m if strict thermodynamic equilibrium were attained. Infrequently, such clusters or droplets would spontaneously form and disappear in the stable liquid, though their concentration should be very small even at the phase transition point T_m. However, for consistency, the same free-energy expressions (VI.14) that suppress such local structures should be utilized both above and below T_m. In principle, this cluster suppression modifies the equilibrium free energy for the stable liquid, and in particular has the formal effect of eliminating an essential singularity that theory predicts will occur exactly at the equilibrium first-order phase transition point [Andreev, 1964; Langer, 1967]. In the strict mathematical sense, this essential singularity would prevent analytic continuation of free-energy functions for the stable liquid across the first-order transition point at $T_m(p)$ into the metastable supercooled liquid regime. However, the singularity has bounded temperature derivatives of all orders and is believed to be numerically virtually undetectable, thus implying that the projection method advocated here is a useful description of the experimental supercooling process. Any error committed is presumably negligible by conventional experimental and simulational standards.

The multidimensional geometry of possible potential energy and potential enthalpy landscapes in principle allows for the existence of many hypersurfaces that have identically the same free-energy functions (at least in the classical limit), but substantially different topographies. In particular, this could be the case for the regions **A** and **Â**. Such a situation implies that conceivably two distinct substances could exhibit identical thermodynamics extending into the supercooled liquid regime but at the same time could also have very different kinetic behaviors, such as distinct glass transition temperatures and different "fragilities." A very simple illustration of this geometric diversity is possible even in a hypothetical one-dimensional "configuration space" [Stillinger and Debenedetti, 2002]. This example composes landscapes out of single-basin modular units that are elementary trigonometric functions defined over the unit interval $0 \le x \le 1$:

$$P(x|+1) = -(9/16)\cos(\pi x) + (1/2)\cos(2\pi x) + (1/16)\cos(3\pi x), \qquad \text{(VI.15)}$$

$$P(x|-1) \equiv P(1-x|+1)$$

$$= (9/16)\cos(\pi x) + (1/2)\cos(2\pi x) - (1/16)\cos(3\pi x).$$

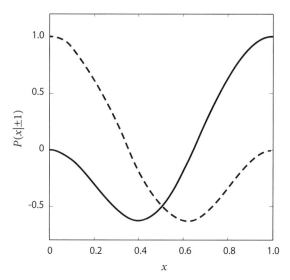

FIGURE VI.7. Plots of the mirror-image modular functions $P(x|+1)$ (solid curve) and $P(x|-1)$ (dashed curve) over their interval of definition $0 \leq x \leq 1$. Reproduced from Stillinger and Debenedetti, 2002, Figure 1.

These modular functions have vanishing slopes at the endpoints $x = 0,1$, equal downward curvatures at those endpoints, and single minima ("inherent structures") inside the unit interval of definition. As the notation indicates, $P(x|+1)$ increases by +1 from left to right, while the mirror-image function $P(x|-1)$ decreases by the same amount from left to right. Figure VI.7 shows plots of these modular functions. It is clear that a steepest descent mapping that was initiated anywhere within the unit interval for either of these functions would converge to the single interior minimum. Furthermore, it is obvious that the intrabasin classical vibrational partition functions is identical for these two modular functions at all temperatures.

By horizontally and vertically translating a sequence of some integer number M of these modular units, it is a straightforward matter to synthesize a model one-dimensional potential function $\Phi(x)$, defined on $0 \leq x \leq M$, that is continuous and twice differentiable. Let this sequence be specified by its string of "Ising spins" $\mu_j = \pm 1$, denoting which of the two choices $P(x|+1)$ or $P(x|-1)$ is used in the jth position. One can impose periodic boundary conditions by initially requiring equal numbers of each spin value (so M must be even):

$$\sum_{j=1}^{M} \mu_j = 0. \tag{VI.16}$$

With this protocol, the explicit form of the model landscape potential in each of the unit intervals is the following:

$$\Phi(x) = P(x|\mu_1) \qquad (0 \leq x \leq 1), \tag{VI.17}$$

$$= P(x - j + 1|\mu_j) + \sum_{k=1}^{j-1} \mu_k \qquad (2 \leq j \leq M, j - 1 \leq x \leq j).$$

Figure VI.8 presents two examples of this construction, Φ_A and Φ_B, both with $M = 60$ and both subject to periodic boundary conditions [Stillinger and Debenedetti, 2002]. This pair of examples shares the further characteristic that the distributions of basins they contain vs. altitude

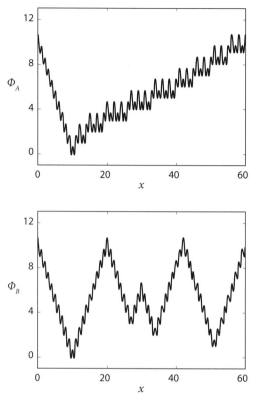

FIGURE VI.8. Two model potential functions in one dimension with identical distributions of 60 inherent structures by height (i.e., by energy) but with distinctly different landscape topographies. Reproduced from Stillinger and Debenedetti, 2002, Figure 3.

are exactly the same. Specifically, the numbers of basins appearing in either Φ_A or Φ_B according to ascending inherent structure energy (reading left to right) is

$$2,4,6,8,8,8,6,6,6,6. \tag{VI.18}$$

This result ensures that the classical canonical configuration integrals for the two cases are identical functions of inverse temperature β:

$$\int_0^M \exp[-\beta\Phi_A(x)]dx \equiv \int_0^M \exp[-\beta\Phi_B(x)]dx. \tag{VI.19}$$

However, the visual appearances of the two topographies clearly are very different. The pattern presented by Φ_A is that of a single broad and deep "valley," whereas that of Φ_B displays several narrower and shallower "valleys" separated by significant barriers. Kinetic redistribution of probability on the former landscape would be relatively rapid at low temperature, but the high internal barriers on the latter landscape would retard redistribution, if not actually trapping the dynamics and thus preventing full relaxation. One easily discovers other sequences of 60 Ising spins summing to zero, beyond those generating Φ_A and Φ_B, that have the same vertical basin/inherent structure distribution, Eq. (VI.18), but that show different distinguishing topographic characteristics.

As the integer M increases, the number of distinguishable model landscapes that can possess identical inherent structure altitude distributions and thus identical classical partition functions also increases. This change expands the possible topographic diversity that they can exhibit. In particular, with sufficiently large M it becomes possible to synthesize model functions whose landscape "roughness" changes qualitatively from the high- to the low-energy portions of the overall topography.

It hardly needs to be mentioned that this class of one-dimensional model potentials is an extreme simplification of the realistic multidimensional potential energy or potential enthalpy landscapes. However, it has been shown [Stillinger and Debenedetti, 2002] that this modular function approach can be generalized to higher dimension, retaining the key mathematical property: Sets of functions can be created for which the classical configuration integrals are identical functions of inverse temperature β but for which the respective multidimensional landscape topographies are distinctly different in ways that would produce qualitative differences in dynamical behavior. The implication that emerges from this landscape modeling is that thermodynamic and kinetic behaviors are not necessarily correlated. Nevertheless, one must keep in mind that this is only a statement about the richness and flexibility of the mathematics of hypersurfaces. Whether the physics of realistic interparticle interactions permits that entire richness and flexibility to occur for real substances remains an important open question.

The broad array of phenomena displayed by real glass-forming substances inevitably invites analysis in terms of their underlying landscape topographies. The promise of important insights that such an approach might produce is reinforced by the simple model illustrated with the two contrasting examples in Figure VI.8. Isochoric (constant-volume) computer simulations that have investigated inherent structures for kinetics and glass formation in binary Lennard-Jones mixtures (with smooth interaction cutoffs at finite pair distances) provide some useful guidance [Sastry et al., 1998, 1999; de Souza and Wales, 2009]. These studies reveal a pattern that is illustrated qualitatively in Figure VI.9. In the high-temperature regime, the mean inherent structure energy per particle $\varphi_a^*(T)$ corresponding to metastable equilibrium is nearly constant, showing only a weak tendency to decline as the temperature is reduced. However, upon entering a lower temperature range $\varphi_a^*(T)$ exhibits a relatively rapid plunge, indicating that the system has begun to occupy a significantly different portion of its configuration space at metastable equilibrium, with a distinctly lower mean altitude by an amount of $O(N)$ for the entire system. This relatively sudden change occurs as the time scale for equilibration within the metastable liquid mixture begins to rise rapidly with cooling. When the glass transition temperature range is encountered and passed upon continued cooling, $\langle\varphi_a\rangle$ is no longer able to keep pace with the declining quantity $\varphi_a^*(T)$ but instead levels off. The result is that the final value observed for $\langle\varphi_a\rangle$, as the system falls out of metastable equilibrium and as $T \to 0$, depends on cooling rate: The slower the cooling rate, the lower the final $\langle\varphi_a\rangle$, as illustrated by the separate curves in Figure VI.9. This is analogous to the behavior of experimental (unmapped) energies or enthalpies indicated in Figure VI.2. Similar results have been reported for isochoric simulations of other model glass formers [Palko and Kieffer, 2004]. The behavior illustrated in the metastably equilibrated portion of Figure VI.9 is consistent with the experimental observations of heat capacity rise as the extent of liquid supercooling increases because it indicates an increasing contribution from $d\varphi_a^*/dT$ or $d\hat{\varphi}_a^*/dT$ to the C_V or C_p expressions, presented, respectively, in Eqs. (V.31)–(V.33) and in Eqs. (V.34)–(V.36).

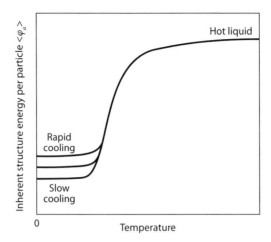

FIGURE VI.9. Temperature behavior of average inherent structure energies per particle, $\langle \varphi_a \rangle$, inferred from isochoric simulations of the binary Lennard-Jones mixtures [Sastry et al., 1998, 1999]. Below the glass transition range, the results depend on the cooling rate; above that range $\langle \varphi_a \rangle = \varphi_a^*(T)$.

The "α," "β" bifurcation of structural relaxation illustrated in Figure VI.5 also occurs as the overall time scale for attaining metastable equilibrium increases substantially. Evidently, it is connected to the declines of $\varphi_a^*(T)$ and $\hat{\varphi}_a^*(T)$ resulting from strong supercooling. The most straightforward interpretation is that the topographical texture of the relevant portion of the energy or enthalpy landscape in set $\mathbf{A}$ or $\hat{\mathbf{A}}$ changes qualitatively from a single to at least a double length-scale character as temperature declines through the bifurcation range. This is illustrated in highly schematic fashion in Figure VI.10, which shows a representative comparison of regions in the Φ or $\hat{\Phi}$ landscape preferentially inhabited at high temperature, VI.10 (a), and at low temperature, VI.10 (b), plotted along representative pathways across the landscapes. The former exhibits uniform roughness, within which structural relaxation kinetics typically only requires thermally activated surmounting of individual transition-state regions that are approximately the same from basin to basin. By contrast, the topography sampled at lower temperature involves a more complex landscape, with organization of basins into larger valleys, or "metabasins," which can act as configurational traps. Whereas the short-range roughness is still present in the low-temperature case (b), the kinetics of full structural relaxation would require surmounting substantially larger barriers that are produced by arrangements of individual basins analogous to that shown in the second example of Figure VI.8 for the elementary one-dimensional model potential. The process that takes the system from the bottom of one metabasin to the bottom of another, even deeper, metabasin requires traversing a sequence of simple basins arranged to create a large net intervening altitude rise, and as a result that process is necessarily relatively slow. The shorter range basin-to-basin kinetics that produces the "β" relaxation empirically retains approximately the same activation energy or enthalpy at all supercooling extents, while the appearance, diversity, and deepening of metabasins as different portions of the multidimensional topography are explored explains the bifurcation and the non-Arrhenius character of the "α" relaxation processes.

It was discussed in Section III.F that individual interbasin transitions statistically involve just particle rearrangements that are localized in three-dimensional space. This is the qualitative na-

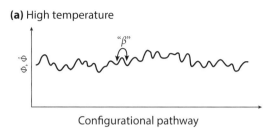

(a) High temperature

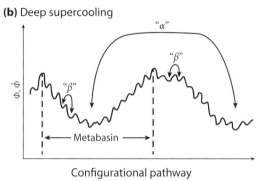

(b) Deep supercooling

FIGURE VI.10. Qualitative distinction between the landscape topographic types accessed by metastably equilibrated liquids (a) at high or slightly supercooled temperatures, and (b) at strongly supercooled temperatures. In (a), the relative homogeneity of the landscape roughness gives rise only to one distinguishable class of structural relaxation processes, whereas in (b) the same processes (now identified as "β" processes) are joined by slower "α" processes that arise from the presence of "metabasins."

ture of "β" relaxation steps. The contrasting "α" relaxation steps require an ordered sequence of these basic interbasin rearrangements, the collection of which is also localized, but over a larger three-dimensional domain. Simulational studies of glass-forming systems have indicated that these sequences tend to displace particles along contorted chains [Donati et al., 1998; Gebremichael et al., 2004] or within neighborhoods that possess particle mobilities higher than that in the surrounding medium [Widmer-Cooper and Harrowell, 2006]. These observations characterize deeply supercooled liquids and their glasses as media displaying "dynamical heterogeneity," a property that evidently underlies the aforementioned failure of the Stokes-Einstein relation, Eq. (VI.7) [Stillinger and Hodgdon, 1994; Swallen et al., 2003].

The metabasin concept, usually introduced only informally in discussions of glass-forming materials, deserves a precise definition in order to allow subsequent detailed analysis of its implications. To begin, recall that inherent structures and their surrounding basins naturally fall into equivalence sets, the members of which differ only by configurational permutation of identical particles. This is a feature that is preserved by the metabasins, which again fall into equivalence classes, each of which contain the same numbers of members because of particle permutation possibilities. The identification of metabasins requires an unambiguous procedure for collecting basins of nonpermutation-equivalent inherent structures into configurationally connected metabasin sets, based upon their locations, shapes, inherent structure energies or enthalpies, and their transition-state interconnections. In particular, the procedure must be generally applicable to the multidimensional configuration space for many-particle systems and must transcend any over-simplification invited by simple one-dimensional landscape representations, such as those in Figs. VI.8 and VI.10.

One possible approach to metabasin identification utilizes the dynamical sequences of inter-basin transitions encountered during molecular dynamics simulations [Heuer, 2008, Section 5.2]. However, the analysis now to be presented is based strictly on the detailed topography of the energy or enthalpy landscape. The key concept to be used for this purpose involves connecting pairs of inherent structures from different equivalence sets with continuous paths of minimal overall altitude variation. These paths begin at an inherent structure, travel to that of a contiguous basin by passing over the lowest simple saddle (transition state) in their shared boundary, and then continue the same process stepwise to end finally at a targeted inherent structure that might be rather remote from the starting point. Pairs of basins, whether contiguous or not, whose nonpermutation-equivalent inherent structures can be connected by such allowed continuous paths of minimal overall altitude variation by definition belong to the same metabasin.

This general basin aggregation process is controlled by a continuously variable energy/enthalpy-rise parameter $\varepsilon \geq 0$. For any chosen ε, the minimal-altitude-change paths passing through inherent structures to be allowed for basin aggregation within $\mathbf{A}$ or $\hat{\mathbf{A}}$ are those whose difference between lowest and highest points does not exceed ε. In order to avoid ambiguity, the ε-dependent aggregation process needs to proceed sequentially, starting with a basin in $\mathbf{A}$ or $\hat{\mathbf{A}}$ that possesses the lowest lying inherent structure. The basins to be included in this initially constructed metabasin are those connected either directly or indirectly to a basin with the lowest lying inherent structure, subject to (a) the ε upper bound constraint, and (b) avoiding connections (either direct or indirect) between pairs of basins that differ only by permutation of identical particles. Once the largest path-connection aggregate (metabasin) containing a lowest inherent structure has been identified for the chosen ε, its other permutation-equivalent metabasins can immediately be identified.

Subsequently, only the remainder basin set in $\mathbf{A}$ or $\hat{\mathbf{A}}$ is considered for the next sequential stage, in which a lowest remaining unaggregated basin serves as the next aggregation seed. This sequential process continues until all basins have been subjected to aggregation into equivalence sets of metabasins. The net result is that the ε-level metabasins provide an effective topographic coarse-graining of the energy or enthalpy landscape, with a height and length scale controlled by the magnitude of ε. Any pair of elementary basins aggregated into the same metabasin must be connected by a path of maximum altitude change $\leq \varepsilon$. Each time ε is changed, it is necessary in principle to redo the entire sequential aggregation process. It is clear that the number of distinct metabasins must decline monotonically as ε increases. The entire initial ($\varepsilon = 0$) collection of individual basins would finally be assigned to a single metabasin equivalence set when ε increases to the macroscopic magnitude $N(\varphi_{\mathrm{max},a} - \varphi_{\mathrm{min},a})$ for isochoric conditions, or $N(\hat{\varphi}_{\mathrm{max},a} - \hat{\varphi}_{\mathrm{min},a})$ for isobaric conditions.

Figure VI.11 schematically indicates how the basin aggregation process operates for an elementary one-dimensional landscape. The maximum allowed altitude variation ε for the paths connecting inherent structures is indicated at the left side of the figure. The construction procedure can in principle result in metabasins whose lowest included inherent structures lie at the boundary with a contiguous metabasin.

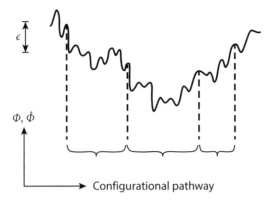

FIGURE VI.11. One-dimensional representation of the metabasin coarse-graining procedure. The basin aggregates formed for the minimal path altitude variation ε shown at the left are delineated by vertical dotted lines and horizontal brackets. By construction, no two basins contained in the same metabasin are related by simple particle permutation.

As already stressed, in the present context it is only the amorphous basins in the projected set **A** or **Â** that are subjected to aggregation into metabasins. However, in principle this is a general algorithm that could equally well be applied to some other projected basin collection relevant to another metastable state. Alternatively, it could just as well be applied to the entire set of basins spanning the full configuration space.

It is natural to classify metabasin equivalence classes by the depths (on a per-particle basis) of the lowest lying inherent structures that they contain. Then as with the basin equivalence classes themselves before aggregation [Section III.C], these metabasin equivalence classes in the large-system limit can asymptotically be enumerated as a function of depth by expressions of the form $\Omega_a(\varphi\,|\,\varepsilon) = \exp[N\sigma_a(\varphi\,|\,\varepsilon)]$ (isochoric conditions) or $\hat{\Omega}_a(\hat{\varphi}\,|\,\varepsilon) = \exp[N\hat{\sigma}_a(\hat{\varphi}\,|\,\varepsilon)]$ (isobaric conditions), where

$$\sigma_a(\varphi\,|\,0) \equiv \sigma_a(\varphi), \tag{VI.20}$$

$$\hat{\sigma}_a(\hat{\varphi}\,|\,0) \equiv \hat{\sigma}_a(\hat{\varphi}).$$

Although two landscapes may possess identically the same enumeration functions σ_a or $\hat{\sigma}_a$ when $\varepsilon = 0$, in analogy to the elementary examples Φ_A and Φ_B shown in Figure VI.8, this degeneracy is usually lifted as ε increases above zero and the metabasin formation process gets underway. In-creasing ε causes the resulting metabasins to become larger and fewer in number; that is, $\sigma_a(\varphi\,|\,\varepsilon)$ and $\hat{\sigma}_a(\hat{\varphi}\,|\,\varepsilon)$ are monotonically decreasing functions of ε. However, this decreasing would be ac-companied by corresponding decreases in "vibrational" free-energy functions $f_{vib,a}(\varphi, \beta, \rho\,|\,\varepsilon)$ and $\hat{f}_{vib,a}(\hat{\varphi}, \beta, p\,|\,\varepsilon)$ that measure intrametabasin configurational displacements.

Provided that the aggregation parameter ε is of order unity, then the depths of the metastable-state thermodynamically dominant metabasins are still identified by the values $\varphi_a^*(\beta, \rho)$ and $\hat{\varphi}_a^*(\beta, p)$, which were originally introduced before aggregation to describe isochoric and isobaric circumstances, respectively. This is the case because $O(1)$ values of ε are negligible in comparison with the energy or enthalpy fluctuations expected in the macroscopic system of interest. This identification is no longer valid as ε rises into its $O(N)$ maximum regime, but this latter extreme situation is not relevant in the following analysis.

For a given ensemble of identically composed systems, let $p_\mu(t|\varepsilon)$ be the occupation probability at time t for the metabasin equivalence class identified by index μ, when the aggregation process has been carried out for parameter value $\varepsilon \geq 0$. When the system is not too far removed from thermal equilibrium in the metastable supercooled liquid, the time dependence of these occupation probabilities in that ensemble can be represented at least approximately by a Markovian "master equation" [Huang, 1963, p. 203; Stillinger, 1990]. In a microcanonical ensemble representation, with conserved total system energy E, the master equation is linear in the probabilities $p_\mu(t|\varepsilon)$, and has the following form:

$$\frac{dp_\mu(t|\varepsilon)}{dt} = \sum_{v \neq \mu} [L_{v \to \mu}(E|\varepsilon)p_v(t|\varepsilon) - L_{\mu \to v}(E|\varepsilon)p_\mu(t|\varepsilon)]. \qquad (VI.21)$$

The E-dependent coefficients $L_{v \to \mu}$ and $L_{\mu \to v}$ are intermetabasin transition rates for the pairs and directions indicated by the subscripts. Positive values for these coefficients in the classical limit imply that the metabasin pair involved shares a common boundary and that the lowest transition state in that boundary lies lower in energy than E. By summing over μ on both sides of Eq. (VI.21), one verifies that the form of this equation automatically preserves the time-independent normalization condition on the full set of metabasin probabilities:

$$\sum_\mu p_\mu(t|\varepsilon) = 1. \qquad (VI.22)$$

The transition rates are required to obey conditions of detailed balance, that is, to ensure that the long-time limiting solution to the master equation is that of microcanonical equilibrium, with no persistent currents around loops of metabasin interconnections. If $p_\mu^{(eq)}(E|\varepsilon)$ is the equilibrium occupation probability for metabasin μ when the total energy is E, then one must have:

$$L_{v \to \mu}(E|\varepsilon) = [p_\mu^{(eq)}(E|\varepsilon)/p_v^{(eq)}(E|\varepsilon)]^{1/2} B_{\mu v}(E|\varepsilon), \qquad (VI.23)$$

$$L_{\mu \to v}(E|\varepsilon) = [p_v^{(eq)}(E|\varepsilon)/p_\mu^{(eq)}(E|\varepsilon)]^{1/2} B_{\mu v}(E|\varepsilon),$$

where $B_{\mu v} \equiv B_{v\mu}$ is a symmetric matrix with respect to the metabasin pair involved [Stillinger, 1990], the only nonvanishing elements of which involve contiguous metabasin pairs. The value of E determines the temperature $T(E)$ of this microcanonical ensemble of isolated systems, independently of the value of the aggregation parameter ε.

As the metabasin complexes enlarge with increasing ε, the transition rates $L_{\mu \to v}$ of course decrease in number, but they also tend to decrease in magnitude. This latter trend is the case because the system on average expends more time exploring the increased configuration space within the interior of each metabasin, and because that larger interior is "rough," the configurational diffusion rate up to and across metabasin boundaries tends to be reduced [Zwanzig, 1988]. Furthermore, the transition barriers to enter neighboring inequivalent metabasins rise roughly in proportion to ε. To some extent, this rate reduction could be offset if the contiguous pairs of expanded inequivalent metabasins shared an increased number of transition states in their mutual boundary.

Master equation (VI.21) can be compactly symbolized in the following form:

$$\frac{d\mathbf{p}}{dt} + \mathbf{L}(\varepsilon) \cdot \mathbf{p} = 0, \tag{VI.24}$$

where $\mathbf{p}$ is a vector whose components are the separate occupancy probabilities $p_\mu(t\,|\,\varepsilon)$, and $\mathbf{L}(\varepsilon)$ is a square matrix composed of the transition rates:

$$[\mathbf{L}(\varepsilon)]_{\mu\nu} = -L_{\nu\to\mu}(E\,|\,\varepsilon) \qquad (\nu \neq \mu), \tag{VI.25}$$

$$[\mathbf{L}(\varepsilon)]_{\mu\mu} = \sum_{\nu\neq\mu} L_{\mu\to\nu}(E\,|\,\varepsilon).$$

For isochoric conditions, the number of components of the vector $\mathbf{p}$ is the total number of distinguishable metabasins of all depths for the amorphous configuration set $\mathbf{A}$, and so is denoted by $\Omega_{tot}(\varepsilon)$. Consequently, matrix $\mathbf{L}(\varepsilon)$ would have size $\Omega_{tot}(\varepsilon) \times \Omega_{tot}(\varepsilon)$. For isobaric conditions with configuration set $\hat{\mathbf{A}}$, these sizes are $\hat{\Omega}_{tot}(\varepsilon)$ and $\hat{\Omega}_{tot}(\varepsilon) \times \hat{\Omega}_{tot}(\varepsilon)$, respectively. The solutions to the master equation generally can be expressed as linear combinations involving the Ω_{tot} or $\hat{\Omega}_{tot}$ eigenvectors $\mathbf{q}^{(n)}$ and the corresponding eigenvalues $\lambda_n \geq 0$ of the matrix $\mathbf{L}$:

$$\mathbf{p}(t) = \sum_n C_n \mathbf{q}^{(n)} \exp(-\lambda_n t), \tag{VI.26}$$

where initial conditions determine the coefficients C_n. The $\mathbf{q}^{(n)}$ are not in general orthogonal to one another. Only one of the eigenvalues vanishes, say $\lambda_0 = 0$, and its eigenvector $\mathbf{q}^{(0)}$ has components that are proportional to the equilibrium occupancy probabilities denoted earlier by $p_\mu^{(eq)}(E\,|\,\varepsilon)$. All other eigenvalues are positive, and with their eigenvectors they describe independent relaxation modes across the metabasin landscape.

A few additional qualitative remarks are in order concerning the properties of the ε-dependent coarse graining of the landscape, the metabasins it produces, and the implications for relaxation distributions. For very small but positive ε, the aggregation process only combines those relatively rare neighboring basin pairs whose inherent structures are very close in the configuration space and therefore close in energy or enthalpy, and whose common boundary contains a very low transition state. These basin pairs constitute the so-called "two-level systems" that are considered in detail in Section VI.C and that contribute to the characteristic low-temperature heat capacity of nonmetallic glasses, Eq. (VI.3). At this early aggregation stage, the enumeration functions $\sigma_a(\varphi\,|\,\varepsilon)$ and $\hat{\sigma}_a(\hat{\varphi}\,|\,\varepsilon)$ have declined very little from their $\varepsilon = 0$ starting values. But as ε increases to become comparable to, and then greater than, the typical transition state barriers for the numerous "β" relaxation processes, the corresponding metabasin enumeration functions $\sigma_a(\varphi\,|\,\varepsilon)$ and $\hat{\sigma}_a(\hat{\varphi}\,|\,\varepsilon)$ have markedly declined in magnitude. The surviving transitions between inequivalent metabasins then involve configuration switches (identified by comparing lowest inherent structures in the metabasins) that include larger numbers of particles but that are still localized in three-dimensional space from the overall macroscopic-system-size (volume V) perspective.

For some purposes, it may be useful to recast metabasin enumeration in terms of a formal division of V into separate three-dimensional volumes $v(\varphi\,|\,\varepsilon)$ [alternatively $\hat{v}(\hat{\varphi}\,|\,\varepsilon)$ for isobaric conditions] in which hypothetically independent binary choices of configurations effectively could be made. By definition of these binary-choice volumes, this convention accounts for the number of metabasins that contribute significantly to relaxation at the relevant portion of the coarse-grained landscape that has been identified with intensive depth parameter φ or $\hat{\varphi}$. For isochoric conditions in the large-system limit, this volume is defined by the expression

$$\Omega_a(\varphi \,|\, \varepsilon) = \exp[N\sigma_a(\varphi \,|\, \varepsilon)] \tag{VI.27}$$

$$\equiv 2^{V/v(\varphi \,|\, \varepsilon)},$$

while the corresponding isobaric version is

$$\hat{\Omega}_a(\hat{\varphi} \,|\, \varepsilon) = \exp[N\hat{\sigma}_a(\hat{\varphi} \,|\, \varepsilon)] \tag{VI.28}$$

$$\equiv 2^{V/\hat{v}(\hat{\varphi} \,|\, \varepsilon)};$$

equivalently, one has ($\rho = N/V$):

$$v(\varphi \,|\, \varepsilon) = \ln 2 / [\rho\sigma_a(\varphi \,|\, \varepsilon)], \tag{VI.29}$$

$$\hat{v}(\hat{\varphi} \,|\, \varepsilon) = \ln 2 / [\rho\hat{\sigma}_a(\hat{\varphi} \,|\, \varepsilon)].$$

Obviously, $v(\varphi \,|\, \varepsilon)$ and $\hat{v}(\hat{\varphi} \,|\, \varepsilon)$ both increase monotonically as the coarse-graining parameter ε increases. However, whereas ε is still of order unity on the molecular energy scale, these volumes themselves remain on the molecular size scale. The volumes $v(\varphi \,|\, 0)$ and $\hat{v}(\hat{\varphi} \,|\, 0)$ are analogous (but not identical) to the Adam-Gibbs concept of "cooperatively rearranging regions" postulated long ago for glass formers [Adam and Gibbs, 1965].

The connection between the master equation eigenvalues λ_n for given ε and measurable distributions of relaxation rates such as represented by the KWW form in Eqs. (VI.9–VI.12) is not straightforward. This situation arises from the high dimensionality of basins and metabasins, each of which possesses $O(N)$ transition states ("exit channels") to neighboring basins or metabasins resulting from localized particle rearrangements. As indicated in Section III.F, this implies that the mean residence time in any basin or metabasin (except at very low temperature) is only of order N^{-1}, thus approaching zero in the large-system limit. The master equation can describe the decay of occupancy probability from a single starting basin or metabasin, using an appropriate linear combination of its eigenvectors for that initial condition. In order for the lifetime in that starting location to be $O(N^{-1})$, some of the master equation eigenvalues λ_n must be $O(N)$. The full spectrum of its eigenvalues thus spans the range from zero (for the equilibrium eigenstate) to an upper limit of macroscopic magnitude proportional to the size of the system. These characteristics apply while the aggregation parameter ε remains $O(1)$.

Let Q represent a quantity of interest that is subject to relaxation, and let Q_μ be defined as its average for a basin or metabasin that is identified by index μ. Examples of such quantities that could monitor the return to metastable equilibrium from an initially prepared disequilibrium state might be (a) the overall dielectric polarization of a sample that should have none after equilibration, (b) the difference in stress components for a sample initially and suddenly subjected to anisotropic external forces, and (c) the differences $\Delta\varphi = \langle\varphi_a\rangle - \varphi_a^*$ or $\Delta\hat{\varphi} = \langle\hat{\varphi}_a\rangle - \hat{\varphi}_a^*$ for an equilibrated sample whose temperature was then rapidly changed. The formal time dependence of Q as described by the master equation would be provided by the following expression:

$$Q(t) = \sum_\mu\sum_n C_n Q_\mu q_\mu^{(n)} \exp(-\lambda_n t), \tag{VI.30}$$

where $q_\mu^{(n)} \equiv (\mathbf{q}^{(n)})_\mu$ is the component of the nth eigenvector for basin or metabasin μ. As before, the set of coefficients C_n is determined by initial disequilibrium conditions and by the weighting of eigenstates according to the specific property Q of interest. The respective time dependences produced by Eq. (VI.30) can vary, for example, yielding decay functions fitted by KWW stretched exponential functions with somewhat different exponents θ and time scales τ.

As ε increases from zero to produce an increasingly coarse-grained description of the energy or enthalpy landscape, the initial effect on the master equation is disproportionately to eliminate the rapid low-barrier relaxations. That is, the small-scale landscape features identified as "β" relaxations first begin to disappear from the relaxation spectrum. However, the slower "α" relaxation processes at this early stage remain largely represented in the spectrum. Continued increase in ε would then begin to project out the more rapid portions of the broad "α" spectrum, while still leaving its slowest components intact. If, for example, the quantity $Q(t)$ at intermediate to long times measured at $\varepsilon = 0$ for a strongly supercooled liquid were to be well described by a KWW stretched exponential function, then for $\varepsilon > 0$ the small-s leading-order behavior of its relaxation function's Laplace transform [Eq. (VI.10)] should continue to be represented by Eq. (VI.12). The previously mentioned observed decrease of the KWW exponent θ for fragile glass formers as T declines toward T_g, which appears in that leading order, indicates that the preferentially occupied portions of the landscape become topographically more diverse as a result of that cooling.

It is important to emphasize that the master equation (VI.21) only describes transitions between distinct metabasin equivalent classes. This is sufficient to represent most relaxation phenomena in supercooled liquids and glass formation. However, there also exist permutational transitions involving small local subsets of identical particles that carry the system between metabasins contained within the same equivalence class. Specifically, these transitions involve simultaneous displacement motion of a small number (e.g., 2, 3, or 4) of identical neighboring particles around a closed path, such that each involved particle adopts the prior position of one of its immediate neighbors along the path, while being replaced by its other immediate neighbor along that path. Such cyclic permutations do not alter the configurational stress pattern within the system, and so are not involved in determining the non-Arrhenius rise of the shear viscosity autocorrelation function as temperature declines, nor in establishing "α" relaxation modes. However, a chronological sequence of distinct cyclic permutations, all of which include a given "tagged" particle, would contribute to self-diffusion of that particle. Although such collective motions would involve surmounting a modest energy or enthalpy barrier and may be facilitated momentarily by radially outward small displacements of surrounding particles, they do not involve the large-scale cooperative motion of many particles that would typically be required as fundamental transitions in shear stress relaxation at low temperature. In other words, these cyclic permutation transitions that contribute to self-diffusion are essentially decoupled from shear viscosity. This distinction may be an important feature in understanding Stokes-Einstein violations that arise from separation between the temperature dependences of shear viscosity and of self-diffusion rate in the deeply supercooled regime for many glass formers.

C. Glasses at Low Temperature

Starting with perfect crystals, the structural changes that require the lowest increase of inherent structure potential energy ($\Delta\Phi$) or potential enthalpy ($\Delta\hat{\Phi}$) typically are the creation of point defects. In the case of atomic or simple ionic crystals, these can be vacancies or interstitials. For ordered alloys, substitutional permutations of the constituent atoms qualify. Molecular crystals can incorporate orientational and conformational anomalies as localized defects. In each of these cases, the creation energy or enthalpy within the perfect crystal has a definite positive magnitude. Consequently, the low-temperature equilibrium concentration of such defects is controlled primarily by the relevant Boltzmann factor $\exp(-\Delta\Phi/k_B T)$ or $\exp(-\Delta\hat{\Phi}/k_B T)$ that vanishes strongly in the $T \to 0$ limit. That in turn implies a similar strong vanishing for any defect-related contribution to thermodynamic properties. In particular, this is true for contributions to the respective equilibrium heat capacities C_V and C_p.

The observation of a roughly linear temperature contribution to heat capacity for nonmetallic glasses near absolute zero, Eq. (VI.3), indicates a qualitative difference in occupied landscape character between glasses and crystals. The existence of long-wavelength propagating phonons in both of these kinds of elastic solids guarantees that a Debye-like T^3 contribution to measured heat capacities should be present for both, but a qualitatively novel mechanism must also be present to dominate this conventional contribution as $T \to 0$. The generally accepted identification of this landscape difference postulates the presence in the amorphous glassy medium of a small concentration of localized bistable degrees of freedom. These have excitation energies or enthalpies that in the large-system limit possess a continuous distribution, including arbitrarily small values [Anderson et al., 1972; Phillips, 1972]. Such bistable degrees of freedom, which interact only weakly with one another because of their large mean separation, presumably involve localized groups of particles that experience a pair of alternative positions of mechanical stability differing only by a small displacement. That these localized displacement modes should exist in a dense amorphous medium is not surprising, given the diversity of possible local particle arrangements and the resulting likelihood of imperfect, incomplete, and frustrated interactions.

Figure VI.12 presents a schematic illustration of the potential energy or potential enthalpy contour along the short one-dimensional pathway in the multidimensional configuration space connecting the nearby pair of inherent structures that determine the two-level system. This pathway is composed of two branches emerging in opposite directions from the saddle point (transition state), both of which are solutions of the steepest descent mapping equations that pass through the respective inherent structures. For the nearby pair of inherent structures that are of interest, the horizontal projection of the transition path has a short length d and is approximately linear in the configuration space. The barrier height b and the inherent structure depth difference $\delta_{IS} > 0$ are similarly small. Box VI.1 presents a simple polynomial approximation for the energy variation $V(x)$ along the transition collective coordinate x that captures the basic qualitative nature of the postulated bistable degrees of freedom.

In the low-temperature regime where these amorphous-phase configurational transitions are important, a quantum mechanical description is mandatory. The time-independent Schrödinger equation for motion in a bistable potential $V(x)$, as illustrated in Figure VI.12 and Box VI.1, may be written as follows:

$$\left[-\left(\frac{\hbar^2}{2m_{eff}} \right) \frac{d^2}{dx^2} + V(x) - E_n \right] \psi_n(x) = 0. \tag{VI.31}$$

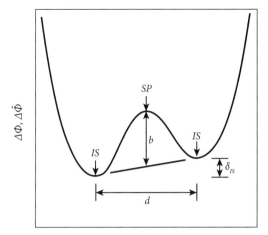

Transition path coordinate

FIGURE VI.12. Typical asymmetric bistable potential that produces a low excitation "two-level system." The transition coordinate indicated as the horizontal axis connects the pair of closely spaced inherent structures (IS) and passes through the intervening simple transition-state saddle point (SP). The distance between the minima, the intervening barrier height measured from a linear interpolation, and the difference in energy or enthalpy of the inherent structures are denoted by d, b, and δ_{IS}, respectively.

Here ψ_n and E_n are the eigenfunctions and eigenvalues for this separated bistable degree of freedom. Equation (VI.31) incorporates an effective mass m_{eff} that is appropriate for the specific bistable coordinate involved. Its value depends upon the number of particles and their individual masses for the collective displacement along the coordinate x between the two closely spaced inherent structures under consideration, but in most cases it is expected to be within an order of magnitude of a single particle mass.

The Schrödinger equation (VI.31) with a bistable $V(x)$ that has the quartic polynomial form shown in Figure VI.12 possesses an infinite number of eigenstates. But to explain the low-T heat capacity anomaly, it is only necessary to consider the ground state ($n = 0$) and the first excited state ($n = 1$) of Eq. (VI.31) and specifically to focus on those cases where the energy separation $\Delta E = E_1 - E_0$ is sufficiently small to be comparable to thermal energy $k_B T$ at very low temperature. The ground-state eigenfunction ψ_0 is nodeless, and the first excited state ψ_1 exhibits a single node located within the barrier region. For ΔE to be small and well separated from other excitations along the same coordinate, it suffices for the bistable potential $V(x)$ to be nearly symmetric about its maximum [i.e., to have a small cubic term $(B/3)x^3$] and to have a sufficiently high barrier between the inherent-structure minima so that both E_0 and E_1 lie below the transition state. In this circumstance, ΔE consists essentially of the sum of the inherent-structure depth difference Δ_{IS}, the difference in ground-state energies of the nearly-harmonic oscillators localized at and measured with respect to the two nearly-equivalent minima, and a small tunneling contribution that tends to split apart the two lowest eigenvalues E_0 and E_1.

In the low-temperature regime where two-level systems play a significant role, the amorphous media that contain them exhibit very small thermal expansions. Consequently, the distinction between isochoric and isobaric conditions for this regime is minor at best. The quantum eigenvalues E_0 and E_1 can be viewed as jointly relevant to both conditions.

BOX VI.1. Elementary Double-Minimum Potential

The simplest representation for a double-minimum potential of the kind illustrated in Figure VI.12 is a quartic polynomial in the displacement coordinate x. Calling this potential $V(x)$, one can write

$$V(x) = (A/4)x^4 + (B/3)x^3 - (C/2)x^2,$$

where $A, C > 0$, and the coefficient B of the asymmetry-producing cubic term can have either sign. Using this form measures energy (or enthalpy) and the displacement coordinate relative to the location of the barrier maximum. Straightforward algebra allows these coefficients to be related to the parameters shown in Figure VI.12:

$$b = (C/12A^2)(B^2 + 3AC),$$

$$d = x_+ - x_- = A^{-1}(B^2 + 4AC)^{1/2},$$

$$\delta_{IS} = (|B|/12A^3)(B^2 + 4AC)^{3/2}.$$

Here the displacement coordinate positions of the two minima have the following values:

$$x_\pm = (2A)^{-1}[-B \pm (B^2 + 4AC)^{1/2}].$$

The curvatures at these minima are the following:

$$V''(x_\pm) = 2C + (B/2A)[B \mp (B^2 + 4AC)^{1/2}],$$

which are always positive for the form chosen for $V(x)$.

For a single chosen two-level system, the thermal mean value of its excitation energy is given by the expression

$$\bar{E}(T) = \frac{\Delta E}{\exp(\Delta E / k_B T) + 1}. \tag{VI.32}$$

Let $P_{tls}(\Delta E)$ stand for the distribution of two-level systems with respect to their quantum mechanical excitation energy ΔE, normalized so that the total number N_{tls} of such qualifying bistable degrees of freedom in the entire N-particle system equals the integral

$$N_{tls} = \int_0^\infty P_{tls}(\Delta E) d(\Delta E). \tag{VI.33}$$

It needs to be emphasized that $P_{tls}(\Delta E)$ describes a glass that has experienced kinetic arrest as it was cooled toward $T = 0$, so that it is both a substance-dependent and a history-dependent quantity. Although the integration limit in Eq. (VI.33) has formally been extended to $+\infty$, the distribution $P_{tls}(\Delta E)$ is only nonvanishing for sufficiently small ΔE cases such that higher eigenvalues of the quantum problem for each bistable system are excluded. Equation (VI.33) leads in turn to the temperature-dependent mean excitation energy for the entire collection of two-level degrees of freedom:

$$\langle \bar{E}(T) \rangle = \int_0^\infty \left[\frac{(\Delta E) P_{tls}(\Delta E)}{\exp(\Delta E/k_B T) + 1} \right] d(\Delta E). \tag{VI.34}$$

At low temperature, the presence of the exponential term in the integrand denominator of Eq. (VI.34) causes virtually the entire contribution to the integral to be confined to very small ΔE. Naturally, this places emphasis on the corresponding leading-order behavior of the distribution $P_{tls}(\Delta E)$ near $\Delta E = 0$. The complexity of the interactions in the amorphous medium and of the two-level degrees of freedom they produce make precise prediction of this leading-order behavior beyond present capability. However, one can reasonably assume a generic scaling form of the type

$$P_{tls}(\Delta E) \cong P_q(\Delta E)^q, \tag{VI.35}$$

where P_q is a positive constant, and the exponent must obey $-1 < q$ in order for integral (VI.33) to converge. One should keep in mind that both exponent q and coefficient P_q can vary according to the substance involved and its cooling history. Insert Eq. (VI.35) into Eq. (VI.34) to obtain

$$\langle \bar{E}(T) \rangle \cong P_q (k_B T)^{q+2} \int_0^\infty \frac{x^{q+1}}{\exp(x) + 1} dx \tag{VI.36}$$

$$= P_q (1 - 2^{-q-1}) \Gamma(q + 2) \zeta(q + 2) (k_B T)^{q+2}.$$

The latter form involves Euler's gamma function $\Gamma(z)$ and Riemann's zeta function $\zeta(z)$.

The heat capacity contribution attributable to the two-level systems follows simply by applying a temperature derivative to the last expression. The result is proportional to T^{q+1}. By comparing this to the leading term in the empirical form shown in Eq. (VI.3), one concludes that

$$u = q + 1. \tag{VI.37}$$

If indeed there are some amorphous solids whose measured low-temperature heat capacities require that $u = 1$, then $q = 0$ and $P_{tls}(0) > 0$.

The heat capacities C_V and C_p for structurally ordered crystals of nonmetallic substances arise only from phonon motions but with inclusion of increasingly important anharmonic corrections as temperature rises toward the melting point. Details are presented in Sections IV.B and IV.C. For these phases, the isochoric temperature variation of C_V/T^3 or analogously the isobaric temperature variation of C_p/T^3 starts at a positive constant in the $T = 0$ limit then declines monotonically toward zero as T increases. But the anomalous heat capacity contribution for amorphous solids, Eq. (VI.3), that is attributable to two-level systems, would cause such plots to diverge to $+\infty$ at $T = 0$. If this were the only difference between the structurally perfect crystalline and the structurally irregular amorphous solids, then the plots would continue to exhibit monotonic decline to zero with increasing T. However, a wide variety of amorphous solids produce C_V/T^3 or C_p/T^3 plots with a broad maximum, usually in the 10 K temperature range [Sokolov et al., 1993]. This maximum is conventionally called the "boson peak," a possibly misleading terminology. Figure VI.13

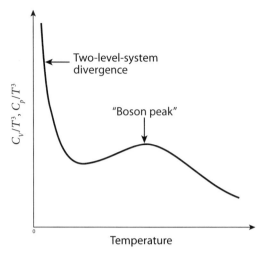

FIGURE VI.13. Schematic representation of heat capacity plots that reveal the so-called "boson peak."

provides a schematic version of this behavior. The strength of the boson peak empirically seems to be related to the "strong" vs. "fragile" classification of glass formers: The "strong" material SiO_2 has a well-developed boson peak, whereas the "fragile" material o-terphenyl has a substantially weaker peak. Other substances lying between these extremes of the conventional strong–fragile scale exhibit peaks with intermediate strength [Sokolov et al., 1993].

It has not yet become clear what characteristic(s) of multidimensional landscapes inhabited by amorphous solids underlie the boson peak phenomenon. A variety of explanations have been advanced. One hypothesis is that the structural disorder present in these materials simply causes a crowding of the purely harmonic vibrational density of states into a relatively low frequency range that would begin to contribute substantially to heat capacity around 10 K [Schirmacher et al., 1998; Xu et al., 2007]. In that view, anharmonicity does not play an important role. Alternatively, an anomalously large number of bistable (or multistable) degrees of freedom possessing relatively large excitation energies ΔE (including those to higher excited states) might be present in an energy range so as to produce the observed phenomena in the 10 K temperature vicinity [Galperin et al., 1991; Lubchenko and Wolynes, 2003]. One must keep in mind that these hypothetical scenarios are not necessarily mutually exclusive.

The thermal conductivity of a finite sample of a dielectric substance in its structurally perfect crystalline state involves its phonons and is controlled in magnitude at a given temperature partly by the density and energy of phonons present, partly by geometry-dependent surface scattering of phonons, and partly by anharmonic processes that limit phonon mean free paths within the sample interior (Section IV.G). The result is that thermal conductivity has a very low value at very low T, passes through a simple maximum, and then declines with increasing T up to the crystal's melting point. Amorphous solids composed of dielectrics provide an intriguing contrast, however. As Figure VI.14 illustrates, thermal conductivity is small at very low T, enters a plateau region, then continues to rise once again with further T increase. This behavior has been observed for a substantial variety of materials, including both inorganic glasses and polymeric organic glasses [Zeller and Pohl, 1971]. Interestingly, the plateau region tends to occur around 10 K, lead-

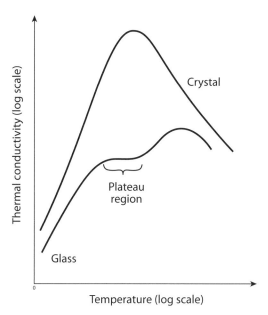

FIGURE VI.14. Typical behavior of thermal conductivity vs. temperature for a dielectric substance in an amorphous solid state, compared to its structurally ordered crystal. Notice that this is a log–log plot.

ing to the suspicion that it may be connected to the boson peak phenomenon. A simple rationale would be that the boson peak excitations are associated with sites in the amorphous medium that can scatter phonons whose frequencies are typical for the 10 K temperature range, but such scattering processes are less effective for the higher frequency phonons that dominate at higher temperature.

D. Ideal Glass Transition Hypothesis

Attention now turns to the behavior of the supercooled liquid properties under the tentative assumption that upon cooling kinetic arrest at a conventional glass transition temperature T_g could somehow be avoided. This assumption might be realized if unusually long times of observation were available (while managing to suppress crystal nucleation and growth), or if some "catalyst" were available to facilitate and hence speed up interbasin transitions in regions $\mathbf{A}$ and $\hat{\mathbf{A}}$ by many orders of magnitude. The objective of such a hypothetical exercise is to provide a critical analysis of the sub-T_g extrapolations such as that illustrated in Figure VI.4 for the entropy. In particular, it is important to assess the validity of the concept of a precise Kauzmann temperature $T_K > 0$ at which the extrapolated supercooled liquid allegedly has its entropy equal to that of the thermodynamically stable crystal and to establish if this is associated with a well-defined "ideal glass transition," a metastable-regime higher order phase transition.

The free-energy expressions shown in Eqs. (VI.14) provide the starting point for the analysis. The basic assumption that kinetic sluggishness can be circumvented implies that these expressions remain valid descriptions of the metastable-state thermal equilibrium for all temperatures down to $T = 0$. Under conventional experimental conditions with 1 atm pressure, the isochoric and isobaric conditions involve descriptions that are numerically close to each other for most glass-forming substances. For the present, the latter representation is favored in the following analysis. Consequently, we need to examine the behavior of the metastable-state enumeration function $\hat{\sigma}_a(\hat{\varphi}, p)$, the vibrational free-energy quantity $\hat{f}_{vib,a}(\hat{\varphi}, \beta, p)$, and the inherent structure enthalpy function $\hat{\varphi}_a{}^*(\beta, p)$, which minimizes the Gibbs free energy at the given β, p.

Accurate measurements of the isobaric heat capacity C_p are available for the organic glass formers diethyl phthalate [Chang et al., 1967] and o-terphenyl [Chang and Bestul, 1972]. These experiments show that well below T_g, where experimentally the glass has substantially ceased all relaxation processes (but where T is still large enough that the contributions of two-level systems and boson peak are negligible), the crystal and the glass exhibit nearly the same values of C_p. Furthermore, this result appears to apply for different glass-forming cooling rates, which influence the $\hat{\varphi}$ value of the kinetically arrested system eventually taken down to $T = 0$. Because the crystal phase is confined to the basins corresponding to perfect (or nearly perfect) configurational order, while the glass is trapped in higher lying disordered basins, the conclusion is that in this sub-T_g regime the vibrational contributions embodied in $\hat{f}_{vib,a}$ are rather insensitive to the enthalpy depth parameter $\hat{\varphi}$, and thus rather insensitive to the extent of configurational order. The implication is that $\hat{f}_{vib,a}$ is a nonsingular function of $\hat{\varphi}$.

This conclusion is reinforced implicitly by computer simulations [Sastry, 2001; Sciortino and Tartaglia, 2001] that have been carried out isochorically for the so-called "binary mixture Lennard-Jones" (BMLJ) model. This model is a variant of one that was originally devised to represent the near-eutectic $Ni_{80}P_{20}$ system [Weber and Stillinger, 1985a, 1985b]. The simulations determined that the constant-volume vibrational free energies $f_{vib,a}$ are close to linear functions of φ at fixed temperature. This is the case over a substantial density range. Because thermal expansion should be a minor attribute of this model's amorphous states at low temperature, the nonsingular φ dependence of $f_{vib,a}$ should also carry over to a nonsingular $\hat{\varphi}$ dependence of $\hat{f}_{vib,a}$ for this model. This means that for any positive temperature

$$-\infty < [\partial \hat{f}_{vib,a}(\hat{\varphi}, \beta, p)/\partial \hat{\varphi}]_{\beta,p} < +\infty. \tag{VI.38}$$

The isobaric criterion for identification of the enthalpic order parameter $\hat{\varphi}_a{}^*(\beta, p)$ is

$$(\partial/\partial \hat{\varphi})_{\beta,p}[\hat{\sigma}_a(\hat{\varphi}) - \beta\hat{\varphi} - \beta\hat{f}_{vib,a}(\hat{\varphi}, \beta)] = 0. \tag{VI.39}$$

This gives rise to Figure VI.15, the isobaric metastable-state version of the graphical construction shown in Figure III.1. Equation (VI.39) requires that for any given T, p there must be equal magnitudes but opposite signs at $\hat{\varphi} = \hat{\varphi}_a{}^*$ for the slopes of the $\hat{\sigma}_a$ curve and for the relevant $-\beta[\hat{\varphi} + \hat{f}_{vib,a}]$ curve, when both are plotted vs. $\hat{\varphi}$. If there were to be an ideal glass transition at $T_K > 0$, it would require slope matching at $\hat{\varphi} = \hat{\varphi}_{\min,a}$. Taking Eq. (VI.38) into account, this would require in turn that

$$0 < [d\hat{\sigma}_a(\hat{\varphi})/d\hat{\varphi}]_{\hat{\varphi}=\hat{\varphi}_{\min,a}} < +\infty. \tag{VI.40}$$

That is, the existence of an ideal glass transition at positive temperature rests upon a finite positive initial slope for the enumeration function $\hat{\sigma}_a$.

The next step in the analysis involves developing directly a physically plausible enumeration for low-lying inherent structures within the amorphous-system configuration sets $\mathbf{A}$ (isochoric

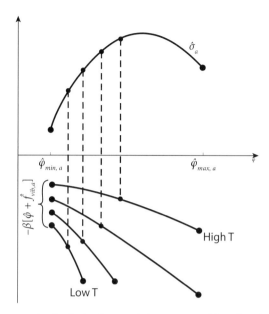

FIGURE VI.15. Graphical construction for the dominant enthalpy parameter $\hat{\varphi}_a^*$ as a function of temperature (at constant pressure). The vertical dashed lines locate positions of equal magnitude, but opposite signs, for the slopes of enumeration curve $\hat{\sigma}_a$ and of the appropriate $-\beta[\hat{\varphi} + \hat{f}_{vib,a}]$ curve.

conditions) or $\hat{\mathbf{A}}$ (isobaric conditions). This analysis can proceed in a manner somewhat similar to that used for the low-temperature two-level degrees of freedom discussed in Section VI.C. But in contrast to that context, history dependence of the substance under consideration is not a relevant issue. Instead, we are now concerned with enumerating all of the lowest inherent structures in the $\mathbf{A}$ or $\hat{\mathbf{A}}$ landscapes and the elementary transitions they can undergo to contiguous basins of nearby inherent structures. These excitations involve localized particle rearrangements within the macroscopic system and are an intrinsic characteristic of amorphous states. They are evidently essentially continuously distributed in the energy or enthalpy difference between the lowest and contiguous inherent structures involved. Because the focus now is just on the sparse excitations expected to dominate the formal metastable-state partition function at low temperature, it is proper to treat these individual localized configurational switches as widely enough separated in the amorphous medium to act essentially independently of one another. The physical circumstances and the form of the analysis are very similar for isochoric and for isobaric conditions at low temperature, so for simplicity only the latter is presented.

For the entire N-body amorphous system, let $NF(\xi)$ represent the density of possible single localized elementary excitations out of the lowest basin equivalence set for which the enthalpic difference between the two inherent structure minima is ξ. If the ξ scale is divided into equal bins corresponding to small increments $\Delta\xi$, then the number of basic excitations within the ith bin whose mean is at $\xi = \xi_i$ is

$$N_i = NF(\xi_i)\Delta\xi. \tag{VI.41}$$

Next let $0 \leq n_i \leq N_i$ be the number of possible elementary configurational excitations in the ith bin that are actually simultaneously excited. Then the enthalpic inherent-structure excitation for the full system involves a sum over the bins:

$$\Delta \hat{\Phi}_{IS} = N \Delta \hat{\varphi} \tag{VI.42}$$

$$= \sum_i n_i \xi_i.$$

The number of ways that this specific excitation magnitude $\Delta \hat{\Phi}_{IS}$ can be attained with the available two-level excitations depends on how many ways for each bin that n_i of the N_i possibilities can be chosen. This is an elementary combinatorial expression, whose logarithm has the following form:

$$\ln \left\{ \prod_i \frac{N_i!}{n_i!(N_i - n_i)!} \right\} \cong \sum_i [N_i \ln N_i - n_i \ln n_i - (N_i - n_i) \ln(N_i - n_i)]. \tag{VI.43}$$

In order to convert these last two expressions into an enumeration function $\hat{\sigma}_a(\Delta \hat{\varphi})$, it is necessary to maximize expression (VI.43) subject to the constraint of a fixed value of expression (VI.42). That constraint requires introduction of a Lagrange multiplier [Korn and Korn, 1968, Chapter 11], to be denoted by Λ. Consequently, one is led to the following variational equation:

$$0 = -\ln n_i + \ln(N_i - n_i) - \Lambda \xi_i, \tag{VI.44}$$

which is equivalent to the result

$$\frac{n_i}{N_i} = \frac{\exp(-\Lambda \xi_i)}{1 + \exp(-\Lambda \xi_i)}. \tag{VI.45}$$

Equation (VI.42) may now be rewritten as follows for the variational maximum:

$$\Delta \hat{\varphi} = \sum_i \left(\frac{N_i}{N} \right) \left(\frac{n_i}{N_i} \right) \xi_i$$

$$= \sum_i [F(\xi_i) \Delta \xi] \left[\frac{\xi_i \exp(-\Lambda \xi_i)}{1 + \exp(-\Lambda \xi_i)} \right]$$

$$\rightarrow \int_0^\infty F(\xi) \left[\frac{\xi \exp(-\Lambda \xi)}{1 + \exp(-\Lambda \xi)} \right] d\xi, \tag{VI.46}$$

where the integral in the last form emerges from the summation by taking the limit $\Delta \xi \rightarrow 0$, which is appropriate for the large-system limit of primary concern. Evaluation of the combinatorial factor in Eq. (VI.43) at its variational maximum, combined with a possible lowest amorphous-state degeneracy measure $\hat{\sigma}_a(\hat{\varphi}_{min,a})$, yields $N \hat{\sigma}_a$. By using the same kinds of transformations that produced Eq. (VI.46) for $\Delta \hat{\varphi}$, one finds

$$\Delta \hat{\sigma}_a(\Delta \hat{\varphi}) = \hat{\sigma}_a(\Delta \hat{\varphi}) - \hat{\sigma}_a(\hat{\varphi}_{min,a})$$

$$= \int_0^\infty F(\xi) \left\{ \ln[1 + \exp(-\Lambda \xi)] + \frac{\Lambda \xi \exp(-\Lambda \xi)}{1 + \exp(-\Lambda \xi)} \right\} d\xi. \tag{VI.47}$$

The last two results may be summarized in the following manner:

$$\Delta\hat{\varphi} = -(d/d\Lambda)I(\Lambda),$$ (VI.48)

$$\Delta\hat{\sigma}_a = [1 - \Lambda(d/d\Lambda)]I(\Lambda),$$

where

$$I(\Lambda) = \int_0^\infty F(\xi)\ln[1 + \exp(-\Lambda\xi)]d\xi.$$ (VI.49)

The next objective in principle is to eliminate Λ between the two expressions in Eqs. (VI.48) so as to produce the desired enumeration function $\hat{\sigma}_a(\Delta\hat{\varphi})$. However, the main focus at present is how this function behaves in the small-$\Delta\hat{\varphi}$ regime, which corresponds to large positive Λ. That in turn puts primary emphasis on the small-ξ portion of the $I(\Lambda)$ integrand because its logarithmic factor then declines rapidly toward zero as the integration variable increases.

The excitation distribution $F(\xi)$ is not known, except in a qualitative and rudimentary way. It may be significantly dependent on the glass-forming substance under consideration. However, a conservative postulate is that its small-ξ leading-order form has a power law form,

$$F(\xi) = F_z\xi^z + \ldots,$$ (VI.50)

where the coefficient $F_z > 0$, and the power z must be restricted by

$$-1 < z,$$ (VI.51)

in order for the $I(\Lambda)$ integral to converge. This postulate then allows $I(\Lambda)$ to be evaluated for large positive Λ by expanding the logarithmic integrand factor into an infinite series:

$$I(\Lambda) \cong F_z\sum_{j=1}^\infty(-1)^{j-1}j^{-1}\int_0^\infty\xi^z\exp(-j\Lambda\xi)d\xi$$ (VI.52)

$$\equiv I_z/\Lambda^{z+1}.$$

Here one has

$$I_z = F_z\Gamma(z+1)\sum_{j=1}^\infty\frac{(-1)^{j-1}}{j^{z+2}},$$ (VI.53)

where as usual $\Gamma(z)$ is the Euler gamma function. Table VI.2 presents several numerical values of the dimensionless quantity I_z/F_z.

By inserting result Eq. (VI.52) into Eqs. (VI.48), one obtains the leading-order expressions

$$\Delta\hat{\varphi} = (z+1)I_z/\Lambda^{z+2},$$ (VI.54)

$$\Delta\hat{\sigma}_a = (z+2)I_z/\Lambda^{z+1}.$$

Upon eliminating Λ between these formulas, one finally concludes that to leading order in $\Delta\hat{\varphi}$, the isobaric enumeration quantity $\Delta\hat{\sigma}_a(\Delta\hat{\varphi})$ has the following form:

$$\Delta\hat{\sigma}_a(\Delta\hat{\varphi}) = \left[\frac{(z+2)I_z^{1/(z+2)}}{(z+1)^{(z+1)/(z+2)}}\right](\Delta\hat{\varphi})^{(z+1)/(z+2)}.$$ (VI.55)

TABLE VI.2. Numerical values of the series appearing in Eq. (VI.53)

z	I_z/F_z
−0.5	1.356188
0.0	0.822467
0.5	0.768536
1.0	0.901543
1.5	1.233034
2.0	1.894066

The coefficient multiplying the right side of this last equation is positive, and the exponent of $\Delta\hat{\varphi}$ lies between 0 and 1 for the stated range of z, Eq. (VI.51). Consequently, the initial slope of $\hat{\sigma}_a(\hat{\varphi})$ above $\hat{\varphi}_{\mathrm{min},a}$ is infinite. Because this result contradicts the earlier Eq. (VI.40) requiring a finite initial slope, the possibility of an ideal glass transition occurring at $T > 0$ is eliminated.

E. Estimating $\sigma_a(\varphi)$, $\hat{\sigma}_a(\hat{\varphi})$ Numerically

It is generally a challenging task to extract the depth-dependent enumeration functions $\sigma_a(\varphi)$ and $\hat{\sigma}_a(\hat{\varphi})$ either from laboratory measurements on real glass formers or from numerical simulations for model systems. However, there can be circumstances that facilitate the estimation of these functions. These circumstances apply most frequently and accurately to fragile glass formers. In particular, for isobaric conditions, suppose that the following conditions apply:

(a) The attainable crystal phase at $T = 0$ is verifiably disorder-free.
(b) The crystal form undergoes no solid-state phase changes from $T = 0$ up to the thermo-dynamic melting point at $T_m(p)$.
(c) Slowly supercooling the liquid phase below $T_m(p)$ results in a relatively sharply defined glass transition at $T_g(p)$.
(d) $C_p^{(crys)}(T)$ has been determined for the crystal phase over the entire range $0 \le T \le T_m(p)$.
(e) The heat of melting $\Delta H_m(p)$ has been determined.
(f) The liquid-phase heat capacity $C_p^{(liq)}(T)$ has been determined in the supercooled regime down to $T_g(p)$ and is a sufficiently smooth function of T to be extrapolated below T_g.
(g) The heat capacity of the glass $C_p^{(glass)}(T)$ has been determined for $0 \le T < T_g(p)$ and is a sufficiently smooth function of T to permit extrapolation up to $T_m(p)$.

Items (a), (b), (d), and (e) in this list imply that the entropy of the crystal $S^{(crys)}$ and of the liquid $S^{(liq)}$ at the melting point can be straightforwardly evaluated. Specifically, on a per-particle basis these are the following:

$$S^{(crys)}[T_m(p)]/Nk_B = \int_0^{T_m(p)} [C_p^{(crys)}(T)/Nk_BT]\,dt; \tag{VI.56}$$

$$S^{(liq)}[T_m(p)]/Nk_B = S^{(crys)}[T_m(p)]/Nk_B + \Delta H_m(p)/Nk_BT_m(p). \tag{VI.57}$$

These entropies of course include both inherent-structure configurational contributions and the average intrabasin vibrational entropy for the phase involved.

Assuming that the crystal remains essentially defect-free up to its melting point [i.e., $\hat{\sigma}^{(crys)}$ $(T_m) \approx 0$], then the melting entropy per particle, the second term of the right side of Eq. (VI.57), is determined by the melting-point increase in $\hat{\varphi}$ plus the mean vibrational entropy increase in passing from the crystal basin to amorphous basins that are relevant at $T_m(p)$. The latter contribution involves both harmonic and anharmonic aspects of the respective vibrational motions and can be extracted from the difference in heat capacities for the crystal and the kinetically arrested glass. Thus, for the liquid at its pressure-p melting point, one can write

$$\hat{\sigma}_a[\hat{\varphi}_a{}^*(T_m)] \cong [\Delta H_m/Nk_B T_m] - \int_0^{T_m} \{[C_p{}^{(glass)}(T) - C_p{}^{(crys)}(T)]/Nk_B T\}dT. \qquad \text{(VI.58)}$$

This expression needs to be supplemented with an expression for $\hat{\varphi}_a{}^*(T_m)$, indicating how much inherent-structure elevation change has occurred in heating the crystal from $T = 0$ up through its first-order melting transition. The required formula parallels Eq. (VI.58) by correcting the contribution from the melting enthalpy per particle by a heat capacity difference integral:

$$\Delta\hat{\varphi}_a{}^*(T_m) = \hat{\varphi}_a{}^*(T_m) - \hat{\varphi}^{(crys)}(T = 0) \qquad \text{(VI.59)}$$

$$\cong \Delta H_m/N - \int_0^{T_m} [C_p{}^{(glass)}(T) - C_p{}^{(crys)}(T)]dT/N.$$

Expressions (VI.58) and (VI.59) can serve as the basis for generalization to $T \neq T_m(p)$, although one must be cautious about how far to carry the generalization away from the melting point. The supercooled liquid isobaric enumeration function and the depth parameter require accounting for both the change in total entropy as T moves away from T_m and the corresponding shift in vibrational entropy. The results are the following:

$$\hat{\sigma}_a[\hat{\varphi}_a{}^*(T)] \cong \hat{\sigma}_a[\hat{\varphi}_a{}^*(T_m)] - \int_T^{T_m} [C_p{}^{(liq)}(T') - C_p{}^{(glass)}(T')]dT'/(Nk_B T'); \qquad \text{(VI.60)}$$

$$\Delta\hat{\varphi}_a{}^*(T) \cong \Delta\hat{\varphi}_a{}^*(T_m) - \int_T^{T_m} [C_p{}^{(liq)}(T') - C_p{}^{(glass)}(T')]dT'/N. \qquad \text{(VI.61)}$$

Accurate calorimetric data at 1 atm are available for the fragile glass former *ortho*-terphenyl [Chang and Bestul, 1972]. These measurements conform to the items (a)–(g) above and have been utilized for estimating the isobaric quantities $\hat{\sigma}_a$ and $\hat{\varphi}_a{}^*$, which in principle result from steepest descent mapping on the space of molecular coordinates and volume [Stillinger, 1998]. Section V.B briefly presented this example, technical details for which are now supplied. This case involves an additional simplification, namely, that the measured heat capacity difference $C_p{}^{(glass)} - C_p{}^{(crys)}$ is sufficiently small to be disregarded numerically (quantized two-level systems that might be present in the glass and contribute to heat capacity at very low temperature are probably too sparse to make significant contributions to $\hat{\sigma}_a$ and $\Delta\hat{\varphi}_a{}^*$). The difference between the measured crystal and liquid heat capacities accurately fits a simple three-term expression ($p = 1$ atm):

$$C_p{}^{(liq)}(T) - C_p{}^{(crys)}(T) = A_2 T^{-2} + A_3 T^{-3} + A_4 T^{-4}, \qquad \text{(VI.62)}$$

$$A_2 = 2.0232 \times 10^7 \text{ J K mol}^{-1},$$

$$A_3 = -4.6172 \times 10^9 \text{ J K}^2 \text{ mol}^{-1},$$

$$A_4 = 3.2346 \times 10^{11} \text{ J K}^3 \text{ mol}^{-1}.$$

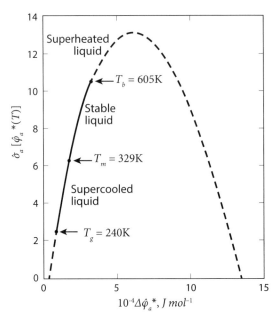

FIGURE VI.16. The isobaric enumeration function for *ortho*-terphenyl plotted against the relative value of the inherent-structure depth parameter. The pressure is 1 atm both for the starting thermodynamic state and for the steepest descent mapping to inherent structures. This curve is based on Eqs. (VI.63) and (VI.64) and has been redrawn from Stillinger, 1998, Figure 2. The specific points on the curve are identified for the glass transition temperature and the melting and boiling temperatures. Dashed portions of the curve locate positions for which the extrapolations may be unreliable.

This in turn leads to the following implicit expressions for the isobaric depth (in J mol^{-1}) and enumeration (in J K^{-1} mol^{-1}) quantities, in which temperature serves as a parameter:

$$\hat{\varphi}_a{}^*(T) = 17{,}191 - A_2(T^{-1} - T_m{}^{-1}) - \frac{A_3}{2}(T^{-2} - T_m{}^{-2}) - \frac{A_4}{3}(T^{-3} - T_m{}^{-3}); \qquad \text{(VI.63)}$$

$$Nk_B\hat{\sigma}_a[\hat{\varphi}_a{}^*(T)] = 52.20 - \frac{A_2}{2}(T^{-2} - T_m{}^{-2}) - \frac{A_3}{3}(T^{-3} - T_m{}^{-3}) - \frac{A_4}{4}(T^{-4} - T_m{}^{-4}). \qquad \text{(VI.64)}$$

Figure VI.16 repeats and supplements information contained in Figure V.2 (in which the enumeration function $\hat{\sigma}_a$ was denoted by the more general metastable-state symbol $\hat{\sigma}_m$). It presents a plot of the isobaric enumeration function vs. the depth parameter, using the expressions displayed in Eqs. (VI.63) and (VI.64). Three points have been identified explicitly on the curve. As indicated in the figure, these correspond to the three distinguished temperatures for the liquid *ortho*-terphenyl, $T_g = 240$ K, $T_m = 329$ K, and the 1 atm boiling point $T_b = 605$ K. They illustrate that within this liquid range, T increases continuously from left to right along the curve.

Notice that the $\hat{\sigma}_a$ curve in Figure VI.16 passes through a single maximum, but that maximum corresponds formally to a superheated state of the 1 atm liquid. This raises the point that for this example the isobaric inherent-structure subset $\hat{\mathbf{A}}$ must be chosen to eliminate not only inherent structures with embedded crystallites but also to eliminate those inherent structures obtained by isobaric mapping of system configurations that contained vapor-filled "bubbles" or larger regions arising from liquid boiling. Note in passing that no similar requirement arises in the isochoric situation. This vapor-nucleated subset of isobaric inherent structures would presumably have an anomalously low number density, a characteristic that could form the basis of the necessary projection operation. However, quantitative details for the projection operation in this superheated regime remain somewhat unclear for the *ortho*-terphenyl example under consideration. Consequently, the curve shown has been dashed in the superheated regime to indicate growing uncertainty with increasing T about its quantitative accuracy. Nevertheless, the overall qualitative shape for the $\hat{\sigma}_a \geq 0$ interval is probably correct. Tracing the curve beyond its maximum nominally requires considering negative T. However, it should be kept in mind that these anomalously high-potential-enthalpy inherent structures can be partially accessed by other means, as mentioned in Chapter III, such as irreversible mechanical working, radiation damage, or rapid film deposition from the vapor.

It is also important to note that the $\hat{\sigma}_a$ curve in Figure VI.16 has been dashed in its very low temperature portion because its simple parametric form is unable to reproduce the expected vertical tangent at its lower extreme that Section VI.D argues is a necessary property. In order for that to emerge, very precise measurements of the crystal and glass heat capacities in the low-temperature limit would have to be incorporated, which were beyond the scope of the experimental determination used for the estimate [Chang and Bestul, 1972]. But in spite of this shortcoming, it is noteworthy that in Figure VI.16 the lowest $\Delta\hat{\varphi}_a^*$ value for which $\hat{\sigma}_a \geq 0$ is itself positive, in line with the requirement that all amorphous inherent structures possess higher energy and/or enthalpy than the crystal.

Analogous $\hat{\sigma}_a(\hat{\varphi}^*)$ calculations have been performed for 1-propanol and for 3-methylpentane, similarly based on calorimetric measurements for these substances [Debenedetti et al., 2001, Figure 13]. These cases have covered ranges of pressure, specifically 0.1–198.6 MPa for the former, and 0.1–108.1 MPa for the latter. In all cases, the $\hat{\sigma}_a$ curves are qualitatively similar to that shown in Figure VI.16 for *ortho*-terphenyl but with quantitative details that demonstrate clear dependence on molecular structure. Over the pressure ranges involved, the variations with pressure are rather small but not negligible.

Glass transition phenomena are not usually investigated in the laboratory under constant-volume (isochoric) conditions. However, that is the data-generating circumstance required in order to form estimates of the isochoric quantities $\Delta\varphi_a^*$ and σ_a. For simplicity in the following, attention is restricted to glass formers whose crystal- and liquid-phase thermal expansions α are positive and whose melting volumes are positive. With this conventional-behavior restriction, and if the fixed volume is that for the system at low T and p, then the constant-volume steepest descent mapping that identifies inherent structures should not have the possibly complicating effect of producing macroscopic voids, but rather the inherent structures fill the given volume uniformly.

Under constant-volume thermodynamic conditions, the first-order melting transition does not occur at a single melting temperature but is stretched out over a nonzero temperature interval of crystal–liquid phase coexistence:

$$T_{m,l}(V) \leq T \leq T_{m,u}(V). \tag{VI.65}$$

Here the lower temperature $T_{m,l}$ locates the onset of melting under equilibrium heating conditions, and the upper temperature $T_{m,u}$ locates its completion. In contrast to isobaric conditions where the entire heat of melting ΔH_m must be added to the system at a fixed temperature, now the heat is added continuously over interval (VI.65). The continuous heat additions involved produce a magnified heat capacity $C_V^{(coex)}(T)$ over that interval, compared to lower or higher temperature, thus exhibiting discontinuities at its endpoints. Once it has been heated isochorically to $T_{m,u}$ or beyond, a good glass former in its uniform liquid state then can be supercooled with determination of $C_V^{(liq)}(T)$ for that metastable liquid. But upon cooling through a glass transition at $T_g(V)$, this constant-volume heat capacity would drop to a smaller quantity $C_V^{(glass)}(T)$. Both of these C_V's should be amenable to extrapolation, respectively, below and above T_g, and then they can be compared with the stable crystal heat capacity, $C_V^{(crys)}(T)$.

With these required changes from the constant-pressure scenario, the isochoric replacements for Eqs. (VI.58) and (VI.59) are as follows:

$$\sigma_a[\varphi_a^*(T_{m,u})] \cong \int_0^{T_{m,l}} [C_V^{(crys)}(T) - C_V^{(glass)}(T)]dT/(Nk_BT) \tag{VI.66}$$
$$+ \int_{T_{m,l}}^{T_{m,u}} [C_V^{(coex)}(T) - C_V^{(glass)}(T)]dT/(Nk_BT);$$

$$\Delta\varphi_a^*(T_{m,u}) = \varphi_a^*(T_{m,u}) - \varphi^{(crys)}(T=0)$$
$$\cong \int_0^{T_{m,l}} [C_V^{(crys)}(T) - C_V^{(glass)}(T)]dT/N \tag{VI.67}$$
$$+ \int_{T_{m,l}}^{T_{m,u}} [C_V^{(coex)}(T) - C_V^{(glass)}(T)]dT/N.$$

Then for other temperatures, the isochoric enumeration and depth function expressions are the following analogs of Eqs. (VI.60) and (VI.61):

$$\sigma_a[\varphi_a^*(T)] \cong \sigma_a[\varphi_a^*(T_{m,u})] - \int_T^{T_{m,u}} [C_V^{(liq)}(T') - C_V^{(glass)}(T')]dT'/(Nk_BT'); \tag{VI.68}$$

$$\Delta\varphi_a^*(T) \cong \Delta\varphi_a^*(T_{m,u}) - \int_T^{T_{m,u}} [C_V^{(liq)}(T') - C_V^{(glass)}(T')]dT'/N. \tag{VI.69}$$

It is also possible to produce estimates of enumeration functions σ_a or $\hat{\sigma}_a$ for models of glass-forming systems in the course of numerical simulations. Some limited results have been reported for the binary mixture Lennard-Jones (BMLJ) model [Sastry, 2001]. Both the roughly inverted parabola shapes of the inferred enumeration functions, as well as their density or pressure dependences, are qualitatively similar to the results mentioned above for glass formers investigated in the laboratory. Simulations using the simple Lewis-Wahnström model for *ortho*-terphenyl [Lewis and Wahnström, 1994] have also yielded information that is consistent with a σ_a qualitatively shaped like an inverted parabola [La Nave et al., 2003]. In contrast to practical laboratory procedures, these simulation studies can take advantage of harmonic-approximation evaluations of intrabasin vibrational contributions, as well as thermodynamic integration paths to ideal gas asymptotic states at infinite temperature or infinitesimal density to provide entropy determinations.

F. Crystal Nucleation and Growth

In order for a macroscopic liquid sample to undergo the first-order freezing transition to its crystal phase at a measurable rate, it is necessary to supercool that sample by a nontrivial amount. Consequently, it is natural to include an examination of at least some basic aspects of the kinetics of the freezing process in the current chapter. The objective is to understand the preferred kinetic pathways that lead the system configuration out of the amorphous landscape denoted by $\mathbf{A}$ (constant-volume conditions) or $\hat{\mathbf{A}}$ (constant-pressure conditions). This involves consideration both of the mechanism whereby the crystal phase makes its first microscopically localized appearance (nucleation) in the macroscopic liquid, as well as the rate of subsequent crystal growth to macroscopic dimensions, both considered as functions of the extent of liquid supercooling. In the following, it is supposed that homogeneous nucleation applies, that is, the starting liquid is free of impurity inclusions or bounding surfaces that could act as heterogeneous nucleation sites. Thus the initiation of the phase change is an intrinsic bulk property of the pure supercooled liquid alone.

After establishing a given initial state of undercooling $T < T_m$, the subsequently observed phenomenological nucleation rate can be denoted by $k_{nuc}(T)$. It implicitly depends upon the pressure, and it has units time^{-1}volume^{-1}. Specifically, it refers to the frequency of spontaneous appearance, per unit time in a unit volume, of identifiable microscopic locations ("embryos") within a macroscopic sample that have begun converting from supercooled liquid to a microscopic growing crystal structure. The typical behavior observed for this rate as a function of temperature is indicated in Figure VI.17, and in conformity with most real laboratory observations it should implicitly be assumed that pressure is held constant. The characteristic behavior shown involves a

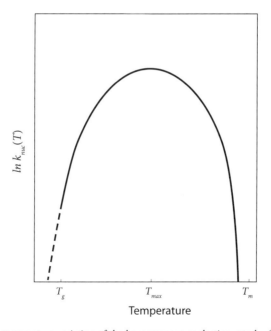

FIGURE VI.17. Qualitative temperature variation of the homogeneous nucleation rate $k_{nuc}(T)$ for supercooled liquids under constant-pressure conditions. The dashed portion of the curve is rendered unobservable because of intervention of the experimental glass transition.

temperature of maximum nucleation rate $T_{max}(p)$ lying between the glass transition range $\approx T_g(p)$ and the thermodynamic melting/freezing temperature $T_m(p)$. On either side of this maximum, the rate plunges rapidly, thus justifying a plot of its logarithm as shown.

The vertical scale and the overall shape of the plot in Figure VI.17 are characteristics that are strongly dependent on the substance under consideration. The maximum rate $k_{nuc}(T_{max})$ is sufficiently large for some cases, such as molten pure alkali halides, that it is very difficult, if not impossible, to form an amorphous glass even by very rapid cooling of a bulk sample: Upon approaching T_{max} from above, nucleation and crystal growth are virtually inevitable. By contrast, good glass formers such as glycerol and salol have much lower nucleation rates in the vicinity of their respective T_{max} values and so remain in an amorphous state as they are brought to low temperatures $T < T_{max}$, even when very slow laboratory cooling rates are used.

Although it suffers from some technical weaknesses, classical nucleation theory identifies at least qualitatively the main ingredients of the crystal nucleation phenomenon [Oxtoby, 1992; Debenedetti, 1996, Section 3.1.5]. These ingredients include (a) identification of a "critical nucleus" that serves as a structural threshold for initiation of the phase change, (b) a time required for spontaneous formation of that critical nucleus from the homogeneous liquid that has been quickly brought to the supercooled state from T_m, and (c) the subsequent growth rate of the critical nucleus to a macroscopic crystal. Classical nucleation theory comes closest to providing a quantitatively accurate description in the small supercooling limit, where the overall rate k_{nuc} tends asymptotically toward zero.

So far as ingredient (a) is concerned, the classical nucleation theory focuses on the formation free energy $W(n)$ that macroscopic thermodynamics would assign to an n-molecule crystallite embedded in the supercooled liquid, where n comprises only a very small fraction of the total number N of molecules present. The theory assumes that the structure and intensive thermodynamic properties of this nucleus are adequately approximated by those of the macroscopic crystal phase at the prevailing undercooling temperature $T = T_m - \Delta T$ and pressure p. The shape of the crystallite is taken to be that which minimizes surface free energy for the crystallite–liquid interface. This minimum is influenced by the fact that the positive surface free energy for crystal–fluid interfaces depends in a nontrivial way on the direction of the normal to the interface relative to the fundamental directions of the underlying crystal structure [Herring, 1951; Woodruff, 1973; Fisher and Weeks, 1983]. In the most direct version of the classical nucleation theory, $W(n)$ is composed simply of a negative bulk contribution proportional to n and a positive interfacial contribution proportional to $n^{2/3}$:

$$W(n) \cong -[\Delta\mu(T, p)]n + \chi\bar{\gamma}(T, p)\mathrm{v}_{crys}(T, p)n^{2/3}. \qquad (VI.70)$$

In this expression, the difference in chemical potentials of the supercooled liquid ("scl") and crystal phases has been denoted by

$$\Delta\mu(T, p) = \mu_{scl}(T, p) - \mu_{crys}(T, p), \qquad (VI.71)$$

the volume per molecule in the crystal is $\mathrm{v}_{crys}(T, p)$, and $\bar{\gamma}(T, p)$ is the liquid–crystal interfacial free energy per unit area averaged over the prevailing shape of the crystallite. The dimensionless numerical factor χ depends upon the geometric shape of the n-molecule crystallite. A sphere would have the minimal value $\chi = (36\pi)^{1/3} \cong 4.8360$, but generally a nonspherical shape would require a larger factor, e.g., $\chi = 2^{2/3} \cdot 3^{7/6} \cong 5.7191$ for a regular octahedron.

As n increases from zero, the formation free-energy function $W(n)$ represented by Eq. (VI.70) at first increases, then passes through a maximum at $n = n^*$, and subsequently declines toward minus infinity. Treating n as a continuous variable, the maximum is predicted to occur at

$$n^*(T, p) = \left[\frac{2\chi\bar{\gamma}(T, p)}{3\Delta\mu(T, p)} \right]^3 \mathrm{v}_{crys}^2(T, p), \qquad (VI.72)$$

and therefore at the prevailing T and p,

$$W(n^*) \cong \frac{4(\chi\bar{\gamma})^3 \mathrm{v}_{crys}^2}{27(\Delta\mu)^2}. \qquad (VI.73)$$

The chemical potential difference $\Delta\mu$ is positive and for small supercooling is proportional to ΔT. Consequently, n^* and $W(n^*)$ both diverge as $\Delta T \to 0$. At least for sizes in the range $n \leq n^*$, embryos should be present in the supercooled liquid at concentrations primarily controlled by the Boltzmann factor $\exp[-\beta W(n)]$. Consequently, the size $n = n^*$ presents a concentration minimum through which the growth to a macroscopic crystal must pass. This "critical nucleus" plays the role of a kinetic bottleneck for the freezing transition, which as $\Delta T \to 0+$ forces $k_{nuc}(T)$ to plunge toward zero, as indicated in Figure VI.17. It is the large size predicted for the critical nucleus when ΔT is small that at least in that limit confers logical consistency on the classical nucleation theory, which utilizes macroscopic characteristics as its conceptual and quantitative input.

Although the classical nucleation theory for freezing of supercooled liquids is phrased in terms of the instantaneous configurations exhibited by the system, it is useful to rephrase the corresponding description in terms of the time sequence of energy or enthalpy basins and their inherent structures. As stressed earlier (Section V.C), an advantage offered by inherent structures is that local patterns of structural geometry in dense many-body media tend to be more clearly identifiable because of removal of intrabasin vibrational smearing. In particular, this point is relevant to identification of local crystalline particle arrangements and to identification of the number n of particles that they contain.

Indeed, such local pattern recognition is central to defining the amorphous-state configurational regions $\mathbf{A}$ and $\hat{\mathbf{A}}$, within which the many-particle system starts before supercooling and nucleation. However, the maximum crystallite size n normally permitted within these regions by their definitions should be small (e.g., $n \approx 10$) and for a given substance should be chosen at the outset to represent the equilibrium liquid accurately. Hence, that choice would have no direct connection to later selections of T and/or p to probe the supercooled regime. The fact that n^* increases without bound as $\Delta T \to 0$ means that successful crystal nucleation events have a threshold (kinetic bottleneck) far removed from the boundaries assigned to regions $\mathbf{A}$ and $\hat{\mathbf{A}}$, and the smaller ΔT is, the farther must be the excursions to pass the nucleation threshold successfully.

If indeed the inherent structures were used to assign n values to embryos present in a supercooled sample, those assignments would of course remain constant throughout the basin containing that inherent structure. The changes in embryo content integers n would only occur at the instant that the system dynamics caused an interbasin transition, i.e., a crossing of a shared basin boundary. However, it must be realized that the majority of interbasin transitions only involve localized rearrangements of $O(1)$ particles that exhibit amorphous arrangements both before and after that transition, hence no change in the set of embryo content integers n. It should also be appreciated that if the n value for an embryo actually does change as a result of an interbasin transition, that change is not necessarily restricted to $\Delta n = \pm 1$. Although it might be a scarce occurrence, the localized particle rearrangement could in principle cause Δn to change by ± 2, or even more. Finally, there could be interbasin transitions whose initial and final inherent structures involve only a restructuring of an embryo without any change in its content integer n.

Consider now an embryo that the dynamics has caused to grow to the critical nucleus size $n = n^*$ for the prevailing T and p. Because its formation free energy $W(n^*)$ is a local maximum with respect to n, there is no net driving force propelling that n either to a smaller or a larger value away from that extremum. However, the fluctuations arising from the many-body dynamics cause n for that critical nucleus to drift away from n^*. A significant fraction of these occasions shows the value of n drifting substantially above n^* to a point where indeed the formation free energy is decreasing enough with n to supply a substantial driving force for uninhibited crystal growth. The nucleation process thus has been completed and the phase-transition process enters the macroscopically observable regime.

The elementary kinetic processes that underlie growth or reduction in embryo sizes naturally slow as temperature declines. In order for a particle on the amorphous medium side of a liquid–crystal interface to add to the crystal, it must have a suitable site on the crystal embryo surface available to accommodate it, and in order to occupy that site it has to be displaced by roughly a nearest neighbor distance. Complex molecules would also typically require reorientation and/or conformational change to fit into the embryo site. These motions, viewed as displacements from one inherent structure to the next, roughly exhibit the same rate decline with reducing temperature as the translational and rotational diffusion constants within the supercooled liquid. The strong reduction in these diffusion constants as T declines toward the temperature range around the glass transition at T_g forces the nucleation rate constant $k_{nuc}(T)$ to decline precipitously, as Figure VI.17 qualitatively indicates.

If a successful nucleation event has occurred, with the n value for the crystallite involved having substantially exceeded the critical nucleus value n^*, then the subsequent crystal growth process releases the heat of freezing per particle Δh. This heat produces at least a temporary rise in temperature in the vicinity of the growing interface that can influence the growth rate itself. That influence depends on whether the temperature of the supercooled medium at the time of the nucleation event is less than or greater than the $T_{max}(p)$ illustrated in Figure VI.17. The corresponding changes would be to increase or to decrease the crystal growth rate, respectively. In the latter case, if the thermal conductivity of the supercooled liquid were anomalously low, the shape of the growing crystallite would be subject to an instability that produces noncompact dendritic forms [Kobayashi, 1993]. This instability thrusts portions of the interface preferentially into regions of the uncrystallized liquid relatively remote from other regions that have been locally heated by the release of the heat of freezing. However, the resulting interfacial area thus produced is larger than for a more compact crystallite with a corresponding increase in interfacial free energy. Longer term, this extended or dendritic form would anneal to reduce that positive interfacial free energy.

The late-stage outcome of the nucleation and crystal growth depends on whether the supercooled sample remained in contact with a thermal reservoir at the initial temperature of supercooling, or whether it was thermally isolated. In the former instance, the entire heat of freezing would eventually be conducted out of the sample and into the reservoir, allowing the crystallization process to go to completion. This would tend to produce at first an imperfect crystalline medium containing a variety of defects that typically would take a very long time to anneal out of the system at the reservoir-fixed temperature. In the latter instance of thermal isolation, the temperature rise resulting from the internal release of the freezing heat could bring the system temperature up to $T_m(p)$ before the entire sample has crystallized. As a consequence, the final state of crystal growth would then exhibit a coexistence of liquid and crystal phases at the thermodynamic melting/freezing point, described in Sections II.H and III.E. A dramatic variant resulting from at least partial thermal isolation is "explosive crystallization," a phenomenon in which the released energy causes extreme acceleration of the crystallization process in thin films of some amorphous materials [Polman et al., 1989; Grigoropoulis et al., 2006].

Many substances at a given pressure can exist stably or metastably in several distinct crystal forms, in other words, they exhibit "crystal polymorphism." Which one of these emerges first from a nucleation event in the supercooled liquid is not necessarily the most stable crystal form at the prevailing T and p. Especially because of differences in the interfacial free energy of these alternate forms in contact with liquid, the critical nucleus free energy $W(n^*)$ could conceivably be lower for a metastable crystal phase than for the stable form. The correspondingly less restrictive kinetic threshold would mean that the metastable form would be the one to appear first and to grow. Later, nucleation of the stable crystal within the macroscopically enlarged metastable form would have to occur to attain thermodynamic equilibrium. This possible scenario justifies mention of the so-called "Ostwald step rule" [Ostwald, 1897; van Santen, 1984]. Formulated on the basis of limited empirical evidence (primarily for precipitation from oversaturated liquid solutions), this rule states that when crystal polymorphism is present, it is the thermodynamically least stable polymorph that nucleates first. Applicability of the Ostwald step rule is especially important in the pharmaceutical industry, whose products frequently exhibit polymorphism with related variation in medical effectiveness [Burley et al., 2007]. In cases that are subject to this Ostwald scenario, it is necessary, as suggested in Section VI.B, to refine the definitions of the amorphous-state regions **A** and **Â** in order to exclude system configurations whose inherent structures contain crystallites of both the stable and metastable polymorphs.

VII.

Low-Density Matter

Because the energy landscape/inherent structure representation is general, it is applicable in principle not only to the dense crystal, liquid, and amorphous glass states but also to lower density forms of matter. These lower density states include dilute vapors, critical and super-critical fluids, and highly porous solids such as aerogels [Fricke, 1988; Mohanan et al., 2005]. A superficial view might suggest that these low-density states could be disregarded in an examination and analysis of condensed matter phenomena; however, that would be inappropriately restrictive. The simple energy landscapes for small isolated clusters provide prototypical characteristics for the much more complicated extended condensed phase cases. Also, some forms of condensed matter can be conveniently prepared by vapor-phase deposition on suitable substrates. Furthermore, continuous thermodynamic paths exist connecting dense liquid states to dilute vapor states, and a complete development of the inherent-structure approach requires understanding quantitatively how this representation evolves along such paths. The fact that irreversible nucleation processes create dense liquid or solid phases out of supercooled or overcompressed vapors constitutes another motivation. What connections, if any, between liquid–vapor critical points on one side and binary liquid mixture critical points and crystalline alloy order–disorder critical points on the other also deserve to be explored. Finally, attaining completeness of the theory requires exploring connections between the energy landscape/inherent structure viewpoint and the vapor-phase virial expansion coefficients [Mayer and Mayer, 1940, Chapter 13; McQuarrie, 1976].

A. Cluster Inherent Structures and Basins

Consider first the case of a cluster of N identical particles isolated in unbounded free space, where for the moment N is a modest number of $O(1)$. Let $\mathbf{x}_1 \ldots \mathbf{x}_N$ represent the position and conformation coordinates (if any) for particles $1 \ldots N$, respectively. Assume that beyond the short range of steric repulsions, these particles experience attractive interactions at all finite distances (although in the case of polyatomic molecules this may involve orientational constraints). This set of N possesses a well-defined collection of bound inherent structures on its potential energy hypersurface $\Phi(\mathbf{x}_1 \ldots \mathbf{x}_N)$. As usual, the isochoric steepest descent mapping process, Eq. (III.2), defines basins belonging to each of the inherent structures. With just a modest finite number N of interacting particles, the total number $\Omega(N)$ of corresponding inherent structures, including those related by permutation, should have a manageable magnitude for exhaustive enumeration. Then because the space is unbounded and allows free translation and rotation of any cluster configuration, the content of all basins must be unbounded. For some purposes, it is useful to remove the center-of-mass degree of freedom and to concentrate just on internal cluster deformations in the resulting subspace. Under this restriction, at least some of the basins remain unbounded because of inclusion of configurations that involve cluster fragmentation displacements of unlimited magnitude. However, other basins subject to fixed center of mass may become bounded, depending on the specific form of Φ.

In the absence of constraints, the dimension of the inherent-structure subspace corresponding to any given Φ minimum depends simply on the number of free translational and rotational degrees of freedom available for the potential-energy-minimized cluster of interest. For a dimer ($N = 2$) comprising a pair of structureless spherical particles in three-dimensional space, there are five such degrees of freedom, three translational and two rotational. The same would be true for any cluster composed of spherical particles if its inherent structure exhibited cylindrical symmetry. All other cases in three-dimensional space possess six such degrees of freedom, three translational and three rotational. Once again, it may be useful to project out and disregard the three unbounded translational degrees of freedom for the center-of-mass motion.

The relative energies and the geometric diversity that can occur with cluster inherent structures can be instructively illustrated, even with the simple model of structureless particles interacting with the pairwise additive Lennard-Jones 12,6 potential [Eqs. (I.35) and (I.37)]. With this choice, the N-particle potential energy function depends just on the set of particle locations $\mathbf{r}_1 \ldots \mathbf{r}_N$ and has the following form:

$$\Phi(\mathbf{r}_1 \ldots \mathbf{r}_N) = \varepsilon \sum_{i=1}^{N-1} \sum_{j=i+1}^{N} \mathrm{v}_{LJ}(r_{ij}/\sigma), \tag{VII.1}$$

$$\mathrm{v}_{LJ}(x) = 4(x^{-12} - x^{-6}).$$

Numerical implementation of steepest descent mapping is straightforward in this case, when N is relatively small. By examining a variety of starting configurations for that mapping, one can be reasonably sure that all distinguishable inherent structures for those N values are uncovered. However, there is no unambiguous criterion for deciding generally that all distinguishable species of inherent structures have been obtained during a search procedure.

Table VII.1 lists all known inherent structures for the Lennard-Jones 12,6 model in three dimensions for cluster sizes in the range $2 \leq N \leq 7$. The table provides values for the reduced potential energy Φ_N/ε, the point group symmetry, a verbal description of the inherent-structure geometry, and the degeneracy d_{IS} enumerating equivalent inherent structures of each geometric type (including inversion) that belong to different (but equivalent) basins. Far more extensive tabulations of inherent structures for this model have been prepared and published [Hoare, 1979; Tsai and Jordan, 1993a], but the difficulty of the enumeration task rises rapidly with particle number N, especially with respect to ensuring completeness. Numerical search over a wide variety of starting configurations with fixed center of mass indicates that the basins for all of the inherent structures listed in Table VII.1 have infinite content. It remains an open question whether for some N values there exist any finite-content basins for clusters of Lennard-Jones 12,6 particles subject to the fixed center-of-mass constraint.

As the number N of particles increases, so too does the range of Φ values for the cluster inherent structures. The sequence of absolute minima, the lowest lying inherent structures, involves surface-minimizing structures that have the potential energy per particle Φ_N/N asymptotically approaching the value for the stable zero-pressure crystal at absolute zero temperature. For the

TABLE VII.1 Inherent structure types for Lennard-Jones 12,6 model clusters in three dimensions ($2 \leq N \leq 7$)

N	Φ_N/ε	d_{IS}	Point Group	Geometric Shape
2	−1.00000	1	$D_{\infty h}$	Symmetric dimer
3	−3.00000	1	D_{3h}	Equilateral triangle
4	−6.00000	2	T_d	Regular tetrahedron
5	−9.10385	20	D_{3h}	Triangular bipyramid
6(a)	−12.71206	30	O_h	Regular octahedron
6(b)	−12.30293	360	C_{2v}	Bicapped tetrahedron
7(a)	−16.50538	504	D_{5h}	Pentagonal bipyramid
7(b)	−15.93504	1,680	C_{3v}	Capped octahedron
7(c)	−15.59321	1,680	C_{3v}	Tricapped tetrahedron
7(d)	−15.53306	5,040	C_2	Four tetrahedra sharing faces in pairs

Lennard-Jones 12,6 potential model, the stable crystal possesses a hexagonal close-packed structure, with potential energy per particle [Stillinger, 2001]:

$$\lim_{N \to \infty} [(\Phi_N/N\varepsilon)]_{lowest} = -8.61107. \tag{VII.2}$$

The crystal producing this energy exhibits a very small spontaneous distortion away from the ideal hexagonal close-packed structure, with shrinkage along the c axis, and expansion within the close-packed planes perpendicular to the c axis. The mean nearest neighbor separation at this lowest lying minimum is 1.0902σ.

The other extreme, associated with the highest lying inherent structures, may correspond to "needle" clusters. Figure VII.1 provides an illustration for 54 particles. These structures maximize surface area while maintaining mechanical stability by consisting of a stack of equilateral triangles, with a 60 degree in-plane twist from one layer to the next along the needle axis [Stillinger and Stillinger, 1990]. The implied upper limit to potential energy per particle is

$$\lim_{N \to \infty} [(\Phi_N/N\varepsilon)]_{highest} = -3.46305. \tag{VII.3}$$

In the middle of long clusters of this kind, the distance between neighbor particles within each equilateral triangle is 1.1191σ, and the distance along the needle axis between the planes of successive triangles is 0.8969σ.

Needless to say, cluster inherent structures depend in energy and geometric characteristics on details of the potential energy function that is applicable. For the purpose of illustrating one aspect of this dependence, consider the following simple generalization of the Lennard-Jones 12,6 pair potential containing a continuous exponent parameter n:

$$v_{LJ}(x) = 4(x^{-2n} - x^{-n}). \tag{VII.4}$$

Even with this extension, the reduced pair potential continues to have a single minimum with the same unit depth, now located at

$$x_{min}(n) = 2^{1/n}. \tag{VII.5}$$

FIGURE VII.1. "Needle" cluster for 54 particles interacting with the Lennard-Jones 12,6 pair potential. This type of extended but mechanically stable arrangement possesses an energy per particle at, or close to, the highest possible value for inherent structures with large numbers of particles interacting with this pair potential. Adapted from Stillinger and Stillinger, 1990, Figure 3.

The curvature at that minimum is a monotonically increasing function of n:

$$v_{LJ}''(2^{1/n}) = 2^{1-(2/n)}n^2. \tag{VII.6}$$

Although it would be inappropriate to use form (VII.4) for modeling extended matter when $n \leq 3$ (energies would be nonextensive in the large-N limit), for present purposes of illustrating the qualitative behavior of isolated cluster potential energy landscapes, it is valid simply to suppose $n > 0$.

The entries in Table VII.1 remain unchanged with use of the extended form in Eq. (VII.4) for $N = 2, 3$, and 4. However, each of the pair distances in each of these three simple clusters is $x_{min}(n)$, thus decreasing as n increases. For $N = 5$, the single type of cluster inherent structure remains a triangular bipyramid, but its reduced energy rises from -10 to -9 as n increases from just above 0 to $+\infty$.

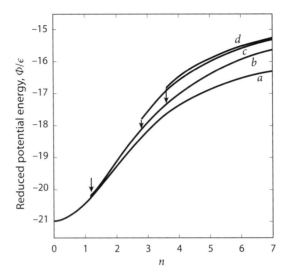

FIGURE VII.2. Variation with n for 7-particle inherent structure energies. The interactions are the Lennard-Jones $(2n),(n)$ pair potential, Eq. (VII.4). The four curves have been labeled in accord with Table VII.1 entries. Downward arrows indicate basin disappearance events as n decreases.

A new feature of n dependence first appears at $N = 6$ and persists with all higher N values. The two distinct inherent-structure types listed in Table VII.1, the regular octahedron and the bicapped tetrahedron, retain their identities for $n > 4$. But as n decreases below approximately 4, the higher energy bicapped tetrahedron ceases to provide local minima of the 6-particle potential energy function, leaving only the regular octahedron as an inherent-structure type. The topographic mechanism for this disappearance as n declines is the continuous approach and mutual annihilation at confluence of the local minimum with a saddle point, leaving neither for $n < 4$. The nature of this process does not imply that the basins for each of the permutation-related bicapped tetrahedron inherent structures ever become finite in extent while n approaches 4 from above, but rather that the basins (subject to the fixed center-of-mass constraint) become arbitrarily "thin" in at least one direction while still retaining infinite extent in other directions in the configuration space.

The four stable cluster structures listed in Table VII.1 for $N = 7$ display a similar behavior as n declines below 6. Figure VII.2 illustrates this in detail with inherent-structure energies plotted vs. n. As n decreases, three of the inherent-structure types and their basins disappear, starting with the highest lying [7(d)], then proceeding downward in energy until only the type 7(a) remains. The downward arrows in the figure locate the n values at which these events occur (approximately 3.6, 2.8, and 1.2), as well as the basin type into which a disappearing basin is absorbed. In particular, one should notice that 7(d) is absorbed into 7(b), bypassing 7(c), which intervenes in energy. Considering the large but different inherent-structure degeneracy factors d_{IS} shown in Table VII.1 for $N = 7$, it is virtually impossible to present a manageable diagram tracing the connectivity fate of every individual basin.

The reduction in number of distinct geometric inherent-structure types, and thus their total number, as the curvature of the attractive well of the potential decreases is apparently a phenomenon that extends well beyond the modest Lennard-Jones illustrations just presented. The Morse potential [Eqs. (I.42)–(I.43)] has a parametrically controllable curvature at its minimum, analogous to the Lennard-Jones exponent n, and this leads qualitatively to the same number reduction behavior that has been documented for even larger cluster sizes N [Braier et al., 1990; Doye and Wales, 1996].

The opposite case of the Lennard-Jones $(2n),(n)$ potential, with large positive n, also has physical relevance, in particular for the modeling of spherical colloid particle aggregation [Anderson and Lekkerkerker, 2002]. In this case, the width of the attractive well becomes narrow compared to the range of strong core repulsion. In the asymptotic limit $n \to +\infty$, the attractive well becomes infinitely narrow, while the repulsive core just inside this attractive surface becomes that of a rigid sphere. Consequently, this limiting case can appropriately be labeled "sticky spheres." The energies of the inherent structures in this limit are simply equal to the number of their sphere pair contacts, and because this feature disregards other geometrical characteristics, the inherent-structure energies develop substantial degeneracies, including those of lowest potential energy [Arkus et al., 2009].

Returning for the moment to the four-particle tetrahedral clusters with finite n, the two inherent structures and their basins differ only in the placing of the numbered particles at the vertices of the tetrahedron. In other words, this pair is related by mirror symmetry. Any configuration within the interior of one of the basins must have a mirror-image partner in the other basin. The boundary separating these basins must then consist of all configurations for which mirror reflection followed by a rigid-body rotation can produce the identity operation, in other words, all coplanar arrangements of the four particles. The coplanar arrangements with lowest potential energy thus locate the saddle points for transitions between this pair of basins. These are simple first-order saddle points, each with just one direction of negative curvature on the Φ hypersurface. In the case of the Lennard-Jones 12,6 potential, the transition-state planar configurations consist of a side-sharing pair of isosceles triangles, with reduced energy:

$$[\Phi_4/\varepsilon]_{sp} = -5.07342. \tag{VII.7}$$

There are six such transition states, depending on how the four coplanar particles are arranged to form the side-sharing triangles. Qualitatively similar results for $N = 4$ also apply to other pair potential choices, in particular, the generalized Lennard-Jones potential, Eq. (VII.4), and the Morse potential, Eqs. (I.42)–(I.43).

As N increases beyond 4, the proliferation of geometrically distinct inherent structures and of the saddle points that connect them calls for a suitable descriptive technique. One such approach starts by defining for each N-particle cluster an available configuration set $\mathbf{A}_N(E)$ that contains all of the configurations that satisfy the condition

$$\Phi(\mathbf{r}_1 \dots \mathbf{r}_N) \le E. \tag{VII.8}$$

In cases involving particles with internal degrees of freedom, the same defining inequality also applies but with $\Phi(\mathbf{x}_1 \dots \mathbf{x}_N)$ replacing $\Phi(\mathbf{r}_1 \dots \mathbf{r}_N)$. This criterion identifies that portion of the entire configuration space that is accessible to the system in classical mechanics, where the total conserved energy is E. When E lies below the global minimum of $\Phi(\mathbf{r}_1 \dots \mathbf{r}_N)$, obviously $\mathbf{A}_N(E)$ is the empty set. As E rises through the value of the Φ global minimum, $\mathbf{A}_N(E)$ at first consists of separate disconnected portions located within the lowest basins. Further rise in E causes analogous additional disconnected portions to appear for higher lying basins. Also, as E exceeds the values corresponding to the potential energies of basin boundary saddle points, connections are made between portions belonging to neighboring basins. Eventually, E rises high enough so that $\mathbf{A}_N(E)$ becomes and remains thereafter a connected set of configurations. For any pair of energies E and E', where $E < E'$, one has the set inclusion condition:

$$\mathbf{A}_N(E) \subseteq \mathbf{A}_N(E'). \tag{VII.9}$$

The changing connectivity of the set $\mathbf{A}_N(E)$ can be used as the basis for an insightful diagrammatic analysis, the "disconnectivity graph" [Becker and Karplus, 1997; Wales et al., 2000; Wales, 2003]. The strategy involves graphically representing energy parameter E vertically and exhibiting a single vertical line at high E, representing full $\mathbf{A}_N(E)$ set connectivity, which subsequently branches in its downward course at each disconnection for the set $\mathbf{A}_N(E)$. The resulting branches terminate at their lower ends at energies equal to those of the corresponding inherent structures. The example of four Lennard-Jones particles, from Table VII.1, leads to the elementary pattern displayed in Figure VII.3. In addition to the energies E associated with the inherent-structure potential minima and the six equivalent saddle points, the figure also indicates the threshold points $E = -3$, -1, and 0 at which particles can escape ("evaporate") to infinity. The lowest of these corresponds to escape of one particle from the remaining three, which are left in the optimal equilateral-triangle configuration. The next is the threshold for a second particle loss to leave a bound dimer. The highest represents the threshold for complete cluster dissociation. It should be mentioned in passing that disconnectivity graphs are not intended to indicate any changes from simple connectivity to multiple connectivity of any portions of $\mathbf{A}_N(E)$, should they occur.

Fully detailed disconnectivity graphs become rapidly more complex as the number of particles involved increases. For many purposes, it is unnecessary to distinguish basins and inherent structures that differ only by inversion or permutation of identical particles. Any information about the possibility of interconversions between equivalent basins would then be suppressed. Consequently, a reduced disconnectivity graph representation suffices in which inversion and permutation equivalent basins are lumped together as single lines, and branching indicates E-dependent separation of those larger basin sets into distinguishable, disconnected parts. Figure VII.4 provides an example of this more economical representation, showing the inversion and permutation reduced disconnectivity for seven identical Lennard-Jones 12,6 particles. It should be noted that the branching (disconnection) points appearing in Figure VII.4 do not exhaust the entire set of simple saddle points connecting basins for distinguishable inherent structures. Higher lying saddles are present on the potential energy landscape for pairs of basin types that are only indirectly connected through a third basin type at lower energy E [Berry et al., 1988].

With this still-modest cluster size $N = 7$, the graph is very simple, exhibiting just elementary one-stage branchings. For larger clusters, the number of branching (saddle) points increases dramatically, even under reduction lumping together equivalent basins, and the branches themselves can display second, third, or higher order branching sequences before terminating at the respective inherent-structure energies. But under no circumstance can a branch rejoin the main stem to produce a loop. Such an occurrence would violate Eq. (VII.9) and would require that $\Phi(\mathbf{r}_1 \ldots \mathbf{r}_N)$ in the inequality (VII.8) be multiple valued over at least a portion of the configuration space, which is obviously inconsistent with the definition of the potential energy function.

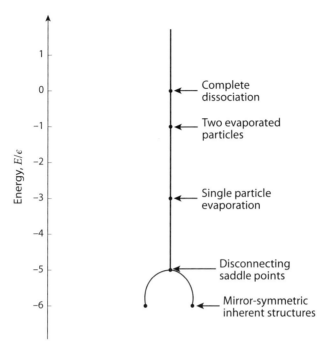

FIGURE VII.3. Disconnectivity graph for four identical particles interacting with the Lennard-Jones 12,6 pair potential.

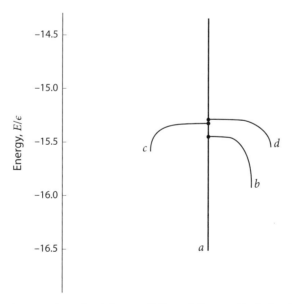

FIGURE VII.4. Inversion and permutation reduced disconnectivity graph for seven identical particles interacting with the Lennard-Jones 12,6 pair potential. Inherent structures and their reduced energies are identified as in Table VII.1. The saddle point energies (branching points) indicated by solid dots occur at reduced energies −15.28, −15.32, and −15.44 [Berry et al., 1988]. Evaporation thresholds are not included, nor are other saddle points that are not directly involved in disconnections.

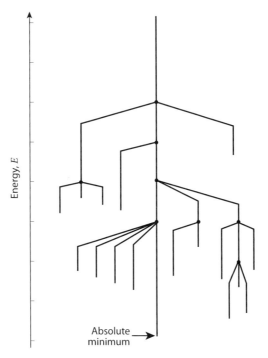

FIGURE VII.5. Schematic version of an energy-coarse-grained disconnection diagram. For this example, the connection/disconnection information is accessed only at equally spaced energies, indicated by the tick marks on the scale at the left. Branchings detected at these discrete energies are highlighted by solid dots. The lower ends of the various branches correspond to the inherent structure potential energies.

The great complexity of even the permutation-reduced disconnectivity graphs describing cluster sizes that have been studied in numerical simulations has motivated the introduction of coarse-grained versions. Though these suppress some of the finer details, they can suffice to convey large-scale topological features of the potential energy landscape [Wales et al., 2000; Wales, 2003]. The coarse graining consists of selecting a finite set of discrete E values at which feasible connections between inherent-structure types are identified. These E values ideally would span the range from just above the lowest lying inherent-structure energy to a sufficiently high value that $\mathbf{A}_N(E)$ is fully connected. Then the data on connections at each successive level, starting from the highest, is used to generate an approximate disconnection graph. Figure VII.5 provides a schematic version of the kind of outcome encountered with this approximate approach. Many specific examples of coarse-grained disconnection diagrams for clusters of various sizes subject to differing potential functions may be found in Wales [2003].

B. Asymptotic Inherent-Structure Enumeration

Entries in Table VII.1 hint at the rise in number of distinguishable inherent-structure types to be encountered as N increases for clusters in free space. However, such small-N enumerations do not identify any qualitative differences that might exist between fixed density circumstances, as discussed in Chapter III, and the vanishing density condition that is part of the present examination. In fact, a rather different type of analysis needs to be brought to bear, leading to distinct conclusions for the two cases in the asymptotic large-N regime. This analysis needs to be presented in a way that clearly distinguishes between the types of inherent structures (from isochoric mapping) occurring in the two regimes: (a) fixed particle number N in the infinite system limit $V \to +\infty$ and (b) low but positive number density $\rho = N/V$ for large-system sizes, where both the number of particles N and the system volume V are regarded as macroscopic. Rather simple qualitative arguments have been developed that provide some guidance for inherent-structure enumeration in both regimes [Stillinger, 2000], the content of which is examined later in this section.

Citing the results of several computer studies helps to set the context in which the two types of enumeration problems need to be confronted. Such studies show that even for direction-independent interactions, a wide variety of mechanically stable inherent structures exists, the great majority of which exhibit vividly noncompact shapes. The needle cluster illustrated in Figure VII.1 provides one simple but extreme example. However, the range of shape diversity for inherent structures in the unbounded space case can be at least partially illustrated by results from potential energy minimizations, starting from expanded, random particle configurations. Figures VII.6 (a)–(c) show some representative examples. These mechanically stable irregular shapes were formed with 400 identical particles confined to two dimensions for visual clarity, where the interactions between all pairs were the Lennard-Jones 12,6 potential without any cutoff [Stillinger and Stillinger, 2006]. The initial conditions from which these structures were created amounted to a randomly generated irregular distribution of the 400 particles within a circular domain, with a number density slightly less than one-twentieth that of the zero-temperature, zero-pressure triangular crystal. No confining boundary was present during the subsequent potential energy minimization because the attractive-pair interactions made surrounding boundaries irrelevant. The transformations from initial to final configurations numerically involved a combination of incremental steepest descent displacements for the initial and intermediate stages of the mapping, followed at the final stage by use of the MINOP procedure, which is designed to find minima of multivariable functions rapidly and to high accuracy [Kaufman, 1999].

The examples presented in Figure VII.6 illustrate several structural features that emerge when boundaries are not present to limit the spatial extent of the inherent structures. First, the shapes are strongly nonconvex, often with thin extended arms that can be rather irregular. Second, arm branching is possible. Third, vacant areas of variable size can be enclosed, producing a "spongy" texture. Fourth, built-in elastic strains can appear that surround vacant areas, which are not simply multivacancies in an otherwise perfect triangular crystalline host arrangement. Examples of this last type of feature tend to show curved lines of particles around the vacant areas. As a result, these vacant areas have associated with them nonvanishing Burgers vectors that quantify the deviation from simple multivacancy character [Ashcroft and Mermin, 1976, Chapter 30; Kittel, 1956, p. 544].

In order to carry out a comparable three-dimensional version of the calculations leading to the structures shown in Figure VII.6, it is reasonable to suppose that the $20^2 = 400$ particles that suffice for two dimensions should be increased to about $20^3 = 8{,}000$. A starting density within a spherical region should have about the same linear expansion as in two dimensions compared to its ground-state crystal structure, i.e., a number density approximately $(1/20)^{3/2} \approx 1/90$ of the stable crystal density. Unfortunately, these requirements place a severe demand on computing

(a)

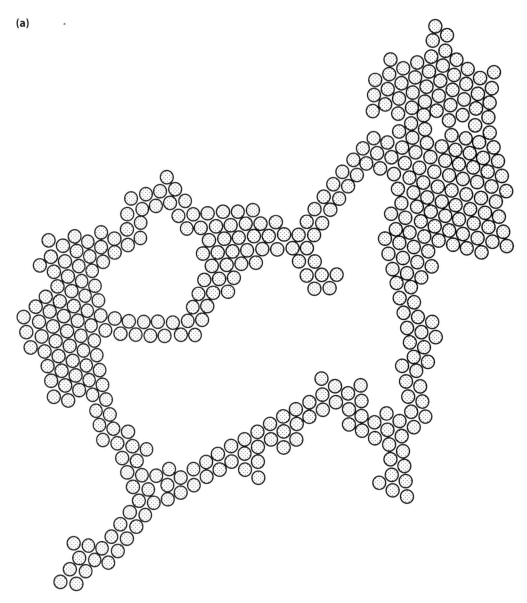

FIGURE VII.6. Three examples of irregularly shaped inherent structures for 400 identical particles confined to two dimensions. The potential energy consists of pairwise Lennard-Jones 12,6 interactions and has the following values (reduced units): (a) −1,010.3099, (b) −930.9967, (c) −969.7488.

resources to carry out an accurate steepest descent mapping for a nontrivial collection of irregular starting configurations. However, if such computations were to be performed, one would expect to observe a diminished tendency for the resulting inherent structures to exhibit short-range crystalline order as they do in the two-dimensional context. Furthermore, one must be prepared to find distinctly noncompact inherent-structure clusters as in the two-dimensional version, and

(b)

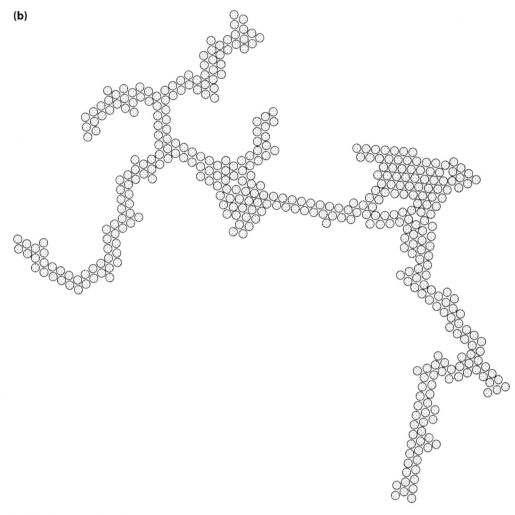

FIGURE VII.6. (*continued*)

one should expect occasionally to observe empty tunnels penetrating into and through the interior of those clusters.

The theoretical process called "diffusion-limited aggregation" (DLA) generates multiparticle clusters whose properties provide both a significant similarity and a contrast to inherent structures formed in free space [Witten and Sander, 1981; Meakin, 1983, 1988]. Both methods create clusters with dramatic spatial extensions for which an overall effective density, defined in terms of the smallest convex enclosing envelope for particle centers, approaches zero as the particle number N becomes very large. This is a circumstance in which the concept of Hausdorff dimension [Hausdorff, 1919], or "fractal" dimension [Mandelbrot, 1977], becomes relevant for the cluster shapes. The basic distinction to be stressed between the DLA process and the steepest descent mapping is that the former involves a sequential process of adding single particles one at a time (via Brownian motion from a random initial source direction) irreversibly to an initial "seed." By

(c)

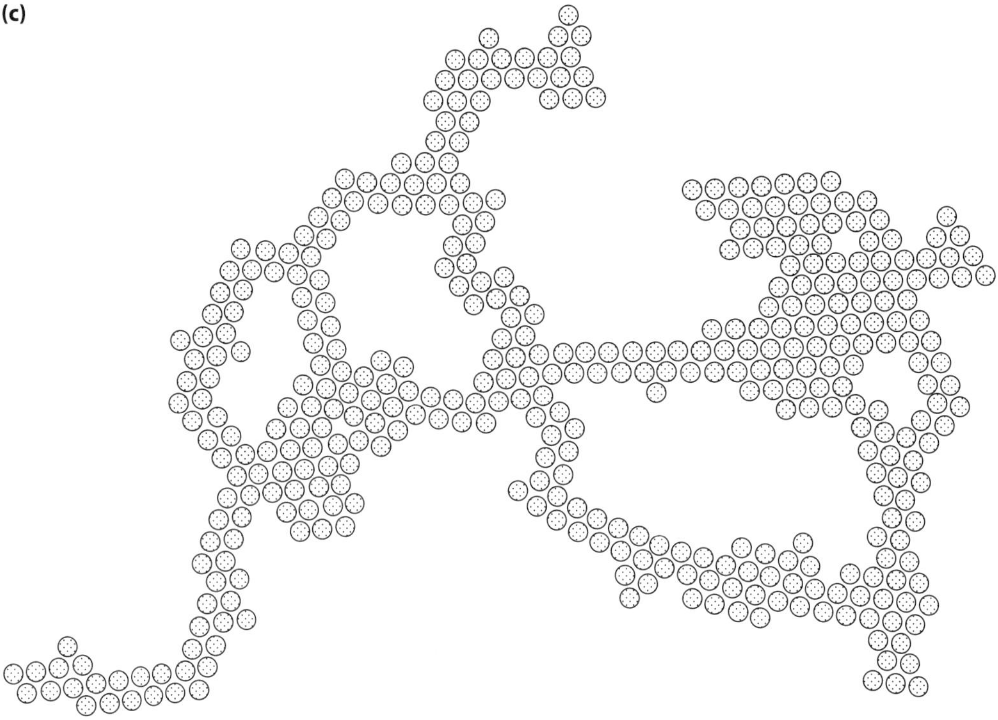

FIGURE VII.6. (*continued*)

contrast, the latter generates inherent structures concurrently, with all particles present both at the outset and throughout the entire mapping operation. Most DLA simulations have been based on a discrete periodic lattice of available particle positions and so cannot generate clusters with strained defect regions described with a nonvanishing Burgers vector. The DLA process tends to create highly dendritic structures that can branch but which diverge radially outward from the location of the initial seed, and which in two dimensions seldom enclose void space disconnected from the exterior. Finally, no potential energy function is present for the DLA process, a fundamental aspect of the inherent-structure approach. An example of a cluster produced by the DLA process in two dimensions appears in Figure VII.7 [Witten and Sander, 1981].

In spite of these obvious distinctions, DLA results help to reinforce the importance of the pattern diversity that must be respected in enumerating inherent structures for clusters that are formed in unbounded space. This is especially significant in view of the fact that for a given number N of included particles, it is computationally far easier to generate DLA clusters than inherent structures. Equivalently, it is feasible to create DLA structures comprising 10^3 to 10^4 particles but currently impractical to carry out an accurate steepest descent mapping for an initially widely dispersed irregular arrangement of the same numbers of particles with a realistic continuous interparticle potential. In this circumstance, it is instructive to note that the large-N DLA clusters have been determined to possess mean radii of gyration $R(N)$ that for large N scale in the following way:

$$R(N) \propto N^f. \tag{VII.10}$$

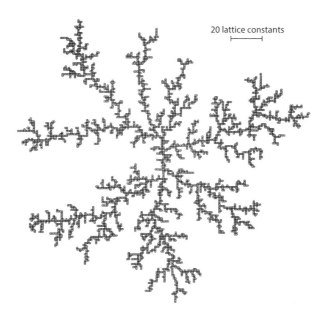

20 lattice constants

FIGURE VII.7. A dendritic cluster produced in two dimensions by the DLA process. Reproduced from Witten and Sander, 1981, Figure 1.

For two-dimensional DLA aggregates, the scaling exponent $f \cong 0.59$, whereas for three-dimensional aggregates, $f \cong 0.40$ [Meakin, 1983]. By contrast, compact arrangements of particles would be characterized by corresponding f values equal to 1/2 and 1/3 for two and for three dimensions, respectively, i.e., $f = 1/D$ for space dimension D. The opposite limiting situation corresponds to linearly extended structures (as exemplified by the "needle" arrangement shown in Figure VII.1) and leads to $f = 1 = 1/D^0$ in spaces of any dimension D. That the DLA exponent f lies between these compact and extended extremes reflects the sparse but spreading extension of DLA structures and allows formal assignment of a fractal dimension $D_f = 1/f$ where $1 < D_f < D$. It is reasonable to expect a similar result for the mean radii of gyration for inherent structures formed by steepest descent from arbitrarily dilute initial conditions, but the corresponding f values and implied fractal dimensions D_f that would result have yet to be determined and may fundamentally depend on the interparticle interactions involved.

The formation of soot particles as a result of various atmospheric chemistry processes, or under carefully controlled laboratory conditions, can produce noncompact geometric patterns similar to those generated by DLA. In a rough sense, these mechanically stable soot particles could be viewed as representatives of inherent structures in unbounded space for the substances out of which they are formed (e.g., carbon). Microscopic observations of soot particles have led to reports of fractal dimensions in the range $1 < D_f < 3$ [Dhaubhadel et al., 2006; Chakrabarty et al., 2009].

An important connection between isochoric inherent-structure characteristics in the low but positive density regime and those for the denser range typical of liquids is introduced in the schematic Figure VII.8. Here, average pressures $p^{(q)}$ for inherent structures, produced by constant-volume mapping, or "quenching," from single-phase fluid states (excluding critical and coexistence

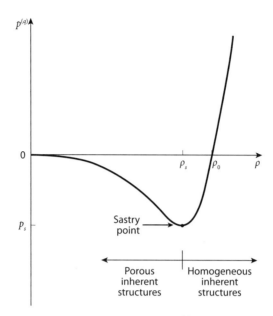

FIGURE VII.8. Schematic plot of average inherent structure pressure $p^{(q)}$ vs. number density ρ for constant volume mapping from homogeneous fluid phases.

regions) have been plotted versus number density. This plot qualitatively summarizes results that have been obtained from simulations for a variety of model substances, including the Lennard-Jones 12,6 single-component system [Sastry et al., 1997], the SPC/E ("simple point charge/ extended") water model [Roberts et al., 1999], the fused salt $ZnBr_2$ [LaViolette et al., 2000], and several low-molecular-weight hydrocarbons [Utz et al., 2001; Shen et al., 2002]. At least for these rather simple model substances, the results as represented in Figure VII.8 are rather insensitive to the temperature of the premapped states. The high-density amorphous inherent structures have positive pressures, though less than the pressures of the premapped states. However, as the density declines, so too does the average pressure of the inherent structures, which crosses the pressure axis at ρ_0 and enters the negative pressure (isotropic tension) regime. Further reduction in density causes the average inherent-structure pressure to pass through a minimum, the "Sastry point," at ρ_S, p_S. This minimum pressure $p_S < 0$, or maximum tension sustainable by inherent structures from the fluid, marks the boundary between homogeneous amorphous arrangements of the participating particles at high density and porous, shredded, or spongy arrangements at low density. The latter consist of connected portions of amorphous particle arrangements separated by void regions, and because of the lower overall density of attractive-particle interactions, the structure is able only to support a reduced tension compared to p_S. The various simulations on which Figure VII.8 are based reveal that ρ_S is less than the triple-point density, and $|p_S|$ can range from 10^1 to 10^2 times the critical pressure p_c for the substance involved.

The available simulations represented qualitatively by Figure VII.8 have all utilized periodic boundary conditions. Interactions between the primary cell and its neighboring periodic image cells permit the existence of negative-pressure inherent structures and of the shredding that arises when $\rho < \rho_S$. An alternative computational scenario could have used impenetrable walls with strong adhesive character to be able to apply traction to the inherent structures, thus also producing negative pressures. In the absence of such wall adhesion, the inherent structures would contract sufficiently to remain unshredded and at zero pressure (in the large-system limit). But with periodic boundary conditions or strongly adhesive walls, when $\rho \ll \rho_S$ the predominant pattern exhibited by the inherent structures is a tenuous connected texture of aggregated particle branches, surrounded and invaded by large void spaces.

For simplicity of presentation, assume that the N particles comprising the system of interest are a single species and spherically symmetric. Whereas the inherent structures are dense and homogeneous for $\rho \geq \rho_S$, the previously derived large-N asymptotic form for the total number Ω_d of distinguishable inherent structures for a fixed volume V remains valid [Eq. (III.15)]:

$$\ln[\Omega_d(N, V)] \sim \sigma_\infty(\rho)N, \tag{VII.11}$$

where neglected terms in the right side may diverge with N, but less rapidly than the linear leading-order term shown. In particular, $\sigma_\infty(\rho_S)$ is a finite positive number that is treated as given for any substance of interest. At this Sastry point density, one may formally express the entire content of all configuration space basins as $(l_S)^{DN}$, where $D = 2,3$ is the space dimension available for the center of each particle and where l_S is a suitable effective "width" of a basin in each Euclidean dimension. In view of the fact that there are in all $N!\Omega_d$ basins tiling the configuration space whose content is V^N, one must have

$$V^N = (l_S)^{DN} N! \exp[\sigma_\infty(\rho_S)N]. \tag{VII.12}$$

Because of the fact that the vast majority of inherent structures at the Sastry point density are spatially uniform, it is natural to presume that length l_S is comparable to the mean nearest neighbor separation in those inherent structures. Applying Stirling's formula for the factorial to this last equation,

$$\sigma_\infty(\rho_S) = \ln\left[\frac{e}{\rho_S l_S^D}\right]. \tag{VII.13}$$

Now consider the effect of reducing the density below that of the Sastry point, say to

$$\rho = x\rho_S, \tag{VII.14}$$

$$0 < x < 1.$$

As a result, the amorphous inherent structures lose their homogeneity and develop an invasive and generally complex surface separating still-dense regions of contacting particles from void space. The mean basin content can still formally be expressed as $[l(x)]^{DN}$ in terms of an appropriate effective length $l(x)$ that varies continuously with density-reduction factor x, subject to $l(1) \equiv l_S$. The same format as shown in Eq. (VII.13) can now be used to express σ_∞ at these lower overall densities in terms of $l(x)$:

$$\sigma_\infty(x\rho_S) = \ln\left[\frac{e}{x\rho_S l^D(x)}\right]. \tag{VII.15}$$

For the purpose of establishing a context in which to interpret expression (VII.15), it is useful to recall that for inverse power potentials, the number of basins and inherent structures is independent of density [Section I.D]. Although this simple class of models includes no attractive forces and thus would not exhibit Sastry points, it is nevertheless a meaningful case because it poses the question of how a length $l(x)$ would have to behave in a circumstance where σ_∞ was independent of the density variation variable x. The answer is immediately obvious from Eq. (VII.15):

$$l(x) = x^{-1/D}l(1).\qquad\qquad (VII.16)$$

A contrasting extreme for isochoric inherent structures, with a hypothetical outcome driven by attractive forces that are present in realistic models, would have the entire excess area ($D = 2$) or volume ($D = 3$) when $0 < x < 1$ consolidated into a single, simply connected, macroscopic void. Because the large-system limit is of interest, this implies that only a negligible fraction of the N particles would be located at the surface of that hypothetical void. The majority of the particles would find themselves aggregated in a macroscopic region occupying the remainder of the system at a local density approaching the zero-pressure value ρ_0 (see Figure VII.8) as $x \to 0$. Consequently, the displacement freedom for each member of this dominating majority would approach the value for a uniform system, without the macroscopic void, at ρ_0. Therefore, this opposite extreme to contrast with Eq. (VII.16) suggests the following:

$$l(x) \equiv x^0 l(\rho_0/\rho_S),\qquad\qquad (VII.17)$$

i.e., a length substantially independent of x for small x.

The actual dependence of distance $l(x)$ on x in the small-x regime should lie between the extremes represented by Eqs. (VII.16) and (VII.17). Because of the shredding patterns expected for inherent structures at densities below ρ_S and the considerable fraction of particles expected to reside at surface sites with enhanced intrabasin displacement freedom, it is sensible to interpolate Eqs. (VII.16) and (VII.17) by the relation

$$l(x) \cong x^{-w} l(\rho_0/\rho_S).\qquad\qquad (VII.18)$$

This line of reasoning implies that exponent w would be subject to the inequalities

$$0 \leq w < 1/D.\qquad\qquad (VII.19)$$

The lower limit 0 has formally been included here to accommodate the interpretation of a logarithmic dependence on x as a result of the relation

$$\lim_{w \to 0} [(x^{-w} - 1)/w] = -\ln x.\qquad\qquad (VII.20)$$

When expression (VII.18) is inserted into Eq. (VII.15), the result is

$$\sigma_\infty(x\rho_S) \cong \sigma_\infty(\rho_S) + D\ln[l(1)/l(\rho_0/\rho_S)] + (Dw - 1)\ln x + O(x).\qquad (VII.21)$$

The coefficient $Dw - 1$ of the third term is negative according to Eq. (VII.19), so the conclusion of this rough estimate is that the inherent-structure enumeration exponent σ_∞ must diverge logarithmically to infinity as density declines to zero (i.e., as $x \to 0$).

The fact that attractive interparticle forces of unbounded range cause inherent structures to be connected single aggregates plays a basic role in the enumeration problem. In order to examine implications of this connectivity, at least in a coarse-grained version, it is useful to divide the D-dimensional system into a simple square ($D = 2$) or simple cubic ($D = 3$) lattice of cells, each of which has a content ω_0. The magnitude of ω_0 is then chosen so that one can reasonably assume that each such cell has just two states: empty, or filled with a single particle. An approximate representation of an inherent structure in this lattice picture is a single cluster comprising all N particles that is fully connected by nearest neighbor contacts between pairs of occupied content-ω_0 cells.

Primary interest concerns large systems, so that the total number M of cells is also large. In addition, the single-occupancy limit for cells requires that M at least equal the number N of particles present. In fact, the dilute regime has

$$M \gg N. \tag{VII.22}$$

Although it is not an essential ingredient of the analysis, it is nevertheless a convenient simplification to suppose that M is an integer that has the form

$$M = 2^{Dm}, \tag{VII.23}$$

where m is itself a positive integer. For purposes of illustration, note that Avogadro's number $N_{Av} = 6.022045 \times 10^{23}$ is a standard measure of a macroscopic amount of matter, and so it is relevant to observe that

$$N = 2^{79} \cong 6.04 \times 10^{23} \tag{VII.24}$$

$$\cong N_{Av}.$$

Choosing for example $m = 29$ in Eq. (VII.23) for $D = 3$ then leads to $M = 2^{87}$, so that the number of particles stated in Eq. (VII.24) dispersed among these M cells would correspond to a dilution factor for the chosen lattice equal to $1/2^8 = 1/256$.

The total number of distinguishable arrangements Ω_0 of particles among cells, regardless of the nearest neighbor-contact connectivity constraint for an inherent structure, is given by the usual combinatorial expression:

$$\ln \Omega_0 = \ln\left[\frac{M!}{N!(M-N)!}\right] \tag{VII.25}$$

$$\cong N\left[\left(\frac{M}{N}\right)\ln\left(\frac{M}{N}\right) - \left(\frac{M-N}{N}\right)\ln\left(\frac{M-N}{N}\right)\right],$$

where Stirling's formula for factorials has been used to simplify the expression. The strong inequality (VII.22) for the dilute regime of primary concern simplifies this even further:

$$\ln \Omega_0 = N\left[\ln\left(\frac{M}{N}\right) + 1 + O(N/M)\right]. \tag{VII.26}$$

The next step is to estimate how the overall connectivity condition reduces this number. The specific choice Eq. (VII.23) for M permits the entire square or cubic array of cells to be subdivided in any of several possible ways into arrays of nonoverlapping identical larger squares or cubes. Each of these larger units would contain 2^{Dj} primary cells (content ω_0), where integer j can be selected from the interval

$$0 \leq j \leq m. \tag{VII.27}$$

The resulting number of larger cell complexes then is $2^{D(m-j)}$. For the illustrative three-dimensional example above, where $M = 2^{87}$, the corresponding range for j would be $0 \leq j \leq 87/3 = 29$. In any case, a wide range of subdivision possibilities would be available for a macroscopic system, whether it is two- or three-dimensional.

The overall demand that enumerated inherent structures correspond to connected single structures on the lattice can be imposed in an ascending order for each j choice. Upon proceeding from level j to the next level $j + 1$, 2^D contiguous units are combined to form a larger similar unit with linear size that has twice the magnitude. At each successive j stage, starting at $j = 0$, it is necessary to ensure that the internal particle distributions within units are consistent with the connectivity requirement. Because not all combinations of connectivity-consistent units at level $j - 1$, when combined to form the next hierarchical level j, meet the requirement, an attrition factor $0 < A_j < 1$ has to be applied for each of the level-j units. This set of attrition factors reduces the indiscriminate combinatorial factor Ω_0 to the desired enumeration quantity Ω_d (or equivalently σ_∞), Eq. (VII.11), as follows:

$$\Omega_d = \Omega_0 \prod_{j=1}^{m} (A_j)^{2^{D(m-j)}} \tag{VII.28}$$

$$= \exp[\sigma_\infty(x\rho_S)N + o(N)].$$

By including the upper limit $j = m$ in the product, periodic boundary conditions are imposed on the allowed configurations.

It is not necessary to exclude j-level cell occupancy patterns merely because the particles contained within that unit are unconnected by face-to-face contacts. The reason is that ultimately they can be externally reconnected within a larger unit. In fact, this possibility allows all occupancy patterns at level $j = 1$ because all primary cells (size ω_0) at this level are at the surface for both $D = 2$ and $D = 3$. Consequently, one must set $A_1 = 1$. However, when $j \geq 2$, there are occupancy patterns within the larger cell complex for which interior occupied cells in that larger complex have no face-sharing connection to surface particles and so must be excluded. Figure VII.9 illustrates this distinction for several occupancy patterns at the $j = 2$ level in two dimensions, showing examples of both permitted and rejected types.

The following argument leads to a reasonable qualitative estimate for each one of the attrition factors A_j, $1 \leq j \leq m$. The number of primary cells at the surface of each j-level grouping, whose particle occupancies can affect A_j, is

$$F(j) = 2^{(D-1)j}D; \tag{VII.29}$$

note that a factor $1/2$ has been included in this expression to avoid double-counting of primary cell face-to-face contacts throughout the full system. Because of particle clusters that would re-

Permitted

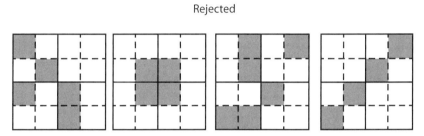

Rejected

FIGURE VII.9. Some two-dimensional occupancy patterns at level $j = 2$. These patterns are classified according to whether they are consistent with the global connectivity requirement that applies to inherent structures and are permitted, or whether they violate that requirement and are rejected. Occupied primary cells are dark, and empty primary cells are light. All of the constituent 2×2 square units in both the permitted and rejected 4×4 examples were permitted by themselves at the $j = 1$ level.

quire contacts across these newly formed lines ($D = 2$) or surfaces ($D = 3$), the logarithm of attrition factor A_j should be proportional to $F(j)$. Furthermore, $\ln A_j$ should also be proportional to the number density (i.e., the primary cell occupancy probability N/M) under consideration, to account for occupied interfacial primary cells that need connection to a large cluster across the interface but fail to find it. Therefore, set

$$\ln A_j \cong -C(D)(N/M)(2^{(D-1)j}D) \qquad (j > 1), \tag{VII.30}$$

where $C(D)$ is a suitable positive constant.

Taking logarithms in Eq. (VII.28), substituting from Eqs. (VII.26) and (VII.30), and dividing the result by N, one obtains the following result applicable in the low-density regime:

$$\sigma_\infty(x\rho_S) \cong -\ln(N/M) + 1 - DC(D)\sum_{j=2}^{m} 2^{-j}. \tag{VII.31}$$

The sum over j in this last expression converges quickly, and for macroscopic systems it is appropriate to replace the upper limit by infinity. Thus, one concludes

$$\sigma_\infty(x\rho_S) = -\ln(x\rho_S\omega_0) + 1 - DC(D)/2. \tag{VII.32}$$

This confirms the earlier conclusion, by comparing with Eq. (VII.21), that σ_∞ diverges logarithmically to plus infinity as density goes to zero. The comparison also indicates that for the exponent w introduced in Eq. (VII.18), one should have

$$w \cong 0, \tag{VII.33}$$

which would be consistent with the logarithmic interpretation, Eq. (VII.20). In any event, it must be remembered that the arguments just presented are only simple approximations and are subject to refinement by more sophisticated analyses.

Divergence of σ_∞ in the low-density limit indicates that the number of distinguishable inherent structures for N attracting particles in infinite space must increase faster with N than as a simple exponential function. The following elementary analysis suggests what alternative asymptotic growth rate may be involved. Suppose that the N-particle inherent structures were to be assembled in sequential order as connected objects, one particle at a time, starting at $N = 1$. At an intermediate stage of this assembly process, when $0 < N' < N$ particles had been emplaced, let $\Omega_d(N', \infty)$ denote the number of distinguishable mechanically stable structures that could have been formed in the infinite space available. As stressed earlier, the majority of these intermediate-stage structures would be distinctly nonconvex and geometrically irregular. Addition of the next particle presumably could occur anywhere along the irregular boundary of the object already formed. The fractal character of that irregular boundary implies that the number of distinct sites for the addition of particle $N' + 1$ would scale with N' substantially as $[K(N')]^r$. Here K is a positive constant, and r is a positive exponent subject to lower and upper limits:

$$(D - 1)/D < r \leq 1. \tag{VII.34}$$

The lower limit corresponds to the compact cluster extreme, and the upper limit corresponds to the fully extended cluster extreme. The consequence of this scaling is that

$$\Omega_d(N' + 1, \infty) \cong K(N')^r \Omega_d(N', \infty). \tag{VII.35}$$

The message conveyed by this last relation is that addition of each particle has a nonvanishing probability of initiating a new family tree of larger inherent structures. This is true even though there always remains a chance that distinct prior stages could converge to a common final inherent structure. Although one recognizes that each addition of an attracting particle inevitably induces small elastic deformation in the substrate structure, this should not undermine the essential validity of Eq. (VII.35).

In the large-N asymptotic regime, one can take logarithms in Eq. (VII.35) and treat N' as a continuous variable. That leads to the following differential relation

$$d\ln\Omega_d(N', \infty)/dN' \cong r\ln N' + \ln K. \tag{VII.36}$$

Upon integrating with respect to N', one finds

$$\ln\Omega_d(N', \infty) \cong rN\ln N + (\ln K - r)N + L, \tag{VII.37}$$

where L is the integration constant. An expression equivalent to the last, invoking Stirling's formula for factorials, is

$$\Omega_d(N, \infty) \cong (N!)^r \exp[-KN + o(N)], \tag{vii.38}$$

where the first factor in the right side dominates the exponential decay factor in the large-N limit.

It could be asked whether the free-space enumeration quantity Ω_d might conceivably rise even faster with N than just as a positive power of $N!$, as indicated in Eq. (VII.38). One hypothetical possibility would involve the function $\exp(BN^y)$ as a factor in $\Omega_d(N, \infty)$, where $B > 0$ and $y > 1$ are N-independent constants. However, detailed analysis indicates that such a rapidly rising function would imply that as N increases, many independent arbitrarily small distortions would convert one inherent structure to another. Such a scenario would apparently violate the qualitative character of acceptable potential energy functions representing real matter.

C. Virial Expansions

The pressure equation of state for a single-component dilute vapor that is an open system in thermal equilibrium can be expressed as a power series in the average number density $\rho = \bar{N}/V$:

$$p/\rho k_B T = 1 + \sum_{n=1}^{\infty} B_n(T)\rho^n. \tag{VII.39}$$

The temperature-dependent quantities $B_n(T)$ are the "virial coefficients." Successive terms in this series provide corrections to the ideal gas pressure, the leading term. For conventional interactions that operate between electrostatically neutral particles, including those possessing permanent dipole moments, virial expansion Eq. (VII.39) has a positive radius of convergence in the complex ρ plane [Ruelle, 1969, Chapter 4]. In the event that the system comprises structureless particles that involve only pairwise additive potentials, and for which classical statistical mechanics applies, the elegant Mayer cluster series approach provides explicit expressions for the virial coefficients in terms of so-called "irreducible cluster integrals" [Mayer and Mayer, 1940, Chapter 13]. However, the present objective concerns more general connections to potential energy landscape topographies, ultimately requiring a less restrictive analysis.

The virial coefficients $B_n(T)$ may formally be connected to particle interactions through the grand partition function $\mathcal{Q}(\beta, \mu)$, Eq. (II.30). The key equations are the following:

$$\ln \mathcal{Q}(\beta, \mu) = \beta p V$$

$$= \exp(\beta\mu)Q_1 + \exp(2\beta\mu)[Q_2 - (Q_1)^2/2] + \exp(3\beta\mu)[Q_3 - Q_2Q_1 + (Q_1)^3/3]$$

$$+ \exp(4\beta\mu)[Q_4 - Q_3Q_1 - (Q_2)^2/2 + Q_2(Q_1)^2 - (Q_1)^4/4] +\dots, \tag{VII.40}$$

and

$$\bar{N} = (\partial \ln \mathcal{Q}/\partial \beta\mu)_\beta$$

$$= \exp(\beta\mu)Q_1 + \exp(2\beta\mu)[2Q_2 - (Q_1)^2] + \exp(3\beta\mu)[3Q_3 - 3Q_2Q_1 + (Q_1)^3]$$

$$+ \exp(4\beta\mu)[4Q_4 - 4Q_3Q_1 - 2(Q_2)^2 + 4Q_2(Q_1)^2 - (Q_1)^4] +\dots. \tag{VII.41}$$

In these expressions, the Q_N are the canonical partition functions for N particles in volume V at temperature $T = (k_B\beta)^{-1}$, described in Sections II.B and II.D, and μ is the chemical potential. By eliminating $\exp(\beta\mu)$ between the last two equations, and identifying the result with the virial series Eq. (VII.39), one obtains expressions for the first few virial coefficients:

$$B_1(T) = -\left(\frac{V}{Q_1}\right)\left[\frac{Q_2}{Q_1} - \frac{Q_1}{2}\right],$$

$$B_2(T) = -\left(\frac{V}{Q_1}\right)^2\left[\frac{2Q_3}{Q_1} - \frac{4(Q_2)^2}{(Q_1)^2} + 2Q_2 - \frac{(Q_1)^2}{3}\right], \qquad \text{(VII.42)}$$

$$B_3(T) = -\left(\frac{V}{Q_1}\right)^3\left[\frac{3Q_4}{Q_1} - \frac{18Q_2Q_3}{(Q_1)^2} + 6Q_3 + \frac{20(Q_2)^3}{(Q_1)^3} - \frac{27(Q_2)^2}{2Q_1} + 3Q_1Q_2 - \frac{(Q_1)^3}{4}\right].$$

In general, the expression for $B_n(T)$ includes canonical partition functions for 1 to $n + 1$ particles, and that expression contains only a single appearance of Q_{n+1}, specifically in its leading term:

$$B_n(T) = -\left(\frac{V}{Q_1}\right)^n\left[\frac{nQ_{n+1}}{Q_1} + \ldots\right]. \qquad \text{(VII.43)}$$

Although the interactions might not be pairwise additive, the expressions for the virial coefficients $B_n(T)$ undergo considerable simplification if the particles involved have no internal degrees of freedom (i.e., are spherically symmetric) and are sufficiently massive that the classical statistical mechanical limit is relevant. One real substance that can be regarded as accurately exhibiting these restrictions is elemental xenon (Xe). Figure VII.10 shows a plot of its second virial coefficient versus temperature. It is qualitatively similar to measured second virial coefficients for other pure substances in that it is positive at high temperature but changes sign and plunges to deeply negative values with decreasing temperature. The temperature T_B at which $B_1(T)$ vanishes is called the "Boyle temperature." The simplified expression for the second virial coefficient in the classical limit, for structureless particles with pair potential Φ_2, is

$$B_1(T) = -(2V)^{-1}\int_V d\mathbf{r}_1\int_V d\mathbf{r}_2\{\exp[-\beta\Phi_2(r_{12})] - 1\}. \qquad \text{(VII.44)}$$

The interaction function Φ_2 for a pair of structureless particles such as xenon atoms exhibits a single inherent structure, a negative minimum, at some scalar separation $r_{min} > 0$. Over most of the two-particle configuration space spanned by the integrations in the $B_1(T)$ expression Eq. (VII.44), the pair is substantially noninteracting ($\Phi_2 \cong 0$). It is the role of the -1 in the integrand to effectively eliminate this majority of the configurations, leaving the small distance configurations in control of the value of the second virial coefficient. At very high temperature (that is, very small β), the integrand makes a negligible contribution for pair separations in the vicinity of r_{min}. The strong steric repulsion at small distances caused by substantial electron cloud overlap forces the pair Boltzmann factor toward zero and thus forces the integrand toward -1. Consequently, the second virial coefficient becomes positive at very high temperatures, just as shown in Figure VII.10. But in the other temperature extreme, with $\beta \to +\infty$, the repulsive core contribution becomes overwhelmed by what then becomes a huge positive Boltzmann factor contribution stemming from the vicinity of the negative-Φ_2 inherent structure at r_{min}. The Boyle temperature T_B locates the precise cancellation between these opposing contributions. Indeed, the qualitative pattern displayed in Figure VII.10 for xenon is also observed experimentally for a wide class of both

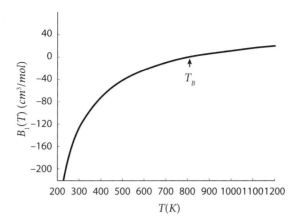

FIGURE VII.10. Second virial coefficient $B_1(T)$ for xenon, from tabular data in Kestin et al., 1984. The Boyle temperature is $T_B \cong 810$ K.

atomic and molecular species, though of course quantitative details such as T_B itself are substance specific [Estrada-Torres et al., 2007]. The Boyle temperature tends to be higher than the critical temperature; for Xe, they are $T_B \cong 810$ K and $T_c \cong 290$ K, and for H_2O, they are $T_B \cong 1,540$ K [Harvey and Lemmon, 2004] and $T_c \cong 647$ K.

The asymptotic low-temperature divergence of $B_1(T)$ for structureless particles can be directly extracted from Eq. (VII.44). In this regime, it suffices to approximate the pair interaction by a two-term Taylor expansion about r_{min} and to approximate the integrand just by the resulting large but narrow Gaussian:

$$B_1(T) = -(1/2\int_0^\infty \{\exp[-\beta\Phi_2(r)]-1\}4\pi r^2 dr$$

$$\sim -2\pi r_{min}{}^2 \exp[-\beta\Phi_2(r_{min})]\int_{-\infty}^\infty \exp[-(\beta/2)\Phi_2''(r_{max})s^2]ds$$

$$= -\frac{(2\pi)^{3/2}r_{min}{}^2}{[\beta\Phi_2''(r_{min})]^{1/2}} \exp[-\beta\Phi_2(r_{min})]. \qquad (VII.45)$$

Analogous considerations apply to the third and higher order virial coefficients. In particular, consider $B_2(T)$ for structureless particles, involving an interaction potential Φ_3 for three identical particles that may include a nonpairwise additive portion. Once again suppose that the classical statistical mechanical limit applies. In that case, the second of Eqs. (VII.42) reduces to the following expression:

$$B_2(T) = -(3V)^{-1}\int_V d\mathbf{r}_1\int_V d\mathbf{r}_2\int d\mathbf{r}_3\{\exp[-\beta\Phi_3(\mathbf{r}_1, \mathbf{r}_2, \mathbf{r}_3)] - \exp[-\beta\Phi_2(r_{12}) - \beta\Phi_2(r_{13})]$$

$$- \exp[-\beta\Phi_2(r_{12}) - \beta\Phi_2(r_{23})] - \exp[-\beta\Phi_2(r_{13}) - \beta\Phi_2(r_{23})]$$

$$+ \exp[-\beta\Phi_2(r_{12})] + \exp[-\beta\Phi_2(r_{13})] + \exp[-\beta\Phi_2(r_{23})] - 1\}. \qquad (VII.46)$$

For simple substances such as xenon, the Φ_3 potential energy function exhibits only a single potential energy minimum (inherent structure), in which the three particles are arranged in an equilateral triangle. In this respect, such substances qualitatively conform to the Lennard-Jones 12,6 model discussed in Section VII.A, but the presence of a three-body contribution in Φ_3 would influence to some extent both the binding energy of the inherent structure, as well as the side length r_{min} of the equilateral triangle. In any case, the negative three-body inherent-structure energy $\Phi_3(min)$ is lower than twice the two-body minimum:

$$\Phi_3(min) < 2\Phi_2(min) < \Phi_2(min) < 0. \tag{VII.47}$$

The result of these inequalities is that the first Boltzmann factor appearing in the integrand of the general classical $B_2(T)$ expression Eq. (VII.46) dominates the low-temperature behavior.

In the vicinity of the three-particle inherent structure, the potential energy surface has a locally quadratic form. It can be expressed in terms of the amplitudes of the three normal modes of vibration for the equilateral triangle, a symmetrical stretch mode (s), and a doubly degenerate asymmetrical stretch mode (u):

$$\Phi_3(\mathbf{r}_1, \mathbf{r}_2, \mathbf{r}_3) \cong \Phi_3(min) + (1/2)[K_s t^2 + K_u(v^2 + w^2)]. \tag{VII.48}$$

Here K_s and K_u are the respective harmonic force constants, and t, v, and w are the corresponding normal mode amplitude coordinates. This suffices to determine the dominating classical low-temperature behavior of $B_2(T)$:

$$B_2(T) \sim -\left(\frac{4\pi^2 r_{min}^3}{3^{1/2}}\right)\left(\frac{2\pi}{\beta K_s}\right)^{1/2}\left(\frac{2\pi}{\beta K_u}\right)\exp[-\beta\Phi_3(min)]. \tag{VII.49}$$

Although it does not apply to the cases of xenon and the other noble gases, it is possible in principle that a sufficiently strong three-body interaction could break the equilateral triangle symmetry of $\Phi_3(min)$. The result could be an isosceles triangle, a symmetric linear structure, or even three-body inherent structures with no symmetry. The asymptotic low-temperature form of $B_2(T)$ would then receive additive contributions from each of these configurations.

It is generally true for n neutral molecules, which may have internal degrees of freedom and exhibit attractions of arbitrary range for one another, that the potential energy $\Phi_n(min)$ of their lowest lying inherent structure(s) in free space is negative. Furthermore, it is clear that as n increases, the $\Phi_n(min)$ forms a monotonically descending sequence, diverging toward $-\infty$:

$$\ldots > \Phi_{n-1}(min) > \Phi_n(min) > \Phi_{n+1}(min) > \ldots. \tag{VII.50}$$

This has the consequence that the virial coefficients of ascending order have stronger and stronger divergences as temperature declines toward absolute zero. To determine the precise limiting low-temperature behavior of $B_n(T)$ in the classical limit, it is necessary to know (a) the cluster symmetry and possible configurational degeneracy of the global minimum structure for Φ_{n+1}, in other words, how many equivalent inherent structures exist at that absolute minimum; (b) the harmonic normal modes of vibration about the global minimum; and (c) the rotational (inertial) moments of the global minimum inherent structure. However, after taking logarithms, the dominant low-temperature behavior for the classical-limit virial coefficients is controlled just by the value of the global minimum inherent-structure potential energy:

$$\ln[-B_n(T)] \sim \beta\Phi_{n+1}(min). \tag{VII.51}$$

In the event that quantum corrections to the classical limit need to be taken into account, the required modification of this last expression simply replaces the lowest inherent-structure potential energy $\Phi_{n+1}(\min)$ with the $(n + 1)$-particle lowest bound state energy eigenvalue $E_0(n + 1)$.

The nth order distribution functions ρ_n or equivalently the correlation functions g_n for the dilute vapor phase of nonelectrolytes in the infinite system limit,

$$\rho_n(\mathbf{r}_1 \dots \mathbf{r}_n; \rho, T) \equiv \rho^n g_n(\mathbf{r}_1 \dots \mathbf{r}_n; \rho, T) \tag{VII.52}$$

also possess series expansions in density with nonzero convergence radii. These expansions can be generated from the formal grand ensemble expression (II.74) in the classical limit. For simplicity, it will continue to be assumed that the particles have no internal degrees of orientational or configurational degrees of freedom. In the case of the pair correlation function, the expansion has the following form:

$$g_2(r_{12}; \rho, T) = \exp[-\beta \Phi_2(r_{12})] + \sum_{j=1}^{\infty} \rho^j g_{2,j}(r_{12}; T). \tag{VII.53}$$

Similarly to Eq. (VII.43) for the virial coefficients, the expression for $g_{2,j}$ involves potential energy functions Φ_l for all orders $2 \leq l \leq j + 2$, representing contributions from clusters up to size $j + 2$. In the low-temperature limit, this maximum cluster size dominates $g_{2,j}$, and this dominance leads to very prominent peaks displayed by this function at the pair distances present in that cluster's ground-state inherent structure.

D. Cluster Melting

The thermodynamic first-order melting/freezing transition is formally an attribute of many-particle systems in the large-system limit. Isolated clusters with the modest sizes ($N \approx 10^1 - 10^3$) that have been considered thus far in this chapter consequently do not qualify to exhibit this phenomenon in its strict interpretation. Nevertheless, numerical studies of clusters subject to controlled internal energy or temperature have been observed to exhibit a precursor of the true melting/freezing transition [e.g., Honeycutt and Andersen, 1987; Wales and Berry, 1990]. Quantitative analysis of this cluster behavior as a time-independent reversible property requires application of suitable constraints that can be implemented in simulations. Experimental studies of clusters are also possible, utilizing for example free jet expansions and electron diffraction [Fargas et al., 1987; Bartell and Huang, 1994], but control of individual cluster conditions tends to be less precise and effective in comparison with numerical simulations for model systems. For analytical purposes in this section, the focus is on a microcanonical ensemble description of cluster properties.

A useful approach to understanding thermal behavior considers each cluster to be an isolated system, undergoing classical dynamics. Its total energy E, center-of-mass momentum $\mathbf{P}$, and angular momentum $\mathbf{L}$ are conserved quantities. With the exception of a zero-measure set of initial conditions, classical dynamics should allow any given cluster to explore ergodically that portion of the phase space available to it subject to its conserved quantities, keeping in mind that E may be below one or more Φ_N disconnection thresholds, as discussed in Section VII.A. A microcanonical ensemble at energy E is composed of a representative collection of the cluster species of interest and includes an appropriately weighted sampling of **P**, **L**, and occupancy of disconnected regions, if any are present.

If the energy E for an N-particle cluster only exceeds its lowest inherent-structure potential energy by a small amount, the classical dynamics preserve that cluster as a bound object. But as indicated in Figure VII.3 for the simple $N = 4$ case, increasing E eventually crosses an evaporation threshold (at the lowest inherent-structure energy for $N - 1$ particles) where the available config-uration space set $\mathbf{A}_N(E)$ begins to contain arrangements with one particle indefinitely far removed from the others. In order to focus attention on the bound N-cluster itself, it is necessary therefore to impose an evaporation-eliminating constraint. The most direct way to implement this con-straint is to introduce a finite-length parameter a_0 and to demand that the network of those pair distances r_{ij} between particle centers which satisfy $r_{ij} \leq a_0$ be at least simply connected. This re-quirement implies that any pair of particles in the full set of N be connected either directly or in-directly by a path of pair distances $\leq a_0$. For the moment, the precise value to be assigned to a_0 is not crucial, but if it is chosen to be in the range 2 to 4 times the nearest neighbor distance in the lowest lying inherent structure, the effect is to eliminate evaporative escapes while not materially affecting the properties of the bound cluster. This connectivity constraint reduces the initial set $\mathbf{A}_N(E)$ of available configurations at total energy E above the evaporation threshold to an included smaller set $\mathbf{B}_N(E)$. But more generally one has

$$\mathbf{B}_N(E) \subseteq \mathbf{A}_N(E). \tag{VII.54}$$

Furthermore, when $E < E'$, the analog of Eq. (VII.9) must be valid:

$$\mathbf{B}_N(E) \subseteq \mathbf{B}_N(E'). \tag{VII.55}$$

Needless to say, $\mathbf{B}_N(E)$ is subject to the same disconnections between basins for different inherent structures at low E as are exhibited by $\mathbf{A}_N(E)$. The extra boundary to the cluster's N-particle dy-namics generated by the connectivity constraint can be treated as a reflecting boundary, which should not destroy the limited ergodicity of a cluster's dynamics within its accessible phase-space region. Application of this nonevaporation constraint has the effect of cutting off remote portions of basins for all of the cluster's inherent structures.

A much simpler and convenient but less discriminating constraint would be to surround the cluster with an impenetrable spherical container that is concentric with the cluster center of mass [Tsai and Jordan, 1993b]. This obviously requires adjusting the radius of that container for the cluster particle number N. As the energy E rises (and evaporation becomes more of an issue), the legitimate cluster configurations become more frequently noncompact. The resulting increase in spatial extension of those cluster configurations requires a corresponding increase in the con-tainer radius to capture all of those legitimate configurations. But then the probability rises that the larger encompassing sphere also contains one or more evaporated particles.

An overall classical ergodicity presumption for an N-particle cluster species implies that the available phase space comprising all possible position and momentum coordinates is uniformly represented in the energy-E microcanonical ensemble. However, this does not lead to a uniform sampling of the coordinate set $\mathbf{B}_N(E)$. Define $P_N(\{\mathbf{x}_i\} | E)$ to be the normalized configurational prob-ability for a cluster configuration specified by the particle coordinates $\mathbf{x}_i$ $(1 \leq i \leq N)$. Under the constant E constraint, any accessible cluster configuration $\{\mathbf{x}_i\}$ has a probability proportional to the measure of the momentum subspace. In the simple case that the N particles of mass m are identical and possess no internal degrees of freedom, this momentum subspace is defined by

$$\sum_{i=1}^{N} (p_i^2/2m) = E - \Phi_N(\mathbf{r}_1 \ldots \mathbf{r}_N), \tag{VII.56}$$

which identifies a hypersphere in that $3N$-dimensional momentum subspace with radius

$$R_p = \{(2m)[E - \Phi_N(\mathbf{r}_1 \ldots \mathbf{r}_N)]\}^{1/2}. \tag{VII.57}$$

The hypersurface of such a hypersphere provides the relevant weight, and it has a hyperarea magnitude proportional to $(R_p)^{3N-1}$. Consequently, the microcanonical configurational distribution for the cluster must have the following form:

$$P_N(\mathbf{r}_1 \ldots \mathbf{r}_N \,|\, E) = C_N(E)[E - \Phi_N(\mathbf{r}_1 \ldots \mathbf{r}_N)]^{(3N-1)/2}, \tag{VII.58}$$

where $C_N(E) > 0$ is a normalizing constant.

In the more general case where the particles might have v internal degrees of freedom and may be composed of atoms with different masses, the momentum hypersphere in Eq. (VII.56) generalizes to a hyperellipse. However, the same type of scaling argument for the corresponding microcanonical configurational probability applies. As a result, the expression in Eq. (VII.58) simply extends to the following form:

$$P_N(\mathbf{x}_1 \ldots \mathbf{x}_N \,|\, E) = C_N(E)[E - \Phi_N(\mathbf{x}_1 \ldots \mathbf{x}_N)]^{[(v+3)N-1]/2}, \tag{VII.59}$$

where now the normalizing constant $C_N(E)$ is modified to account for the internal degrees of freedom.

The temperature $T(E)$ to be assigned to the N-particle cluster when its microcanonical ensemble energy is E follows from the kinetic energy average for that microcanonical ensemble. For the basic case of identical structureless spherically symmetrical particles, and assuming that $\mathbf{P}$ and $\mathbf{L}$ are unconstrained, the definition of $T(E)$ in the classical limit follows the usual connection:

$$T(E) = (2/3Nk_B)\left\langle \sum_{i=1}^{N}(p_i^2/2m) \right\rangle_E \tag{VII.60}$$

$$\equiv (3mk_B)\langle p_j^2 \rangle_E, \qquad (1 \le j \le N),$$

where $\langle \ldots \rangle_E$ stands for the microcanonical ensemble average. Presuming that this temperature function has been independently evaluated over a substantial range of E values above the lower limit set by the lowest lying inherent structure, a plot of E vs. $T(E)$ can be analyzed for evidence of a precursor to the macroscopic system melting/freezing transition.

Figure VII.11 provides generic views of an E vs. $T(E)$ plot for a cluster of modest size, as well as the corresponding slope plot, emerging from a microcanonical ensemble representation. For the temperature range indicated in this plot, it has been assumed that no evaporation occurs. At low temperature, $E(T)/N$ at first rises linearly with a slope characteristic of harmonic motion near the basin bottoms for the set of permutation-related lowest inherent structures. As T rises, intrabasin anharmonicity can produce an increasing slope, as can the interbasin transitions that become possible as E begins to exceed the saddle point potential energies involved. Presuming that a sufficiently large collection of excited-inherent-structure basins become accessible over a relatively

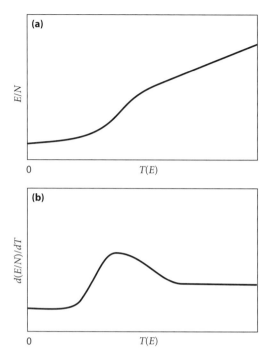

FIGURE VII.11. Qualitatively typical results from a microcanonical ensemble representation of a cluster of microscopic size, $N \approx 10^1 - 10^3$. Panel (a) illustrates the relation between the assigned temperature T and the ensemble fixed energy per particle E/N. Panel (b) presents the corresponding plot of the slope, with a broad maximum identifiable as a smeared-out version of a heat of transition.

small E increment, the per-particle slope $N^{-1}(dE/dT)$ displays a maximum before settling down to a smaller high-T behavior. This maximum invites interpretation as a smeared-out version of the heat of transition for the macroscopic system melting/freezing first-order phase change. In at least a rough sense, one would then expect the maximum to sharpen and become higher with increasing cluster size N, to approach a Dirac delta function in the $N \to \infty$ limit, with a strength determining the thermodynamic heat of melting (at $p = 0$). The cluster configuration probability function would reflect the structural changes as E passes through this zone of maximum slope, and an even more vivid indicator would be the occurrence-frequency-weighted collection of inherent structures and energies obtained by steepest descent mapping from an unbiased selection of energy-E cluster configurations.

This seemingly straightforward picture of melting/freezing precursors needs to be elaborated with a few complicating details, some aspects of which have been mentioned in this chapter. A notable feature is that for small to intermediate N, the lowest lying inherent structure may not represent a local portion of the stable crystal structure for the substance under consideration. In the case of the Lennard-Jones 12,6 pair potential model for which the stable macroscopic crystal at low temperature and pressure is hexagonal close packed [Kihara and Koba, 1952; Stillinger, 2001], Table VII.1 shows that this discrepancy exists even for $N = 7$. Furthermore, the same pair potential model yields several lowest energy inherent structures exhibiting icosahedral symmetry that is incompatible with that of the macroscopic crystal. These layered structures are called

"Mackay icosahedra" and are composed of the following "icosahedral magic numbers" of particles [Mackay, 1962]:

$$N(n) = (1/3)(10n^3 + 15n^2 + 11n + 3) \qquad\qquad (VII.61)$$

$$= (1/3)(2n + 1)(5n^2 + 5n + 3), \qquad (n > 0).$$

It is believed that these provide the global inherent-structure minima for the Lennard-Jones 12,6 pair interaction model, at least for the following cluster sizes [Romero et al., 1999; Doye and Calvo, 2002]:

$$N(1) = 13,$$

$$N(2) = 55,$$

$$N(3) = 147,$$

$$N(4) = 309. \qquad\qquad (VII.62)$$

Furthermore, there are many other N values with this interaction where the lowest lying inherent structures consist of modified icosahedra or fused portions of icosahedral units [Wales, 2003, Section 8.3].

As might be expected from the N-dependent diversity of global-minimum inherent-structure energies, the corresponding spectra of excited-state inherent-structure energies also is not constrained to have a simple N dependence. One possibility is that the apparent width of the melting precursor region shown schematically in Figure VII.11 might not diminish in width smoothly, let alone monotonically, as N increases. Another possibility for some model potentials is that there might be a precursor solid-state transition at E and T values below the range of precursor melting. Finally, it needs to be stressed that not all interactions produce a thermodynamically stable liquid phase, but just a vapor, as occurs in the case of buckminsterfullerene C_{60} [Hagen et. al., 1993; Doye and Wales, 1996]. In such cases, with clusters prevented from evaporating, there would by definition be no legitimate melting precursor phenomenon.

Numerical studies involving stepwise excitation of single members of a microcanonical ensemble can produce hysteresis effects, especially in the neighborhood of the precursor melting range. This can occur whether or not the cluster center-of-mass momentum $\mathbf{P}$ and total angular momentum $\mathbf{L}$ are held fixed. If such a cluster system begins at energy E just above a global minimum of the potential Φ_N, its dynamics would be trapped in that global minimum's basin, but a sufficiently long dynamical run to guarantee anharmonicity-produced ergodicity at that E could adequately sample the basin, yielding an accurate $T(E)$ via Eq. (VII.60) or its polyatomic particle generalization. But the cluster dynamics cannot escape that basin until E has been incrementally stepped up far enough to exceed the lowest basin disconnection value of Φ_N for the cluster. At that point, the dynamics would have failed to sample the basin for the excited state or states for which the disconnection is relevant, a situation that amounts to a failure to attain overall ergodicity. Further stepwise increases in conserved energy E suffer the same type of disconnection-related non-ergodicity for all higher lying inherent structures and their enclosing basins. Eventually, E could be ramped up sufficiently so that all disconnection thresholds would have been exceeded, with the allowed configuration set $\mathbf{B}_N(E)$ then connected and dynamically fully available for ergodic sampling.

An inverse situation applies to stepwise lowering of E in a dynamical sequence for a single cluster. Upon descending below a disconnection level, the configuration of the system can become trapped in a subset of $\mathbf{B}_N(E)$ and not be any longer representative of the microcanonical ensemble at lower E. In particular, the chance that the cluster finally ends up in one of its global minimum basins can be rather small, especially for large N.

These difficulties that tend to plague numerical simulations can be surmounted in a variety of ways. One way is just to forego the conceptually useful microcanonical ensemble and to replace it with the canonical ensemble in a Monte Carlo implementation. However, this replacement eliminates access to details of the cluster dynamical evolution, including rates of interbasin transitions emerging from classical dynamics.

E. Interacting Clusters

The descriptions of cluster characteristics in Sections VII.A, VII.B, and VII.D concerned isolated clusters. The exception was the presentation in Section VII.C of virial coefficients and other related density expansions for thermal equilibrium conditions. Whereas the virial coefficient description of the vapor phase is formally exact, the physical content of those coefficients, especially those of high order in the intermediate- to high-temperature range, is somewhat obscure. We now return to examine the description of the imperfect vapor phase in terms of interacting physical clusters, with the aim of restoring at least some aspects of physical insight. This physical cluster viewpoint was advocated long ago by Frenkel and Band, who presented some of the key concepts involved [Frenkel, 1939a, 1939b; Band, 1939a, 1939b]. It is possible now to convey a more thorough version of that approach to representing the vapor phase and to connect it with the inherent-structure formalism.

The cluster connectivity distance a_0 that was introduced previously to eliminate evaporative dissociation of isolated clusters is useful in this more general context. In fact, this distance can serve as the basis for identification of distinct clusters in a nonzero-density imperfect vapor phase. Figure VII.12 illustrates the elementary concept involved, namely, that two particles are consigned to the same "physical cluster" if either their associated diameter-a_0 spheres overlap or if they are indirectly connected by a path of such sphere overlaps. For the moment, we shall assume that all the particles are the same species and are either spherically symmetric or have a high symmetry (e.g., tetrahedral) with the associated sphere centered at the particle's centroid. This criterion divides the N particles that are present in any given configuration into n_1 monomeric clusters, n_2 dimeric clusters, ..., where

$$\sum_{s=1}^{N} s n_s = N. \tag{VII.63}$$

Here the interest centers about the large-system case, where the N particles reside in a volume V and where both N and V have macroscopic magnitudes. Of course, the division of the N particles among cluster sets of distinct sizes s changes as time progresses, and the particle dynamics modifies the system configuration.

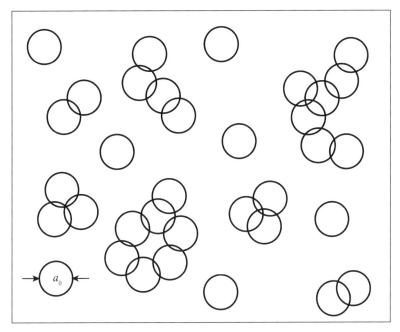

FIGURE VII.12. Identification of distinct "physical clusters" in a moderate-density fluid according to overlaps of diameter-a_0 particle-associated spheres. In this schematic illustration, the numbers n_j of j-particle clusters are $n_1 = 7$, $n_2 = 2$, $n_3 = 2$, $n_4 = 1$, and $n_7 = 2$.

In the event that the N particles are apportioned between coexisting vapor and liquid phases in their own separate connected macroscopic subvolumes, there is a range of sphere diameters a_0 such that one macroscopic cluster spans the domain occupied by the liquid, whereas the vapor portion consists of many disconnected microscopic clusters. In other words, for that range of a_0, the vapor is below the percolation threshold for connectivity, while the higher density liquid is above that percolation threshold [Stauffer and Aharony, 1992]. The range of a_0 values for which this distinction applies shrinks as the difference in number densities of the coexisting phases declines. If the vapor and liquid phases are in thermodynamic equilibrium with one another, that number density difference continuously approaches zero as the temperature rises to the critical temperature T_c. Hence the lower and upper a_0 limits converge to a common value a^* at the critical point. This situation is illustrated in Figure VII.13, where percolation threshold loci for $a_0 < a^*$ and $a_0 > a^*$ are superposed on a portion of a typical pressure–density phase diagram, including coexistence and critical regions. Henceforth, we shall suppose that the unique choice $a_0 = a^*$ has been imposed [Stillinger, 1963].

The grand ensemble offers a convenient context for examining the equilibrium statistics of connected clusters as they interact in an imperfect gas. As a starting point, we recall the two fundamental relations for that equilibrium ensemble from Section II.C (cited in Section VII.C for virial coefficients). These are now expressed in a form that takes advantage of the classical limit for the particles currently under consideration, which are treated as structureless for notational simplicity.

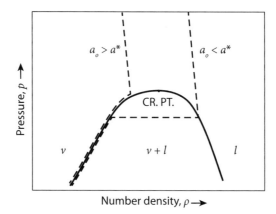

FIGURE VII.13. Percolation thresholds (dashed curves) with alternative choices for the connectivity distance a_0, displayed in a portion of the pressure (p) vs. number density (ρ) plane that includes vapor (v), liquid (l), and coexistence ($v + l$) regions. System-spanning connectivity percolation exists to the left of a dashed curve. In order for the percolation threshold to pass through the liquid–vapor critical point (CR. PT.), one must have the unique choice $a_0 = a*$.

The first relation is the grand partition function:

$$\mathfrak{Q}(\beta, y) = \sum_{N=0}^{\infty} (y^N/N!) \int \ldots \int \exp[-\beta \Phi_N(\mathbf{r}_1 \ldots \mathbf{r}_N)] d\mathbf{r}_1 \ldots d\mathbf{r}_N \qquad \text{(VII.64)}$$

$$\equiv \exp(-\beta\Omega),$$

where the integrals span all configurations inside the fixed macroscopic system volume V and where y is the absolute activity:

$$y = (2\pi m k_B T/h^2)^{3/2} \exp(\mu/k_B T). \qquad \text{(VII.65)}$$

Here the symbols m, h, and μ have their conventional meanings, namely particle mass, Planck's constant, and chemical potential, respectively. The absolute activity has the low-density asymptotic property

$$\lim_{\rho \to 0}(y/\rho) = 1. \qquad \text{(VII.66)}$$

We shall suppose that periodic boundary conditions apply to the system volume V, and because it therefore possesses no surfaces one can make the following identification with the pressure equation of state for the imperfect vapor phase [cf. Eq. (II.32)]:

$$-\Omega = pV. \qquad \text{(VII.67)}$$

If indeed one or more bounding surfaces were present, this last expression would have to be supplemented with extra terms representing the reversible isothermal work of inserting those surfaces.

The second fundamental relation provides the mean number of particles in the system subject to the given values of $\beta = 1/k_B T$ and y. It is generated by an isothermal isochoric derivative of $-\beta\Omega$ with respect to $\ln y$:

$$\langle N \rangle = \left(\frac{\partial}{\partial \ln y} \right)_{\beta, V} \ln[\mathcal{Q}(\beta, y)]$$

$$= \exp(\beta\Omega)\sum_{N=1}^{\infty} N(y^N/N!)\int\ldots\int\exp[-\beta\Phi_N(\mathbf{r}_1\ldots\mathbf{r}_N)]d\mathbf{r}_1\ldots d\mathbf{r}_N \qquad \text{(VII.68)}$$

$$\equiv \sum_{s=1}^{\infty} s\langle n_s \rangle.$$

The last form simply resolves the grand ensemble average of N according to the cluster size distribution, Eq. (VII.63).

Having introduced a cluster identification criterion that is based on the overlap distance a^*, it is now possible to transform the fundamental relations (VII.64) and (VII.68) so as to exhibit explicitly their underlying dependences on the cluster distributions that may be present instantaneously. Note first that the number of substantially different ways to permute N particles among n_1 monomeric clusters, n_2 dimeric clusters, $\ldots$, n_s clusters of s particles, $\ldots$ is given by the combinatorial factor

$$N!/\left[\prod_{s=1}^{N} n_s!(s!)^{n_s}\right]. \qquad \text{(VII.69)}$$

Consequently, the grand partition function can be expanded into a sum over all possible sets $\{n_s\}$ of cluster sizes present in the system

$$\exp(-\beta\Omega) = \sum_{\{n_s\}} \int_{\{n_s\}}\ldots\int\exp[-\beta\Phi(\{n_s\})]\prod_{t=1}^{\infty}(1/n_t!)\prod_{\alpha=1}^{n_t}(y^t/t!)\prod_{k(\alpha)=1}^{t} d\mathbf{r}_{k(\alpha)}. \qquad \text{(VII.70)}$$

Here the index α has been introduced to enumerate clusters of the same size n_t, $k(\alpha)$ is an identifier for particles within a given cluster, and the integrations are restricted configurationally to preserve the specified set $\{n_s\}$ of internally connected clusters. It needs to be realized that the potential energy function for each cluster division, denoted here simply by $\Phi(\{n_s\})$, generally can include interactions between different clusters. The correspondingly transformed version of Eq. (VII.68) is the following:

$$\langle N \rangle = \exp(\beta\Omega)\sum_{\{n_s\}} \int_{\{n_s\}}\ldots\int\exp[-\beta\Phi(\{n_s\})]\prod_{t=1}^{\infty}(1/n_t!)\prod_{\alpha=1}^{n_t}(ty^t/t!)\prod_{k(\alpha)=1}^{t} d\mathbf{r}_{k(\alpha)}. \qquad \text{(VII.71)}$$

The original version of the Frenkel-Band theory assumed that intercluster interactions could be disregarded, whether those interactions were from the potential energy function or were due to the nonoverlap restriction. In other words, the clusters were regarded as independent of one another, an assumption justifiable in the low-density-limit regime. In that circumstance [to be indicated now by superscript (0)], the multiple integral terms in the last two rather complicated expressions reduce to simple products of the following configurational partition functions for the individual clusters ($s \geq 1$):

$$Z_s^{(0)}(\beta) = (1/s!)\int_{(s)}\ldots\int\exp[-\beta\Phi_s(\mathbf{r}_1\ldots\mathbf{r}_s)]d\mathbf{r}_1\ldots d\mathbf{r}_s. \qquad \text{(VII.72)}$$

As a result, Eqs. (VII.70) and (VII.71) would undergo drastic simplifications to the following forms:

$$-\beta\Omega = \beta p V \qquad \text{(VII.73)}$$

$$= \sum_{s=1}^{\infty} Z_s^{(0)}(\beta)y^s,$$

and

$$\langle N \rangle = \sum_{s=1}^{\infty} s Z_s^{(0)}(\beta) y^s. \tag{VII.74}$$

In order to take proper account of cluster interference effects that arise for denser vapors, it is useful to introduce a probability function that describes the actual configurational preferences for a cluster of size s. Specifically, let

$$P_s(\mathbf{r}_1 \dots \mathbf{r}_s) d\mathbf{r}_1 \dots d\mathbf{r}_s \tag{VII.75}$$

stand for the probability that incremental volume elements $d\mathbf{r}_1 \dots d\mathbf{r}_s$ simultaneously contain the centers of serially numbered particles $1, \dots, s$ that comprise a single connected cluster of size s. As a result of this definition one has

$$\langle n_s \rangle = \int_{(s)} \dots \int P_s(\mathbf{r}_1 \dots \mathbf{r}_s) d\mathbf{r}_1 \dots d\mathbf{r}_s. \tag{VII.76}$$

The function P_s can be expressed as a grand ensemble average, written as a sum over cluster populations, as in Eqs. (VII.70) and (VII.71):

$$P_s(\mathbf{r}_1 \dots \mathbf{r}_s) = (y^s/s!) \exp(\beta\Omega) \sum_{\{n_t\}} n_s \int_{(n_t)} \dots \int \exp[-\beta\Phi(\{n_t\})] \prod_{t=1}^{\infty} (1/n_t!) \prod_{\alpha=1}^{t} (y^t/t!) \prod_{k(\alpha)=1}^{n_t - \delta_{s,t}} d\mathbf{r}_{k(\alpha)}. \tag{VII.77}$$

This expression accounts for the possibility that any one of n_s clusters of size s that happen to be present in the system could occupy the incremental volume elements at $\mathbf{r}_1 \dots \mathbf{r}_s$ but only in one serially ordered permutation among those numbered positions.

Subject to the cluster nonoverlap restriction, all clusters represented in Eq. (VII.77) but one are free to move throughout the volume V. The exception is the one occupying the fixed positions $\mathbf{r}_1 \dots \mathbf{r}_s$. So far as the remaining movable clusters are concerned, the presence of the fixed s-cluster acts as a localized externally applied potential, not only because it cannot be overlapped by the movable clusters, but also because its contribution to the overall potential energy function can extend beyond its exclusion zone. Because of that effective external potential, the grand partition function for just those movable clusters deviates from that in its absence and can be denoted by [Stillinger, 1963]:

$$\exp[-\beta\Omega_s(\mathbf{r}_1 \dots \mathbf{r}_s)]. \tag{VII.78}$$

This expression permits Eq. (VII.77) to be expressed in a more compact form:

$$P_s(\mathbf{r}_1 \dots \mathbf{r}_s) = (y^s/s!) \exp[\beta\Omega - \beta\Omega_s(\mathbf{r}_1 \dots \mathbf{r}_s) - \beta\Phi_s(\mathbf{r}_1 \dots \mathbf{r}_s)]. \tag{VII.79}$$

The amount of reversible isothermal work W_s that must be expended in order to place this effective external potential within the system is the difference

$$W_s(\mathbf{r}_1 \dots \mathbf{r}_s; \beta, y) = \Omega_s(\mathbf{r}_1 \dots \mathbf{r}_s; \beta, y) - \Omega(\beta, y), \tag{VII.80}$$

which is equivalent to a cavity free energy for reversible insertion of a fixed-configuration s-cluster within the macroscopic open system. As a result of this identification, Eq. (VII.76) now can be rewritten

$$\langle n_s \rangle = Z_s(\beta, y) y^s, \tag{VII.81}$$

where

$$Z_s(\beta, y) = (1/s!) \int_{(s)} \ldots \int \exp[-\beta \Phi_s(\mathbf{r}_1 \ldots \mathbf{r}_s) - \beta W_s(\mathbf{r}_1 \ldots \mathbf{r}_s; \beta, y)] d\mathbf{r}_1 \ldots d\mathbf{r}_s \qquad (\text{VII.82})$$

supplants the free-cluster quantity $Z_s^{(0)}(\beta)$, Eq. (VII.72). The mean number of particles in the system is now expressed by

$$\langle N \rangle = \sum_{s=1}^{\infty} s Z_s(\beta, y) y^s. \qquad (\text{VII.83})$$

The grand partition function and therefore the pressure p for the imperfect vapor can be obtained from Eq. (VII.83) by integrating that expression with respect to $\ln y$ as required by the first line of Eqs. (VII.68). The result may be written as follows:

$$\beta p V = \sum_{s=1}^{\infty} Z_s^*(\beta, y) y^s. \qquad (\text{VII.84})$$

The coefficient functions in this series are the following:

$$Z_s^*(\beta, y) = (1/s!) \int_{(s)} \ldots \int \exp[-\beta \Phi_s(\mathbf{r}_1 \ldots \mathbf{r}_s) - \beta W_s^*(\mathbf{r}_1 \ldots \mathbf{r}_s; \beta, y)] d\mathbf{r}_1 \ldots d\mathbf{r}_s, \qquad (\text{VII.85})$$

which contain modified cavity free energy functions W_s^* for the connected clusters:

$$\exp[-\beta W_s^*(\mathbf{r}_1 \ldots \mathbf{r}_s; \beta, y)] = (s/y^s) \int_0^y (y')^{s-1} \exp[-\beta W_s(\mathbf{r}_1 \ldots \mathbf{r}_s; \beta, y')] dy'. \qquad (\text{VII.86})$$

This completes the formal description of the thermal equilibrium state of the imperfect vapor in terms of properly interacting Frenkel-Band connected clusters, at least for structureless particles. If the particles were not spherically symmetric but contained intramolecular degrees of freedom, the formalism just presented could be generalized simply by replacing position vectors $\mathbf{r}_1 \ldots \mathbf{r}_N$ in the various expressions with the appropriately extended configurational vectors $\mathbf{x}_1 \ldots \mathbf{x}_N$. Furthermore, an extended version of the theory for polyatomic molecules offers the possibility of generalizing the use of a single diameter-a^* overlap criterion to the use of several overlap-criterion spheres per molecule whose intramolecular positions conform to that molecule's instantaneous shape.

F. Metastability Extensions

Figure VII.13 schematically showed the conventional vapor–liquid coexistence region possessed by a single-component macroscopic system at thermal equilibrium in the density–pressure plane. In analogy to the experimental possibility of attaining long-lived metastable undercooled liquids that reside within a liquid–crystal coexistence region, it is also experimentally possible to produce metastable vapor and liquid states within their own coexistence region. This naturally raises questions about how those extensions might respectively be defined in terms of their own configuration space projection operations analogous to that described in Chapter VI for supercooled liquid states.

For a single-component macroscopic system, let $T_{lv}(p)$ be the pressure-dependent equilibrium coexistence temperature for liquid and vapor phases, where the pressure p lies above the triple-point pressure and below the critical-point pressure: $p_{tp} \leq p \leq p_{cp}$. Then starting with the system in an equilibrated homogeneous vapor phase at some intermediate point along the $T_{lv}(p)$ locus, it is possible experimentally to lower the temperature isobarically by a modest amount to generate an undercooled homogeneous vapor with sufficiently long lifetime to exhibit reproducible properties of its metastable state. Alternatively, it is possible to increase the pressure isothermally by a modest amount to yield a homogeneous overcompressed vapor. However, as indicated qualitatively by classical nucleation theory [Abraham, 1974], the average lifetime of a metastable phase decreases strongly with increasing depth of penetration into a coexistence region, so there is a limit for reproducible observation that has historically been identified as a "spinodal" state [Hansen and McDonald, 1976, p. 347].

Analogously, starting with a homogeneous liquid sample along the $T_{lv}(p)$ locus, followed by modest isobaric temperature increase, the liquid would enter a metastable superheated state. Alternatively, an isothermal pressure reduction would generate a decompressed metastable liquid. As in the case of the metastable vapor states, the average lifetime of these liquids declines as the depth of penetration into the coexistence region increases, again empirically suggesting a "spinodal" limitation for reproducible measurements.

These observable metastable-state phenomena invite possible identification of inherent-structure basin projection operations for quantitative description of any homogeneous metastable states that can exist at least temporarily within the equilibrium vapor–liquid coexistence region. For the metastable vapor states, it is necessary to rule out initial configurations exhibiting droplets or substantial regions of the liquid that have formed by nucleation and subsequent growth. Under both isobaric and isochoric mappings to minima, these "density anomalies" should remain identifiable and available to trigger projection elimination. Metastable liquid states need to be free of substantial voids or vapor-filled bubbles, which analogously should remain identifiable after either isochoric or isobaric mapping to inherent structures. Extensive numerical simulation studies are necessary to determine the size limits of liquid droplets and of voids above which the configurations should be eliminated. In view of the fact that the equilibrium vapor and liquid phases become indistinguishable upon approaching the critical point, it appears necessary for the two projection operations to be defined so that the configuration sets that they retain also should be indistinguishable upon approaching the critical point.

At least for the undercooled or overcompressed vapor metastable extension, an alternative approach might be considered that does not rely on either isochoric or isobaric mapping to acceptable inherent structures. The basic strategy is to use the grand ensemble interacting cluster series Eqs. (VII.83) and (VII.84) for $\langle N \rangle$ and βpV, respectively, with the same particle overlap distance a^* as before but with β, y choices corresponding to metastability. However, an upper limit s_{max} would be imposed on the allowed sizes of clusters defined by that a^*. The metastable extension relations thus become simply the following:

$$\beta pV \cong \sum_{s=1}^{s_{max}} Z_s^*(\beta, y)y^s, \tag{VII.87}$$

and

$$\langle N \rangle \cong \sum_{s=1}^{s_{max}} sZ_s(\beta, y)y^s, \tag{VII.88}$$

where $Z_s(\beta, y)$ and $Z_s^*(\beta, y)$ are defined as before but for the cluster-size-constrained medium. For the β, y range of interest, the choice of s_{max} should be as large as possible while not exceeding the "critical nucleus" size that homogeneous nucleation theory assigns on the basis of maximum cluster free energy as a function of s [Abraham, 1974]. This alternative description of vapor undercooling deserves eventually to be quantitatively evaluated in comparison with the steepest descent mapping approach mentioned above.

If a similar mapping-free strategy were to be formulated for the metastable liquid, relying on the same overlap distance a^*, the allowed interacting physical clusters would have to include a macroscopically large one (or several) spanning the entire system because the density exceeds the a^* percolation threshold. Smaller physical clusters could also be present. However, the system-spanning large cluster(s) would have to be configurationally constrained so as not to allow the presence of large regions within which it was (or they were) absent, while only small "vapor" clusters occupied such regions; otherwise, this would constitute voids or bubbles of vapor that would have to be disallowed in the metastable liquid phase.

VIII.

The Helium Isotopes

T
he stable helium isotopes ^{3}He and ^{4}He present crystal and liquid phases whose properties are dominated by quantum effects. These effects arise in part from the low atomic masses and relatively weak interatomic attractions that are involved. But additionally, the intrinsically quantum mechanical classification of these isotopes as fermions and bosons, respectively, produces unusual characteristic differences. The striking observable phenomena include ^{3}He and ^{4}He liquid ground states for modest pressures, as well as superfluidity [Atkins, 1959; Wilks and Betts, 1987; Dobbs, 2000]. These quantum-dominated helium isotopes also exhibit an unusual equilibrium phase behavior explained below that is not present for the heavier noble gases, namely, "inverse melting," whereby constant-pressure heating causes the liquid to freeze into a higher entropy crystal. Those and other unusual properties of the helium isotopes form a sufficiently vivid collection to justify devoting this entire chapter to presenting their empirical details and related theoretical explanations, especially regarding implications for the inherent-structure, energy-landscape perspective.

The two helium isotopes both occur naturally on the Earth but with widely different concentrations. In the terrestrial atmosphere, the lighter isotope ^{3}He comprises only a fraction 1.37×10^{-6} of the total amount of the element [Lide, 2003, p. **9**-93]. Somewhat higher (but quite variable) fractions are present in various geological formations within the planetary mantle, but the lighter isotope still remains distinctly the rarer of the two [Aldrich and Nier, 1948].

A. Experimental Phase Diagrams

Figure VIII.1 presents a portion of the phase diagram in the T,p plane for the lighter isotope ^{3}He. This view shows the normal liquid phase (in the "fluid" region) as well as a body-centered cubic (bcc) crystal phase at moderate pressure that gives way to the hexagonal close-packed (hcp) crystal structure at higher pressure. However, the temperature and pressure scales used for those features obscure the liquid–vapor coexistence curve and its critical point. They also obscure the existence of additional phases that occur at very low temperature, specifically two superfluid phases (A and B) and an antiferromagnetically ordered bcc crystal phase. Figure VIII.2 renders these latter features visible by utilizing a millikelvin (mK) temperature scale. Below T approximately equal to 1 mK, the ^{3}He vapor-pressure curve (indistinguishable from the horizontal axis in Figure VIII.2) directly separates the superfluid B phase and the vapor phase, with no intervening normal fluid phase.

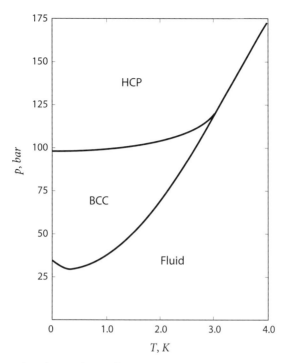

FIGURE VIII.1. Equilibrium phase diagram for pure ^{3}He, showing the normal fluid, bcc crystal, and hcp crystal regions. The liquid–vapor coexistence curve and its critical point at 3.32 K occur at low pressures on this scale and have been suppressed for visual clarity.

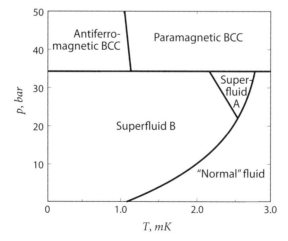

FIGURE VIII.2. Equilibrium phase diagram for pure ^{3}He with a millikelvin temperature scale to show clearly the two superfluid phases and the paramagnetic and antiferromagnetic nuclear-spin-ordered bcc phases occurring in this very low temperature regime.

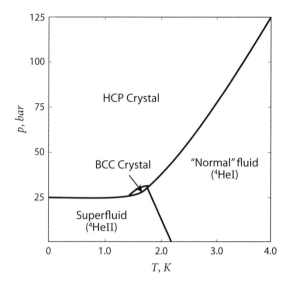

FIGURE VIII.3. Equilibrium phase diagram for pure ^{4}He.

The substantially different T,p phase diagram for the heavier isotope ^{4}He appears in Figure VIII.3. In this case, a second low-temperature scaling is unnecessary. For this heavier isotope, the liquid region is divided into two distinguishable portions, conventionally denoted as "I" and "II," at higher and lower temperature, respectively. The former is a "normal" liquid, and the latter exhibits superfluid behavior.

Because the liquid–vapor coexistence curves for the two isotopes occur at pressures too low to be conveniently represented in Figures VIII.1 and VIII.3, as an alternative Table VIII.1 presents the critical-point parameters for the two isotopes, which locate the upper termination points of those coexistence curves. Once again, the results display notable differences between the isotopes that stem from the strong underlying quantum effects. As temperature falls below an isotope's T_c, its vapor pressure declines monotonically, vanishing at $T = 0$ because the ground state of each isotope is a quantum many-body bound state in free space.

The obvious differences in phase behavior observed for the two isotopes have virtually no precedent with other elements. As indicated earlier, these distinctions stem in part from the small, but substantially unequal, masses $m_3 < m_4$ of the two isotopes. The differences are also attributable to the fact that the lighter isotope obeys Fermi statistics because of its nuclear spin 1/2 (measured in units $\hbar$) whereas the heavier isotope with vanishing nuclear spin obeys Bose-Einstein statistics.

TABLE VIII.1. Critical point parameters for the pure helium isotopes

	^{3}He	^{4}He
Temperature T_c, K	3.32	5.20
Pressure p_c, bar	1.15	2.27
Number density ρ_c, mol/cm^3	0.0137	0.0169

Although representing a departure from experimental reality, it would nevertheless be theoretically informative to have sets of properties for *four* "heliums" in which each mass m_3 and m_4 could be separately represented by both fermionic and bosonic species. Comparison of properties would then permit an explicit separation of light-mass effects from fermion–boson effects. More generally, one might imagine being able to vary the masses of each of the fermionic and bosonic species at will and comparing the ways that they individually approach classical behavior in their respective positive-temperature, high-mass asymptotic limits.

In relation to this general mass variation issue, it may be worth noting that molecular hydrogen H_2 offers a stable particle with only half the mass of a ^{4}He atom. Specifically, *para*-H_2 has both its electron spins and its proton spins paired, and at low temperature only the $J = 0$ rotational state of the diatomic molecule is populated. Consequently, it masquerades as a spherically symmetric boson. But in spite of its small mass, the zero-pressure ground state of *para*-H_2 is not liquid as with the helium isotopes, but is an hcp crystal [Silvera, 1980]. This indicates that stronger attractive interactions are present in *para*-H_2 compared to the heliums. It nevertheless raises the question about whether placing crystalline *para*-H_2 under isotropic tension at $T = 0$ might produce a metastable liquid, perhaps even a superfluid.

A brief list of the most striking characteristics presented by the phase diagrams in Figures VIII.1–VIII.3 would include the following observations:

(1) Both pure isotopes remain in a fluid state at $T = 0$, unless the pressure exceeds a positive threshold that is approximately 34.4 bar for ^{3}He [Greywall, 1985], 25.3 bar for ^{4}He [Swenson, 1950]. The number densities of the coexisting liquid and crystal phases under these $T = 0$ conditions, respectively, are $\rho_{liq} = 0.0395$ mol/cm^3 and ρ_{crys}(bcc) $= 0.0415$ mol/cm^3 for ^{3}He [Huang and Chen, 2005], and $\rho_{liq} = 0.0430$ mol/cm^3 and ρ_{crys}(hcp) $= 0.0472$ mol/cm^3 for ^{4}He [Swenson, 1950].

(2) The bcc crystal structure appears as a stable phase for both isotopes. But as comparison of Figures VIII.1 and VIII.3 indicates, this is the spatial structure observed for solid ^{3}He over a substantial temperature–pressure range, including $T = 0$, while it exists only in a narrow region along a small $T > 0$ portion of the melting curve for ^{4}He [Grilly and Mills, 1962; Donohue, 1982]. The dominant crystal structure for this heavier isotope is hcp, to which the lighter isotope also reverts under elevated pressure. It should additionally be noted that the face-centered cubic (fcc) crystal structure appears for both isotopes in the kbar pressure range and at higher temperatures beyond the ranges covered by Figures VIII.1 and VIII.3 [Arms et al., 2003].

(3) Although superfluid phases exist for both isotopes, the temperature ranges involved are very different. The A and B superfluids for ^{3}He appear only below $T \approx 0.0025$ K, roughly three orders of magnitude less than the $T \approx 2$ K location of the ^{4}He boundary between its normal (helium I) and superfluid (helium II) phases. The helium I–helium II boundary at its lower end intersects the liquid–vapor curve at $T = 2.17$ K and at its upper end intersects the melting curve at $T = 1.76$ K.

(4) This latter ^{4}He phase boundary locates a higher order phase transition between the helium I and II fluids. It has historically been called the "lambda line," denoted by $T_\lambda(p)$, because of the reported logarithmic divergence at its lower end exhibited by the liquid heat capacity $C_{sat}(T)$ along the saturated vapor pressure curve [Buckingham and Fairbank,

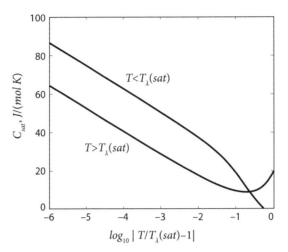

FIGURE VIII.4. Liquid ^{4}He heat capacity $C_{sat}(T)$ along a portion of the saturated vapor curve that contains the lambda transition. Redrawn from Ahlers, 1971, Figure 11.

1961; Wilks and Betts, 1987, p. 169]. Figure VIII.4 presents that measured heat capacity $C_{sat}(T)$. The isochoric and isobaric heat capacities $C_V(T)$ and $C_p(T)$ have also been measured across the full length of the lambda line $T_\lambda(p)$, and although these heat capacities appear also to display strong positive singularities upon crossing the lambda line, there is some indication that they may remain bounded and can be described in leading order by a small positive power of $|T - T_\lambda(p)|$ [Ahlers, 1973, 1980].

(5) The ^{3}He transitions indicated in Figure VIII.2 from normal fluid to either superfluid A or B are second order, with simple heat capacity discontinuities but no latent heat. However, a latent heat has been observed for the A–B transition line [Alvesalo et al., 1981].

(6) The slope dp_m/dT of the ^{3}He melting curve $p_m(T)$ is clearly negative over the temperature range $0 < T < 0.316$ K, then passing through a minimum and changing sign upon further temperature rise [Wilks and Betts, 1987, p. 15; Huang and Chen, 2005, Table VI]. Consequently, adding heat isobarically to a liquid sample that is initially at a point along that lower temperature portion of the melting curve causes it to crystallize, a counterintuitive phenomenon called "inverse melting" [Stillinger et al., 2001; Stillinger and Debenedetti, 2003]. A similar $p_m(T)$ negative slope leading to a minimum at approximately 0.8 K also occurs for ^{4}He, but this is a very shallow minimum by comparison to that for ^{3}He [Wilks and Betts, 1987, p. 16]. This inverse melting phenomenon must conform to the thermodynamic Clausius-Clapeyron equation:

$$\frac{dp_m}{dT} = \frac{S_l - S_c}{V_l - V_c}, \tag{VIII.1}$$

relating the slope to the ratio of entropy and volume changes between the coexisting liquid (l) and crystal (c) phases. For both isotopes in their inverse melting ranges, the volume decreases discontinuously upon crystallization. Consequently, the spatially uniform and disordered liquid at low positive temperature must possess a lower entropy than the spatially ordered solid, a seemingly paradoxical conclusion. It should also be noted that because the entropy difference approaches zero at $T = 0$, the enthalpy difference $H_l - H_c$ also vanishes at $T = 0$ for both isotopes at their respective crystallization pressures.

In view of the strong differences in the phase diagrams for the helium isotopes in pure form, it is natural to inquire about the properties of their mixtures at low temperatures. As Figure VIII.5 illustrates, the liquid-state mixtures are far from ideal and indeed show phase separation below 0.87 K. Increasing the mole fraction x_3 of the lighter isotope from zero causes the lambda transition temperature to decline and to join a tricritical point at $x_3 = 0.67$, which serves as the uppermost point of the phase separation region. The $T = 0$ behavior shows an unusual asymmetry, with ^{3}He dissolving in liquid ^{4}He up to $x_3 = 0.0648$, while ^{4}He is virtually insoluble in liquid ^{3}He [Wilks and Betts, 1987, Chapter 10].

The field of helium cryophysics historically presents a repeated examination of the possibility of "supersolid" behavior for solid ^{4}He at sufficiently low, but positive, temperature. This presumes that some version of Bose-Einstein condensation could occur analogous to that in the liquid below its $T_\lambda(p)$ locus. A large scientific literature considering this unusual behavior has emerged, involving both theoretical [e.g., Chester, 1970; Leggett, 1970] and experimental [e.g., Kim and Chan, 2004; Choi et al., 2010] efforts. However, sufficiently compelling arguments do not yet exist to incorporate a distinguished supersolid phase region in the ^{4}He equilibrium phase diagram, Figure VIII.3 [Kim and Chan, 2012].

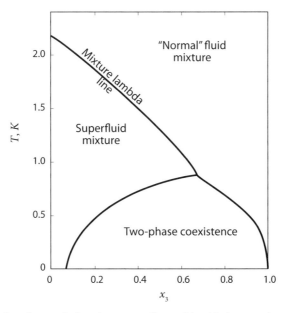

FIGURE VIII.5. Solution phase diagram for liquid mixtures of ^{3}He and ^{4}He. The horizontal coordinate x_3 is the ^{3}He mole fraction. The pressure is low, just large enough to eliminate occurrence of vapor phases.

B. Ground-State Wave Functions

The basic tool for analyzing the quantum phenomena that determine observable properties of the helium isotopes is the many-body Schrödinger equation. In particular, understanding the properties of its ground states is the logical starting point. The spin-free case of bosonic ^{4}He is conceptually simpler than that of the fermionic spin-1/2 case of ^{3}He and so is considered first.

The time-independent N-atom wave equation with Hamiltonian operator $\mathbf{H}^{[4]}$ satisfied by the ^{4}He ground-state eigenfunction $\psi_0^{[4]}$ and eigenvalue $E_0^{[4]}$ is the following:

$$(\mathbf{H}^{[4]} - E_0^{[4]})\psi_0^{[4]}(\mathbf{r}_1,....\mathbf{r}_N) = 0, \qquad (\text{VIII.2})$$

$$\mathbf{H}^{[4]} = -(\hbar^2/2m_4)\sum_{j=1}^{N}\nabla_j^2 + \Phi(\mathbf{r}_1,....\mathbf{r}_N).$$

As before, m_4 represents the mass of one of the heavier-isotope atoms, and Φ is the Born-Oppenheimer electronic ground-state potential function for the nuclear motion. The eigenfunction $\psi_0^{[4]}$ is symmetric under all permutations of the particle positions $\mathbf{r}_1...\mathbf{r}_N$, and it is conventional to suppose that this function is everywhere non-negative. If the volume V available to the N atoms is equal to or greater than the ground-state liquid volume at $p = 0$, then $E_0^{[4]} < 0$, i.e., the ground state is a bound state. However, reducing V forces neighboring atoms into closer contact, thus causing $E_0^{[4]}$ to rise and eventually become positive as the kinetic energy contribution and eventually the potential energy contribution are driven upward.

Under most temperature and pressure conditions that have been applicable in experiments on liquid and crystalline helium (both isotopes), the N-atom potential function Φ appears to be dominated by isotropic pair interactions. Therefore. for present purposes it is assumed that

$$\Phi(\mathbf{r}_1...\mathbf{r}_N) = \sum_{j=2}^{N}\sum_{l=1}^{j-1} \mathrm{v}^{(2)}(r_{jl}). \qquad (\text{VIII.3})$$

However, it should be kept in mind that if a helium sample is subjected to extreme compression, forcing the atoms' electron clouds to overlap strongly with one another, this pairwise additivity is significantly violated. Such compression eventually would produce a metallic state [Young et al., 1981].

The scientific literature devoted to helium contains many approximate determinations of the interatomic pair potential $\mathrm{v}^{(2)}(r)$ [e.g., Bruch and McGee, 1970; Aziz et al., 1979, 1987; Bich et al., 1988; Aziz and Slaman, 1991; Tang et al., 1995]. Box VIII.1 presents details for one of these [Aziz et al., 1979], also shown graphically in Figure VIII.6. Other more recent determinations are qualitatively quite similar and in fact differ quantitatively by rather small amounts for pair distances r, which are relevant for the helium ground state at moderate pressures.

The squared modulus of any ^{4}He eigenfunction is a non-negative function of particle positions and thus nominally could be interpreted as a suitably normalized Boltzmann factor for a classical ensemble of the same number of particles. In particular, this is true for the ground state for which one can write $[C_0(N) > 0]$:

$$[\psi_0^{[4]}(\mathbf{r}_1,....\mathbf{r}_N)]^2 = C_0(N) \exp[-\beta_{eff}\Phi_{eff}(\mathbf{r}_1...\mathbf{r}_N)], \qquad (\text{VIII.4})$$

where trivially the normalization constant for system volume V is

$$C_0(N) = [\textstyle\int_V d\mathbf{r}_1....\int_V d\mathbf{r}_N[\psi_0^{[4]}(\mathbf{r}_1,....\mathbf{r}_N)]^2]\{\int_V d\mathbf{r}_1....\int_V d\mathbf{r}_N \exp[-\beta_{eff}\Phi_{eff}(\mathbf{r}_1,....\mathbf{r}_N)]\}^{-1}, \quad (\text{VIII.5})$$

BOX VIII.1. Helium Interatomic Pair Potential

One of the available estimates for the distance variation of the interaction potential $v^{(2)}(r)$ for a pair of helium atoms comes from Aziz et al. (1979). It involves a mathematical form originated by Ahlrichs et al. (1976) but with shifts in the magnitudes of the included parameters. The potential is scaled with well-depth energy ε and with distance r_m that locates its single minimum

$$v^{(2)}(r) = \varepsilon v(r/r_m).$$

The reduced interaction function $v(x)$ is assigned the following form:

$$v(x) = A\exp(-ax) - \left[\frac{C_6}{x^6} + \frac{C_8}{x^8} + \frac{C_{10}}{x^{10}}\right]F(x),$$

where

$$F(x) = \exp\left[-\left(\frac{D}{x} - 1\right)^2\right] \quad \text{for } x < D,$$

$$= 1 \quad \text{for } x \geq D.$$

The scaling parameters have the values

$$\varepsilon/k_B = 10.8 \text{ K},$$

$$r_m = 2.9673 \text{ Å},$$

and the dimensionless parameters are

$$A = 5.448504 \times 10^5,$$

$$a = 13.353384,$$

$$C_6 = 1.3732412,$$

$$C_8 = 0.4253785,$$

$$C_{10} = 0.178100,$$

$$D = 1.241314.$$

The role of the modulating function $F(x)$ is to continuously switch off at small separations the inverse-power attraction terms it multiplies; its form involves a discontinuous second derivative at $x = D$, but that feature should have no noteworthy physical implication.

The value of this pair potential at vanishing pair separation is finite but very large:

$$v^{(2)}(r = 0)/k_B = \varepsilon A/k_B$$

$$\cong 5.88 \times 10^6 \text{ K.}$$

Presumably, this should diverge to $+\infty$ because of the coulombic repulsion between the nuclei. But whether or not that is the case, the ground-state and low-lying excited-state eigenfunctions of the N-atom Schrödinger equation (VIII.2) would exhibit extremely small values whenever any $r_{kl} = 0$ in spite of the quantum mechanical tunneling phenomenon, rendering essentially irrelevant the precise value assigned to $v^{(2)}(0)$.

The distance r_0 at which $v^{(2)}(r_0) = 0$ is $r_0 \cong 0.88919 \, r_m \cong 2.6385$ Å. It should be noted that the distance ratio r_0/r_m is close to the corresponding value for the standard Lennard–Jones 12,6 pair potential, namely $r_0/r_m = 2^{-1/6} \cong 0.89090$.

The curvature of the reduced potential at its unit-depth minimum has the value

$$v''(x = 1) \cong 74.9.$$

This is only slightly larger than the corresponding value for the Lennard–Jones 12,6 pair potential [Eq. (I.39)] when the latter is distance-scaled so that its minimum is also at $x = 1$, specifically,

$$(2^{1/6})^2 v_{LJ}''(x = 2^{1/6}) = 72.$$

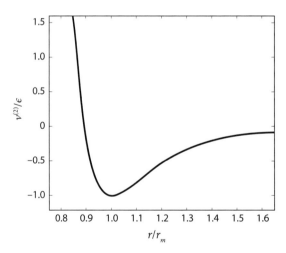

FIGURE VIII.6. Distance dependence of the approximate helium interatomic pair potential $v^{(2)}$ determined by Aziz et al., 1979. This plot is presented in reduced form with $\varepsilon/k_B = 10.8$ K and $r_m = 2.9673$ Å.

and where formally one can set $\beta_{eff} = 1/k_B T_{eff}$. The absolute magnitudes of β_{eff} and of Φ_{eff} separately in Eq. (VIII.4) are irrelevant; only the dependence of their product on the particle configuration $\mathbf{r}_1 \ldots \mathbf{r}_N$ matters. Note that expressions (VIII.4) and (VIII.5) remain valid if an arbitrary constant is added to $\beta_{eff} \Phi_{eff}(\mathbf{r}_1 \ldots \mathbf{r}_N)$, which simply produces a compensating change in $C_0(N)$. It should be kept in mind that changing the system volume V can cause nontrivial changes in the form of Φ_{eff}, in contrast to a conventional classical canonical ensemble situation, where the functional forms of the contributing interparticle potential functions do not change with V. A representation of an excited-state eigenfunction similar to that shown in Eq. (VIII.4) for the ground state would entail an effective potential in the Boltzmann factor diverging logarithmically to $+\infty$ at those N-particle configurations where that excited-state eigenfunction possessed a node. In principle, $\Phi_{eff}(\mathbf{r}_1 \ldots \mathbf{r}_N)$ defines a multidimensional landscape, with its own basin tiling and inherent structures; however, that has at best only an indirect bearing on the relevant helium system landscape defined by the proper interaction function $\Phi(\mathbf{r}_1 \ldots \mathbf{r}_N)$.

The fact that strong interatomic repulsion requires the ^{4}He ground-state wave function $\psi_0^{[4]}$ to decline markedly and nearly vanish when any one of the $N(N-1)/2$ pair distances approaches zero could be replicated in the Boltzmann factor (VIII.4) if Φ_{eff} incorporated contributions that are positive and large for those configurations. This qualitative attribute by itself would be satisfied if Φ_{eff} consisted of a sum of atom-pair terms $v_{eff}(r_{jl})$, each one of which becomes large and positive at small separation r_{jl}. That approximation would generate the specific form

$$\psi_0^{[4]}(\mathbf{r}_1 \ldots \mathbf{r}_N) \cong [C_0(N)]^{1/2} \exp[-\beta_{eff}/2) \sum_{j=2}^{N} \sum_{l=1}^{j-1} v_{eff}(r_{jl})] \tag{VIII.6}$$

$$\equiv [C_0(N)^{1/2} \prod_{j=2}^{N} \prod_{l=1}^{j-1} b(r_{jl}),$$

where the $b(r_{jl})$ are individual pair Boltzmann factors for effective pair interaction v_{eff} at temperature $2T_{eff}$. This product-form approximation to the ^{4}He ground-state eigenfunction is called a "Jastrow" wave function [Jastrow, 1955], now to be denoted generically by $\psi_0^{[4,J]}$. Regarded as a variational trial function with the exact Hamiltonian operator $\mathbf{H}^{[4]}$ in Eq. (VIII.2), the Jastrow form produces an upper bound to the exact ground-state energy $E_0^{[4]}$:

$$E_0^{[4,var]} = \langle \psi_0^{[4,J]} | \mathbf{H}^{[4]} | \psi_0^{[4,J]} \rangle / \langle \psi_0^{[4,J]} | \psi_0^{[4,J]} \rangle \tag{VIII.7}$$

$$\geq E_0^{[4]}.$$

At least one such variational calculation has been reported for the homogeneous liquid phase of ^{4}He, for which the $b(r_{ij})$ factors were assigned the form $\exp[-(2.6\text{Å}/r_{ij})^5]$ [McMillan, 1965]. The results obtained for this Jastrow approximation were in rough agreement with observed properties of the ^{4}He liquid state at $T = 0$, but those results indicated the need for better approximations to $\psi_0^{[4]}$.

The time-independent Schrödinger equation (VIII.2) for the ^{4}He ground state can be trivially rearranged to the following form:

$$\Phi(\mathbf{r}_1 \ldots \mathbf{r}_N) = E_0^{[4]} + \left(\frac{\hbar_2}{2m_4}\right) \left(\frac{\sum_{j=1}^{N} \nabla_j^2 \psi_0^{[4]}(\mathbf{r}_1 \ldots \mathbf{r}_N)}{\psi_0^{[4]}(\mathbf{r}_1 \ldots \mathbf{r}_N)}\right). \tag{VIII.8}$$

This identity in principle offers a way to generate the exact potential energy function (with possible ambiguity of a configuration-independent additive constant) whose bosonic ground-state eigenfunction is the given $\psi_0^{[4]}$. In particular, it can formally supply an N-body interaction $\Phi^{(J)}$ for which the exact ground-state eigenfunction has a chosen Jastrow product form $\psi_0^{[4,J]}$. In other words, by working "backwards" using Eq. (VIII.8), one can deduce the precise form of a "Jastrow Hamiltonian." By substituting Eq. (VIII.6) into Eq. (VIII.8), one finds for such a case that

$$\Phi^{(J)}(\mathbf{r}_1....\mathbf{r}_N) = E_0^{(J)} + \sum_{i<j} v^{(2,J)}(r_{ij}) + \sum_{i<j<k} v^{(3,J)}(\mathbf{r}_i, \mathbf{r}_j, \mathbf{r}_k). \qquad (VIII.9)$$

Here $E_0^{(J)}$ is the ground-state eigenvalue [not to be confused with the variational energy $E_0^{[4,\text{var}]}$ in Eq. (VIII.7)] for the Jastrow Hamiltonian. The two-body and three-body contributions in this expression have forms that can be exhibited in terms of the Jastrow pair quantities from Eq. (VIII.6), which are written for simplicity as

$$u(r) \equiv (\beta_{eff}/2)v_{eff}(r). \qquad (VIII.10)$$

The results are the following:

$$v^{(2,J)}(r_{ij}) = \left(\frac{\hbar^2}{m_4}\right)\left\{[u'(r_{ij})]^2 - u''(r_{ij}) - \left(\frac{2}{r_{ij}}\right)u'(r_{ij})\right\}, \qquad (VIII.11)$$

and

$$v^{(3,J)}(\mathbf{r}_i, \mathbf{r}_j, \mathbf{r}_k) = \left(\frac{\hbar^2}{m_4}\right)[u'(r_{ij})u'(r_{jk})\cos\theta_{ijk} + u'(r_{ik})u'(r_{kj})\cos\theta_{ikj} + u'(r_{ji})u'(r_{ki})\cos\theta_{jik}], \qquad (VIII.12)$$

where $\cos\theta_{ijk}$ stands for the angle subtended at particle j by the line segments from it to particles i and k.

The fact that a Jastrow product wave function $\psi_0^{[4,J]}$ formally involves strong three-body interactions at short range points to one of its principal physical shortcomings as a description of the $T = 0$ phases of ^{4}He. Those three-body interactions could be canceled by generalizing the Jastrow approximation Eq. (VIII.6) to include in its exponent a sum of carefully chosen three-body correction terms $v_{eff}^{(3)}(\mathbf{r}_i, \mathbf{r}_j, \mathbf{r}_k)$. However, inserting that extended Jastrow form into Eq. (VIII.8) would then generate additional contributions to the corresponding implied system potential energy that are four-body and five-body terms. By extension, this suggests that in order for the purely pairwise additive helium potential, Eq. (VIII.3), to emerge from Eq. (VIII.8), it would be necessary for the exact ground-state wave function to contain contributions of all possible orders in its exponent. This can symbolically be indicated as follows:

$$\psi_0^{[4]}(\mathbf{r}_1....\mathbf{r}_N) = [C_0(N)]^{1/2}\exp\left[-\sum_{i<j}w^{(2)}(i, j) - \sum_{i<j<k}w^{(3)}(i, j, k) - \sum_{i<j<k<l}w^{(4)}(i, j, k, l) - ....\right]. \qquad (VIII.13)$$

At this point, it is useful to recall that in a classical canonical ensemble the potential of mean force (i.e., the configuration-dependent free energy) for a group of N solute particles, averaged over the degrees of freedom of a solvent medium containing them, generally contains contributions of orders 2, 3, ..., N that are analogs of the $w^{(2)}$, $w^{(3)}$, ..., $w^{(N)}$ terms in Eq. (VIII.13). Alternatively, a classical canonical ensemble of N polyatomic molecules, if averaged over all molecular degrees of freedom except the position of a specific atom in each molecule, would result in a potential of mean force (free energy) for the remaining N positions that likewise would generally involve two-body, three-body, ..., N-body contributions. These observations suggest a conceptual strategy to create families of trial wave functions to represent the ^{4}He ground state with greater physical precision than can be attained with a simple Jastrow product function.

One specific implementation of this strategy involves the "shadow" wave function approximation $\psi_0^{[4,s]}$ for the ^{4}He ground state [Vitiello et al., 1988; MacFarland et al., 1994]. This takes the following form:

$$\psi_0^{[4,s]}(\mathbf{r}_1....\mathbf{r}_N) = [C_0^{(s)}(N)]^{1/2} \exp\left[-\sum_{i<j}^N u^{(s)}(r_{ij})\right] \int d\mathbf{s}_1....\int d\mathbf{s}_N \left[\prod_{k=1}^N l(|\mathbf{r}_k - \mathbf{s}_k|)\right] \exp\left[-\sum_{l<t}^N \tilde{u}^{(s)}(s_{lt})\right], \quad \text{(VIII.14)}$$

including Jastrow-like functions both outside and within the multiple integral. Each particle at location $\mathbf{r}_k$ is linked to its own "shadow" particle at $\mathbf{s}_k$ by a short-ranged integrand factor $l(|\mathbf{r}_k - \mathbf{s}_k|)$. By integrating over shadow particle degrees of freedom, a set of many-particle contributions for the N "real" particles is generated, as schematically indicated in Eq. (VIII.13).

The square of the shadow wave function is the approximation to the N-particle probability distribution function, which as remarked earlier can be interpreted as a classical canonical distribution:

$$[\psi_0^{[4,s]}(\mathbf{r}_1....\mathbf{r}_N)]^2 = C_0^{(s)}(N) \exp\left[-2\sum_{i<j}^N u^{(s)}(r_{ij})\right] \int d\mathbf{s}_1 d\mathbf{s}'_1....\int d\mathbf{s}_N d\mathbf{s}'_N$$

$$\times \left[\prod_{k=1}^N l(|\mathbf{r}_k - \mathbf{s}_k|) l(|\mathbf{r}_k - \mathbf{s}'_k|)\right] \exp\left\{-\sum_{l<t}^N [\tilde{u}^{(s)}(s_{lt}) + \tilde{u}^{(s)}(s'_{lt})]\right\}. \quad \text{(VIII.15)}$$

This description amounts to a set of N flexible triatomic molecules, the middle atoms of which are at locations $\mathbf{r}_1....\mathbf{r}_N$ and the terminal atoms of which are respectively at $\mathbf{s}_1....\mathbf{s}_N$ and at $\mathbf{s}'_1....\mathbf{s}'_N$. The l functions are present to control the distribution of the two intramolecular bond lengths. A terminal atom in one triatomic molecule can interact intermolecularly only with terminal atoms of its own kind (unprimed with unprimed, primed with primed). Figure VIII.7 provides a schematic illustration of these triatomic species. The integrations indicated in Eq. (VIII.15) over all $2N$ shadow particles create a substantially wider range of functional forms than would be possible with just a simple Jastrow approximation [the pre-integral factor in Eq. (VIII.14)].

In some shadow wave function calculations, the intramolecular bond factor l has been chosen to be a simple Gaussian form centered at zero separation, based upon a path integral representation of the quantum mechanical problem [Reatto and Masserini, 1988; Ceperley, 1995]. Several alternative functional forms have been examined for the pair functions u and $\tilde{u}$, with shape optimization assigned by minimizing the corresponding variational approximations to the ground-state energy [Reatto and Masserini, 1988; MacFarland et al., 1994]. The conclusion is that the shadow wave function family provides a significant improvement over the simple Jastrow approximation for energies and structures of the liquid and crystal ground states of ^{4}He. In particular, the Jastrow approximation tends to predict unphysically strong localization of the particles in the crystal phase, a shortcoming that is rectified by the optimized shadow wave function approach. In particular, the hcp crystal at its melting pressure experimentally has an unusually large value for the Lindemann ratio of mean-square atomic displacement to nearest neighbor separation [Section IV.I], a phenomenon that Jastrow functions apparently are unable to reproduce.

Whereas it may suffice for homogeneous liquid or crystal phases to approximate $\beta_{eff}\Phi_{eff}$ in the Boltzmann factor of Eq. (VIII.5) with short-range functions, this would not be appropriate for

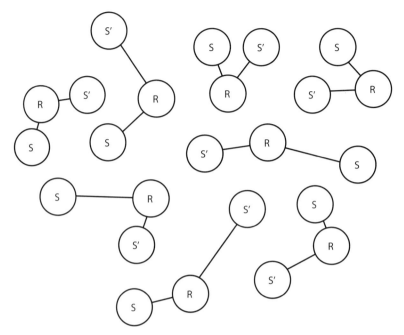

FIGURE VIII.7. The effective triatomic species involved in the probability distribution provided by the square of the shadow wave function. The intramolecular bond factors l have been represented as solid line segments. The only intermolecular interactions present are those acting between pairs of the same species: "real" particles R, shadow particles S, and shadow particles S'.

N-atom bound states in a very large or unbounded volume. In such cases, the ^{4}He atoms' ground state consists of a liquid droplet within that volume. If one imagines continuously displacing one of the atoms away from the remaining $N-1$ atoms localized within the droplet, the ground-state wave function $\psi_0^{[4]}(\mathbf{r}_1\ldots\mathbf{r}_N)$ would have to decline exponentially (in leading order) with that displacement distance at a rate determined by the one-atom binding energy to the remainder droplet. Short-range effective interactions of the types that have been used in Jastrow and shadow wave function approximations cannot produce this functional behavior.

The singlet reduced-density matrix for the ^{4}He ground state is

$$\rho_{1,N}(\mathbf{r}_1\,|\,\mathbf{r}_1') = N\!\int_V d\mathbf{r}_2.\ldots\int_V d\mathbf{r}_N\psi_0^{[4]}(\mathbf{r}_1, \mathbf{r}_2.\ldots\mathbf{r}_N)\psi_0^{[4]}(\mathbf{r}_1', \mathbf{r}_2.\ldots\mathbf{r}_N). \qquad \text{(VIII.16)}$$

Its diagonal elements ($\mathbf{r}_1 = \mathbf{r}_1'$) provide the spatial distribution of particle number density, which is the position-independent constant N/V for both liquid and crystal phases when periodic boundary conditions apply. In the density region for the ^{4}He homogeneous liquid ground state, as the two positions $\mathbf{r}_1$ and $\mathbf{r}_1'$ recede from one another within a macroscopic volume V, the off-diagonal elements of the singlet reduced density matrix approach a constant fraction of the diagonal value N/V:

$$\rho_{1,N}(\mathbf{r}_1\,|\,\mathbf{r}_1') = f_0 N/V \qquad (0 < f_0 < 1). \qquad \text{(VIII.17)}$$

The intensive quantity f_0 measures the off-diagonal long-range order (ODLRO) of the $T = 0$ liquid, a final consequence of Bose-Einstein condensation (BEC) that first appears as a warm liquid sample is cooled across the lambda line [Glyde, 1994, Sections 7.2 and 7.3].

When viewed from the standpoint of the equivalent classical canonical distribution function, the process of separating the positions $\mathbf{r}_1$ and $\mathbf{r}_1'$ from one another to detect ODLRO in the ground-state $\rho_{1,N}$ has a useful interpretation. This can be described as the splitting of a particle into a pair of "semiparticles," which are then separated spatially from one another. Each semiparticle then interacts with the remaining $N-1$ intact particles with the same interactions as before, but reduced by factor 1/2. In terms of notation introduced in Eq. (VIII.13), the semiparticle at $\mathbf{r}_1$ experiences the following $2k_B T_{eff}$-reduced interactions with its surroundings:

$$\sum_{i=2}^{N} w^{(2)}(\mathbf{r}_1, \mathbf{r}_i) + \sum_{i=3}^{N}\sum_{j=2}^{i-1} w^{(3)}(\mathbf{r}_1, \mathbf{r}_i, \mathbf{r}_j) + \sum_{i=4}^{N}\sum_{j=3}^{i-1}\sum_{k=2}^{j-1} w^{(4)}(\mathbf{r}_1, \mathbf{r}_i, \mathbf{r}_j, \mathbf{r}_k) + \dots, \qquad \text{(VIII.18)}$$

with a similar expression for the distant semiparticle at $\mathbf{r}_1'$. The value of f_0 in this equivalent classical description is determined by the reversible work ΔW_{eff} that must be expended isothermally and isochorically to replace one particle (i.e., conjoined semiparticles) in the system with a pair of widely separated semiparticles:

$$f_0 = \exp(-\beta_{eff}\Delta W_{eff}). \qquad \text{(VIII.19)}$$

For an ideal gas of bosons in its ground state, $f_0 = 1$. This value may be approached experimentally in Bose-Einstein-condensed dilute vapors of some metallic elements [Fukuhara et al., 2007; Kraft et al., 2009].

Penrose and Onsager were the first to pursue this approach to the ODLRO [Penrose and Onsager, 1956], and using a hard-sphere approximation for both a ^{4}He atom and each of its two semiparticles, they concluded that for the ground state of liquid ^{4}He at low pressure,

$$f_0 \approx 0.08. \qquad \text{(VIII.20)}$$

A later estimate based on a Jastrow wave function yielded a somewhat higher value for the low-pressure liquid [McMillan, 1965]:

$$f_0 \approx 0.11. \qquad \text{(VIII.21)}$$

Calculations based on shadow wave functions have also been utilized for this quantity; one such approximate wave function denoted by "M+A(S)" has produced the values [MacFarland et al., 1994]

$$f_0 = 0.078 \pm 0.001 \qquad (N/V = 0.0218/\text{Å}^3), \qquad \text{(VIII.22)}$$

$$f_0 = 0.031 \pm 0.003 \qquad (N/V = 0.0262/\text{Å}^3),$$

where the number densities indicated in parentheses refer to vanishing pressure and to the freezing pressure, respectively. Finally, it should be mentioned that the direct wave function numerical construction technique entitled the Green's function Monte Carlo (GFMC) method has also been used to estimate the ground-state ODLRO, yielding the value [Kalos et al., 1981]

$$f_0 = 0.090 \pm 0.003 \qquad (N/V = 0.02185/\text{Å}^3). \qquad \text{(VIII.23)}$$

In the case of the hcp crystal phase of ^{4}He at $T = 0$, the singlet reduced-density matrix $\rho_{1,N}(\mathbf{r}_1 | \mathbf{r}_1')$ would be expected to exhibit ODLRO if that phase exhibited some form of Bose-Einstein condensation, possibly with an accompanying "superflow" characteristic. But as background, the estimated reduction in the liquid-phase f_0 indicated in Eq. (VIII.22) that results from compression, along with the discontinuous density increase upon freezing [Swenson, 1950], would seem to imply that any nonzero f_0 for the hcp crystal would be very small. Indeed, a path integral Monte Carlo (PIMC) calculation for a small defect-free hcp crystal with periodic boundary conditions finds no indication of ODLRO [Clark and Ceperley, 2006].

The separating semiparticles view of how $\rho_{1,N}(\mathbf{r}_1 | \mathbf{r}_1')$ varies away from its diagonal supports the vanishing of f_0 for perfect ^{4}He crystals. As positions $\mathbf{r}_1$ and $\mathbf{r}_1'$ move away from one another, one or both of the semiparticles are forced to leave the immediate vicinity of a crystal site. This leaves behind a semivacant or a fully vacant crystal site to which the semiparticle or semiparticles are normally bound, a situation that does not apply to the amorphous liquid phase. But as indicated above in connection with particle displacement from a ground-state droplet in a large volume, displacing one or both semiparticles from a binding site would cause one or both wave functions which $\rho_{1,N}$ incorporates to decline essentially exponentially with displacement. That in turn would cause $\rho_{1,N}(\mathbf{r}_1 | \mathbf{r}_1')$ to decline exponentially with increasing $|\mathbf{r}_1 - \mathbf{r}_i'|$, thus forcing f_0 to vanish.

In comparison with the situation for the heavier isotope ^{4}He, analysis of the ground-state eigenfunctions for macroscopic samples of ^{3}He is substantially more challenging. This challenge stems from the nonvanishing nuclear spins $s = \pm 1/2$ (in the basic unit $\hbar$) of this lighter isotope and their associated magnetic moments, as well as the Fermi-Dirac statistics that these spins force the atoms to obey. Fortunately, it is an accurate assumption that in configuration space the electronic-ground-state potential energy function Φ is the same for the two isotopes, and so for ^{3}He it should again be adequately approximated by the same pairwise-additive sum shown in Eq. (VIII.3).

The relevant Hamiltonian operator $\mathbf{H}$ for ^{3}He systems includes a spatial operator $\mathbf{H}^{[3]}$, a mass-adjusted version of the form $\mathbf{H}^{[4]}$ shown in Eq. (VIII.2) for ^{4}He, plus an additional operator $\mathbf{J}$ that represents nuclear magnetic dipole–dipole interactions:

$$\mathbf{H} = -(\hbar^2/2m_3)\sum_{j=1}^{N}\nabla_j^2 + \Phi(\mathbf{r}_1....\mathbf{r}_N) + \mathbf{J}(s_1....s_N, \mathbf{r}_1....\mathbf{r}_N) \qquad \text{(VIII.24)}$$

$$\equiv \mathbf{H}^{[3]} + \mathbf{J}.$$

Here $s_1....s_N$ stand for the nuclear spin components along some chosen fixed axis for the particles located, respectively, at $\mathbf{r}_1....\mathbf{r}_N$. The eigenfunctions of $\mathbf{H}$ depend explicitly on both the atomic positional coordinates and the respective nuclear spin components in such a way that the exchange antisymmetry condition is satisfied. In particular, for the ^{3}He fermionic ground-state eigenfunction $\psi_0^{[3]}$ (to be distinguished from the ^{4}He bosonic ground-state eigenfunction $\psi_0^{[4]}$), this requires for all particle pairs j,k that

$$\psi_0^{[3]}(....\mathbf{r}_j, s_j....\mathbf{r}_k, s_k....) = -\psi_0^{[3]}(....\mathbf{r}_k, s_k....\mathbf{r}_j, s_j....). \qquad \text{(VIII.25)}$$

This exchange antisymmetry property also applies to all ^{3}He excited states.

The nuclear magnetic dipole–dipole interaction $\mathbf{J}$ amounts to a very weak perturbation to the configurational operator $\mathbf{H}^{[3]}$. This perturbation plays a key role in the existence of the A and B superfluid phases, as well as in the antiferromagnetism of the bcc crystal. But as Figure VIII.2 vividly illustrates, these phases only appear when T is reduced to the mK range, thus presenting direct evidence for the weakness of $\mathbf{J}$. The primary objective in this chapter (detailed in Section VIII.D) is to examine the inherent-structure mappings from the spin-independent configurational distribution for the N atoms. That distribution and the resulting inherent structures should be relatively insensitive to the effects of $\mathbf{J}$ even as T approaches absolute zero and the ^{3}He system traverses the A, B, and antiferromagnetic bcc phases. For that reason, the eigenfunction problem is considered just for the unperturbed configurational operator $\mathbf{H}^{[3]}$ but of course subject to fermion symmetry constraints.

If a pair of particles possessing antiparallel nuclear spin components are interchanged, the antisymmetry required by Eq. (VIII.25) is supplied by the nuclear spin factor in the wave function, so the spatial part is symmetric under that interchange. Because the operator $\mathbf{H}^{[3]}$ is spin independent, this sign change of the nuclear spin factor generates no energy penalty (with neglect of $\mathbf{J}$). However, if the nuclear spin components of an exchanging pair are parallel, the situation is reversed. In that case, the nuclear spin factor is symmetric under pair interchange, and the spatial part must change sign. This latter situation forces the spatial part of the wave function to include a nodal hypersurface (i.e., a configuration set on which $\psi_0^{[3]} = 0$) between the pre-interchange and post-interchange configurations. The resulting increase in spatial variation of $\psi_0^{[3]}$ produces an increased contribution to kinetic energy. Thus the larger the number of parallel pairs of nuclear spins, the larger is the number of nodal hypersurfaces with a resulting net increase in energy. Consequently, the ^{3}He ground state as described by $\mathbf{H}^{[3]}$ contains the minimum total number of parallel spin pairs. For simplicity, we can assume that the number N of atoms is an even integer, so that minimizing the number of parallel pairs leads to equal numbers of up and down nuclear spins:

$$\sum_{j=1}^{N} s_j = 0. \tag{VIII.26}$$

Even subject to this constraint, the spatial component of $\psi_0^{[3]}$ contains a large number, $(N/2)$ $[(N/2) - 1]$, of exchange-generated nodal hypersurfaces, distinguishing it vividly from the nodeless ^{4}He ground-state wave function. In this case of equal numbers of up and down nuclear spins, the $\mathbf{H}^{[3]}$ eigenvalue corresponding to $\psi_0^{[3]}$ is denoted by $E_0^{[3]}$.

In view of the foregoing discussion, it is useful to exhibit a generic representation for $\psi_0^{[3]}$ that explicitly separates the sign changes due, respectively, to the nuclear spin exchanges and to the spatial configuration exchanges. This can be initiated by considering first the spin subspace in which particles $1 \leq j \leq N/2$ all have down spins ($s_j = -1/2$), while the remainder $N/2 + 1 \leq j \leq N$ have up spins ($s_j = +1/2$). Let $\alpha_-(s)$ and $\alpha_+(s)$, respectively, denote the $s = -1/2$ and $s = +1/2$ single-spin eigenfunctions:

$$\alpha_-(-1/2) = \alpha_+(+1/2) = 1, \tag{VIII.27}$$

$$\alpha_-(+1/2) = \alpha_+(-1/2) = 0. $$

Then this N-spin subspace component of $\psi_0^{[3]}$ being considered first can be represented as follows:

$$F(\mathbf{r}_1 \ldots \mathbf{r}_{N/2} \,|\, \mathbf{r}_{N/2+1} \ldots \mathbf{r}_N) \alpha_-(s_1) \ldots \alpha_-(s_{N/2}) \alpha_+(s_{N/2+1}) \ldots \alpha_+(s_N). \tag{VIII.28}$$

Here the configurational function $F(....|....)$ for the N particles is antisymmetric under pair exchange of any two position vectors preceding the vertical bar, as well as for pair exchange of any two position vectors following the vertical bar:

$$F(\mathbf{r}_1...\mathbf{r}_i...\mathbf{r}_j...\mathbf{r}_{N/2}|\mathbf{r}_{N/2+1}...\mathbf{r}_N) = -F(\mathbf{r}_1...\mathbf{r}_j...\mathbf{r}_i...\mathbf{r}_{N/2}|\mathbf{r}_{N/2+1}...\mathbf{r}_N), \qquad \text{(VIII.29)}$$

$$F(\mathbf{r}_1...\mathbf{r}_{N/2}|\mathbf{r}_{N/2+1}...\mathbf{r}_i...\mathbf{r}_j...\mathbf{r}_N) = -F(\mathbf{r}_1...\mathbf{r}_{N/2}|\mathbf{r}_{N/2+1}...\mathbf{r}_j...\mathbf{r}_i...\mathbf{r}_N).$$

These possibilities account for all antisymmetries arising from position exchanges involving particles with the same nuclear spin assignment. But in addition, one should take this configurational function to be symmetric under exchange of any pair of particle positions across the vertical line:

$$F(\mathbf{r}_1...\mathbf{r}_i...\mathbf{r}_{N/2}|\mathbf{r}_{N/2+1}...\mathbf{r}_j...\mathbf{r}_N) = F(\mathbf{r}_1...\mathbf{r}_j...\mathbf{r}_{N/2}|\mathbf{r}_{N/2+1}...\mathbf{r}_i...\mathbf{r}_N). \qquad \text{(VIII.30)}$$

The full wave function $\psi_0^{[3]}$ requires a sum over N-spin subspaces satisfying Eq. (VIII.26), of terms corresponding to expression (VIII.28). Including this case illustrated in Eq. (VIII.28), the total number of such spin-space components contained in $\psi_0^{[3]}$ is $N!/[(N/2)!]^2$, one for each of the distinct assignments of spin eigenfunctions α_- and α_+ to the particles $1....N$. Let ξ be a running index to label these distinct assignments, each of which has equal numbers of up and down spins, Eq. (VIII.26). For any ξ, one can formally designate the corresponding product of single-particle spin eigenfunctions in the following manner:

$$\alpha_-(s_{\xi(1)})...\alpha_-(s_{\xi(N/2)})\alpha_+(s_{\xi(N/2+1)})...\alpha_+(s_{\xi(N)}). \qquad \text{(VIII.31)}$$

Here by convention the $N/2$ α_-'s continue to precede the $N/2$ α_+'s, as in expression (VIII.28), but the spin variables on which they depend have been switched according to what ξ requires. To be specific, these switches involve spin exchanges between pairs of particles, one particle i in the range $1 \le i \le N/2$, and one particle j in the remaining range $N/2 + 1 \le j \le N$. It needs to be emphasized that if two or more such pair interchanges are involved, there are several ways and several orders in which those interchanges can be applied; however, Eqs. (VIII.29) can be used to show that these produce an identical outcome, so only one needs to be identified with a given index ξ. For each subscript pattern denoted by ξ, the minimum number of such pair interchanges required is denoted by $n(\xi)$. With these notation conventions, $\psi_0^{[3]}$ adopts the following generic form:

$$\psi_0^{[3]}(\mathbf{r}_1, s_1....\mathbf{r}_N, s_N) = [\tilde{C}_0(N)]^{1/2}\sum_{\xi}(-1)^{n(\xi)}F(\mathbf{r}_{\xi(1)}....\mathbf{r}_{\xi(N/2)}|\mathbf{r}_{\xi(N/2+1)}...\mathbf{r}_{\xi(N)})$$

$$\times \alpha_-(s_{\xi(1)})....\alpha_-(s_{\xi(N/2)})\alpha_+(s_{\xi(N/2+1)})....\alpha_+(s_{\xi(N)}). \qquad \text{(VIII.32)}$$

Here $\tilde{C}_0(N)$ is a normalization constant. It can be verified that this last expression satisfies the general antisymmetry condition Eq. (VIII.25).

Since the operator $\mathbf{H}^{[3]}$ is independent of nuclear spin, the function F itself must be an eigenfunction of that spatial Hamiltonian operator with $E_0^{[3]}$ as its eigenvalue:

$$(\mathbf{H}^{[3]} - E_0^{[3]})F(\mathbf{r}_1....\mathbf{r}_{N/2}|\mathbf{r}_{1+N/2}....\mathbf{r}_N) = 0. \qquad \text{(VIII.33)}$$

This can simply be restated in a variational format:

$$E_0^{[3]} = \min_{(F)} \left[\frac{\int d\mathbf{r}_1 \dots \int d\mathbf{r}_N F \mathbf{H}^{[3]} F}{\int d\mathbf{r}_1 \dots \int d\mathbf{r}_N F^2} \right], \tag{VIII.34}$$

where the minimization is to be carried out over all test functions F that obey the configurational exchange symmetry and antisymmetry conditions that have been specified earlier. By rewriting Eq. (VIII.33) in the same manner that previously led to Eq. (VIII.8) for the heavier isotope, one obtains

$$\Phi(\mathbf{r}_1 \dots \mathbf{r}_N) = E_0^{[3]} + \left(\frac{\hbar_2}{2m_3} \right) \left(\frac{\sum_{j=1}^{N} \nabla_j^2 F(\mathbf{r}_1 \dots | \dots \mathbf{r}_N)}{F(\mathbf{r}_1 \dots | \dots \mathbf{r}_N)} \right). \tag{VIII.35}$$

This provides a basic consistency requirement between the form of the spatial function F and the eigenvalue $E_0^{[3]}$ on one side and potential energy function Φ on the other. Specifically, this concerns the behavior of F in the immediate configurational neighborhood of one of its $(N/2)[(N/2) - 1)]$ nodal hypersurfaces. For any N-atom configuration in which all $r_{ij} > 0$, the function F must be continuous and differentiable. Therefore, both the numerator and denominator of the latter right-side factor in Eq. (VIII.33) must vanish exactly on those hypersurfaces to avoid unphysical divergences of Φ. That is, both F and its net curvature measure $\nabla^2 F$ generated by the multidimensional Laplacian operator:

$$\nabla^2 \equiv \sum_{j=1}^{N} \nabla_j^2 \tag{VIII.36}$$

must continuously approach zero and simultaneously change sign as a nodal hypersurface is crossed.

A functional form for F that trivially exhibits the necessary nodal hypersurfaces and which automatically obeys the curvature condition imposed by Eq. (VIII.35) at those hypersurfaces is provided by the ideal gas limit $\Phi \equiv 0$, involving double occupancy of all free-particle states within a Fermi surface. The corresponding function F can be denoted by $F_{id}(\mathbf{r}_1 \dots | \dots \mathbf{r}_N)$. As is the case for the ^{3}He system of interest, this would incorporate $N/2$ particles of each spin component but of course would not exhibit the effects of strong particle repulsion at small pair separation as required by the helium interaction potential. It is then tempting to consider variational wave functions for ^{3}He that utilize a product form for F, specifically F_{id} multiplied by a non-negative Jastrow function or a shadow function of the kinds used for bosonic ^{4}He, Eqs. (VIII.6) and (VIII.14), respectively. Though such approximations can be used to produce a variational upper bound to the ground-state energy $E_0^{[3]}$, the curvature conditions required by Eq. (VIII.35) would not be satisfied. Furthermore, one should expect the positions of the nodal hypersurfaces in the multidimensional configuration space for ^{3}He to be displaced from the F_{id} positions, and indeed these nodal hypersurfaces may possess a topology that is distinct from that of the ideal fermion gas.

It should not escape notice that the wave function generic form presented in Eq. (VIII.32) for the ground state can also represent excited states in the same spin space. Of course, the specific form of the configuration-space function $F(\dots | \dots)$ must change. In particular, at least one additional nodal hypersurface must be included to ensure orthogonality to the ground-state eigenfunction. Nevertheless, the same pair exchange symmetry and antisymmetry conditions that apply to the ground-state function $F(\dots | \dots)$ must also apply to its excited-state replacement.

In spite of the nodal surface reconfiguring produced by particle interactions, the energy eigenvalue spectrum of the low-temperature ^{3}He liquid can be mapped at least approximately onto that of the degenerate ideal Fermi gas. This is the basis of the Landau theory of Fermi liquids [Wilks and Betts, 1987, Chapter 3; Glyde, 1994, Chapter 13]. The Landau formalism replaces individual particles in the degenerate ideal gas with an equal number of "quasiparticles" created by the interactions. This set of quasiparticles also possesses a Fermi surface, the presence of which controls quasiparticle excitations. Although the Landau approach does not make ab initio quantitative predictions about excitation spectra, it does provide a conceptual framework within which various low-temperature experimental measurements on normal liquid ^{3}He can be interpreted to yield those spectra.

Without committing to a specific candidate family of ground-state wave functions, the ^{3}He system can be numerically investigated by the "Green's function Monte Carlo" (GFMC) procedure [Lee et al., 1981; Panoff and Carlson, 1989]. But because this procedure does not produce a closed-form wave function for examination, much of the subtle physics involved remains out of site of the investigators invoking this method. It should also be mentioned that the GFMC method is substantially more computer intensive for fermion systems than for boson systems with the same number of particles because of the wave function sign alternations that are present for the former but not for the latter.

Although the perturbation $\mathbf{J}$ has been disregarded for the later inherent-structure considerations, the following observations should be noted in passing. The sign alternations forced upon the ^{3}He ground-state wave function by the Fermi-Dirac statistics play a fundamental role in the millikelvin-range superfluid A and B states indicated in Figure VIII.2. This superfluidity involves atom pairing analogous to that producing the Bardeen-Cooper-Schrieffer (BCS) superconducting state of the electron gas [Bardeen et al., 1957]. But in contrast to the antiparallel-spin, zero-angular-momentum (s-state) electron pairs in the BCS state, the ^{3}He pairs in the A and B phases have parallel spins and are subject to binding in a rotational p-state. Such a p-state pairing intrinsically incorporates a spatial node. This situation causes the A and B phases to exhibit distinctive order parameters, thus departing from the isotropic liquid category [Wilks and Betts, 1987, Chapter 9; Vollhardt and Wölfle, 1990, Chapter 6]. The benefit of p-state pairing is that it helps to avoid the small-distance repulsive part of the pair potential $v^{(2)}(r)$ while still taking advantage of its larger distance attractive portion and magnetic dipole interactions. In this connection, it should be recalled that in Section I.A it was remarked that the BCS mechanism for superconductivity fundamentally involves a violation of the Born-Oppenheimer separation of electronic and nuclear motions. That is not the case for the pairing mechanisms in the superfluid states of liquid ^{3}He, which owe their properties in part to the presence and effects of the Born-Oppenheimer interaction potential $\Phi(\mathbf{r}_1 \ldots \mathbf{r}_N)$.

The presence of off-diagonal long-range order (ODLRO) in superfluid ^{4}He suggests that an analogous phenomenon should also appear in the superfluid A and B phases for ^{3}He. However, these latter cases do not exhibit an identifying attribute in the singlet density matrix. Instead, ODLRO appears as an off-diagonal property of the pair density matrix because of the fact that superfluidity is tied fundamentally to collective pairing processes [Yang, 1962].

C. Phonon Excitations

Phonon excitations (i.e., quantized sound waves) represent an important class of low-lying excitations for both helium isotopes in their liquid and crystal phases. The class of phonons to be considered first is that occurring in superfluid ^{4}He (helium II), specifically longitudinal phonons, with a description following the Feynman approach [Feynman, 1954, 1972, Chapter 11]. This begins with the nodeless ground-state wave function $\psi_0^{[4]}(\mathbf{r}_1 \ldots \mathbf{r}_N)$ for number density $\rho = N/V$ within the homogeneous liquid range and recognizes that this ground state automatically incorporates embedded ground states for independent harmonic phonon modes. Included among the N-particle configurations whose relative probabilities are given by $(\psi_0^{[4]})^2$ are those containing wavelike density fluctuations. These fluctuations are measured by Fourier components of the overall density distribution, which amount to collective density variables $\rho(\mathbf{k})$ for chosen wave vectors $\mathbf{k}$ that are consistent with the system's periodic boundary conditions [Section I.D(6)]:

$$\rho(\mathbf{k}) = \sum_{j=1}^{N} \exp(i\mathbf{k}\cdot\mathbf{r}_j). \tag{VIII.37}$$

The mean-square value of this quantity's magnitude in the ground state, on a per-particle basis, is given by

$$\langle \psi_0^{[4]} | \rho^*(\mathbf{k})\rho(\mathbf{k}) | \psi_0^{[4]} \rangle / N \langle \psi_0^{[4]} | \psi_0^{[4]} \rangle$$

$$= 1 + \sum_{j=1}^{N} \sum_{l(\neq j)=1}^{N} \langle \psi_0^{[4]} | \exp[i\mathbf{k}\cdot(\mathbf{r}_j - \mathbf{r}_l)] | \psi_0^{[4]} \rangle / N \langle \psi_0^{[4]} | \psi_0^{[4]} \rangle$$

$$= 1 + (N-1)\langle \psi_0^{[4]} | \exp(i\mathbf{k}\cdot\mathbf{r}_{12}) | \psi_0^{[4]} \rangle / \langle \psi_0^{[4]} | \psi_0^{[4]} \rangle$$

$$= 1 + \rho \int [g_2(r_{12}) - 1] \exp(i\mathbf{k}\cdot\mathbf{r}_{12}) d\mathbf{r}_{12}$$

$$\equiv S(k). \tag{VIII.38}$$

The last two lines refer to the infinite system limit, with ρ the number density in that limit. Here $g_2(r_{12})$ is the pair correlation function from the diagonal part of the pair density matrix:

$$\rho_{2,N}(\mathbf{r}_1, \mathbf{r}_2 | \mathbf{r}_1, \mathbf{r}_2) = \rho^2 g_2(r_{12}), \tag{VIII.39}$$

and $S(k)$ is the structure factor [Eq. (V.22)] for liquid helium II in its ground state.

The first excited state of a simple harmonic oscillator amounts to multiplying its ground state by the first power of the oscillator variable involved. In this respect, the phonon excitation for wave vector $\mathbf{k}$ should be obtained in good approximation upon multiplying $\psi_0^{[4]}$ (which includes the phonon ground state) by $\rho(\mathbf{k})$, i.e., the unnormalized wave function for liquid helium II containing a single phonon excitation would be

$$\psi_{\mathbf{k}}(\mathbf{r}_1 \ldots \mathbf{r}_N) = \left[\sum_{j=1}^{N} \exp(i\mathbf{k}\cdot\mathbf{r}_j) \right] \psi_0^{[4]}(\mathbf{r}_1 \ldots \mathbf{r}_N). \tag{VIII.40}$$

It is easy to show that two wave functions of the form Eq. (VIII.40) with distinct nonvanishing wave vectors are orthogonal to one another. The energy $E_{\mathbf{k}}$ of $\psi_{\mathbf{k}}$ can be directly evaluated as follows:

$$E_{\mathbf{k}} = \langle \psi_{\mathbf{k}} | \mathbf{H}^{[4]} | \psi_{\mathbf{k}} \rangle / \langle \psi_{\mathbf{k}} | \psi_{\mathbf{k}} \rangle$$

$$= \langle \psi_0^{[4]} | \rho^*(\mathbf{k})\rho(\mathbf{k})\mathbf{H}^{[4]} | \psi_0^{[4]} \rangle / \langle \psi_0^{[4]} | \rho^*(\mathbf{k})\rho(\mathbf{k}) | \psi_0^{[4]} \rangle$$

$$+ (\hbar^2/2m_4)\left\langle \psi_0^{[4]} | \sum_{j=1}^{N} [\exp(-i\mathbf{k}\cdot\mathbf{r}_j)\mathbf{k}]\cdot[\exp(i\mathbf{k}\cdot\mathbf{r}_j)\mathbf{k}] | \psi_0^{[4]} \right\rangle / \langle \psi_{\mathbf{k}} | \psi_{\mathbf{k}} \rangle. \tag{VIII.41}$$

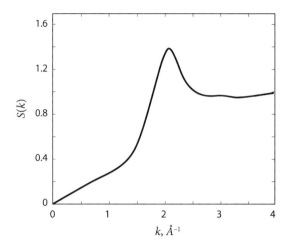

FIGURE VIII.8. Structure factor $S(k)$ for liquid helium II at $T = 1.1$ K and the corresponding vapor pressure. The curve shown was adapted from Figure 12 in Cowley and Woods, 1971.

The first term on the right of the last form is simply the ground-state energy E_0[4]. The remaining term has undergone integration by parts and represents the phonon excitation energy $\varepsilon_{\mathbf{k}} \equiv \varepsilon(k)$. Therefore,

$$\varepsilon(k) = \frac{\hbar^2 k^2}{2 m_4 S(k)}, \tag{VIII.42}$$

where Eq. (VIII.38) has been used to introduce the structure factor in the denominator. The phonon group velocity (longitudinal sound speed) $c(k)$ is determined by the following derivative:

$$c(k) = \frac{1}{\hbar} \frac{\partial \varepsilon(k)}{\partial k}. \tag{VIII.43}$$

The structure factor $S(k)$ is experimentally accessible by both neutron scattering [Cowley and Woods, 1971] and X-ray diffraction [Wirth and Hallock, 1987]. Figure VIII.8 presents this function for liquid helium II at 1.1 K and the corresponding vapor pressure, inferred from both types of experiments. Because the density of excitations in the liquid should be low in this low-temperature state, the $S(k)$ result shown should be close to that for the ground state at the same low pressure. The isothermal compressibility relation presented earlier for classical liquids [Eqs. (V.21) and (V.23)]:

$$\rho k_B T \kappa_T = 1 + \rho \int [g_2(r) - 1] d\mathbf{r} \tag{VIII.44}$$

$$\equiv S(0)$$

is equally valid in the quantum liquid regime with the appropriate g_2 from Eq. (VIII.39), and because κ_T remains finite as $T \to 0$, one must have $S(0)$ vanishing in the ground state. In particular, the linear behavior indicated in Figure VIII.8 for $S(k)$ near $k = 0$ should apply to the ground state, with

$$S(k) = \frac{\hbar k}{2m_4 c_0} + O(k^2), \qquad\qquad\qquad \text{(VIII.45)}$$

where c_0 stands for the long-wavelength limit of the sound speed $c(k)$. The phonon excitation energies in the long-wavelength regime therefore can be expressed

$$\varepsilon(k) = \hbar c_0 k + O(k^2). \qquad\qquad\qquad \text{(VIII.46)}$$

It is worth noting that the small-k linear behavior noted in Eq. (VIII.45) for the structure factor implies a large-r property for the pair correlation function $g_2(r)$. If $g_2(r) - 1$ were exponentially damped as r increased, the Fourier integral appearing in the next to last line of Eq. (VIII.38) would generate a power series in k^2 for $S(k)$. However, the linear term appearing in Eq. (VIII.45) discounts that possibility and instead requires for large r that $g_2(r) - 1$ exhibit an algebraic decay, the dominating behavior of which is

$$g_2(r) - 1 \sim -\left(\frac{\hbar}{2\pi^2 m_4 \rho c_0}\right)\left(\frac{1}{r^4}\right). \qquad\qquad \text{(VIII.47)}$$

The form of this result has been long known as a basic property of the liquid helium ground state [Reatto and Chester, 1967]. It can be extracted from the Fourier integral expression for $S(k)$ in Eq. (VIII.38). This negative algebraic tail stems from the phonon zero-point fluctuations incorporated in the boson ground-state wave function $\psi_0^{[4]}(\mathbf{r}_1 \ldots \mathbf{r}_N)$.

The maximum exhibited by $S(k)$ for $k \approx 2.0$ Å^{-1} in Figure VIII.8 arises largely from the first-neighbor peak in $g_2(r)$ at $r \approx 3.1$ Å. This maximum is large enough and sufficiently narrow to produce a relative minimum in the phonon energy expression, Eq. (VIII.42). This is illustrated in Figure VIII.9. Although the plot refers to phonons in the helium II ground state at low pressure, it is qualitatively relevant for the entire pressure interval up to freezing.

Neutron scattering experiments that measure the $\varepsilon(k)$ dispersion curve [Cowley and Woods, 1971] provide a key comparison for the theoretical result (solid curve) indicated in Figure VIII.9. Although such measurements (dashed curve in Figure VIII.9) confirm the linear rise for small k, followed by a subsequent downturn producing a minimum at $k \approx 2.0$ Å^{-1}, the height and depth of the experimentally determined maximum and minimum, respectively, are about a factor of 2 lower in energy than those emerging from Eq. (VIII.42). This quantitative discrepancy should not be surprising. Although phonon wave functions that have the form shown in Eq. (VIII.40) are physically plausible for large wavelengths (small k), they begin to lose credibility when the wavelength approaches the typical nearest neighbor distance in the liquid medium. But that wavelength regime is where the measured $\varepsilon(k)$ develops and passes through its local maximum and subsequent minimum, the "roton" region. This quantitative failure can be connected to the imprecise assignment of wave function nodal hypersurface positions and shapes, discussed in Box VIII.2.

Feynman and Cohen (1956) introduced a qualitatively distinct wave function family for excitations lying in the roton region. These excitations replace the individual $\exp(i\mathbf{k}\cdot\mathbf{r}_j)$ terms summed in Eq. (VIII.40) with terms for each particle j having a "smoke ring" character. The intuitive justification was that short-distance repulsive interatomic interactions required a "backflow" phenomenon. Each of these modified terms exhibits localized irrotational circulation around a closed vortex loop whose diameter is roughly comparable to the mean nearest neighbor distance

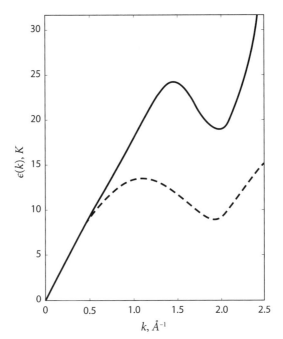

FIGURE VIII.9. Plot of the excitation energy $\varepsilon(k)$ for helium II according to Eq. (VIII.42), shown as a solid curve. The experimentally determined $\varepsilon(k)$ has been indicated by the dashed curve. Adapted from Wilks and Betts, 1987, Figure 11.4.

in the liquid helium II medium. The flow through and around the vortex loop obeys the Onsager quantization condition for vortex lines [Onsager, 1949; Feynman, 1972, p. 336]. Each vortex loop individually presents a locus of wave function singularity, a property that has no analog in the simple phonon wave functions in Eq. (VIII.40). As k increases from zero and approaches the region of the dispersion curve maximum, the predominant helium II excitations presumably change continuously from simple phonon to roton character, with appropriately repositioned nodal hypersurfaces. One should notice that the group velocity expression in Eq. (VIII.43) states that for the k interval between the maximum and the subsequent roton minimum in the dispersion curve, this velocity is opposite in direction to wave vector $\mathbf{k}$ (i.e., negative sound velocity). Numerical study of the roton wave function approximation significantly improves the extent of agreement with the experimentally determined dispersion curve in the vicinity of the minimum, but it is limited in precision in part by the necessity of approximating three-atom and four-atom correlation functions in the ground-state liquid [Feynman and Cohen, 1956].

The phonon wave function in Eq. (VIII.40) can be extended to the case of two simultaneous excitations with distinct small wave vectors, say $\mathbf{k}$ and $\mathbf{k}'$. Provided that these are both long-wavelength excitations, they are sufficiently weakly interacting with each other to be substantially independent. Thus the corresponding wave function involves multiplying the ground-state wave function by a pair of configurational factors [Feynman, 1954, Eq. (24)]:

$$\psi_{\mathbf{k},\mathbf{k}'}(\mathbf{r}_1\ldots\mathbf{r}_N) = \left[\sum_{j=1}^{N}\exp(i\mathbf{k}\cdot\mathbf{r}_j)\right]\left[\sum_{l=1}^{N}\exp(i\mathbf{k}'\cdot\mathbf{r}_l)\right]\psi_0^{[4]}(\mathbf{r}_1\ldots\mathbf{r}_N). \qquad \text{(VIII.48)}$$

BOX VIII.2. Wave Function Nodal Hypersurface Geometry

The wave function form (VIII.40) nominally describing a single phonon excitation actually presents an orthogonal pair of real wave functions:

$$\psi_{\mathbf{k}}(\mathbf{r}_1\ldots\mathbf{r}_N) = \psi_{\mathbf{k},c}(\mathbf{r}_1\ldots\mathbf{r}_N) + i\psi_{\mathbf{k},s}(\mathbf{r}_1\ldots\mathbf{r}_N),$$

where

$$\psi_{\mathbf{k},c}(\mathbf{r}_1\ldots\mathbf{r}_N) = \left[\sum_{j=1}^{N}\cos(\mathbf{k}\cdot\mathbf{r}_j)\right]\psi_0^{[4]}(\mathbf{r}_1\ldots\mathbf{r}_N),$$

$$\psi_{\mathbf{k},s}(\mathbf{r}_1\ldots\mathbf{r}_N) = \left[\sum_{j=1}^{N}\sin(\mathbf{k}\cdot\mathbf{r}_j)\right]\psi_0^{[4]}(\mathbf{r}_1\ldots\mathbf{r}_N).$$

These two functions differ only by the phase of their trigonometric factors [...] as the N-atom system undergoes uniform translation. Those factors determine the nodal hypersurfaces across which the wave functions change sign in the $3N$-dimensional configuration space.

When $\mathbf{k}$ is sufficiently small that its wavelength $k/2\pi$ equals many interatomic spacings in the liquid helium, small errors in the placing of a nodal surface incur negligible energy error in the corresponding energy expression Eq. (VIII.41). This is true because the wave function remains close to zero over a substantial distance from the nodal hypersurface, and any error in placement of that nodal hypersurface has very little weight in determining the energy. By implication, the simple phonon energy expression Eq. (VIII.42) should be asymptotically correct in the small-k regime.

This fortunate simplification vanishes as the $\mathbf{k}$ increases to where its wavelength becomes comparable to the average nearest neighbor spacing in the liquid. This is the interval encompassing the experimentally measured $\varepsilon(k)$ maximum and the subsequent roton minimum. Now approximate wave functions of the type (VIII.40) vary spatially more rapidly, so nodal positioning errors receive substantially greater weights in the energy expression (VIII.42). This situation therefore requires more elaborate wave function forms to predict $\varepsilon(k)$ accurately through the roton regime.

The energy is the sum of individual phonon excitation energies:

$$\varepsilon(k, k') = \varepsilon(k) + \varepsilon(k'). \tag{VIII.49}$$

This wave function multiplication procedure of course extends to the case of larger numbers of simultaneous phonon single excitations in helium II, with an additive total excitation energy. However, it must be kept in mind that as the concentration of excitations rises, so too do their interactions, an effect that eventually undermines the accuracy of the simple wave function multiplication representation and additive excitation energy.

Only single excitations of each harmonic phonon mode in the liquid have been considered thus far. However, any true harmonic oscillator individually possesses an infinite sequence of excited states that are equally spaced in energy. With respect to small-k phonon states, there exist further excited states beyond those represented by the product-form wave functions in Eqs.

(VIII.40) and (VIII.48). These would require factors that involve the identifying $\mathbf{k}$ and a symmetrical combination of the N atom positions in a functional form that corresponds directly to harmonic oscillator higher excited-state wave functions [e.g., Pauling and Wilson, 1935, Chapter III].

The long-wavelength phonons that are the dominant excitations in helium II at very low temperature are directly analogous to the acoustic phonons present in insulating solids at very low temperature. In both cases, the excitation energies are linear in $|\mathbf{k}|$ and are proportional to the acoustic sound speed. As a result, the low-temperature isochoric heat capacity expression derived in Eq. (IV.32) can be adapted to helium II, bearing in mind that only longitudinal phonons are present, not transverse modes as in solids:

$$\frac{C_V}{Nk_B} = \left(\frac{2\pi^2}{15\rho(\hbar c_0)^3}\right)(k_B T)^3 + O(T^4). \qquad \text{(VIII.50)}$$

As T rises to its value $T_\lambda(\rho)$ at the lambda line (now regarded for convenience as a function of density rather than pressure), accounting quantitatively for all thermodynamically relevant excitation modes becomes difficult. This difficulty stems partly from the nonlinearity of the dispersion relation for increasing k, including the roton minimum region, but also from the significant extent of interaction between the excitations that are present at increasing concentration. Nevertheless, it is possible to establish a connection between the observed isochoric heat capacity singularity at the lambda line and the distribution of energy eigenvalues. An appropriate representation for the density of eigenvalues in the macroscopic system-size regime would have a form exponential in N. Specifically, one can introduce the expression

$$\exp[N\zeta(\eta,\rho)]d\eta \qquad \text{(VIII.51)}$$

to stand for the number of eigenvalues lying in the interval $N(\eta \pm d\eta/2)$ for the Hamiltonian $\mathbf{H}^{[4]}$ of the N atoms. Here the function $\zeta(\eta,\rho)$ is a non-negative function of the intensive energy and density properties η and ρ and is defined by the asymptotic large-system limit. The minimum value of η is $\eta_0(\rho) = E_0^{[4]}/N$, the ground-state energy per atom at density ρ, but it has no upper bound. In this representation, the canonical partition function takes the form

$$\exp(-\beta F_N) = \int_{\eta_0}^{\infty} \exp\{N[\zeta(\eta,\rho) - \beta\eta]\}d\eta. \qquad \text{(VIII.52)}$$

The large-N asymptotic limit of this last expression is of interest. In that case, the value of its integral is dominated by the contribution of the integrand in the immediate neighborhood of its very large maximum at $\eta = \eta^*(\beta,\rho)$. That maximum is determined by the following condition:

$$\left(\frac{\partial \zeta}{\partial \eta}\right)_{\eta^*,\rho} = \beta. \qquad \text{(VIII.53)}$$

In leading order in N, the Helmholtz free energy per atom then becomes

$$\beta F_N(\beta,\rho)/N = -\zeta[\eta^*(\beta,\rho),\rho] + \beta\eta^*(\beta,\rho). \qquad \text{(VIII.54)}$$

This large-N format reduction is directly analogous to that described in Section III.D leading up to Eq. (III.29) for the classical Helmholtz free energy.

Successive β derivatives applied to the free energy yield the constant volume heat capacity:

$$\frac{C_V}{Nk_B} = -\beta^2\left(\frac{\partial^2 \beta F_N/N}{\partial \beta^2}\right)_\rho$$

$$= -\beta^2\left(\frac{\partial \eta^*}{\partial \beta}\right)_\rho \tag{VIII.55}$$

$$\equiv \frac{1}{k_B}\left(\frac{\partial \eta^*}{\partial T}\right)_\rho.$$

By comparing this last result with Eq. (VIII.50), one finds that the leading low-temperature behavior of η^* is the following:

$$\eta^*(\beta, \rho) = \eta_0(\rho) + \left(\frac{\pi^2}{30\rho(\hbar c_0)^3}\right)(k_B T)^4 + O(T^5). \tag{VIII.56}$$

But as T rises through the lambda line at $T_\lambda(\rho)$, the constant-volume heat capacity exhibits a strong singular maximum, the dominating behavior of which reportedly can be described by small fractional powers of the temperature displacement from $T_\lambda(\rho)$ [Ahlers, 1973, 1980]. This can be expressed in the following way for helium I and helium II liquids, respectively:

$$\frac{C_V}{Nk_B} \approx A_I - B_I[T - T_\lambda(\rho)]^{q_I} \qquad (T > T_\lambda), \tag{VIII.57}$$

$$\approx A_{II} - B_{II}[T_\lambda(\rho) - T]^{q_{II}} \qquad (T < T_\lambda).$$

Here all six quantities $A_I, \ldots, q_{II}$ are positive. In order for these forms to arise from η^*, it is necessary for that function to possess the following singular behavior in the vicinity of $T_\lambda(\rho)$:

$$\eta^*(\beta, \rho) - \eta^*(\beta_\lambda, \rho) \approx k_B\left[A_I(T - T_\lambda) - \left(\frac{B_I}{1 + q_I}\right)(T - T_\lambda)^{1+q_I}\right] \qquad (T > T_\lambda),$$

$$\approx k_B\left[A_{II}(T - T_\lambda) + \left(\frac{B_{II}}{1 + q_{II}}\right)(T_\lambda - T)^{1+q_{II}}\right] \qquad (T < T_\lambda). \tag{VIII.58}$$

Such singular behavior must arise from cooperative interactions between the excitations that are present in this temperature range.

Above the $T = 0$ freezing pressure for liquid helium II, the crystalline ground state is hexagonal close packed (hcp), as indicated in Figure VIII.3. The unit cell for this structure has a basis of two atoms. Consequently, the phonon spectrum for the crystal has six branches, three "acoustic" and three "optic" branches, where the former includes one longitudinal and two transverse phonon branches. At very low positive temperature, only these low-frequency acoustic modes would be excited in the crystal, and they would cause the heat capacity C_V to exhibit a T^3 behavior. Unlike the crystal, however, the helium II superfluid cannot support transverse acoustic modes, only longitudinal phonon modes. This reduced set of excitations also produces a T^3 contribution to C_V [Eq. (VIII.50)], but with a correspondingly reduced positive coefficient. The result is that the entropies of the coexisting phases (both of which vanish at $T = 0$) increase at different rates as T rises; that of the hcp crystal increases faster than that of the liquid. As mentioned earlier, the Clausius-Clapeyron equation [Eq. (VIII.1)] relates the temperature derivative of the equilibrium melting pressure p_m to the ratio of differences in molar entropies and in molar volumes of the crystal (c) and liquid (l) phases at coexistence. Because the molar volume of the liquid is substantially larger than that of the hcp solid, the sign of the melting curve slope dp_m/dT is controlled by the entropy difference, which is negative because of the discrepancy in phonon excitation densities. This is the primary physical source of the helium II inverse melting behavior in the range $0 < T < 0.8$ K.

Even for small wave vectors $\mathbf{k}$, constructing accurate wave functions for solid ^{4}He with a single phonon excitation is considerably more challenging than what has been described above for liquid helium II. Qualitatively, this task requires accounting for the quantized translational fluctuations in the periodic long-ranged crystal order. The amplitudes and directions of these excitation-induced fluctuations must conform to the symmetry and basis set of the crystal, specifically the hcp case, which possesses direction-dependent elastic constants. Self-consistent approximations for this problem have been proposed and examined in the published literature [Glyde, 1994, Chapter 5]. Presenting the complicated technical details of that approach is beyond the scope of this brief chapter.

Proper description of longitudinal-phononlike excitations in liquid ^{3}He requires a more elaborate analysis than that just presented for helium II. Specifically, it is not correct simply to invoke the analog of Eq. (VIII.40), which involved multiplying the ground-state wave function by the collective density fluctuation function $\rho(\mathbf{k}) = \sum \exp(i\mathbf{k}\cdot\mathbf{r}_j)$. This would generally produce configurational functions $F(....|....)$ violating the boundedness condition at their nodal hypersurfaces that would emerge from the excited-state version of Eq. (VIII.35). The alternative is to utilize the Landau formalism for Fermi liquids. It is important to realize that the low-T heat capacity of the normal ^{3}He is proportional to T [Greywall, 1983], a characteristic of a degenerate Fermi fluid. This stands in clear distinction to a T^3 heat capacity contribution expected from a phonon excitation spectrum [cf. Eq. (VIII.50)], thus presenting yet another qualitative difference between the two isotopes. Consequently, any phononlike excitations in normal ^{3}He liquid at low T must represent a small fraction of the available excitations.

In the Landau representation, ordinary longitudinal sound propagation at low temperature arises from the hydrodynamics of quasiparticle excitations present in the medium. It requires that the temperature be sufficiently far above absolute zero to have generated a collection of quasiparticle excitations and that the sound frequency ω be low enough that collisions between the quasiparticle excitations in the liquid occur sufficiently rapidly to be able to attain equilibrium during the sound wave oscillation period. That is, one requires $\omega\tau \ll 1$, where τ is the mean time between collisions experienced by a single quasiparticle. However, if the sound frequency is sufficiently high, collisions would occur infrequently over the sound wave period. This latter circumstance corresponds to the opposite circumstance $\omega\tau \gg 1$. Experimentally, this alternative involves a measurable increase in sound velocity compared to ordinary sound and has been called "zero sound." The transition from the former ordinary sound to zero sound can be produced in the laboratory by lowering the temperature while holding the exciting frequency ω constant because of the marked increase in mean free path and thus time τ between collisions for the quasiparticle excitations. Measurements at low pressure indicate that the speed of zero sound is several percent higher than that of ordinary sound [Abel et al., 1966].

Although the inverse melting phenomenon in ^{3}He is confined to a lower temperature range ($T < 0.32$ K) than that in ^{4}He ($T < 0.8$ K), this unusual phenomenon is more striking in the lighter isotope case as measured by the amount that the melting pressure declines from its value at $T = 0$ to its minimum. The mechanism involved differs from the ^{4}He case, which stemmed from the relative densities of phonon excitations in the liquid and crystal phases. For ^{3}He, the nuclear spins play a central role. In the bcc crystal above ≈ 10 mK, the nuclear spins are substantially uncorrelated, thus contributing $R \ln 2$ to the molar entropy. However, the coexisting liquid phase described by the Landau model as a Fermi fluid whose quasiparticles are constrained by a Fermi surface does not possess the same freedom of nuclear spin reorientation and thus has a lower molar entropy than the crystal below 0.32 K [Atkins, 1959, Section 8.7; Wilks and Betts, 1987, Section 9.1]. The strong inverse melting behavior in ^{3}He provides the basis for operation of the "Pomeranchuk refrigerator" [Betts, 1974].

Below their respective critical temperatures, the helium isotopes can display liquid–vapor interfaces. A basic class of collective motions at these interfaces involves quantized versions of the capillary waves discussed in Section V.H. Although little attention thus far has been devoted to the ^{3}He interface, the quantized waves on liquid ^{4}He have been called "ripplons" [Cole, 1970, 1980; Gould and Wong, 1978]. These are the interfacial analogs of bulk-liquid phonons. Their quantitative influence on the intrinsic interfacial profile remains an open question.

D. Inherent Structures

In order to examine the nature of isochoric inherent structures and their basins for the condensed phases of both helium isotopes, it should be sufficient to represent the multidimensional potential energy Φ as a pairwise-additive form, Eq. (VIII.3). The steepest descent mapping operation on the Φ hypersurface for many-atom configurations is the same as that used for classical many-particle systems. However, the distribution of initial configurations to which the mapping is applied is a quantum mechanical configuration probability function $P_N(\{\mathbf{r}_j\}, t)$ that was introduced in Eq. (II.10). Generally time dependent, this probability function for ^{4}He is determined by the diagonal elements of that system's density matrix in the configurational representation. In the ^{3}He case, this probability also involves the diagonal elements of the density matrix with a configurational representation but incorporates a sum over spin variables. For present purposes, attention is restricted to the equilibrium canonical ensemble for both isotopes, so that P_N depends on the temperature and particle number density but is independent of time. With this restriction, for bosonic ^{4}He we can write

$$P_N^{[4]}(\{\mathbf{r}_j\}, \beta) = [Q_N^{[4]}(\beta)]^{-1} \sum_n \exp(-\beta E_n^{[4]}) \, |\psi_n^{[4]}(\{\mathbf{r}_j\})|^2, \qquad \text{(VIII.59)}$$

where the n summation covers all eigenstates, $Q_N^{[4]}$ is the canonical partition function, and $\beta = 1/k_B T$ as usual. The corresponding expression for fermionic ^{3}He requires summation over its nuclear spins s_j, and thus is the following:

$$P_N^{[3]}(\{\mathbf{r}_j\}, \beta) = [Q_N^{[3]}(\beta)]^{-1} \sum_{\{s_j\}} \sum_n \exp[-\beta E_n^{[3]}(\{s_j\})] \, |\psi_n^{[3]}(\{\mathbf{r}_j, s_j\})|^2. \qquad \text{(VIII.60)}$$

It is important to recognize that even as $T \rightarrow 0$ where Eqs. (VIII.59) and (VIII.60) only include the respective ground states, the distributions of initial atom arrangements specified by $P_N^{[4]}$ and $P_N^{[3]}$ in principle span essentially the entire $3N$-dimensional configuration space. The zero-measure exceptions include $r_{ij} = 0$ particle coincidences where the interaction diverges to $+\infty$, and in the ^{3}He case possibly at some special configurations arising from wave function nodal surfaces [as illustrated by the $F(\dots|\dots)$ terms shown in the ground-state eigenfunction, Eq. (VIII.32)]. In any case, it is important to realize that for both isotopes all inherent structures are accessible in principle from the ground states, although many may have extremely small relative occurrence probabilities. Because of quantum mechanical tunneling, even those inherent structures lying higher in potential energy than the ground-state energies have nonzero probabilities of occurrence.

As background information, it is useful to have available the density dependences of Φ evaluated at the perfect bcc and hcp structures. The Aziz et al. (1979) pair potential discussed in Box VIII.1 has been used for this purpose, with results presented graphically in Figure VIII.10. The arrows locate the experimentally measured densities of the $T = 0$ crystals for the two isotopes at their depressurization melting points. For both isotopes, the crystals exhibit far lower number densities than those at the respective Φ/N minima, and the magnitudes of Φ/N at those observed densities are less than half the values at the minima. In the case of ^{4}He, the potential energy for the stretched but structurally perfect hcp crystal has been reduced to about 41% of its maximum binding strength. For the lighter isotope ^{3}He, the effect for its bcc crystal structure is even more substantial; the corresponding result is approximately 32%. Quantum mechanical kinetic energy produces profound dilation and binding reduction on crystals of both isotopes.

Because of the fact that both the perfect bcc and hcp structures possess reflection symmetry, they must be configurations at which the multidimensional gradient of the potential energy vanishes, $\nabla \Phi = 0$, a statement of the obvious fact that no particles experience a net force in those configurations. Using the Aziz et al. (1979) potential, the harmonic normal mode spectrum for

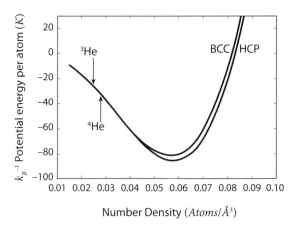

FIGURE VIII.10. Crystal structure energies per atom vs. atom number density for the perfect bcc and hcp structures. These are the crystal forms exhibited respectively by ^{3}He and ^{4}He in the vicinity of their $T = 0$ melting pressures. The potential energies have been computed using the additive atomic pair potential derived by Aziz et al. (1979). The densities of the crystals at their ground-state melting points have been indicated by arrows. For unit conversion, note that 1 atom/Å^3 is equivalent to 1.66057 mol/cm^3.

the hcp crystal structure has been calculated for ^{4}He at its quantum mechanically dilated $T = 0$ melting point density, located by an arrow in Figure VIII.10 [Hodgdon and Stillinger, 1995]. Every one of the $\mathbf{k} \neq 0$ phonons in the first Brillouin zone was found to have $\omega^2(\mathbf{k}) < 0$, i.e., all harmonic normal modes had imaginary frequencies indicating instability. In other words (aside from the three overall translational degrees of freedom), the perfect hcp configuration at the measured crystal density is a local maximum of Φ. An earlier approximate calculation based on the Lennard-Jones 12,6 pair potential reached the same conclusion [De Wette and Nijboer, 1965]. Because the dilation effect is even stronger for the lighter isotope ^{3}He, the same local Φ maximum characteristic should also be expected for the corresponding bcc structure.

The imaginary-frequency phonons exhibited by the hcp structure at low density can be restored to real frequencies by compression. This does not happen simultaneously for all phonons, but sequentially. As a result, the relative maximum exhibited by Φ at the $T = 0$ melting point converts stepwise through saddle points of decreasing order (in other words, decreasing the number of imaginary frequencies) until it finally becomes a local minimum (inherent structure). This process involves reducing the hcp nearest neighbor distance from 3.65 Å to 3.30 Å, corresponding experimentally to compressing the crystalline ^{4}He phase to about 1,300 bar [Hodgdon and Stillinger, 1995]. This compressed state is still considerably less dense than that needed to produce the greatest binding potential energy for the perfect hcp structure, specifically involving a nearest neighbor distance of about 2.90 Å.

Although a precise closed-form expression is not available for $\psi_0^{[4]}$ describing the low-pressure hcp ground-state crystal of ^{4}He, its localization around the ideal crystallographic configurations [numbering $(N - 1)!$ when periodic boundary conditions apply] suggests a convenient way to sample qualitatively the most probable types of inherent structures that underlie this phase. Specifically, the N ^{4}He atoms can be given sets of small random displacements off of the local Φ maximum to serve as a set of initial conditions for steepest descent mapping. Despite the fact that the maximum is characterized harmonically by only $3N - 3$ normal-mode downward curvatures, the collection of random initial displacements can initiate mapping to a far larger number of distinct inherent structures, namely, a number rising exponentially with N in accord with the total expected number of inherent structures, as described in earlier chapters. In other words, each hcp maximum is a common boundary point shared by an exponentially large number of Φ basins. It is currently unknown whether all inherent-structure types for this expanded hcp structure can be accessed this way, although it is conceivable that some rare inherent structures lie higher in Φ than the hcp maxima and would thus be inaccessible by this simple sampling procedure. So far as structurally similar but permutationally distinct inherent structures are concerned, these would be accessed primarily by correspondingly permuting the labeled atoms within the basic hcp periodic structure before implementing the steepest descent mapping.

Using a modest $8 \times 8 \times 8$ unit cell with periodic boundary conditions, this hcp plus random displacement procedure has been used to generate inherent structures for the ^{4}He crystalline ground state at its melting pressure [Hodgdon and Stillinger, 1995]. The displacement magnitudes used were about one-twentieth of the nominal nearest neighbor spacing 3.65 Å, as well as about one-third of that spacing; no significant differences in statistical outcomes for these two alternatives were detected. Figure VIII.11 graphically presents the atomic configuration at one of the randomly small-displaced hcp initial conditions (a), as well as the inherent structure that results from its steepest descent mapping (b). The figure also shows an atypical inherent structure that

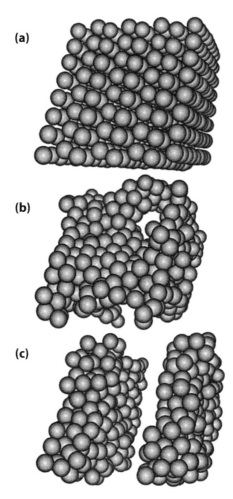

FIGURE VIII.11. Specific examples of (a) premapping hcp configuration with small random atom displacements, (b) typical postmapping inherent structure, and (c) an atypical inherent structure resulting from an initial displacement set biased toward opening a planar crack. The fixed density corresponds to the experimental ground-state crystal at its melting pressure, at which the nominal nearest neighbor spacing is 3.65 Å. Reproduced from Hodgdon and Stillinger, 1995, Figure 3.

emerged from a biased (and therefore atypical) set of initial displacements designed to predispose toward a planar crack in the hcp configuration (c).

The spatial patterns presented by the inherent structures from the $T = 0$ melting point crystal show a strong resemblance to those discussed in Chapter VII for low-density classical many-particle systems. Specifically, they show the atoms aggregated into amorphous packings that are penetrated by empty pores or voids, thus presenting a kind of "aerogel" texture. Qualitatively, this is the type of inherent structure expected upon mapping to minima a classical canonical ensemble for a fluid below its Sastry point density (Figure VII.8). In the present ^{4}He application, it is noteworthy that the initial crystalline long-range order essentially completely disappears as the mapping moves the atoms away from their initial near-hcp positions to their final resting places. This is true both for the typical inherent structures exemplified by Figure VIII.11(b), as well as the special planar crack case illustrated in Figure VIII.11(c). Furthermore, the mapping operation causes the binding potential energy quantity Φ/Nk_B to drop from about –35 K to an inherent-structure average of about –65 K. The amorphous packing character of the atomic arrangements in these inherent structures has been verified by calculating the atomic pair distribution function [Hodgdon and Stillinger, 1995], which resembles that for glassy arrangements of classical Lennard-Jones particles [Rahman et al., 1976] but differs substantially from that of their crystals, which show fully segregated coordination shells.

An analogous calculation could be (but has not yet been) carried out to reveal the dominant types of inherent structures for the lighter ^{3}He isotope at its $T = 0$ melting pressure. This would start with the perfect bcc lattice at the appropriate density, add small random displacements to the N atoms, then apply the steepest descent mapping to Φ minima. As observed for the ^{4}He crystal, the expected outcome would again be porous inherent structures involving amorphous atom aggregates. However, the lower number density involved at the zero-pressure melting point for ^{3}He (see Figure VIII.10) implies that the fraction of empty space in the resulting inherent structures would be correspondingly larger.

As the $T = 0$ pressure rises above the respective melting-point thresholds for the ground states of the two isotopes, the inherent structures would reflect the increase in atomic number density. At first, this would result in a decrease in the fraction of the system volume occupied by empty pore space, while the atoms would continue to display amorphous packing. But upon crossing the quantum mechanical version of the Sastry point at which empty pore space vanishes, subsequent $T = 0$ compression should yield inherent structures whose atomic packings develop short-range orders that increasingly resemble the coordination structure of the bcc or hcp crystals for ^{3}He or ^{4}He, respectively.

The question naturally arises about how periodic hcp and bcc crystal structures at relatively low pressures can arise from, or at least be related to, collections of inherent structures that individually are disordered. This situation, arising from strong quantum effects in the helium isotopes, reverses the trend in the classical limit, where mapping to inherent-structure minima typically increases configurational order. With respect to the hcp phase of ^{4}He at its $T = 0$ melting-point pressure, the measured binding energy per atom is only [Edwards and Pandorf, 1965]

$$E_0^{[4]}/Nk_B \cong -6.0 \text{ K}. \tag{VIII.61}$$

This is considerably smaller in magnitude than the potential energy per atom for the hcp lattice at the same density, as shown in Figure VIII.10, specifically –35 K. Consequently, one can conclude from the Schrödinger Eq. (VIII.2) that at any one of the $(N - 1)!$ perfect hcp configurations (with periodic boundary conditions) the ^{4}He ground-state non-negative wave function obeys the following inequality:

$$\nabla^2 \psi_0^{[4]} < 0, \tag{VIII.62}$$

where ∇^2 represents the $3N$-dimensional Laplacian operator. Because the atoms are equivalent at these configurations, this inequality implies that $\psi_0^{[4]} > 0$ exhibits a local (if not global) maximum at the hcp configurations, in spite of the fact that the potential energy function also exhibits a maximum there. Of course, the same would be true for the configurational probability given by

---

BOX VIII.3. One-Dimensional Analog

The situation for ^{4}He in its ground-state hcp phase at low to moderate pressure is roughly analogous to that displayed by a symmetrical one-dimensional quartic oscillator possessing two equivalent off-center minima, which are its trivial "inherent structures" at depth $V_{IS} = 0$. The potential energy function for this simple analog possesses the elementary form $(\bar{V}, \bar{x} > 0)$:

$$V(x) = \bar{V}(x^2 - \bar{x}^2)^2.$$

If the minima at $\pm \bar{x}$ are widely separated [and the barrier $V(0) = \bar{V}\bar{x}^4$ correspondingly high], the ground-state eigenfunction $\psi_0(x) > 0$ for a mass-m particle subject to this $V(x)$ would exhibit a symmetric pair of distinct maxima, respectively, near $\pm \bar{x}$. However, as the two $V(x)$ minima approach each other moving symmetrically toward $x = 0$, the barrier while still positive eventually would decline below the ground-state eigenvalue, at which point the eigenfunction would start to display a single maximum at $x = 0$. In this situation, the off-center inherent structures can be viewed as cooperating to produce the single wave function maximum. Of course, sampling the square of that wave function for initial conditions in steepest descent mappings would identify the two off-center inherent structures at $x_{IS} = \pm \bar{x}$ with equal probability 1/2.

$[\psi_0^{[4]}]^2$ that supplies the starting points for steepest descent mapping to inherent structures. Consequently, members of the large collection of dominant inherent structures, in spite of their disorder, are not too far removed from the hcp configuration. An elementary analog appears in Box VIII.3 to help justify this claim.

As pressure is reduced below the melting threshold value for the ^{4}He ground state, the reduction in atomic number density resulting from the phase change implies that the dominant inherent structures underlying the superfluid ground state would be located farther from the hcp configurations than in the crystal-phase case. Consequently, these inherent structures would display an even larger fraction of empty pore space. Assuming that the pores become more numerous rather than merely wider, this implies that a substantially larger fraction of the atoms would be found at the interfaces between the pore and the amorphous atom-packing subvolumes. Consequently, the average inherent-structure energy per atom should rise upon melting.

If the pressure exerted on the ^{4}He ground state is just high enough to stabilize the hcp crystal phase, the configurational-space displacements of the many contributing inherent structures from the hcp geometry inevitably have an enhancing effect on the root-mean-square deviation of the atoms from their nominal crystal site locations. Experimental observations indeed show this to be the case [Hansen and Pollock, 1972; Moroni and Senatore, 1991]. The dimensionless ratio of the root-mean-square deviation to the nominal nearest neighbor distance in the crystal is the basis of the Lindemann melting criterion that was briefly discussed in Section IV.I. The fact that this ratio for ^{4}He low-temperature, low-pressure melting exceeds Lindemann criterion melting values for classical models of structureless particles by 50% or more is a clear indication of the unusual quantum inherent-structure situation relevant to the former and missing for the latter. By substantially increasing the pressure on the ^{4}He ground-state crystal to reduce the fraction of its imaginary harmonic normal modes and eventually eliminate them altogether, the Lindemann dimensionless ratio should exhibit a corresponding reduction. Analogous remarks should apply as well to the ^{3}He bcc crystal.

It is worth noting that the peaking of the ^{4}He probability at hcp crystal configurations enforced by the Schrödinger equation, and away from the less symmetric inherent structures, has an analog among crystal phases of some of the heavier elements. This analog is entropy driven and is exhibited specifically by the group IVB elements Ti, Zr, and Hf [Souvatzis et al., 2008]. The bcc structures observed for these elements at elevated temperatures owe their presence to anharmonic vibrational motions. At least some of the harmonic normal modes evaluated at the bcc structures of these elements with appropriately accurate potential functions are found to have imaginary frequencies. Consequently, inherent structures would involve broken bcc symmetry. However, the combination of thermally driven anharmonic vibrational displacements that are biased on average toward the bcc configuration lead to that structure as the observed phase at thermal equilibrium. A similar entropy-driven situation has also been cited as the reason that Ca under elevated pressure displays a simple cubic (sc) crystal phase [Errea et al., 2011; Di Gennaro et al., 2013].

Considering the geometric complexity and diversity of the inherent structures underlying the low-pressure liquid and crystal phases of ^{4}He in its ground state, it seems unlikely that low-energy excitations above the ground state for this isotope, at fixed number density, would produce any qualitative change in the inherent-structure distributions. Specifically, this should apply in the presence of low numbers of phonons in both crystal and liquid phases, as well as roton excitations in the liquid. As noted earlier, under sufficiently high compression, the hcp crystal becomes its own inherent structure as all imaginary-frequency harmonic modes have become squeezed out, and phonons (if not too highly excited) then simply reduce to intrabasin oscillations. Analogous remarks should apply to ^{3}He in its bcc crystal phase. However, it has yet to be determined if Landau-liquid excitations attributed to liquid ^{3}He at low temperature produce distinctive signatures of the inherent-structure distribution.

The diversity and unusual nature of the properties exhibited by condensed phases of the stable helium isotopes present many opportunities for the steepest descent mapping and inherent-structure formalism to enhance understanding of those phenomena. But at the time of this writing, the approach has had rather limited quantitative application to these substances. This situation invites listing cases for which identification of underlying inherent structures would likely yield significant further insights into the distinctive quantum behaviors of ^{3}He and ^{4}He. In addition to opportunities implied in the preceding text, such a list could include the following:

(A) It would be valuable to determine the way that liquid ^{4}He inherent-structure energies vary along paths that cross the lambda transition line between the normal (He I) and superfluid (He II) phases, to determine if the average of those energies as a function of temperature exhibits a singularity of the type underlying the measured thermodynamic heat capacity.

(B) Are there clear geometric differences between the inherent structures mapped isochorically from liquids of the two isotopes that have identical number densities?

(C) For the ground-state ^{3}He bcc and ^{4}He hcp crystals taken to sufficiently high pressure so that all harmonic normal modes have positive frequencies, do the resulting inherent structures exhibit substantial concentrations of point defects (vacancies, interstitials) or other structural defects?

(D) Both isotopes have ground states for fixed numbers of particles in free space that amount to liquid droplets. It would be useful to know how the mean radii and moments of inertia of those droplets change as a result of steepest descent mapping to inherent structures.

(E) The nonzero solubility of ^{3}He in ^{4}He at low pressure raises questions about how these isotopes would be distributed spatially inside a free droplet containing both. How would that relative distribution be affected upon transforming to the inherent-structure distribution?

IX.

Water

T he undeniable importance of water in the physical and biological sciences, in engineering and technology, and in a wide variety of human affairs requires that this substance receive unparalleled attention. Understandably, it has served as the focus of a vast and diverse body of research effort, with a proportionately huge published literature. Its special role and unusual properties, both as a pure substance and as a solvent medium, deserve careful examination from the landscape/inherent-structure approach developed in this text. This chapter provides basic aspects of that examination.

Because of the fact that the constituent nuclei are light, quantum effects on the properties of water are non-negligible. This leads to significant differences in measurable properties for distinct isotopic variants of the water molecules, whether in isolation, in clusters, or in condensed phases. Table IX.1 lists the six usually encountered hydrogen and oxygen isotopes. In the following sections, when specific isotopic compositions are not mentioned, the default interpretation is that properties discussed refer to the natural abundance of terrestrial water, overwhelmingly dominated by "light water" $(^1H)_2(^{16}O)$.

A. Molecular Interactions

The structure of the water molecule and the interactions that occur between neighboring water molecules are fundamental characteristics that underlie all of the equilibrium and nonequilibrium properties possessed by macroscopic condensed phases of this substance. Figure I.4 shows the stable geometry for the three nuclei comprising an isolated H_2O molecule. This is the unique minimum on the ground-electronic-state Born-Oppenheimer potential surface. Its nonlinear geometry has C_{2v} symmetry and is identical for other isotopic variants of the water molecule (HDO^{16}, D_2O^{16}, HTO^{18}, …). The vibrational frequencies for excursions of the nuclei away from this single-molecule inherent structure have been exhaustively studied and allow detailed characterization of the intramolecular force field, including both harmonic terms as well as anharmonic contributions [Eisenberg and Kauzmann, 1969, Chapter 1].

The normal modes of vibration of the common light water molecule $H_2^{16}O$ have the following frequencies (i.e., energies of its distortional quantum excitations) [Eisenberg and Kauzmann, 1969, Chapter 1]:

$$v_1 = 3{,}656.65 \text{ cm}^{-1} \qquad \text{(symmetrical stretch)},$$

TABLE IX.1. Nuclear isotopes available to form water molecules[a]

Isotope	% Natural Abundance	Lifetime	Nuclear Spin[b]
^{1}H(H)	99.9885	∞	1/2
^{2}H(D)	0.0115	∞	1
^{3}H(T)	0.0000	12.33y	1/2
^{16}O	99.757	∞	0
^{17}O	0.038	∞	5/2
^{18}O	0.205	∞	0

[a]Lide, 2006, p. **11**–50.
[b]In units of $\hbar = h/2\pi$.

$$v_2 = 1{,}594.59 \text{ cm}^{-1} \qquad \text{(symmetrical bend)}, \tag{IX.1}$$

$$v_3 = 3{,}755.79 \text{ cm}^{-1} \qquad \text{(asymmetric stretch)}. \tag{}$$

In the harmonic approximation, the sum of zero-point kinetic energies for these normal modes is

$$(h/2)(v_1 + v_2 + v_3) \cong 12.88 \text{ kcal/mol}. \tag{IX.2}$$

Needless to say, the corresponding values would be smaller for more massive isotopic variants of the water molecule. Aside from a small additional contribution caused by rotational motion, this represents the irreducible quantum incremental energy that the molecule possesses above its inherent-structure potential energy minimum. This may be compared with the difference in potential energy $\Delta\Phi_{dissoc}$ on the Born-Oppenheimer ground-state potential hypersurface between the inherent structure and the dissociation limit involving widely separated neutral O–H and H fragments, the former at its own minimum. That difference has been estimated to be [Stillinger, 1978]

$$\Delta\Phi(dissoc) = \Phi[\text{OH}] + \Phi[\text{H}] - \Phi[\text{H}_2\text{O}] \tag{IX.3}$$

$$\cong 126 \text{ kcal/mol}.$$

An interacting pair of water molecules in the electronic ground state also exhibits only a single type of inherent structure. It is illustrated in Figure IX.1. The structure shown is based on a potential energy function for the six nuclei comprising flexible molecules that utilizes a large

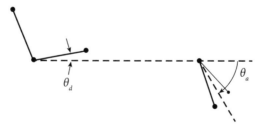

FIGURE IX.1. Geometric arrangement of nuclei at the single type of inherent structure exhibited by a pair of water molecules. This is based on the flexible-molecule potential energy hypersurface determined by Leforestier et al., 2002.

database of spectroscopic information and quantum calculations [Leforestier et al., 2002]. The two participating molecules each retain nearly the same intramolecular geometry as that of an isolated molecule, Figure I.4. But their positions in the dimer are inequivalent, with one molecule acting as a "donor" of a hydrogen toward the oxygen of the other "acceptor" molecule. The six-nucleus configuration at this inherent structure possesses a plane of symmetry that contains the donor molecule, as well as the angle bisector of the acceptor molecule. The distance between the oxygen nuclei is 3.034 Å, and the two angles defined in Figure IX.1 have values $\theta_d = 9.67°$ and $\theta_a = 62.3°$. The potential energy of this nearly linear single-hydrogen-bond inherent structure, compared to twice that for a monomer at its minimum (i.e., the inherent binding energy of the dimer), is

$$\Delta\Phi(\text{dimer}) = \Phi[(H_2O)_2] - 2\Phi[H_2O] \qquad \text{(IX.4)}$$

$$= -5.145 \text{ kcal/mol.}$$

This water dimer inherent structure is 24-fold degenerate because of the 4! distinct but equivalent ways that the four hydrogen nuclei could be assigned to the locations shown within the dimer. There exists a variety of transition pathways connecting these 24, passing over saddle points on the potential energy hypersurface. Transitions that retain the initial intramolecular chemical bonds have relatively low barriers. The lowest of these also retains the near-linear hydrogen bond but rotates the acceptor molecule through ±180° about its own symmetry axis, thereby interchanging the positions of its own two hydrogens. The next higher type of intact-molecule barrier exchanges the roles of donor and acceptor. A pathway that passes through a "bifurcated hydrogen bond" transition configuration to interchange the two donor-molecule hydrogens must rise higher still between the equivalent inherent structures [Keutsch et al., 2003a]. Saddle points on the potential energy hypersurface corresponding to disruption of intramolecular chemical bonds and exchange of hydrogens between the molecules lie considerably higher yet in energy.

The isolated water molecule has a substantial permanent dipole moment 1.855 D oriented along the bisector of the H–O–H bond angle [Dyke and Muenter, 1973]. Consequently, dipole–dipole interactions dominate the water pair interaction at large oxygen–oxygen separation r_{OO}. It is possible to regard r_{OO} as a controllable parameter in the pair potential. Upon continuously increasing r_{OO} from the inherent-structure value mentioned above, the tendency toward formation of a single near-linear hydrogen bond at short separation diminishes, then suddenly gives way to a configuration with the two molecular dipole moments collinear and the molecular planes perpendicular. One rough estimate based on the Hartree-Fock approximation places this singular symmetry-changing distance at $r_{OO} \approx 7.73$ Å [Stillinger and Lemberg, 1975].

The water trimer $(H_2O)_3$ has been the frequent object of both experimental and theoretical research attention [Keutsch et al., 2003b]. Its potential energy function reveals a lowest lying inherent structure that displays three strained hydrogen bonds. The arrangement of the three participating molecules in this structure displays no symmetry and thus is chiral. The spatial pattern is illustrated in Figure IX.2. It exhibits a cyclic structure of three hydrogen bonds, with each molecule acting both as donor and as acceptor. Notice that each dimer is distorted compared to the optimal arrangement shown in Figure IX.1. The three pendant hydrogen atoms that are not directly involved in the hydrogen bonds lie well out of the plane containing the oxygen nuclei. Two are on one side of that plane, the third on the opposite side. Flipping one of the former two across

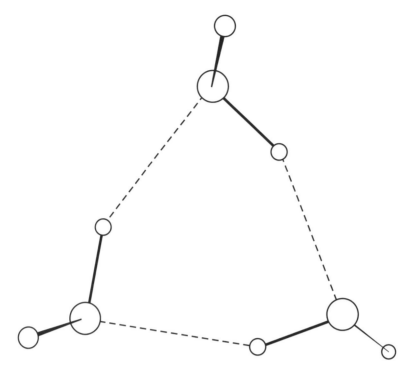

FIGURE IX.2. Arrangement of three water molecules at one of the six equivalent lowest energy inherent structures for intact molecules. The view depicted is from a location directly above the plane defined by the three oxygen nuclei.

the plane has a low barrier and merely converts one of the six equivalent lowest inherent structures to another one. A quantum mechanical study of this cyclic trimer utilizing rigid water molecules [Mas, et al., 2003, Table I] concluded that its inherent-structure binding energy compared to three isolated water molecules is –12.91 kcal/mol. This binding energy includes –10.73 kcal/mol as the sum of the three dimer interactions in the trimer geometry, and an additional –2.18 kcal/mol that is an intrinsic three-molecule interaction. This same study found that the oxygen pair separations in the trimer are 2.78 Å, substantially shorter than the separation in the optimal isolated dimer. As indicated above, if the three water molecules are treated as intact participants, there are only 6 equivalent inherent structures, but allowing exchange of hydrogens increases this degeneracy to 240.

Other higher energy inherent structures also exist for the water trimer. The type requiring the lowest excitation energy apparently has the same cyclic structure of three roughly linear hydrogen bonds but with all three pendant hydrogen nuclei on the same side of the plane containing the oxygens. This arrangement has been estimated to lie approximately 0.72 kcal/mol above the trimer global minimum energy [Mas et al., 2003, Table IV].

Significantly, the higher lying trimer inherent structures include examples where the hydrogen-bonding roles of the participating water molecules can change qualitatively. The two cases already discussed have each molecule simultaneously acting as both a hydrogen donor and an acceptor. However, Figure IX.3 shows structures with two near-linear hydrogen bonds, with a central molecule that in one case, (a), serves as a double donor of hydrogens to its neighbors, and in the other

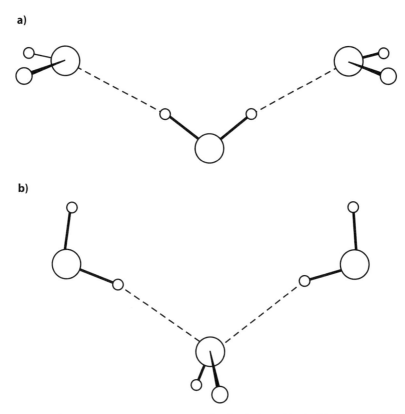

FIGURE IX.3. Trimer inherent structures containing a central molecule that is (a) a double donor, and (b) a double acceptor of hydrogens.

case, (b), serves as a double acceptor of hydrogens from its neighbors. Both of these structures exhibit C_{2v} symmetry. The angle requirements for these alternative geometries place the two end molecules too far apart to form a third hydrogen bond. In contrast to the global-minimum cyclic trimer, the three-body (nonadditive) contribution to the binding is positive for these two structures. The double donor case (a) has potential energy –8.681 kcal/mol, comprising –9.101 kcal/mol from the dimers, and +0.421 kcal/mol from three-molecule nonadditivity; the corresponding values for the double acceptor (b) are –8.723 kcal/mol total, comprising –8.909 kcal/mol from the dimers and +0.186 kcal/mol from three-molecule nonadditivity [Mas et al., 2003, Table VI]. This multiplicity of water trimer inherent structures is a premonition of structural and energy diversity possessed by inherent structures in extended aqueous condensed phases.

Spectroscopic techniques have been developed that allow experimental determination of the structures of stable water clusters involving up to $n = 6$ molecules [Pérez et al., 2012]. However, performing accurate and comprehensive quantum calculations for water clusters $(H_2O)_n$ rapidly becomes impractical as the number n increases above 3. Not only does the number of nuclear configurational coordinates increase in proportion to n, but the number of electronic basis functions typically required for high accuracy increases exponentially with n. This practical computational barrier to direct quantum calculation places a premium on the possibility to develop useful semi-empirical effective pair potentials, which for suitably limited ranges of external parameters (e.g., temperature and density) might provide reasonably accurate approximations for the total n-molecule potential energy function Φ as a simple sum of pair terms:

$$\Phi(\mathbf{x}_1 \ldots \mathbf{x}_n) = \sum_{i<j} \mathrm{v}^{(2)}(\mathbf{x}_i, \mathbf{x}_j) + \sum_{i<j<k} \mathrm{v}^{(3)}(\mathbf{x}_i, \mathbf{x}_j, \mathbf{x}_k) + \ldots + \mathrm{v}^{(n)}(\mathbf{x}_1 \ldots \mathbf{x}_n) \quad \text{(IX.5)}$$

$$\cong \sum_{i<j} \mathrm{v}_{e\!f\!f}(\mathbf{x}_i, \mathbf{x}_j).$$

Here the $\mathbf{x}_i$ stands for the full set of coordinates describing molecule i. Once chosen, $\mathrm{v}_{e\!f\!f}$ would be treated as independent of temperature and density. This seems to be a rational strategy for water,

---

BOX IX.1. Effective Pair Potentials

The concept of an effective pair interaction $\mathrm{v}_{e\!f\!f}$ does not uniquely lead to a precise definition. In fact, the choice of this function would depend on the specific application or applications for which it is intended as a computational model. Suppose that one wishes to investigate a set of properties $P_1 \ldots P_q$ of the substance of interest (water in the present case). This set could include both bulk thermodynamic and transport properties, as well as surface characteristics. The logical framework for determining $\mathrm{v}_{e\!f\!f}$ in principle involves minimizing a suitable objective function, such as the following non-negative net error quantity:

$$O\{\mathrm{v}_{e\!f\!f}\} = \sum_{i=1}^{q} w_i (\Delta P_i)^2. \quad \text{(IX.6)}$$

In this expression, ΔP_i represents the difference between the experimentally measured value of property P_i and its value predicted by the model-system pair potential $\mathrm{v}_{e\!f\!f}$. The $w_i > 0$ are relative weights attached to each of the properties considered and depend on the intended application. In principle, the minimization of objective function $O\{\mathrm{v}_{e\!f\!f}\}$ should be carried out over a well-defined flexible function family from which $\mathrm{v}_{e\!f\!f}$ is to be selected. In practice, such an idealized optimization procedure is too complex and cumbersome to be carried out in full detail. Instead, typically a finite sequence of trial approximations to a "best" $\mathrm{v}_{e\!f\!f}$ is generated, using previous results in that sequence to suggest an appropriate choice for the next approximation.

One version of this variational selection procedure would be to reproduce exactly some small set of experimental properties at a single state point, say, at the temperature–density pair T^*, ρ^*. Even if this were achieved, the constructed $\mathrm{v}_{e\!f\!f}$ generally would not exactly reproduce the same small set of experimental properties at a shifted state point $T^* + \Delta T$, $\rho^* + \Delta \rho$. The utility of a given $\mathrm{v}_{e\!f\!f}$ might then be judged by how small these deviations remain over the ΔT, $\Delta \rho$ ranges of interest.

Replacing the exact potential function Φ with a sum of effective pair interactions must be expected to modify the multidimensional potential landscape topography. Suppose that the scientific objective is to replicate optimally the topography of that portion of the landscape that is thermodynamically most relevant at a given temperature and density. This suggests focusing attention on the Boltzmann factor $\exp(-\beta\Phi)$, realizing that any finite extremum of Φ (minimum, saddle point, maximum) is also an extremum of the Boltzmann factor, and vice versa. Hessian eigenvalues change sign under this transformation, and so in particular minima become maxima. The goal thus becomes finding an optimal approximation to $\exp(-\beta\Phi)$ over the available configuration space in the form of an effective Boltzmann factor $\exp(-\beta\Sigma\mathrm{v}_{e\!f\!f})$. Some consequences of this topographic approach have been explored in general, and specifically for water [Stillinger, 1970, 1972].

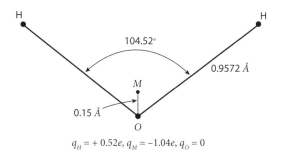

FIGURE IX.4. Rigid water molecule geometry assumed for the TIP4P effective pair potential.

in view of the fact that three-body nonadditivity appears to be a modest correction to two-body energies, and that even higher order nonadditivities are smaller still [Goldman et al., 2005]. Indeed, the published literature offers a large number of candidates for effective pair potentials to describe water in various states of aggregation [Guillot, 2002]. The general strategic situation that underlies selection of an effective pair interaction, whether for water or some other substance, provides the subject of Box IX.1.

One of the more widely used effective pair potentials is designated the "TIP4P" potential [Jorgensen et al., 1983]. This case assumes that the individual water molecules are rigid, the geometry of which appears in Figure IX.4. For any pair of molecules, the potential consists of a sum of 10 terms that depend on distances between force centers located in the different molecules. One of these terms is a Lennard-Jones 12,6 potential acting between the oxygens, while the other nine are simple coulombic charge–charge interactions:

$$v_{TIP4P} = \varepsilon v_{LJ}(r_{OO}/\sigma) + \sum_{i=1}^{3}\sum_{j=1}^{3} q_i q_j/r_{ij}. \tag{IX.7}$$

The Lennard-Jones parameters have been assigned the values

$$\varepsilon = 0.15504 \text{ kcal/mol}, \tag{IX.8}$$

$$\sigma = 3.1536 \text{ Å} .$$

As indicated in Figure IX.4, electrostatic charges have been placed at the hydrogens ($q_H = +0.52e$) and at a point M ($q_M = -1.04e$) that is 0.15 Å ahead of the oxygen, along the angle bisector. Using classical Monte Carlo simulations, these parameters were chosen to fit both thermodynamic and pair correlation properties of liquid water at 25°C and 1 atm pressure, a specific application of the general strategy indicated in Eq. (IX.6). One outcome of this fit is that the point charges used produce a molecular dipole moment 2.166 D, significantly larger than the experimentally determined value 1.855 D for an isolated water molecule. This dipole enhancement can be rationalized as an average polarization effect acting between neighbors in the liquid. In view of the fact that classical statistical mechanics was the parameter-fitting tool, a shifted set of TIP4P parameter values would be expected if liquid D_2O were the substance of interest instead of H_2O. It should be noted that in the bulk system limit, the TIP4P effective pair potential produces a water-phase diagram surprisingly similar to that observed experimentally (see Section IX.B) but with significant temperature shifts [Sanz et al., 2004]. However, a more recent reparameterization of the TIP4P potential, designated TIP4P/2005, partially improves the predicted phase diagram up to moderate pressures [Abascal and Vega, 2005].

The reader should be warned that the TIP4P potential formally has no lower bound since it contains unphysical divergences to $-\infty$ if a positively charged hydrogen nucleus moves into coincidence with a negatively charged "M" site on another molecule. However, such anomalous configurations are not normally accessed in most calculations because reaching them from widely separated starting configurations of a dimer requires surmounting a very high positive energy barrier. In most applications, this peculiarity may simply be ignored. However, if circumstances such as high kinetic energies are involved that might conceivably surmount the high energy barrier and visit these "dangerous" pair configurations, it would be possible to modify the very short range part of the unlike-charge Coulomb terms to a nondivergent form. This would not substantially influence the potential at other, more conventional, dimer configurations.

Despite its origin based on properties of bulk liquid water, the TIP4P effective pair interaction appears able semiquantitatively to represent low-lying inherent structures for isolated clusters of small numbers of water molecules. In the case of the dimer, it produces a hydrogen-bonded structure of the type shown in Figure IX.1 but with $r_{OO} = 2.75$ Å, $\theta_a = 46°$, and $\Delta\Phi(\text{dimer}) = -6.24$ kcal/mol [Jorgensen, et al., 1983, Table II]. Because a rigid-molecule model is assumed, this effective pair interaction must implicitly contain a contribution from interaction-caused shifts in the zero-point energies of participating molecule vibrational modes. Presumably, the TIP4P representation of inherent structures becomes more accurate as the number n of water molecules increases because of the greater relative contributions of nonadditive interactions, as are present in the extended liquid.

The TIP4P effective pair interaction provides a convenient tool to generate approximations to $(H_2O)_n$ inherent structures for $n \geq 3$. While producing estimates of the geometries for the low-lying inherent structures, this effective interaction may invert the ordering of cases lying close in energy.

Nevertheless, its application to clusters reveals general patterns of hydrogen-bonding behavior for water molecules that also emerge from other effective pair interactions and from quantum calculations (to the extent that they are feasible). Specifically, roughly linear single hydrogen bonds, as seen in Figures IX.1–3, are a consistent pattern to connect close neighbors. For $n = 4,5$, the global minima display cyclic structures analogous to that of the trimer, with each water molecule acting simultaneously as a hydrogen donor and as a hydrogen acceptor [Wales and Hodges, 1998]. In the subsequent size range $6 \leq n \leq 21$, published calculations show that roughly linear single hydrogen bonds connecting neighbors is a continuing theme, but the topology of those bonds becomes more complex, with polygons of hydrogen bonds sharing vertices and edges. This produces examples of water molecules engaging in three or four hydrogen bonds simultaneously, but no more [Wales and Hodges, 1998]. Water molecules engaging in three hydrogen bonds can be a single donor and a double acceptor or a double donor and a single acceptor. Those involved in four hydrogen bonds invariably are double donors and double acceptors. The hydrogen-bond polygons present in the cluster inherent structures exhibit both even and odd numbers of edges.

Thus far, the hydrogen-bond concept has been treated informally but with strong support from cluster inherent-structure properties. However, having a formal and precise definition of "hydrogen bond" is desirable for several reasons. In the case of large aggregates of water molecules, whether they are at potential energy minima or not, there can occur pairs of molecules whose hydrogen-bonding status would be ambiguous in the absence of an exact assignment criterion. Furthermore, an unambiguous identification of hydrogen bonds is a starting point for determination of the topological characterization of extended hydrogen-bond networks, including such attributes as distribution of water molecules by numbers of hydrogen bonds in which they participate and the distribution of polygons by their number of sides. Because the hydrogen-bond concept intrinsically recognizes energy stabilization for collections of molecules, a natural way to construct a hydrogen-bond criterion compares the interaction energy for any pair of molecules with a physically motivated cutoff value $V_{hb} < 0$ [Rahman and Stillinger, 1973; Jorgensen et al., 1983]. All pairs of molecules whose interactions lie below V_{hb} are declared to be hydrogen-bonded, while those pairs with interaction above V_{hb} are not. In model systems where interactions are represented by effective pair potentials, the criterion is obvious and simple to state for the presence of a hydrogen bond connecting molecules i and j:

$$V_{hb} \geq v_{eff}(\mathbf{x}_i, \mathbf{x}_j). \tag{IX.9}$$

However, the following more elaborate statement is required for the case of the exact interactions that include contributions of all orders up to n, the number of molecules present:

$$V_{hb} \geq v^{(2)}(\mathbf{x}_i, \mathbf{x}_j) + (1/3)\sum_k{}' v^{(3)}(\mathbf{x}_i, \mathbf{x}_j, \mathbf{x}_k) + (1/6)\sum_{k<l}{}' v^{(4)}(\mathbf{x}_i, \mathbf{x}_j, \mathbf{x}_k, \mathbf{x}_l) + \ldots \tag{IX.10}$$
$$+ [2/n(n-1)v^{(n)}(\mathbf{x}_i,\ldots,\mathbf{x}_z).$$

The primed summations exclude the molecule pair i,j of interest. This expression is based on an equal division of every $v^{(l)}$ among all $l(l-1)/2$ pairs that it involves. A specific value of the cutoff potential, developed in connection with the TIP4P effective interaction, is the following [Jorgensen et al., 1983]:

$$V_{hb} = -2.25 \text{ kcal/mol}. \tag{IX.11}$$

At a basic level, the use of a binary distinction between hydrogen-bonded and not-hydrogen-bonded for a pair of water molecules contradicts the fact that interactions are continuous functions of configurational variables. Nevertheless, it is a very helpful pedagogical device that is a physically motivated coarse-grained description of a given configuration of molecules. In particular, it highlights vivid topological distinctions between crystal and liquid phases of water and provides a convenient device for describing complex structural relaxation processes in a relatively simple and straightforward manner.

B. Phase Diagram

Considering the nonspherical shape of the water molecules and the distinctive directionality of the hydrogen bonds that they form with one another, it should come as no surprise that this substance displays a rather complex phase diagram. In this respect, it is a far more interesting and challenging substance than the noble elements heavier than He (i.e., Ne, Ar, Kr, Xe). Figure IX.5 presents the arrangement of thermodynamically stable phases in an elevated-pressure portion of the temperature–pressure plane to exhibit the diversity of known crystalline phases (ice

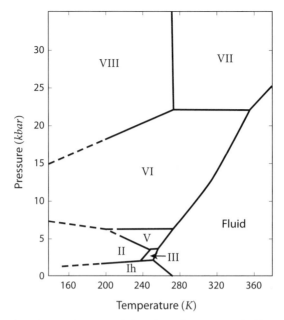

FIGURE IX.5. Equilibrium phase diagram for H_2O, including elevated pressure. Dashed lines are extrapolations of phase boundaries to low temperature, where phase transitions become too sluggish to observe directly.

polymorphs). These ice phases are conventionally distinguished by Roman numerals. At the extremes of low temperature and high pressure, experimental difficulties prevent precise determination of the locations of phase boundaries, so Figure IX.5 must be viewed as an incomplete representation of the complete phase behavior of water. The temperature and pressure scales in Figure IX.5 are such that the liquid–vapor coexistence and its critical point are not shown. These additional characteristics appear in the supplementary Figure IX.6.

The first-order phase transition boundaries $T(p)$ appearing in Figures IX.5 and IX.6 have slopes that must obey the thermodynamic Clausius-Clapeyron equation:

$$\frac{dT(p)}{dp} = \frac{\Delta H}{T(p)\Delta V},$$ (IX.12)

in which ΔH and ΔV are the isobaric reversible enthalpy and volume changes across the transition, respectively [cf. Eq. (IV.126)]. The ice I–liquid melting transition stands out as the only transition boundary in the phase diagram with an obviously negative slope, and this is because of the well-known fact that at coexistence the liquid is denser than the crystal ($\Delta V < 0$), while melting requires heat input ($\Delta H > 0$). The phase boundaries between ices shown in Figure IX.5 that are horizontal or nearly so (Ih–III, III–V, V–VI, VI–VII) have $\Delta V \approx 0$, and the nearly vertical VII–VIII boundary has $\Delta H \approx 0$.

No region appears in Figure IX.5 labeled "IV." Ice IV is a metastable crystalline form of water [Bridgeman, 1935]. It can be formed and has been observed to persist within the stable region of ice V.

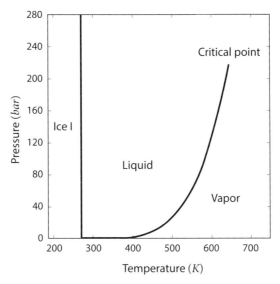

FIGURE IX.6. Region of the H$_2$O phase diagram showing the ice I–liquid–vapor triple point, the liquid–vapor coexistence curve, and the liquid–vapor critical point.

Table IX.2 lists some basic structural characteristics of the stable ice phases. For more complete details about the crystal structures of the various ices, the reader should consult the available specialized sources [Whalley and Davidson, 1965; Eisenberg and Kauzmann, 1969; Fletcher, 1970]. Unit cells for these crystals and the numbers of water molecules they contain have been assigned on the basis of oxygen atom positions alone. As Table IX.2 indicates, several of the ices exhibit proton (hydrogen atom) disorder; the water molecules remain intact and substantially undistorted, but they can be arranged orientationally in a variety of ways with an indeterminacy that formally would require arbitrarily large unit cells as the periodically replicated molecular arrangements. In all of the ices, the water molecule orientations lead to formation of hydrogen bonds to neighbors, though in some of the cases these bonds display significant angular distortions

TABLE IX.2. Crystal symmetries of thermodynamically stable ices

Ice Type	Crystal System	Molecules per Unit Cell	Proton Order
Ih	Hexagonal	4	No[a]
II	Rhombohedral	12	Yes
III	Tetragonal	12	No
V	Monoclinic	28	No
VI	Tetragonal	10	No
VII	Cubic	2	No
VIII	Cubic	2	Yes[b]

[a]A cubic modification of Ih observed at low temperature, also proton disordered, is designated "ice Ic."
[b]This is a proton ordered form of ice VII.

away from linearity because of the elevated external pressure, so as to permit higher packing density overall.

The equilibrium phases appearing in Figures IX.5 and IX.6 do not exhaust all possible equilibrium phases of water. This is especially evident from the presence of proton-disordered ices with associated positive configurational entropy. Upon approach to $T = 0$, the zero-entropy ground state should dominate thermodynamically, exhibiting some form of proton order. Therefore, at sufficiently low temperature at least all of the hydrogen-disordered ices listed in Table IX.2 in principle should transform to hydrogen-ordered forms [Salzmann et al., 2006]. Although precisely determined regions of stability for these or any other additional forms of ice have not been absolutely determined, some reliable experimental evidence of their existence and location in the phase plane has been accumulated. The following additional phases, with their attached Roman numeral designations, are now recognized:

(a) Ice IX is a metastable form of ice III [Whalley et al., 1968].

(b) Ice X is an ordered form of ices VII and VIII, in which pressures estimated to be above 490 kbar compress all hydrogen bonds to the point where they are symmetrical, i.e., protons are located at the bond midpoints so no distinct H_2O molecular units are present [Lee et al., 1992].

(c) Ice XI is the low-temperature ferroelectric phase that is the proton-ordered form of ice Ih [Tajima et al., 1982]; see Section IX.C.

(d) Ice XII is a metastable, proton-disordered phase existing within the ice V region of the equilibrium phase diagram [Lobban et al., 1998; Koza et al., 2000].

(e) Ice XIII is a proton-ordered form of ice V [Salzmann et al., 2008].

(f) Ice XIV is a proton-ordered form of the metastable ice XII [Salzmann et al., 2006].

(g) Ice XV is a proton-ordered form of ice VI that is stable below $T \approx 130$ K [Salzmann et al., 2009].

The vapor–liquid–ice Ih triple point for H_2O occurs at $T = 273.16$ K, $p = 0.0061$ bar. The coexisting-phase mass densities at this triple point are 4.847×10^{-6} g/cm^3 for the vapor, 0.9998 g/cm^3 for the liquid, and 0.9164 g/cm^3 for ice Ih. The molar enthalpy changes at this triple point are $\Delta H_m = 1.4363$ kcal/mol for melting ice I to liquid and $\Delta H_{lv} = 10.767$ kcal/mol for vaporization of the liquid. These enthalpy changes correspond to entropy changes per molecule $\Delta S/Nk_B$, respectively, equal to 2.469 and 14.54. Replacing H_2O with heavy water D_2O causes the vapor–liquid–ice Ih triple-point temperature to rise to 276.98 K. Other triple points in the phase diagram also increase in temperature when H_2O is replaced by D_2O but not by proportional amounts [Eisenberg and Kauzmann, 1969, Table 3.5], so that isotopic substitution thermodynamically does not have a simple rescaling character.

The liquid–vapor critical point for H_2O occurs at $T_c = 647.3$ K and $p_c = 221.2$ bar. In the case of heavy water H_2O, these are displaced slightly to $T_c = 644.1$ K and $p_c = 218.6$ bar. It is noteworthy that this H → D isotope shift in temperature is opposite in direction to that of the triple point.

Strong confinement of water can significantly modify the hydrogen-bond patterns that the bulk ice phases present. In particular, this modification is particularly evident in the case of water confined to the interior of carbon nanotubes with variable internal diameters. Computer simulation has been applied to this circumstance and has utilized the inherent-structure pattern clarification property to reveal a complex "phase diagram" with unusual spatial patterns of hydrogen bonds [Takaiwa et al., 2008].

C. Ices Ih and Ic

The usual form of crystalline water present in the Earth's environment is hexagonal ice Ih. A cubic modification, ice Ic, has been detected in the upper atmosphere, where it apparently forms by vapor sublimation at low temperature [Whalley, 1983]. Presumably, the latter is metastable with respect to the former. Laboratory preparations of ice Ic created at low temperature spontaneously and irreversibly revert to ice Ih when heated at ordinary pressure to about 190 K [Dowell and Rinfret, 1960]. The latent heat of the transformation apparently is very small, less than 1.5 cal/g [Beaumont et al., 1961]. The local structures of both Ih and Ic are similar, with each water molecule linked to four nearest neighbors by nearly linear hydrogen bonds. The geometric patterns of hydrogen bonding are illustrated in Figure IX.7. In the geometrically ideal forms of each of these structures, pairs of directions from any oxygen to two of its nearest neighbors form the tetrahedral angle $\arccos(-1/3) \cong 109.47°$. One can identify closed circuits of hydrogen bonds in each of these structures, jumping from the oxygen position of one water molecule to that of a nearest neighbor, and so forth, until the jumps return to the starting position. For both Ih and Ic, the shortest closed circuits have six sides, and all closed circuits have even numbers of sides. In both structures, all even-number circuit lengths ≥ 6 are present. In ice Ic, all hexagons of hydrogen bonds have "chair" form, whereas in ice Ih, half are "chair" form and half are "boat" form. The ice Ih structure consists of two interpenetrating hexagonal close-packed (hcp) lattices of oxygen atoms, in contrast to the ice Ic structure, whose oxygen atom arrangement consists of two interpenetrating face-centered cubic (fcc) lattices.

The measured distance between nearest neighbor oxygens in ice Ih at 77 K and 1 atm is 2.75 Å [Fletcher, 1970, Chapter 2]. Thermal expansion causes this to increase slightly to about 2.76 Å at the melting temperature 273 K. The density of ice Ic is very close to that of Ih at the same temperature and pressure conditions, so their closely related structures involve essentially the same oxygen–oxygen nearest neighbor distances. Because the water molecules in both of these ices retain their individual molecular identities, with O–H covalent bond lengths that are close on average to those of the free molecules, single hydrogen atoms forming each hydrogen bond in the crystals must reside asymmetrically along those bonds, displaced about 0.4 Å one way or the other from the hydrogen-bond midpoint. Each water molecule thus has two hydrogens close to it (its own) pointing toward neighbor oxygens and two more remote hydrogens pointing toward it from neighbor molecules. In other words, each intact water molecule has six possible orientations at its crystal site, with its hydrogens occupying two of the four available hydrogen-bonding directions. These "ice rules" for the positions of the hydrogen atoms are incomplete configurational constraints for both Ih and Ic, thus allowing for the presence of hydrogen disorder, as reported in the last column of Table IX.2. The intermolecular interactions that give rise to these ice rules also influence the average molecular dipole moment, which has been estimated to be 2.73 D in ice Ih [Onsager, 1973], substantially larger than the free molecule value 1.855 D [Dyke and Muenter, 1973]. It should be noted in passing that if the locations of all hydrogens in an ice sample were simultaneously reversed along their bonds, the resulting configuration would also satisfy the "ice

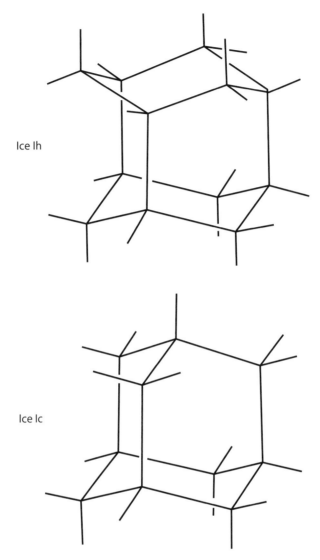

FIGURE IX.7. Hydrogen bond networks in ices Ih and Ic. Positions of the oxygen nuclei are the network nodes, from each of which four bonds emanate to connect to nearest neighbor oxygens.

rules." Furthermore, quasielastic neutron scattering (QENS) experiments on ices Ih and Ic observe concerted hydrogen switches around oriented closed paths of hydrogen bonds, taking the samples from one "ice rule" structure to another [Bove et al., 2009].

A full listing of the macroscopic three-dimensional hydrogen-bond networks that water molecules can form should include those that appear in gas hydrates, known as "clathrates" [Davidson, 1973]. These provide an informative contrast to ices Ih and Ic and are created by cocrystallizing any of a wide variety of sparingly soluble gases with water under moderate pressure. A well-known example is methane hydrate, whose unit-cell composition is $8CH_4 \cdot 46H_2O$. Its crystal structure is one of the conventionally named "structure I" clathrates. As in the ices Ih and Ic, the water molecules are each hydrogen-bonded to four nearest neighbors, donating two hydrogens and accepting two others along those hydrogen-bond directions. But unlike ices Ih and Ic, the three-dimensional networks involved are arranged to form two kinds of face-sharing polyhedra, regular dodecahedra (12 pentagonal faces) and tetrakaidecahedra (2 planar hexagons and 12 pentagons). Single CH_4 molecules occupy both of these types of polyhedra. The symmetry of the resulting crystal is cubic [Pauling, 1960, pp. 469–472]. The ice rules and hydrogen disorder are also present for the clathrate-structure water networks. The mass density of the empty water network in this clathrate is 0.79 g/cm^3, considerably less than that for ice Ih at its 1 atm melting point, namely 0.92 g/cm^3. Closed hydrogen-bond circuits on the methane clathrate network include paths with 5, 6, 8, 9, 10, … bonds.

Other gases besides methane that form the same clathrate structure I with water are Ar, Kr, Xe, N_2, and CO_2. Larger molecules such as sulfur hexafluoride (SF_6) and propane (C_3H_8) are too large to fit inside the polyhedra of the methane hydrate structure. However, they do cocrystallize with water in alternative forms, with water arranged in fully hydrogen-bonded networks that possess even lower mass densities [Davidson, 1973; Mao et al., 2007].

It is a reasonable presumption that each of the ice Ih and Ic structures satisfying the ice rules corresponds to a single well-defined inherent structure (both isochoric and isobaric) for the many-molecule system. Pauling devised an approximate, but surprisingly accurate, enumeration method for these ice-rule inherent structures [Pauling, 1935]. This estimate is based on elementary properties established by the ice rules. The fourfold coordination present throughout ices Ih and Ic implies that an N-molecule crystal contains $2N$ hydrogen bonds between nearest neighbors. Two off-center positions are nominally available for the single hydrogen residing along each of these bonds. If nothing else mattered, this would lead to 2^{2N} distinct arrangements for the set of $2N$ hydrogens. However, the ice rules demand that only intact water molecules be present, with each oxygen having two of the four hydrogens as its nearby covalently bonded partners, the other two instead belonging covalently to neighbor oxygens. This implies that only six of the $2^4 = 16$ hydrogen arrangements around each oxygen are acceptable. The Pauling approximation assumes that this same attrition factor applies independently at each oxygen vertex, leading to the following approximate inherent-structure enumeration that applies to both three-dimensional structures:

$$\Omega_{ice,3D} \cong 2^{2N}(6/16)^N \qquad \text{(IX.13)}$$

$$= (3/2)^N.$$

This Pauling enumeration estimate can be compared to an exact result that is available for a two-dimensional "ice" model. This model has fourfold coordination by nearest neighbors just as in ices Ih and Ic, but it is two-dimensional and places its analogs of oxygen atoms at the sites of a square lattice [DiMarzio and Stillinger, 1964]. Figure IX.8 illustrates this "square ice," showing that its analogs of hydrogen atoms sit asymmetrically along the nearest neighbor bonds. In order to conform to hydrogen bonding on the square lattice, the "water molecules" freely adopt configurations in which the angle between their intramolecular O–H bonds can either be $\pi/2$ or π. By using a transfer matrix method, Lieb has obtained the following exact result for this square ice lattice model in the large-system limit [Lieb, 1967]:

$$\Omega_{ice,2D} = [(4/3)^{3/2}]^N \qquad \text{(IX.14)}$$

$$\cong [1.53960]^N.$$

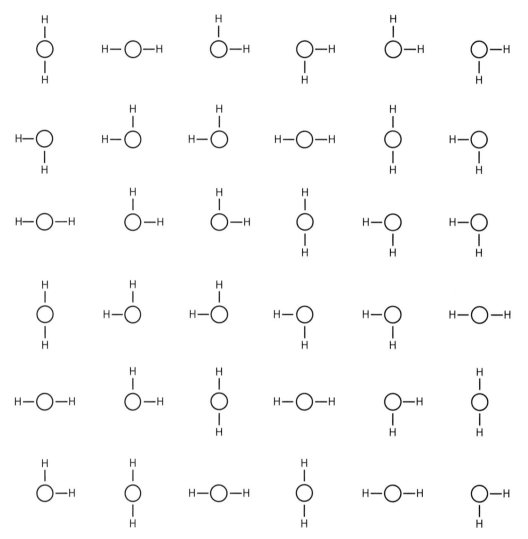

FIGURE IX.8. The square ice model. The configuration shown obeys periodic boundary conditions left-to-right and top-to-bottom.

The Pauling approximation would assign the same approximate value $(3/2)^N$ to this model as it does to the three-dimensional ices Ih and Ic, so evidently its use of independent attrition factors for each oxygen site is a bit overly restrictive.

Although "square ice" was originally created simply as a theoretical model, it was later discovered that it can be realized in the laboratory under special circumstances [Wang et al., 2006]. This realization involves a square lattice of single-domain magnetic islands, the magnetic directionalities of which can be observed using a scanning probe. In addition, it has been proposed that "square ice" could also be realized by colloids suspended in optical trap lattices [Libál et al., 2006].

Unfortunately, the transfer matrix method cannot be adapted to three-dimensional structures to produce a corresponding exact enumeration of all ice-rule configurations. Nevertheless,

Nagle has devised an approach that can generate systematic corrections to the Pauling approximation [Nagle, 1966]. This leads to the improved estimate:

$$\Omega_{ice,3D} = [1.50685 \pm 0.00015]^N, \tag{IX.15}$$

which applies to both forms Ih and Ic. The fact that this result is considerably closer to Pauling's approximation Eq. (IX.13) than is the square-ice value Eq. (IX.14) arises from the fact that the smallest closed cycles of hydrogen bonds in square ice are shorter than those in ices Ih and Ic, and it is these closures that underlie the nonindependence of the attrition factors.

There is no intrinsic reason to suppose that the energies and vibrational free energies for all of the ice-rule inherent structures are identical. However, experimental calorimetry implies that at least the majority are indeed close enough to remain about equally populated until the ice temperature is sufficiently low to inhibit kinetic rearrangements. Heat capacity measurements allow one to infer that indeed a residual entropy remains when pure ice Ih is cooled toward absolute zero, with the value [Giauque and Stout, 1936]

$$S_{res} = 0.82 \pm 0.05 \text{ cal/mol deg}. \tag{IX.16}$$

This can be compared with the formal three-dimensional enumeration results:

$$S_{res} = R \ln \Omega_{ice,3D}$$

$$\cong 0.8060 \text{ cal/mol deg (Pauling)} \tag{IX.17}$$

$$\cong 0.8145 \pm 0.0002 \text{ cal/mol deg (Nagle)}.$$

The inherent-structure enumeration problem generated by the ice rules in three dimensions also arises in a rather different context, namely describing the degeneracy of magnetic moment directions in "spin ice" [Bramwell and Gingras, 2001; Castelnovo et al., 2008]. This involves some crystalline materials with a cubic pyrochlore structure. Examples are holmium titanate ($Ho_2Ti_2O_7$) and dysprosium titanate ($Dy_2Ti_2O_7$). The rare earth ions in these crystalline materials reside within vertex-sharing oxygen tetrahedra. Each of those shared oxygens has associated with it a magnetic moment vector with two possible orientations that are analogous to the pair of hydrogen positions in ice Ic and that are subject to the analog of the ice rules: Each tetrahedron has two moments pointing in and two pointing out.

Although hydrogen-atom positional disorder evidently can be kinetically frozen into pure ice, this misses a low-temperature ordering transition for Ih that strict equilibrium includes. This transition can be induced in the laboratory by doping the ice with potassium hydroxide (KOH) to enhance the rates of configurational transitions at low temperature. Using such an approach, it has been established that disordered ice Ih at 1 atm undergoes a first-order phase transition as it is cooled through 72 K [Tajima et al., 1982]. The resulting structure exhibits ferroelectric ordering of the hydrogens, and, as mentioned in Section IX.B, has been named "ice XI." The fully ordered form is illustrated in Figure IX.9 [Leadbetter et al., 1985]. The oxygen atom positions in ice XI are only slightly displaced from those of ice Ih. The behavior of the phase boundary between ices Ih and XI as pressure rises has not been determined directly, but because the volume change of the transition is evidently small, it is probably insensitive to pressure. It is currently unknown if there is a similar hydrogen ordering transition at low temperature for the metastable ice Ic.

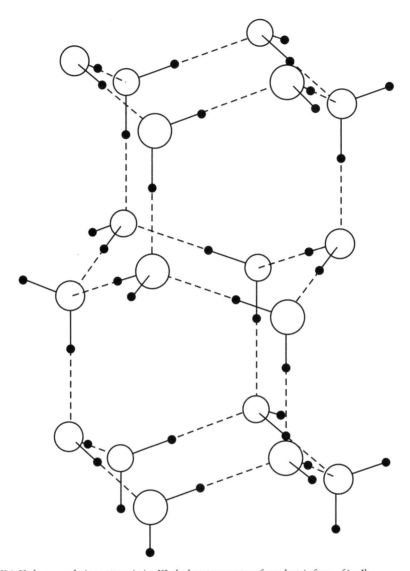

FIGURE IX.9. Hydrogen ordering pattern in ice XI, the low-temperature ferroelectric form of ice Ih.

The inherent structures in ices Ih and Ic that obey the ice rules do not exhaust all possible inherent structures for intact water molecules in those crystalline phases. Additional arrangements include those containing "Bjerrum defects" [Bjerrum, 1952]. These are local orientational defects that entail violation of the ice rules, that nominally place two, or no, hydrogens along directions from an oxygen to a nearest neighbor oxygen. They are conventionally named Bjerrum "D-defects" and "L-defects," respectively (shortened from the German words "doppelt" and "leer"). A neighboring D,L defect pair can be created by misorienting one water molecule in the ice crystal structure. This can then be followed by other neighboring water molecule reorientation transitions that have the effect of moving the defects apart. These transitions between distinct inherent structures and their basins are indicated schematically in Figure IX.10. The hopping motions of

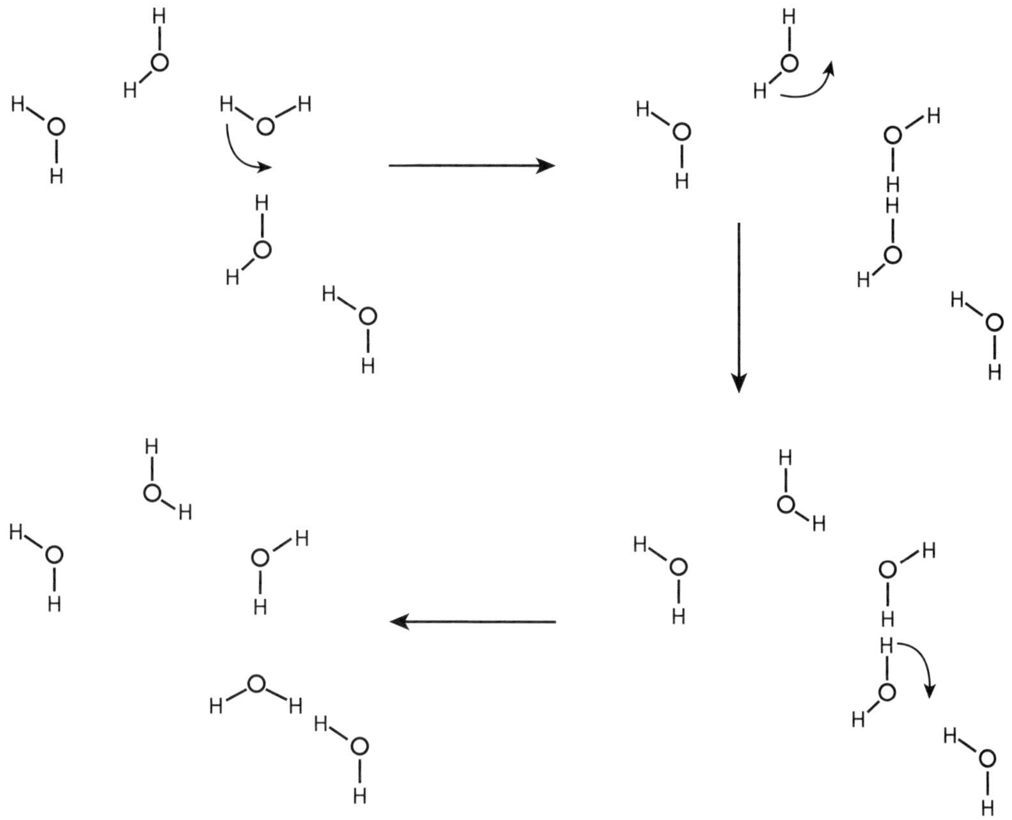

FIGURE IX.10. Creation and hopping separation of a *D,L* pair of Bjerrum defects in ice.

the defects throughout the ice lattice have the effect of transforming the nondefective parts of the crystal from one ice-rule structure to others. In particular, the changing directions of water molecule dipole moments along the pathway of a Bjerrum defect pilgrimage can produce dielectric relaxation.

The unnatural state of a nearest neighbor pair of oxygens with either two crowded hydrogens (*D*-defect) or a missing hydrogen (*L*-defect) involves a substantial energy increment over that of the parent ice-rule structure. In particular, the expected strong repulsion between the two hydrogens at the *D*-defect must produce a significant local distortion, forcing those hydrogens off of the nominal bond line and affecting the positions of the two immediately involved oxygens. Let E_{DL} be the average energy required to create a *D,L* pair of defects that are widely separated in an otherwise ice-rule-obeying structure at 1 atm. Various doped-ice (Ih) experiments have estimated this energy to be in the range 0.69–0.79 eV, while a density functional theory (DFT) quantum mechanical calculation has yielded the somewhat larger value 1.153 ± 0.04 eV [de Koning, et al., 2006]. Because of the magnitude of this creation energy E_{DL}, the thermal equilibrium value of the equal concentrations of the *D* and *L* defects in pure ice is small. At $-10°C$, these concentrations have been estimated to be 7×10^{15} cm^{-3} [Fletcher, 1970, p. 156], or about one defect of either kind for every 4×10^6 water molecules.

A pair of unusual, apparently related, properties exhibited by ice Ih deserve mention. Accurate lattice constant measurements [Röttger et al., 1994, 2012] reveal first that both H_2O and D_2O ice at ordinary pressure possess negative thermal expansion below approximately 73 K. Second, those measurements establish that the isothermal H → D isotope replacement causes the ice to expand over its full temperature stability range 0 K to 273 K. Evidently, the hydrogen atom vibrational motions with substantial quantum zero-point contributions underlie these observations. The negative thermal expansion at low temperature indicates that a temperature decrease and consequent vibrational amplitude reduction produce a crystal density decrease. Increasing the hydrogen isotope mass also reduces vibrational amplitude and creates a density decrease. This latter phenomenon is the reverse of that expected for a crystal composed of structureless particles such as noble gas atoms, where isotope mass increase causes density increase [Herrero, 2002]. These unusual ice Ih properties have been demonstrated in carefully designed quantum calculations [Pamuk et al., 2012].

D. Anomalous Liquid Thermodynamic Properties

The strongly directional hydrogen-bonding interactions that operate between neighboring water molecules create the wide diversity of crystal structures that this substance displays in the temperature–pressure plane. The same directional interactions exist in the liquid, and it is no surprise that they exert a strong and characteristic influence on observable thermodynamic properties. Indeed, the results of those interactions include the presence of several anomalies when viewed in the context of other "more conventional" liquids. This section discusses the most prominent of those water thermodynamic anomalies, covering both the stable liquid as well as its metastable extensions under supercooling and stretching. In a final analysis, the anomalous liquid water properties emerge from that portion of the potential energy or enthalpy landscape occupied by the liquid and the basin shapes and depth distributions that those landscape regions include.

(1) DENSITY MAXIMUM

It is common knowledge that ice floats on liquid water because of buoyancy arising from a negative volume of melting. That property is a bit unusual among all pure substances, but it is hardly unique. Nevertheless, even among those substances that exhibit a negative volume of melting, water (including all its isotopic variants) is especially noteworthy for the fact that at constant pressure its molar volume continues to decrease for a short temperature interval above its melting temperature T_m. Figure IX.11 provides a plot of the molar volume of water at 1 atm pressure, showing values for the equilibrium ice Ih and liquid phases. The plot also includes results of a metastable supercooling extension for the liquid, from its freezing point down to –34°C [Angell, 1982, Figure 7]. Below this supercooling range, crystal nucleation rates become too high to permit reliable density measurements on bulk liquid water samples.

The density maximum (i.e., molar volume minimum) shown in Figure IX.11 occurs at T_{max} (1 atm) = 4.0°C, at which point the mass density is 1.0000 g/cm³. The molar volume reduction between the melting point T_m and T_{max} is only 2.34×10^{-3} cm³/mol, far smaller than the molar volume reduction on melting of ice Ih, 1.62 cm³/mol. As pressure rises above 1 atm, the profile

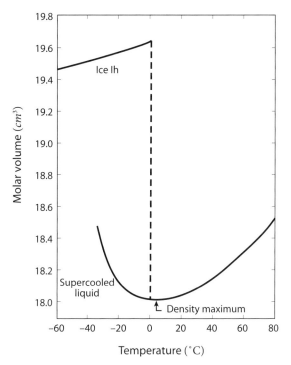

FIGURE IX.11. Molar volume of ordinary (light) water at 1 atm pressure.

of the density maximum becomes even less distinctive and shifts slightly to lower temperature, reaching 0°C at approximately $p = 1$ kbar. However, this remains well within the equilibrium liquid range because the ice Ih melting temperature at this pressure has declined to $T_m = -10°C$. Further pressure increase causes the density maximum eventually to disappear, apparently merging with a lower temperature shallow density minimum to produce a point of inflection at $T \cong -5°C$, $p \cong 1.7$ kbar [Eisenberg and Kauzmann, 1969, Figure 4.16(a)].

The temperature-dependent molar volume of water exhibits a small but significant isotope effect. This arises from quantum corrections to the classical limit of the underlying statistical mechanics for the bulk water system. Table IX.3 contains the T_m, T_{max}, and molar volumes at the respective density maxima for several isotopic water variants at 1 atm. The results shown in the table demonstrate the complexity of quantum effects present in water. First, the upward shifts in T_m and in T_{max} are not proportional to one another for a given increase in mass of either the hydrogens or the oxygen of the constituent molecules. Second, the shift in molar volume at the density maximum depends upon which element has its mass increased: Adding one mass unit to each of the hydrogens ($H \rightarrow D$) raises that molar volume, in contrast to the effect of adding two mass units to the oxygen ($^{16}O \rightarrow {}^{18}O$), which reduces the molar volume. Indeed, this variable-sign isotope effect may control the respective molar volumes over the entire 1 atm stable liquid range from melting to boiling [Eisenberg and Kauzmann, 1969, Figure 4.13].

The sharply rising molar volume displayed in Figure IX.11 as water supercools toward −34°C invites a simple extrapolation that would suggest equality of the ice Ih and supercooled liquid

TABLE IX.3. Isotope effects on liquid water density maxima at 1 atm (equilibrium melting temperatures are included for comparison)[a]

Isotopic Species	Melting Temperature (T_m, °C)	Density Maximum Temperature (T_{max}, °C)	Molar Volume at T_{max} (cm^3)
$H_2{}^{16}O$	0.0	4.0	18.0153
$D_2{}^{16}O$	3.8	11.2	18.1085
$T_2{}^{16}O$	4.5[b]	13.4	18.1398
$H_2{}^{18}O$	—	4.2	17.9912
$D_2{}^{18}O$	—	11.4	18.0844[c]

[a]Entries based on data presented in Eisenberg and Kauzmann, 1969.
[b]Xiang, 2003, Table 2.
[c]Linear estimate based on values for $H_2{}^{16}O$, $D_2{}^{16}O$, and $H_2{}^{18}O$.

molar volumes in the temperature range around –45°C. However, that appears to be unwarranted. By utilizing nanometer-range cylindrical pores as containers, it is possible experimentally to carry small water samples to considerably lower temperature without having them experience crystal nucleation. By means of small-angle neutron scattering for $D_2{}^{16}O$ in such confinement, density measurements have been performed on liquid samples down to 160 K (–113°C) [Liu et al., 2007]. The experimental measurements imply the presence of a 1 atm density minimum at 210 ± 5 K (–64 ± 5°C), at which the molar volume is approximately 19.24 cm^3/mol, less than that of ice Ih at the same temperature. Further cooling causes density to rise again, increasing somewhat the difference with ice Ih. Some caution needs to be exercised in interpreting this result because all of the D_2O sample examined is close to a confining surface and thus subject to configurational perturbation. Nevertheless, it is tempting to suppose that this density minimum and its higher temperature maximum are the 1 atm versions of the minima and maxima discussed above that converge and disappear at elevated pressure. Some support for this notion comes from classical molecular dynamics simulations, using the "TIP5P-E" and "ST2" model potentials, both of which show low-temperature supercooled liquid density minima and higher temperature density maxima [Liu et al., 2007].

The existence of a liquid-phase density maximum evidently arises from structural competition between the tendency for hydrogen bonding to produce open networks at low temperatures and the entropy-driven tendency for the system to occupy more orientationally disordered and compact molecular arrangements. It should not be surprising then that solutes present in the aqueous medium could affect this competition, producing a shift in the temperature of maximum density. Both soluble organic solutes as well as simple electrolytes have been shown to produce a shift [Cawley et al., 2006], though with varying effectiveness. For the solutes thus far examined, the density maxima have always been shifted to lower temperature compared to pure water, indicating primarily a solute-induced disruption of open hydrogen-bonding networks.

In order to provide a broader context for these density characteristics, Table IX.4 presents experimentally measured mass densities and equivalent molar volumes for several regular water networks. The entries show densities spanning more than a factor of two that are attainable by mechanically stable three-dimensional patterns of hydrogen bonds (with varying extents of deviation from linearity). In the cases of ices III, VI, and VIII, the values shown correspond to samples quenched to low temperature and brought to ambient pressure.

TABLE IX.4. Mass densities and molar volumes for some water networks occurring in some ices and clathrate hydrate crystals [Stillinger and Weber, 1983a]

Structure	Temperature, K	Mass Density g/cm^3	Molar Volume cm^3/mol
Ice Ih	273	0.9164	19.66
Ice Ic	143	0.934	19.29
Ice III[a]	83	1.14	15.80
Ice VI[a]	98	1.31	13.75
Ice VIII[a]	100	1.491	12.08
Class I clathrate (Cl$_2$ hydrate)	277	0.7904	22.79
Class II clathrate (tetrahydrofuran + H$_2$S hydrate)	253	0.7844	22.97
Br$_2$ hydrate	263	0.7446	24.19
(CH$_3$)$_3$CNH$_2$ hydrate	243	0.7012	25.69

[a]High-pressure ice phases were quenched to low temperature and then decompressed to 1 atm.

(2) HEAT CAPACITIES

At ambient pressure 1 atm, the isobaric heat capacity C_p is a rather flat function of T over the equilibrium liquid range. Its value at the melting point 0°C is 18.15 cal/mol deg, approximately twice that of ice Ih at the same temperature. At 35°C, it has declined approximately 0.2 cal/mol deg, to a shallow minimum. Subseqent heating to the boiling point 100°C causes it to rise by about the same amount to return very close to its freezing point value. By contrast, supercooling the liquid produces a distinctive rise in C_p. Experiments using emulsified samples (to reduce crystal nucleation and growth) show that this metastable-state C_p approaches 30 cal/mol deg as T declines toward −40°C [Angell et al., 1973].

Recalling the thermodynamic identity that connects isobaric and isochoric heat capacities C_p and C_V to isothermal compressibility κ_T and thermal expansion α_p [Guggenheim, 1950, Chapter IV]

$$C_p = C_V + \frac{\alpha_p{}^2 TV}{\kappa_T}, \tag{IX.18}$$

where

$$\alpha_p = \frac{1}{V}\left(\frac{\partial V}{\partial T}\right)_p, \tag{IX.19}$$

$$\kappa_T = -\frac{1}{V}\left(\frac{\partial V}{\partial p}\right)_T,$$

it is clear that these two heat capacities along each isobar are equal at the density maximum and minimum just discussed because that is precisely where $\alpha_p = 0$. Within the 1 atm equilibrium liquid regime, C_V declines monotonically with increasing temperature, from about 18.1 cal/mol deg where it coincides with C_p at 4°C, to about 16.2 cal/mol deg at the 100°C boiling point. As water at 1 atm is cooled below 4C, through the freezing point, and into the supercooled state, C_V

rises above C_p. But as Eq. (IX.18) requires, these quantities again approach one another and attain equality at the temperature of minimum density.

(3) ISOTHERMAL COMPRESSIBILITY

The positive thermal expansion exhibited by "conventional" liquids results in a steady reduction in density as temperature rises under isobaric conditions. This density reduction then leads to a corresponding steady increase in isothermal compressibility. Except at high pressure, water deviates significantly from this simple behavior. Below $T = 46°C$ on the 1 atm isobar, κ_T decreases as temperature rises. After passing through a minimum at 46°C, it rises continuously as T increases to the 100°C boiling point [Eisenberg and Kauzmann, 1969, Figure 4.15]. In line with similar behavior noted for the molar volume and for C_p, the isothermal compressibility at 1 atm rises strongly as water is supercooled toward –30°C. Measurements of κ_T are not available below this temperature for the supercooled state, and in particular do not extend to the region of the previously mentioned density minimum that has been reported to exist around –64°C.

(4) LIQUID WATER IN TENSION

In addition to being rendered metastable by supercooling below the equilibrium freezing/melting temperature $T_m(p)$, liquid water can also be brought into a state of tension, i.e., $p < 0$. Figure IX.12 indicates schematically this larger domain of metastable states of liquid water in the T,p plane. Four qualitatively distinct cases are shown, whereby an equilibrium liquid sample at point A could be rendered metastable along distinct paths in the T,p plane. Simple supercooling at positive pressure is represented by path $A \rightarrow B$, where the nucleation event that would conceivably put the metastable liquid in jeopardy would be the spontaneous formation of an ice Ih seed. Path $A \rightarrow C$ is a representative superheating path, taking the liquid across the equilibrium vapor pressure curve but remaining at positive pressure. The relevant nucleation event for $A \rightarrow C$ is formation of a critical-nucleus vapor bubble. Path $A \rightarrow D$ can also be classified as producing superheated liquid but now in a state of tension, $p < 0$. Finally, $A \rightarrow E$ generates a liquid sample in tension that is simultaneously metastable, both with respect to ice Ih nucleation and to vapor bubble nucleation, so paradoxically it is both supercooled and superheated.

The basin projection procedure that must be applied to the inherent-structure representation of the partition function in order to describe metastable states has been described in Section III.G. In the case of metastable liquids produced by process $A \rightarrow E$, that projection operation needs to be extended to reject basins whose inherent structures display either or both local regions with ice Ih-like, or open bubblelike, characteristics. This type of projection extension is not restricted to this specific case. Even for water, it would apply elsewhere in the T,p plane. In the neighborhood of the ice Ih–ice II–fluid triple point, description of the isobarically supercooled fluid would require elimination of those basins whose inherent structures exhibit portions of either or both ices.

In order to place a liquid sample experimentally in a state of tension, it is possible to surround that sample with a container material to which the water molecules at the surface are strongly attracted, and then to arrange for the container volume to increase by a controlled amount. Alternatively, a substantially fixed-volume container with similarly attracting walls could be fully loaded

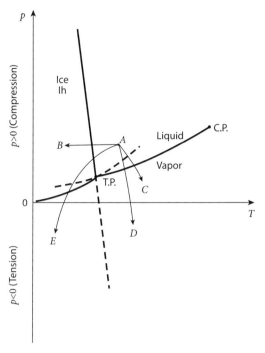

FIGURE IX.12. Schematic diagram showing distinct routes for liquid water to pass from an equilibrium state to alternative types of metastable states (not drawn to scale). Dashed lines indicate extrapolations of phase boundaries. The liquid–vapor–ice Ih triple point is indicated by "T.P.," the liquid–vapor critical point by "C.P."

with hot liquid, which then cools with reduction in pressure. It is even possible to rely on centrifugal force in a spinning liquid sample to create the state of tension [Winnick and Cho, 1971].

Using a Pyrex container with a design originated by Meyer, and employing the hot-liquid-filling tactic, Henderson and Speedy were able to put liquid water into tension, reaching the neighborhood of $p = -200$ bar [Henderson and Speedy, 1987a]. This involved samples in the temperature range from approximately 50°C down to approximately –15°C, including samples of H_2O, D_2O, and their equimolar mixture. Referring to Figure IX.12, these processes included both paths of types $A \rightarrow D$ and $A \rightarrow E$. The results clearly show that $T_{max}(p)$ increases as the tension magnitude increases:

$$dT_{max}(p)/dp < 0, \qquad\qquad (IX.20)$$

thus continuing the trend observed at positive pressures. The same investigators established that the ice Ih melting line also continues to have a negative slope in the $p < 0$ region:

$$dT_m(p)/dP < 0, \qquad\qquad (IX.21)$$

but $T_{max}(p)$ remains greater than $T_m(p)$ [Henderson and Speedy, 1987b].

It has been reported that small water droplets confined to cavities within inorganic crystals such as quartz, calcite, and fluorite can be brought into states of significantly greater tension [Zheng et al., 1991]. This has apparently allowed considerable extension of the determination of the

TABLE IX.5. Static dielectric constant for liquid water at 1 atm[a]

$T(°C)$	$\varepsilon(0)$, H_2O	$\varepsilon(0)$, D_2O
0	87.90	—
5	85.90	85.48
10	83.95	83.53
15	82.04	81.62
20	80.18	79.76
25	78.36	77.94
30	76.58	76.16
35	74.85	74.43
40	73.15	72.74
45	71.50	71.08
50	69.88	69.47
55	68.30	67.90
60	66.76	66.36
65	65.25	64.86
70	63.78	63.39
75	62.34	61.96
80	60.93	60.56
85	59.55	59.19
90	58.20	57.86
95	56.88	56.55
100	55.58	55.28

[a]Weast, 1976, p. E-61.

$T_{max}(p)$ curve into the $p < 0$ regime. In particular, Zheng et al. (1991) reported that a density maximum has been detected at 42°C with the water at tension −1400 bar.

(5) STATIC DIELECTRIC CONSTANT

Compared to many other insulating liquids of roughly similar molecular weight or size (e.g., CH_4, NH_3, Ne, N_2, O_2, H_2S), water exhibits a large static dielectric constant $\varepsilon(0)$ in its stable liquid range. These large values owe their existence to the substantial static dipole moment of the isolated water molecule, to hydrogen-bond interactions that tend to align dipole moments of neighboring molecules, and to various interactions that increase the average molecular dipole moment above that of the isolated molecule. The static dielectric constant specifies the thermodynamic behavior of water in the presence of applied electric fields that are small enough to remain in the linear response regime. Table IX.5 presents measured liquid-phase $\varepsilon(0)$ values for both H_2O and D_2O at ambient pressure. The heavier species D_2O consistently exhibits smaller $\varepsilon(0)$ values than the lighter H_2O, but interestingly neither species has any extremum analogous to those displayed by the molar volumes or isothermal compressibility.

Similar monotonic decline of $\varepsilon(0)$ with increasing temperature also occurs along the vapor–pressure curve from the triple-point temperature T_t to the critical-point temperature T_c. At its $T_c = 374.15°C$, H_2O has a static dielectric constant that is no greater than 9.74 [Eisenberg and Kauzmann, 1969, p. 191].

E. Shear Viscosity and Self-Diffusion

The anomalous liquid thermodynamic properties discussed in Section IX.D emerge from static sampling of the basin distribution for the relevant portion of the landscape at the given T and p. By contrast, the shear viscosity η and the self-diffusion constant D fundamentally involve time-dependent attributes of the many-particle systems, specifically the rates of executing kinetic transitions between basins. However, the properties η and D depend, respectively, on different characteristics of interbasin transitions. Whereas self-diffusion in pure substances is intrinsically a single-molecule property, shear viscosity is a many-body flow property.

Figure IX.13 presents an Arrhenius plot of the liquid water shear viscosity for 1 atm pressure. This plot includes measurements for both thermodynamically stable and supercooled states. The resulting curve displays a strong upward curvature, especially in its low-temperature portion. The slope increases by a factor of approximately 4.7 over the range shown. This can be interpreted as a change in effective enthalpy of activation for viscous flow from about 2.8 kcal/mol at the boiling point to about 13 kcal/mol at the supercooling limit shown. Such distinctly non-Arrhenius behavior is qualitatively similar to that exhibited by η for the fragile glass-forming materials discussed in Chapter VI. This similarity indicates that the region of the potential enthalpy landscape inhabited by liquid water substantially changes its topographical character as the liquid is cooled and

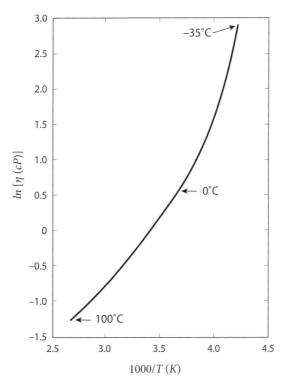

FIGURE IX.13. Arrhenius plot for the shear viscosity of liquid water at 1 atm pressure. Results are included for both the stable liquid temperature range [Weast, 1976, p. F-51], as well as for the supercooled liquid down to –35°C [Angell, 1982, Table III].

then supercooled. In particular, it strongly suggests that cooling causes the multidimensional configuration point to occupy a more rugged, metabasin-characterized section of the landscape.

Just as the low-pressure ices Ih and Ic exhibit open structures resulting from the directionality of hydrogen bonding, a similar but somewhat diminished tendency toward structural openness characterizes the low-pressure inherent structures for the liquid. Increasing the pressure has the effect of crushing that openness, thus necessarily disrupting the preferred hydrogen-bond directionality. An experimental observation bearing out this tendency is the fact that at low temperature, increasing the pressure isothermally from 1 atm initially causes the shear viscosity η to decrease. At 2.2°C, this decrease occurs until the pressure reaches approximately 1,500 bar, at which point it passes through a minimum and then increases with further pressure rise [Bett and Cappi, 1965]. The net reduction at the minimum compared to 1 atm is approximately 8%. This nonmonotonic behavior of η as a function of pressure is suppressed at higher temperature and does not appear above 50°C, at least within the stable liquid range of pressures. Although accurate measurements are apparently not available for supercooled water, elementary extrapolation of $\eta(T, p)$ from the equilibrium liquid indicates that the anomalous pressure-induced reduction of shear viscosity persists and even magnifies in the supercooled regime.

The Stokes-Einstein relation [Eq. (V.143)] provides a traditional, if only approximate, connection between the self-diffusion constant D and the shear viscosity η:

$$D = \frac{k_B T}{C \eta R_{eff}}.$$ (IX.22)

Recall that the underlying model for this relation is that the diffusing particle acts as a radius R_{eff} sphere embedded in an incompressible viscous-fluid medium. The numerical value of the dimensionless constant C depends on the applicable hydrodynamic boundary condition; it is 4π or 6π, respectively, for "slip" or "stick" at the sphere surface. Of course, the water molecule is far from a rigid sphere, and the strongly directional hydrogen bonds that it forms with its discrete neighbors hardly resemble the presence of a structureless hydrodynamic medium. Consequently, one might suspect that predictions based on Eq. (IX.22) would be well off the mark.

States of liquid water for which both D and η measurements are available allow for a rough test of the Stokes-Einstein relation. In particular, the measured values can be combined so as to predict the effective hydrodynamic radius R_{eff}, which then can be compared with the known molecular dimensions. Trivial rearrangement of Eq. (IX.22) produces the following expressions for the two alternative boundary conditions:

$$R_{eff}(stick) = k_B T/(6\pi D \eta)$$ (IX.23)

$$= 2R_{eff}(slip)/3 \, .$$

Table IX.6 shows implied values for these hydrodynamic radii that follow from D and η measurements for a temperature range encompassing both equilibrium and supercooled liquid states. It is somewhat surprising that the radii are more nearly independent of temperature in the upper portion of the range shown, as compared to the temperature variations of D and η. The $R_{eff}(stick)$ values are suspiciously small compared, say, to half the oxygen–oxygen separation in a well-formed hydrogen bond (≈ 1.4 Å), though the $R_{eff}(slip)$ values conform somewhat better in this respect.

TABLE IX.6. Hydrodynamic radii implied by the Stokes-Einstein relation for liquid water at 1 atm

T (K)	D (10^{-5} cm^2 s^{-1})[a]	η(cP)[b,c]	R_{eff} (stick) (Å)	R_{eff} (slip) (Å)
298.2	2.38	0.890	1.03	1.55
285.3	1.69	1.23	1.00	1.51
273.5	1.12	1.77	1.01	1.52
263.7	0.750	2.55	1.01	1.51
261.5	0.675	2.80	1.01	1.52
258.7	0.590	3.18	1.01	1.52
255.8	0.509	3.73	0.99	1.48
254.0	0.467	4.14	0.96	1.44
251.8	0.421	4.84	0.91	1.36
250.0	0.364	5.50	0.91	1.37
248.2	0.343	6.45	0.82	1.23
246.3	0.281	7.63	0.84	1.26
244.4	0.234	9.10	0.84	1.26
242.5	0.200	11.1	0.80	1.20

[a]Angell, 1982, Table II.
[b]Weast, 1976, p. F-51.
[c]Angell, 1982, Table III, with interpolation.

However, this latter observation is surely more coincidence than direct support for a literal interpretation of the Stokes-Einstein model as applied to liquid water.

The decline of both R_{eff}'s as the liquid becomes more and more deeply supercooled represents a decoupling of D and η, just as noted in Chapter VI for fragile glass formers. Self-diffusion rates do not decrease as fast with decreasing temperature as might be inferred from the non-Arrhenius temperature dependence of shear viscosity. Although the increasing ruggedness of the inhabited landscape upon isobaric cooling influences both D and η, these properties evidently sample that ruggedness in distinctly different ways.

At least in the temperature range 0°C–100°C, the self-diffusion constant increases isothermally with increasing pressure, up to approximately $p = 1$ kbar [Angell, 1982, Figure 20]. This is another indication that compressing and crushing the irregularly hydrogen-bonded water medium acts to free individual water molecules from their confining neighbors.

F. Liquid Pair Correlation Functions

A basic descriptor of molecular order in the condensed phases of water is the pair distribution function ρ_2. The four hydrogen nuclei plus two oxygen nuclei included in a pair of water molecules present 18 position coordinates that in general are all required to serve as variables for ρ_2. However, the isotropy and uniformity of an extended fluid phase reduces these 18 to 12 by eliminating the six degrees of freedom that correspond to rigid-body translations and rotations of a dimer. These 12 degrees of freedom are the same set needed for the water molecule pair potential $v^{(2)}$ discussed in Section IX.A. For some purposes, these twelve variables can be reduced even further to six, with an obvious partial loss of information, by recognizing the intrinsic stiffness of the molecules and integrating over intramolecular vibrational deformations. One natural choice for these six remaining configurational variables would be three polar coordinates that locate the oxygen nucleus of a second molecule relative to the first one held in a fixed orientation plus three Euler angles to specify the orientation of that second molecule relative to the first.

Let $\mathbf{x}_{12}$ denote the six-component vector of relative configuration for a pair of rigid water molecules, and suppose that the system has a given temperature and number density $\rho = \rho_O = \rho_H/2$. Then one can write

$$\rho_2(\mathbf{x}_{12}) = \rho^2 g_2(\mathbf{x}_{12}) \qquad \text{(IX.24)}$$

where g_2 is the pair correlation function for the water medium in the infinite system limit. If the phase under consideration is an isotropic fluid, one has the statistical independence of the two water molecules when they are widely separated:

$$\lim_{R_{12} \to \infty} g_2(\mathbf{x}_{12}) = 1, \qquad \text{(IX.25)}$$

where R_{12} is a scalar distance between fixed points in each water molecule.

The principal experimental methods for determining local order in condensed-matter systems are X-ray and neutron diffraction. Establishing the full pair distribution function ρ_2 for water would require more information than these techniques can provide. However, they are capable of determining radial distributions for each of the three kinds of nuclear pairs in water. The corresponding nuclear pair distribution functions and their correlation functions at given temperature and density depend only on scalar distance and are the following:

$$\rho_{OO}(r_{12}) = \rho_O^2 g_{OO}(r_{12}),$$

$$\rho_{OH}(r_{12}) \equiv \rho_{HO}(r_{12}) = \rho_O \rho_H g_{OH}(r_{12}), \qquad \text{(IX.26)}$$

$$\rho_{HH}(r_{12}) = \rho_H^2 g_{HH}(r_{12}).$$

The three nuclear pair correlation functions g_{OO}, g_{OH}, and g_{HH} appearing here each approach unity at large separation r_{12}. The *OH* and *HH* pair functions contain contributions from both intramolecular and intermolecular pairs of nuclei.

The limited amount of structural information carried even by the full pair correlation function $\rho_2(\mathbf{x}_{12})$ implies that it alone cannot characterize the condensed phases of water in the detail that might reasonably be desired. As examples, it cannot provide the size distribution of closed hydrogen-bond polygons, or even the distribution of water molecules according to the number of hydrogen bonds in which they participate for the thermodynamic state of interest. Nevertheless, $\rho_2(\mathbf{x}_{12})$ information provides constraints on such spatial distributions and thus can serve in principle as a consistency check on proposals for those distributions.

One outstanding exception to this limitation involves the isothermal compressibility κ_T. Recall that this thermodynamic quantity is related to density fluctuations of long wavelength in the many-molecule system. Presuming that the water molecules substantially all remain intact, the intramolecular chemical bonds force the oxygens and the hydrogens to participate together in such long-wavelength density fluctuations. Consequently, even the radial pair correlation functions contain the relevant information for determining κ_T. Specifically, one has the following equivalent expressions for water [cf. Eq. (V.21)]:

$$\rho k_B T \kappa_T = 1 + \rho \int [g_{OO}(r) - 1] d\mathbf{r}$$

$$= \rho \int [g_{OH}(r) - 1] d\mathbf{r} \qquad\qquad (IX.27)$$

$$= (1/2) + \rho \int [g_{HH}(r) - 1] d\mathbf{r},$$

where the integrations in the latter two forms include the intramolecular contributions.

One might note in passing that the three nuclear pair correlation functions suffice to describe all of the classical-limit thermodynamic properties for a specific primitive but analytically convenient family of water models. These are the so-called "central force models," in which the full potential energy hypersurface for the interacting flexible water molecules is approximated by a sum of radial atom pair interactions [Lemberg and Stillinger, 1975; Rahman et al., 1975; Duh et al., 1995]. The pair interaction functions are chosen to form stable H_2O molecular units that exhibit intramolecular vibrations and that are capable of dissociating at elevated temperature.

The substantial difference in neutron scattering lengths of H and D nuclei offers an opportunity to determine the three radial pair correlations for water, in both liquid and solid states. Separate diffraction experiments at a chosen temperature and pressure can be carried out on samples of pure light water H_2O, pure heavy water D_2O, and their mixtures in which random H, D exchanges spontaneously occur. Assuming that isotope effects on spatial distributions of nuclear pair distances are negligible, the resulting diffraction data can then be processed to extract separate contributions from oxygen–oxygen, oxygen–hydrogen, and hydrogen–hydrogen pairs [Soper, 2000; Botti et al., 2002]. Figure IX.14 presents an example of the inferred radial pair correlation functions for liquid water at ambient conditions. These functions present several notable features:

(a) A small-distance, strong, narrow peak stands out in the cross-correlation function $g_{OH}(r)$. Its center is close to $r = 1$ Å, and contains all intramolecular $O - H$ pairs.

(b) Although not quite as narrow, prominent, or separated as the first $g_{OH}(r)$ peak, $g_{HH}(r)$ also presents an intramolecular peak centered close to $r = 1.6$ Å. The difference in widths correlates with the stiffness of covalent $O - H$ bonds with respect to stretching, vs. the relative softness of the intramolecular bending degree of freedom which directly affects the $H - H$ separation.

(c) The first $g_{OO}(r)$ peak around $r = 2.8$ Å occurs in the distance range expected for directly hydrogen-bonded water molecule pairs, but it is not isolated from the remainder of the distribution of $O - O$ pairs at larger separation.

(d) The second $g_{OH}(r)$ peak around $r = 1.8$ Å occurs where one would expect to find pairs involved in a hydrogen bond as a donated hydrogen and a recipient oxygen.

(e) The second $g_{HH}(r)$ peak centered near $r = 2.3$ Å is consistent with the geometry of $H - H$ pairs, one of which is a donated hydrogen atom in a near-linear hydrogen bond, the other a pendant hydrogen in the recipient water molecule.

(f) The broad second peak exhibited by $g_{OO}(r)$ around $r = 4.5$ Å is centered near where one would expect an oxygen pair from a sequence of three hydrogen-bonded water molecules, with an angle between the two hydrogen bonds distributed around the tetrahedral angle $\theta_t = \cos^{-1}(-1.3) \cong 109.47°$. However, the considerable breadth of this feature, and the resulting overlap with the first $g_{OO}(r)$ peak and with the larger distance remainder of

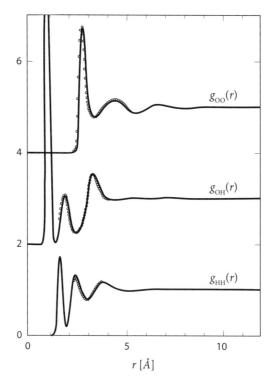

FIGURE IX.14. Radial pair correlation functions for liquid water at $T = 298$ K, $p = 1$ bar. The pairs of closely-spaced curves (solid line, open circles) represent two different ways of processing the neutron diffraction data. Reproduced from Soper, 2000, Figure 6.

    the distribution, indicates that the characteristic hydrogen-bond network is very defective in comparison with the ice Ih ideal.

(g) The short-range order appearing in each of the three radial correlation functions has substantially disappeared at $r = 10$ Å and beyond.

If it were possible experimentally to map onto inherent structures the water system configurations that have contributed to the results shown in Figure IX.14, there is no doubt that the various features of the pair correlation functions would sharpen substantially. In this respect, water would qualitatively behave similarly to nonhydrogen-bonded liquids. However, one would expect the intrabasin vibration-free results to reveal even more vividly the intrinsic disorder of the hydrogen-bonding geometry that pervades the liquid water medium.

Computer simulation studies directed at understanding liquid water have a substantial historical record that includes results for the three radial correlation functions. Examples from that history include classical simulations based upon a variety of effective pair potentials, such as the "ST2" potential [Stillinger and Rahman, 1974], the "MCY" potential [Lie et al., 1976], the "TIP4P" potential [Jorgensen et al., 1983], and the "SPC/E" potential [Berendsen et al., 1987]. Although these simulation results quantitatively differ somewhat from experimental determinations and differ quantitatively somewhat among themselves, they are in satisfying qualitative agreement with the features (a)–(g) listed above. To the extent that such classical simulations are reliable indicators of overall hydrogen-bond structure in liquid water, they support the view that a space-filling, irregular, defective hydrogen-bond network is present in the liquid at ordinary temperatures and pressure. Many published water simulations have examined the effects on pair correlation functions caused by variations in temperature and pressure. The qualitative view of irregular and defective hydrogen-bonding networks is also supported by simulations based on more sophisticated representations of the water molecule interactions, such as the "ab initio" method that computes overall interaction "on the fly" [Sharma et al., 2007], as well as "path integral" approaches that incorporate quantum effects on the nuclear degrees of freedom [Stern and Berne, 2001].

The incomplete picture of local structure in liquid water that is conveyed by the three radial correlation functions g_{OO}, g_{OH}, and g_{HH} can be supplemented by another independent method of presenting molecular pair correlation function information [Svishchev and Kusalik, 1993]. This involves construction and graphical representation of three-dimensional probability contours for oxygen nucleus occurrences in the neighborhood of a spatially fixed water molecule. Once again, this involves projecting the full six-dimensional pair correlation function for rigid molecules, or the full twelve-dimensional pair correlation function for flexible molecules, onto a lower dimensional space. This alternative approach conveys an independent view from that of the three radial correlation functions. Nevertheless, it again confirms the presence of an irregular and defective space-filling hydrogen-bond network as the basic structural character of liquid water, at least under conditions that do not involve extremely high temperature and/or pressure.

Although construction of inherent structures for the liquid is not directly possible experimentally, it is a procedure that has frequently been carried out numerically as part of computer modeling efforts [Tanaka and Nakanishi, 1991; Stillinger and Head-Gordon, 1993; Liu et al., 2005]. These published examples have uniformly invoked isochoric (constant-volume) mapping to potential energy minima. Just as has been observed for numerical simulations of "simple" liquids, mapping to water model inherent structures markedly enhances the image of local order in the many-molecule medium. As expected, the distinguishing features of the radial pair correlation functions undergo significant narrowing and height enhancement, thus reducing interpretation ambiguity. The image-enhanced results clearly reinforce the picture of irregularity and defectiveness in the hydrogen-bond network. Figure IX.15 provides a specific example of the image-sharpening characteristic of isochoric mapping to inherent structures. The case shown is a $T = 298.15$ K liquid sample at the experimental density, modeled using molecular dynamics with the TIP4P effective pair potential [Tanaka and Nakanishi, 1991]. It is clear that the intermolecular features have undergone considerable clarification upon removal of intrabasin vibrational displacements, but typically no long-range order has appeared in the mapped pair correlation functions.

G. Ionic Dissociation

Under ambient conditions ($T \approx 298$ K, $p \approx 1$ bar), the vast majority of the water molecules in the equilibrated liquid remain chemically intact as they vibrate, rotate, translate, and experience time-varying interactions with neighbors. This situation continues to apply upon lowering the temperature, leading either to supercooling or to freezing into ice Ih. Theoretical investigation of the majority of water properties for N molecules under these circumstances can then justifiably be concerned only with that portion of the entire $9N$-dimensional configuration space in which each oxygen remains bonded to two specific hydrogens.

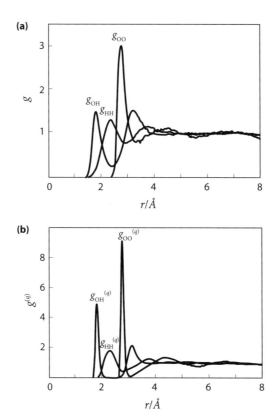

FIGURE IX.15. Intermolecular portions of radial pair correlation functions for liquid water at $T = 298.15$ K and the corresponding experimental density, simulated using the TIP4P effective pair potential. Panel (a) shows results for the thermally equilibrated liquid, and panel (b) shows the radial pair correlation functions averaged over inherent structures obtained by mapping configurations from the thermally equilibrated run. Reproduced from Tanaka and Nakanishi, 1991, Figures 1 and 2.

Although it is very improbable under those ambient conditions, it is nevertheless possible for a local energy fluctuation to cause reversible dissociation of a water molecule into ionic fragments:

$$H_2O \leftrightarrow H^+ + OH^-.$$

The resulting hydrogen cation H^+ and hydroxide anion OH^- can diffuse away from one another and recombine with other dissociation-product ions of opposite charge. Alternatively, the following reversible atom exchange reactions can conceivably occur:

$$H^+ + H_2O \leftrightarrow H_2O + H^+,$$

$$OH^- + H_2O \leftrightarrow H_2O + OH^-.$$

The net result of these processes is to open up the full $9N$-dimensional configuration space and all its basins for occupancy. These basins include all those for intact water molecules with all possible atom permutational exchanges. But additional basins exist for which the inherent structures exhibit distinct local features ("point defects") that can be identified as hydrogen and hydroxide ions that have resulted from dissociation.

Experimentally determining the equal concentrations of H^+ and OH^- ions in pure water traditionally has relied on measuring the voltage of electrochemical cells. These cells are based in part on the redox properties of H^+. The device often used for this purpose is the "glass electrode" [Eisenman, 1967], which in fact can be used to determine hydrogen cation concentrations in a wide variety of aqueous liquids, not just pure water. For most practical situations, its output is interpreted in terms of "pH", a quantity related to the hydrogen cation concentration in moles/liter, $c(H^+)$, by the formula:

$$pH = -\log_{10}[c(H^+)].$$

(IX.28)

A conventional alternative measure of the extent of ion formation at thermal equilibrium is the "dissociation constant" for water, expressed as a ratio of "activities" for the species:

$$K_w = \frac{a(H^+)a(OH^-)}{a(H_2O)}.$$

(IX.29)

The usual convention for defining these activities sets $a(H_2O)$ equal to unity for undissociated water, and the ion activities are reckoned relative to those of an ideal solution, so in view of the very small extent of dissociation around ambient conditions, this last expression reduces to

$$K_w \approx c(H^+)c(OH^-) = [c(H^+)]^2.$$

(IX.30)

Table IX.7 presents the experimentally measured ion concentrations for the liquid at 1 atm. The low values listed justify the claim that water is a very weak acid. They also justify neglect of

TABLE IX.7. Temperature dependence of the concentration of dissociation-produced ions in liquid H_2O at 1 atm[a]

$T(°C)$	$c(H^+) = c(OH^-)(10^{-7}$ mol/L$)$
0	0.337
5	0.430
10	0.540
15	0.671
20	0.825
25	1.004
30	1.212
35	1.445
40	1.708
45	2.004
50	2.340
55	2.701
60	3.101

[a]Concentrations calculated from results listed in Robinson and Stokes, 1959, Appendix 12.2.

the dissociation process for analysis of many thermodynamic and kinetic properties, whether by laboratory measurement, by analytical theory, or by numerical simulation.

It may be useful to provide alternative ways to interpret the concentrations listed in Table IX.7. One is simply to calculate a mean distance of separation l between neighboring ions of either kind by means of the formula

$$l = \{V/[N(H^+) + N(OH^-)]\}^{1/3} \tag{IX.31}$$

$$\equiv [2c(H^+)]^{-1/3},$$

where $N(H^+) = N(OH^-)$ is the expected number of ions of the indicated type in a liquid sample of volume V. The 25°C entry in the table leads to the result $l = 2022$ Å. Another relevant length is the Debye electrolyte shielding distance κ^{-1} for these monovalent ions. For this case

$$\kappa^2 = \frac{8\pi c(H^+)e^2}{\varepsilon(0)k_B T}, \tag{IX.32}$$

involving the proton charge e and the static dielectric constant $\varepsilon(0)$ of pure water. At the same 25°C, 1 atm liquid state, this yields $\kappa^{-1} = 9590$ Å. The ions produced by dissociation in this thermodynamic state form a very dilute and thus weakly interacting electrolyte. By comparing these two characteristic distances, one sees that the shielding ion atmosphere according to the Debye theory of electrolytes encompasses many local ions of both charges.

Some experimental results are available for water ionization at elevated temperature and pressure [Tödheide, 1972, Section 5]. By following the saturation (vapor-pressure) curve of the liquid from the triple point to the critical point, the degree of dissociation at first rises but then passes through a maximum at around 240°C and decreases monotonically to the critical point, a feature attributable to diminished ion solvation caused by the substantial decrease in density along that path. Alternatively, raising the temperature while holding the liquid density fixed inevitably causes the degree of dissociation to rise without interruption, well into the supercritical regime. Under extremely high T and p, the majority of the water molecules may dissociate into ions, even before entering the regime of electronic excitation.

Under the same temperature and pressure conditions as for light water H_2O, heavy water D_2O exhibits a significantly lesser extent of ionization. Qualitatively at least, this is to be expected because of the lesser contribution to quantum kinetic energy within an intact molecule with the heavier hydrogen isotope. At 25°C and 1 atm, for example, $c(D^+) = c(OD^-)$ is only about 37% of that for light water [Covington et al., 1966].

Ionic dissociation is not confined to the liquid state but exists as well in ice Ih. However, accurate experimental determination of the dissociation constant in ice poses a more difficult experimental challenge in comparison with the liquid. Available estimates indicate that near its melting point, ion concentrations in H_2O ice Ih are approximately two orders of magnitude less than extrapolation of equilibrium liquid values (such as those in Table IX.7) into the supercooled regime would indicate [Fletcher, 1970, p. 150]. Presumably, ion concentrations in D_2O ice Ih at the same temperature and pressure would be even less. It should be noted in passing that the "spin ice" magnetic analog of proton-disordered ice Ic has magnetic monopole excitations that correspond to ionic defects in the latter [Castelnovo et al., 2008].

TABLE IX.8. Thermodynamic property changes caused by ionization in liquid light water (H$_2$O) at $T = 298$ K, 1 atm[a]

Property	Value
ΔH^0	13.34 kcal/mol
ΔS^0	−19.31 cal/mol deg
$\Delta C_p^{\,0}$	−53.5 cal/mol deg
ΔV^0	−22.13 cm^3/mol

[a]Values taken from Hepler and Woolley, 1973, Table IV.

The variations of the dissociation constant K_w with respect to temperature and pressure lead to calculations for changes in basic thermodynamic properties as a result of ionic dissociation. In the large-system limit, ionic dissociation of one mole of intact water molecules at $T = 298$ K and $p = 1$ atm leads to standard enthalpy (ΔH^0), entropy (ΔS^0), heat capacity ($\Delta C_p^{\,0}$), and volume (ΔV^0) changes listed in Table IX.8. The positive enthalpy change ΔH^0 includes a very large energy, estimated by quantum calculations to be 426 kcal/mol [Newton and Ehrenson, 1971], that must be expended to break apart an isolated intact water molecule into well-separated bare and static ionic fragments H$^+$ and OH$^-$. Most of this is regained as each of those ions is solvated by surrounding water. In addition, the changes in entropy ΔS^0 and volume ΔV^0 indicate that those electrostatically charged ionic fragments produce a major local disruption in the irregular hydrogen-bond network, compressing it and reducing its configurational freedom. This is a molecular-scale version of the macroscopic electrostriction phenomenon that dielectric materials experience when they are subjected to strong static electric fields [Landau and Lifshitz, 1960, Section 12]. The negative volume change at least partly rationalizes the steep reduction in dissociation upon freezing the water because the rather rigid ice Ih structure is largely incompatible with such a local compression. The negative change in heat capacity $\Delta C_p^{\,0}$ indicates that the solvation structures around the H$^+$ and OH$^-$ ions are relatively rigid, effectively resisting thermal excitation around room temperature as compared with the water medium away from ions.

One approach to understanding the solvation of H$^+$ and OH$^-$ ionic fragments, and therefore the Table IX.8 entries, is to examine structures and energies of simple gas-phase clusters H$^+$(H$_2$O)$_n$ and OH$^-$(H$_2$O)$_n$. Figure IX.16 provides two elementary examples, illustrating the structures of the singly and doubly hydrated proton clusters at their electronic ground-state potential minima (cluster inherent structures) as predicted by the Hartree-Fock approximation [Newton and Ehrenson, 1971].

The monohydrate, Figure IX.16(a), is usually called the "hydronium ion," although "oxonium ion" is also used. This monohydrate is pyramidal with C_{3v} symmetry. Consequently, there is no distinguishing the "extra" proton from those supplied by the hydrating molecule. The binding energy of a proton to a water molecule to produce a hydronium ion, as defined by the energies of the respective inherent structures, has been estimated from quantum mechanical calculations to be approximately 177 kcal/mol [Kollman and Bender, 1973]. The resulting four-atom cluster H$_3$O$^+$ is isoelectronic with the neutral ammonium molecule NH$_3$, which it resembles geometrically. The barrier height for inversion of the H$_3$O$^+$ pyramid ($C_{3v} \rightarrow D_{3h} \rightarrow C_{3v}$) has been estimated

(a)

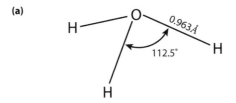

(b)

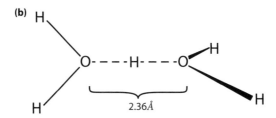

FIGURE IX.16. Stable geometries (inherent structures on the Born-Oppenheimer ground-state potential surface) for (a) a monohydrated proton (H_3O^+) [Kollman and Bender, 1973], and (b) a dihydrated proton ($H_5O_2^+$) [Newton and Ehrenson, 1971].

to lie in the 2–3 kcal/mol range, which is substantially less than the corresponding value for ammonia, 5.78 kcal/mol [Kollman and Bender, 1973].

The proton dihydrate ($H_5O_2^+$) inherent structure shown in Figure IX.16(b) exhibits symmetry D_{2d}. It consists of a short symmetrical hydrogen bond connecting the oxygens of the two minimally distorted water molecules. The structure is analogous to that of the isoelectronic bifluoride anion FHF$^-$, which also contains a symmetrical hydrogen bond [Pauling, 1960, Section 12-3], likewise short compared to the hydrogen-bond length between a neutral pair of neighboring HF molecules [Dyke et al., 1972]. The pendant water molecules of the dihydrate occupy perpendicular planes and retain O–H covalent bonds close in length and angle to those of an isolated water molecule. In this cluster, the centered hydrogen is naturally identified as the excess ion, in contrast to the ambiguity presented by the monohydrate. Holding the other six nuclei fixed, this proton experiences a flat potential energy as it is displaced along the axis connecting the oxygens [Newton and Ehrenson, 1971].

Experimentally determined crystal structures are known that exhibit localized H_3O^+ or $H_5O_2^+$ cationic units. The former has been observed both in hydrogen chloride monohydrate ($HCl \cdot H_2O$) [Yoon and Carpenter, 1959] and in sulfuric acid dihydrate ($H_2SO_4 \cdot 2H_2O$) [Taesler and Olovsson, 1969]. The latter occurs in hydrogen chloride dihydrate ($HCl \cdot 2H_2O$) [Lundgren and Olovsson, 1967] and in perchloric acid dihydrate ($HClO_4 \cdot 2H_2O$) [Olovsson, 1968].

Both the proton monohydrate and the dihydrate structures are capable of donating their outward-pointing hydrogens to additional neighboring water molecules in a formation of near-linear hydrogen bonds. The former can thus add three additional waters to form an $H_9O_4^+$ entity, at the center of which an H_3O^+ unit is still recognizable. However, quantum mechanical calculations for this cluster indicate that the central hydronium unit has become planar because of weak repulsions among the three added water molecules [Newton and Ehrenson, 1971]. This proton tetrahydrate, along with its incompleted trihydrate precursor $H_7O_3^+$, have been observed in crystals of hydrogen bromide tetrahydrate ($HBr \cdot 4H_2O$), inevitably somewhat perturbed in structure

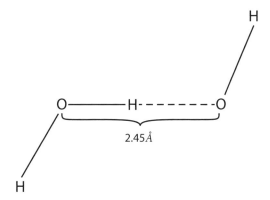

FIGURE IX.17. Inherent structure predicted for the hydroxide monohydrate cluster $H_3O_2^-$ [Newton and Ehrenson, 1971].

by crystal forces [Lundgren and Olovsson, 1968]. The analogous result of forming hydrogen bonds to additional water molecules with the four external hydrogens of the dihydrate cation would be $H_{13}O_6^+$. The lowest inherent structure for this cluster has not yet been the object of high-accuracy quantum mechanical calculations. However, it is worth pointing out that such a cationic entity has also been observed crystallographically in the solid phase of the cage compound $[(C_9H_{18})_3 (NH)_2Cl]^+Cl^-$ [Bell et al., 1975]. These facts provide some additional background for the question of what local geometry or geometries dominate the arrangement of water molecules neighboring an excess H^+ in the liquid or ice phases.

Figure IX.17 illustrates the inherent structure of the isolated hydroxide monohydrate anion $H_3O_2^-$, based on the Hartree-Fock approximation for its electronic ground state. This cluster is planar, with a single short interior hydrogen bond that is slightly longer than the corresponding single hydrogen bond in $H_5O_2^+$ shown in Figure IX.16(b). The bonding hydrogen sits in one of two equivalent asymmetric positions between the oxygens, separated by 0.23 Å. Displacing that hydrogen from one of these positions to the other involves surmounting a low potential energy barrier estimated to be 0.13 kcal/mol [Newton and Ehrenson, 1971]. The two oxygens are available to act as hydrogen acceptors from neighboring water molecules if this $H_3O_2^-$ unit were to be incorporated in a larger hydrating cluster or even an extended liquid water medium.

Under the influence of an applied electric field, the apparent conductances of hydrated H^+ and of OH^- ions in liquid water substantially exceed those of other univalent inorganic ions. Table IX.9 illustrates this behavior with values extrapolated to zero concentration. The reason for the discrepancy is that H^+ and OH^- have available to them sequences of kinetic exchange transitions that do not apply to other ions dissolved in water. The charge-hopping processes involved have traditionally been called the "Grotthuss mechanism" [de Grotthuss, 1806], but its satisfactory molecular-level description is a more recent contribution [Hückel, 1928; Bernal and Fowler, 1933]. Figure IX.18 indicates schematically how an excess proton can be exchanged stepwise along a chain of properly oriented hydrogen bonds by the Grotthuss mechanism, following which the chain orientation has been reversed. Figure IX.19 presents the corresponding sequence for a "proton hole," i.e., a hydroxide anion. Once again the hydrogen-bond orientations are reversed by passage of the ionic excitation.

Several comments are in order. First, it is important to acknowledge that substantial quantum effects (enhanced kinetic motion and tunneling) must be involved in the dynamics of transfer

TABLE IX.9. Limiting equivalent conductances for ions in water[a,b]

$T(°C)$	$\lambda^0(H^+)$	$\lambda^0(OH^-)$	$\lambda^0(Li^+)$	$\lambda^0(Na^+)$	$\lambda^0(K^+)$	$\lambda^0(Cl^-)$	$\lambda^0(Br^-)$
0	225	105	19.4	26.5	40.7	41.0	42.6
5	250.1	—	22.8	30.3	46.8	47.5	49.3
15	300.6	165.9	30.2	39.8	59.7	61.4	63.2
18	315	175.8	32.8	42.8	63.9	66.0	68.0
25	349.8	199.2	38.7	50.10	73.50	76.35	78.1
35	397.0	233.0	48.0	61.5	88.2	92.2	94.0
45	441.4	267.2	58.0	73.7	103.5	108.9	110.7
55	483.1	301.4	68.7	86.9	119.3	126.4	127.9
100	630	450	115	145	195	212	—

[a]Robinson and Stokes, 1959, p. 465.
[b]Conductances λ^0 in cm^2/ohm·equiv.

motion for light protons, but somewhat less so for deuterons. Consequently, the limiting equivalent conductances $\lambda^0(D^+)$ and $\lambda^0(OD^-)$ in D_2O should be less than the light-water values shown in Table IX.9. Second, the fact that $\lambda^0(H^+) > \lambda^0(OH^-)$ must stem at least in part from the observation that the hydrogen bond within $H_5O_2^+$ is shorter and symmetrical [Figure IX.16(b)] compared to that in the asymmetrical $H_3O_2^-$ [Figure IX.17]; these would respectively be the local geometric intermediates for the transfer steps involved in the Grotthuss mechanisms illustrated in Figures IX.18 and IX.19. Note, however, that the ratio $\lambda^0(H^+)/\lambda^0(OH^-)$ trends toward unity with increasing temperature while both conductances are rising, apparently because of the increasing imperfection of the hydrogen-bond network of the liquid. Third, it was noted earlier that the relatively rigid ice Ih structure reduces the ionization probability compared to that of supercooled liquid. Furthermore, the tendency of the relatively rigid ice structure to hold oxygens farther apart than in the $H_5O_2^+$ and H_3O^- transfer units would by itself retard ion mobility. However, the immediate

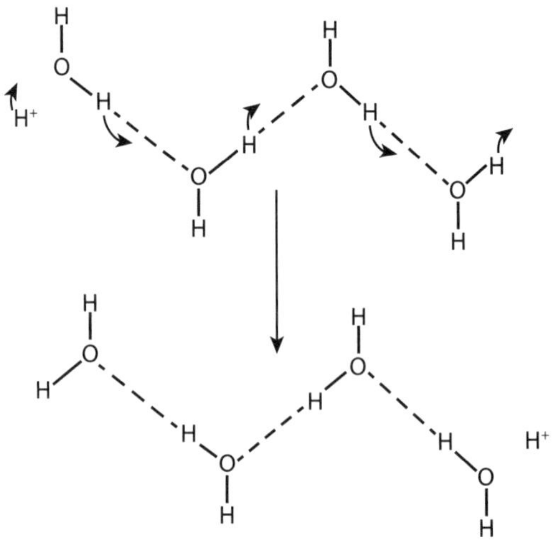

FIGURE IX.18. Sequence of proton shifts along a chain of oriented hydrogen bonds that underlies the anomalously large value of $\lambda^0(H^+)$ in liquid water. This is the so-called Grotthuss mechanism.

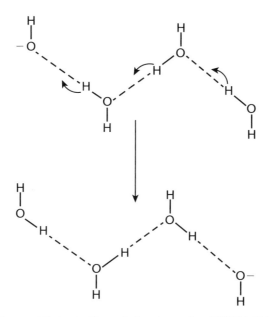

FIGURE IX.19. Sequence of proton shifts involved in producing a large value of $\lambda^0(OH^-)$. This is the "proton hole" version of the Grotthuss conduction mechanism.

availability of many hydrogen-bond pathways in ice Ih compared to the liquid would act in the opposite direction.

The published literature contains a large number of claims and discussions about what would typically be the local liquid-phase surroundings of the H^+ and OH^- ions. In other words, it is concerned with what a molecular-level snapshot would reveal about the solvation geometry for these species that interact strongly with the surrounding undissociated water molecules. For the solvated proton, two prominent proposals are the "Eigen ion" and the "Zundel ion." The former postulates that the dominating solvation structure is that of the $H_9O_4^+ \equiv (H_3O^+)(H_2O)_3$ in which the excess proton is not uniquely distinguished [Eigen, 1964]. The latter suggests that at least a significant structural contributor is the $H_5O_2^+$ unit, with the distinguished excess proton located along the identifiable short central hydrogen bond [Zundel, 1976]. In both cases, these arrangements would interact with surrounding water molecules by additional hydrogen bonds and consequently would experience some structural perturbation compared to the corresponding gas-phase clusters. A similar binary categorization does not exist for the hydrated hydroxide anion, perhaps because a "proton hole" does not have a cluster structure analogous to that of the hydronium H_3O^+.

In principle, distinguishing among the Eigen, Zundel, or any other distinctive structural alternatives for a solvated proton requires a precise configurational criterion for deciding which is the applicable category and precisely where that ion resides, given the positions of all nuclei in that ion's neighborhood. The same situation applies to the hydrated hydroxide anion. A general strategy for this purpose can be based on the collection of inherent structures for the dissociating water system. That is, the ion pattern recognition problem can be reduced first to that of the inherent structures. Then any system configuration within a given basin is categorized by the assignment attached to its inherent structure. This strategy eliminates any source of ambiguity that arises from intrabasin vibrational deformation of the set of nuclei.

At least within a classical dynamical description, a thermally induced dissociation of a water molecule involves a continuous motion of the ionic fragments away from one another. However, it is clear that the inherent structures that follow this process view it in a discontinuous manner. The elementary charges of the H^+ and the OH^- products of dissociation have a strong electrostatic attraction for one another and so do not reside close to one another in a mechanically stable inherent-structure configuration. The initial step in dissociation has to be viewed as a special large-amplitude intrabasin vibration. Strong electrostatic repulsions likewise imply that solvated pairs of H^+ cations or pairs of OH^- anions are not found close to one another in any inherent structure. A relevant open problem is to determine what is the smallest cluster inherent structure $H^+(H_2O)_n(OH^-)$ within which separated and distinguishable hydrogen and hydroxide ions both exist.

The limiting ion conductances listed in Table IX.9 illustrate an additional feature arising from the Grotthus mechanism for H^+ and OH^- transport that distinguishes these from other simple ions. In spite of the fact that crystallographic radii increase along the sequence Li^+, Na^+, and K^+, as well as from Cl^- to Br^- [Pauling, 1960, p. 514], the entries in Table IX.9 at all temperatures show that the mobility increases in the same order. This runs counter to the qualitative expectation based upon mobility of a sphere in a viscous fluid medium, which should be inversely proportional to the sphere radius, whether "stick" or "slip" boundary conditions apply. The explanation involves the strength of binding of the first shell of water molecules around the ion involved, which is greater the closer those water molecules can approach to the ion involved. In particular, this ion–water binding would be especially strong for the very small ion Li^+. Consequently, that ion and its immediate shell of solvating water molecules would have to move as a rather rigid object with a correspondingly enhanced hydrodynamic radius. This hydrodynamic enhancement effect naturally diminishes with increasing ion crystallographic radius, thus allowing for greater mobility of the larger monatomic ions. But of course, such considerations do not come into play when a Grotthus mechanism applies.

Before determining the geometric statistics for H^+ and OH^- hydration, whether in liquid water or in ice Ih, an in-principle algorithm should be proposed to identify which of the hydrogen nuclei are the centers of cations and which of the oxygen nuclei are at the positions of the anions. The following procedure accomplishes that goal on a conceptually simple basis [Stillinger, 1978]. In a system that aside from dissociation nominally contains N water molecules, let $\mathbf{r}_1 \ldots \mathbf{r}_N$ be the positions of the oxygen nuclei in an inherent structure of interest, and let $\mathbf{r}_{N+1} \ldots \mathbf{r}_{3N}$ be the positions of the hydrogen nuclei. These positions define a total of $M = 2N^2$ oxygen–hydrogen distances:

$$l(i,j) = |\mathbf{r}_i - \mathbf{r}_j| \qquad (1 \leq i \leq N, N+1 \leq j \leq 3N). \tag{IX.33}$$

Next, order these lengths by increasing magnitude:

$$l(i_1, j_1) \leq l(i_2, j_2) \leq \ldots \leq l(i_M, j_M). \tag{IX.34}$$

In a macroscopic system, the vast majority of these scalar distances far exceed molecular dimensions and so are irrelevant for this algorithm. However, interest focuses on those $l(i,j) < L$, where the length L should be chosen as a physically motivated O–H bond-stretch dissociation threshold. A reasonable choice for L is one-half of the hydrogen-bond length between nearest neighbor oxygens in the ice Ih crystal at $T = 0$ K and 1 atm:

$$L = 1.375 \text{ Å}. \tag{IX.35}$$

After imposing this length criterion, the remaining distance subset contains, but is not equivalent to, identifiable chemical bonds whose assignment locates the H_2O, H^+, and OH^- species in the inherent structure under examination.

The following iterative procedure now needs to be implemented:

(1) Formally consider as bonded the oxygen–hydrogen pair that exhibits the shortest pair distance in the ordered list. Remove that distance from the list.
(2) Remove from the list all distances $l(i_\mu, j_\nu)$ that involve an oxygen i_μ that has already been bonded twice and/or a hydrogen j_ν that has been bonded once. Proceed to consider the assignment of the smallest remaining distance in the correspondingly edited list.
(3) Return to step (1) if any distances remain in the list.

The outcome of this procedure is that all oxygens are bonded to one or two specific hydrogens, corresponding respectively to OH^- and H_2O species. Each hydrogen participates in no bonds or just one, corresponding, respectively, to H^+, or to incorporation in OH^- or H_2O. Here the possibility of a totally unbonded oxygen (O^{2-}) can be disregarded as too high in energy and reactive to be encountered in an inherent structure. Having completed the steps of this "pattern recognizing" algorithm, it is subsequently possible to analyze the statistics of solvation for each of the three identified species.

H. Surface Properties

At its melting temperature, and for a substantial interval of lower temperature, ice Ih in equilibrium with its vapor exhibits a liquidlike noncrystalline layer at its surface. This can be viewed as a phenomenon inverse to that noted earlier for some of the liquid alkanes and a few other liquids, which possess two-dimensional crystalline order at their liquid–vapor interfaces for a short temperature range above their triple points [Section V.H]. The presence of this mobile surface layer on ice was first recognized and analyzed by Michael Faraday [Faraday, 1860]. More recently, various forms of surface spectroscopy have been turned to the task of determining the temperature-dependent width of the liquidlike layer [Elbaum et al., 1993; Bluhm et al., 2002]. The results from a variety of experimental methods exhibit considerable quantitative discrepancy but agree qualitatively on the presence of a mobile layer above ≈ 225 K [Rosenberg, 2005]. Evidently, the liquidlike film grows in width as T approaches T_m for all exposed crystal surfaces, but in the absence of adsorbed foreign substances, that width remains finite at T_m. It should be expected that at any given $T \leq T_m$, the width of the liquidlike film would depend on which crystal surface was involved.

The presence of a disordered layer at the surface of ice Ih is another manifestation of the directionality and fourfold coordination preference for water molecules in bulk. In the top surface layer, the molecules are missing one or more immediate neighbors and are unable to satisfy that ideal tetrahedral hydrogen-bonding geometry. Because of the presence of fewer hydrogen bonds to hold them in place, the surface molecules are able to vibrate and librate more freely than those in bulk ice, producing stretched and bent hydrogen bonds. Such an intrinsically bond-deficient two-dimensional layer evidently would also have substantially lower barriers to lateral diffusion, compared to the crystal interior. Furthermore, the presence of disorder in an outermost layer would in turn affect underlying layers to some extent. The implication of these observations for the system's relevant inherent structures is that they also must exhibit a disordered surface zone, increasing in width as temperature approaches the bulk melting point. Molecular dynamics simulations of the ice surface reproduce the liquidlike layer phenomenon and clearly illustrate the orientational and translational disorder within the layer [Carignano et al., 2005].

Because the disordered surface remains limited in width even as $T \to T_m$, it would be inaccurate to identify the properties of its outer surface with those of the equilibrium liquid–vapor interface at $T = T_m$. The underlying ice Ih crystal can be expected to exert some influence on the molecular distribution in the liquidlike layer. Furthermore, the finite width of that layer would not permit full development of capillary wave surface fluctuations that the free liquid–vapor interface exhibits.

The planar surface tension $\gamma(T)$ for liquid water in contact with its saturated vapor has been measured from the supercooled regime [Floriano and Angell, 1990] up to the critical temperature $T_c = 647.3$ K (374.1°C) [Schmidt, 1969, p. 172]. Figure IX.20 displays this function of temperature. These data have all emerged from terrestrial-surface experiments. None are available for the liquid–vapor interface subject to substantially different gravitational fields or their centrifugal equivalents. The plot indicates that the second derivative $\gamma''(T)$ is negative at low temperature but changes sign upon approach to the critical point. The point of inflection $\gamma''(T_{infl}) = 0$ occurs at $T_{infl} \approx 250$°C. The available data do not permit a reliable extrapolation toward $T = 0$, but Figure IX.20 roughly suggests that if the corresponding measurements could be made, they would show $\gamma(0) > 100$ dyne/cm.

Upon approach to the critical temperature, the slope $\gamma'(T)$ increases to zero. This reflects the existence of a positive critical exponent μ for surface tension (not to be confused with the symbol for chemical potential):

$$\gamma(T) \cong \gamma_0 (T_c - T)^\mu, \qquad (\gamma_0 > 0). \tag{IX.36}$$

The value that has been assigned to μ for classical liquids including water is [Rowlinson and Widom, 1982, Table 9.1]

$$\mu \approx 1.26. \tag{IX.37}$$

Thermodynamic identities convert the $\gamma(T)$ function for water to values for its interfacial surface excess per unit area of energy, E_s, and entropy, S_s [cf. Eq. (V.107)]:

$$E_s(T) = \gamma(T) - T\gamma'(T); \tag{IX.38}$$

$$S_s(T) = -\gamma'(T). \tag{IX.39}$$

Consequently, both of these quantities also vanish at the critical point, but only with the small leading-order exponent $\mu - 1 \approx 0.26$. However, in the low-temperature limit $\gamma \to E_s$, which in

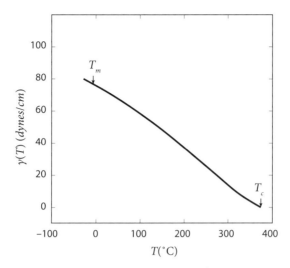

FIGURE IX.20. Surface tension of liquid water in contact with its saturated vapor.

principle would give an estimate of the interface correction to the average inherent-structure energies for the deeply supercooled liquid.

Because of their nonspherical symmetry, water molecules located within an interface between two distinct phases, whether involving ices, the liquid, or the vapor, tend to exhibit anisotropic orientational probabilities. Consider specifically the liquid–vapor interface. Suppose coordinate z measures the distance normal to the nominally planar interface (i.e., its Gibbs dividing surface) located at $z = 0$, with the liquid phase at $z < 0$ and the vapor phase at $z > 0$. Then the singlet density function for the water molecules, assumed for the moment to be nondissociating, may be expressed as $\rho_1(z, \Omega)$, where Ω comprises the intramolecular degrees of freedom. If the molecules can be treated as rigid rotors, Ω amounts to the set of three Euler angles that specify orientation of the molecule relative to a fixed orientation. In the case of the liquid–vapor interface, large positive or large negative z refers to the interior of a bulk phase for which ρ_1 loses its dependence both on z and on Ω.

Because of the broken spatial symmetry within the interfacial zone, the water molecule dipole and higher multipole moments whose orientations are described by Ω do not average to zero when z is small, i.e., the interface would be expected to exhibit a surface polarization and nonvanishing average electric field. This is a phenomenon that in principle applies to liquid–vapor interfaces of all substances whose molecules possess permanent dipole moments. Associated with that surface polarization is a difference in mean electrostatic potential between the interior of the liquid (l) and vapor (v) phases separated by the interface:

$$\Delta \bar{\psi}_{vl} = \bar{\psi}_l - \bar{\psi}_v. \tag{IX.40}$$

This potential difference generally results from contributions of all molecular multipole moments that are anisotropically arranged within the interface. It must be realized that the interfacial polarization and electric field occur spatially where the average magnitude of molecular multipole moments change from that of the bulk liquid (enhanced by neighbor interactions) to that of the vapor (closer to those of an isolated molecule). Upon approach to the critical point, the distinction between liquid and vapor phases vanishes, the broken symmetry disappears, and so $\Delta\bar{\psi}_{vl} = 0$ at the critical point.

Experimental determination of the potential difference $\Delta\bar{\psi}_{vl}$ for any planar liquid–vapor interface of a pure substance is still an elusive goal. However, for the case of water around room temperature, several classical molecular dynamics computer simulation studies have been directed toward determination and analysis of this quantity. Using relatively small clusters of molecules ($N = 64$ and 125) interacting with the TIP4P effective pair potential, Brodskaya and Zakharov (1995) concluded that in the temperature range 290–300 K the liquid-phase potential was below that of the vapor, with $\Delta\bar{\psi}_{vl} \cong -0.65$ V. Sokhan and Tildesley (1997), using the SPC/E effective pair potential for systems with $N = 500$ and 1,000, obtained $\Delta\bar{\psi}_{vl} \cong -0.55$ V at $T = 298$ K. Wick et al. (2006) carried out simulations based on a polarizable four-site water-molecule model, with $N = 1,000$, concluding that $\Delta\bar{\psi}_{vl} \cong -0.48$ V at 298 K, and -0.51 V at 323 K. Because of the modest size of the interfaces produced in these simulations, their respective interfaces would involve rather little contribution from thermally excited capillary waves, compared to a macroscopic interface in a laboratory setting. However, it is unlikely that this distinction would substantially affect the potential difference $\Delta\bar{\psi}_{vl}$. The rough interpretation for these negative values of $\Delta\bar{\psi}_{vl}$ is that water molecules in the interface prefer to point their positively charged hydrogens toward the vapor, leaving their negatively charged oxygens closer to the liquid interior.

The spontaneous water molecule dissociation process that at subcritical temperatures is weak in the interior of the liquid phase, and that is essentially undetectable in the interior of the vapor, is perturbed by the presence of the interface. On the basis of an elementary continuum electrostatic picture, one would conclude that ions solvated within a high-dielectric-constant liquid phase would be repelled from the planar interface with the low-dielectric-constant vapor by their electrostatic point images. This simple viewpoint concludes that a point ion with charge q_i within the liquid ($z < 0$) experiences a single-particle potential energy function:

$$U_{image}(z) = \frac{(\varepsilon_l - \varepsilon_v)q_i^2}{4\varepsilon_l(\varepsilon_l + \varepsilon_v)|z|}, \tag{IX.41}$$

where ε_l and ε_v are the static dielectric constants of the liquid and vapor, respectively. This effective repulsion and the ion depletion it implies for the interface region has formed the basis for theory of the surface tension of dilute electrolytic solutions [Onsager and Samaras, 1934; Buff and Stillinger, 1956].

In view of that ion image effect, it is perhaps surprising that quantum mechanical calculations have established that H^+ cations have an energy preference to accumulate at the outer boundary of the $(H_2O)_{48}H^+$ cluster, as well as in a thin local region at the planar liquid surface [Buch et al., 2007; Swanson et al., 2007]. This is an adsorption phenomenon not shared by the OH^- anions. The calculations show that the surface-localized H^+ cations are present substantially as H_3O^+ hydronium units, oriented so that its three OH bonds participate in hydrogen bonds with water molecules of the liquid phase, thus exposing its oxygen in the direction of the vapor phase. The electrical double layer formed by these surface-confined protons and the diffuse arrangement of charge-compensating hydroxides farther into the liquid [perhaps by a distance comparable to the liquid-phase Debye length κ^{-1}, Eq. (IX.32)] contributes its own surface potential to $\Delta\bar{\psi}_{vl}$, reinforcing and thus increasing the negative value inferred for undissociated water molecules. The precise amount of this reinforcement has not been determined because of present ignorance about the surface density of surface protons, but in any case this contribution should diminish in magnitude upon replacing H_2O with more weakly dissociating D_2O.

I. Solvent Characteristics

The diversity of phenomena involving water as a solvent medium is enormous. Within that broad context, this section has a limited objective. Several specific aqueous solutions and solution phenomena are examined briefly to illustrate some basic principles that indeed have wider applicability. These principles are linked to the underlying energy/enthalpy landscape topographies for water and to the resulting distributions of basins, their inherent structures, and the measurable properties that they produce.

(1) HYDRATED ELECTRONS

The potential energy surface for N water molecules, in the ground-state Born-Oppenheimer approximation, involves $10N$ electrons equally distributed between up and down spin states. Experimental procedures are available for injecting an extra electron into a water system with which it then strongly interacts. The result for the dynamics and statistical mechanics of the N water molecules then requires knowledge of the corresponding $(10N + 1)$-electron Born-Oppenheimer ground-state energy surface. The comparison with properties in the absence of that extra electron by definition constitutes the properties of a hydrated electron.

While isolated from any other water molecules, an electron plus a single water molecule are insufficiently attractive to form a bound state. However, the water dimer anion $(H_2O)_2^-$ does exhibit a quantum mechanical bound state [Naleway and Schwartz, 1972]. Single extra electrons can also bind quantum mechanically to isolated clusters of $N \geq 3$ water molecules, the results of which have been subject to spectroscopic examination [Paik et al., 2004; Asmis et al., 2007].

The intrinsic equivalence of all electrons with the same spin makes it inappropriate to consider the excess electron as a distinguishable species. However, it is legitimate to inquire about the electron distribution in space when $10N + 1$ electrons are present in an N-water-molecule system. For small water clusters with the extra electron, perhaps in the range $4 \leq N \leq 12$ and spatially arranged at their inherent structures, the excess electron density that is clearly not part of the molecular covalent bond distribution appears to reside at the cluster surface, where one of the molecules of the hydrogen-bonded cluster (a double H-bond acceptor) orients its own hydrogens outward from that cluster [Jordan, 2004]. For larger N, such surface-localized excess charge configurations can still be expected to exist as inherent structures, associated with similar outwardly oriented surface molecules. However, for sufficiently large N, one can expect to have inherent structures in which the excess charge is localized within the cluster interior, at a cavity in the hydrogen-bond network of the water molecules. Such interior states for large N are the ones that approach the structures of hydrated electron states observed in bulk water. The interior cavities are thought to have diameters in the range 4–5 Å [Jordan, 2004] and may exhibit inward-pointing OH covalent bonds from surrounding water molecules.

An excess electron solvated within the bulk liquid water presents a roughly spherical localized distribution, usually displaying a radius in the range 0.22 to 0.24 nm. It is capable of being driven spectroscopically into an excited state, a process that can crudely be described as an "s" to "p" transition, in analogy to atomic excitations. This process is observable as a broad absorption around 1.7 eV [Jordan, 2004]. The spectral broadening stems largely from the diversity of geometric arrangements of the water molecules surrounding the cavity. The transition carries the system onto a higher Born-Oppenheimer hypersurface, formally presenting its own collection of inherent structures and transition states for nuclear motion, but from which spontaneous de-excitation returns the system to its ground-state hypersurface.

An inverse to the solvated electron case would be the removal of an electron, leaving $10N - 1$ for the N water molecules. One might initially be tempted to call this a "solvated hole" state, borrowing terminology from solid-state physics for a missing electron. However, the expected outcome is not a mirror image of the solvated electron but would probably entail the loss of one of the N water molecules by the following process:

$$H_2O - e^- \rightarrow (H_2O)^+ \rightarrow H^+ + OH, \tag{IX.42}$$

yielding a hydrogen ion and a hydroxyl free radical, both solvated by the surrounding intact water molecules.

(2) AMMONIUM FLUORIDE IN ICE Ih

When they are cooled to a sufficiently low temperature, most aqueous solutions at ordinary pressures exhibit nucleation and growth of essentially pure ice Ih crystals, leaving the dissolved solute behind and thus increasing its concentration in the coexisting liquid. This solute exclusion thermodynamically induces a freezing-point depression, familiar in the action of "de-icing" substances. It also forms the basis of "zone refining" technology that is widely used for materials purification [Pfann, 1966]. These phenomena involve the simple principle that most solutes in the liquid phase do not readily fit into the crystal structure of the pure solvent.

A notable exception to this behavior occurs with ammonium fluoride, NH_4F. It has been pointed out that this ionic substance in pure form displays a hexagonal crystal structure similar to that of ice Ih [Lonsdale, 1946]. The heavy atoms (N, F) occupy sites closely analogous to the oxygens in ice Ih, with four neighbors and nearest neighbor distances only about 0.1 Å less than that in ice. Furthermore, the tetrahedral symmetry of the ammonium cation NH_4^+ naturally lends itself to hydrogen bonding to its four neighboring fluoride anions. This close isomorphism between crystalline ammonium fluoride and ice Ih lends itself naturally to solid-state solutions of the former in the latter. Indeed, that has been observed in the laboratory, with several mole percent of the water molecules replaced in their normal ice Ih sites by NH_4^+ or by F^- [Labowitz and Westrum, 1961].

A dilute solid solution of ammonium fluoride in ice Ih would typically have the ions widely separated and each one surrounded by four water molecules as nearest neighbors. The resulting local pattern of hydrogen bonds is indicated in a two-dimensional schematic representation in Figure IX.21. Because the ammonium cation acts as a quadruple hydrogen donor to its neighboring water molecules, while the fluoride anion acts as a quadruple hydrogen acceptor, their presence perturbs the "ice rules" degeneracy that applies to the pure ice Ih crystal. A straightforward extension of the Pauling approximation for that degeneracy that was presented in Section IX.C provides an estimate of the perturbation. First, with the incorporation of the two ions, there now

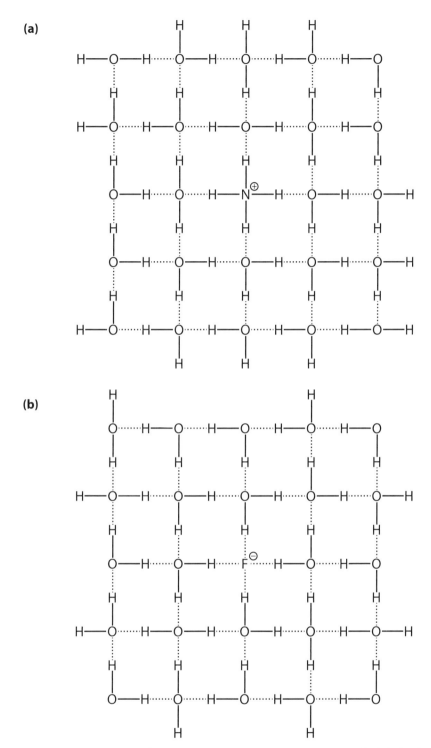

FIGURE IX.21. Schematic representation of (a) ammonium cations and (b) fluoride anions, each of which substitutes for a single water molecule in the ice Ih lattice.

remain only $2N - 8$ hydrogen bonds along which the hydrogens have a pair of possible locations, compared with $2N$ in the pure ice crystal. In assigning attrition factors to the water molecules as Pauling did, one now has to distinguish the eight first neighbors of the incorporated ions. For each of these eight, simple counting shows that only $3/8$ of the hydrogen location assignments along the three outer bonds correspond to intact water molecules. This is the same attrition factor as for pure ice Ih. Consequently, the pure ice degeneracy $(3/2)^N$ is reduced to

$$2^{2N-8}(3/8)^{N-2} = (1/36)(3/2)^N. \tag{IX.43}$$

That is, with each ammonium fluoride ionic unit incorporated and dispersed, the ice rule degeneracy is reduced by a factor approximately equal to $1/36$.

(3) CARBON DIOXIDE

When it is in its ground state and isolated, the carbon dioxide molecule (CO_2) at its potential energy minimum is linear. Its carbon atom resides at the center and is flanked by equivalent oxygens, each at distance 1.159 Å [Pauling, 1960, p. 267]. This centrosymmetric structure necessarily has no dipole moment, but it exhibits an axial quadrupole moment arising from an electron distribution that causes the carbon atom to act as though it had a partial positive charge, and the oxygens a compensating negative charge. The thermodynamic and kinetic properties of aqueous CO_2 solutions have direct relevance to subjects as diverse as the carbon distribution in the terrestrial biosphere and the economic health of the beverage industry. These properties are influenced by a key reversible chemical reaction between carbon dioxide and water that produces carbonic acid:

$$CO_2 + H_2O \leftrightarrow H_2CO_3. \tag{IX.44}$$

Quantum mechanical calculations verify that carbonic acid is a locally stable molecule on its own energy hypersurface, but it lies approximately 9 kcal/mol higher than the sum of energies of the isolated CO_2 and H_2O precursors. However, an energy barrier of approximately 44 kcal/mol must be surmounted to reverse the gas-phase reaction (IX.44), forming carbonic acid, which is high enough to allow experimental isolation of pure H_2CO_3 in solid form at low temperature [Loerting et al., 2001b].

The presence of water molecules immediately neighboring a carbonic acid molecule can lower the energy barrier for decomposition of the latter. Figure IX.22 indicates qualitatively how this catalyzed decomposition can proceed with a single or a pair of added H_2O molecules. The quantum mechanical calculations indicate that these one and two water molecule chains reduce the decomposition barrier to approximately 27 kcal/mol and 24 kcal/mol, respectively [Loerting et al., 2001b]. This phenomenon has substantial kinetic consequences for the spontaneous decomposition of pure solid H_2CO_3 at low temperature. The initial slow decomposition process for the anhydrous material produces water molecules, which then act to speed up the subsequent chemical reaction.

In liquid water at room temperature, the low concentration of carbonic acid that forms when gaseous carbon dioxide dissolves does not have only the decomposition route available to it. The carbonic acid can subsequently ionize by loss of a proton, yielding a bicarbonate anion:

$$H_2CO_3 \leftrightarrow H^+ + HCO_3^-. \tag{IX.45}$$

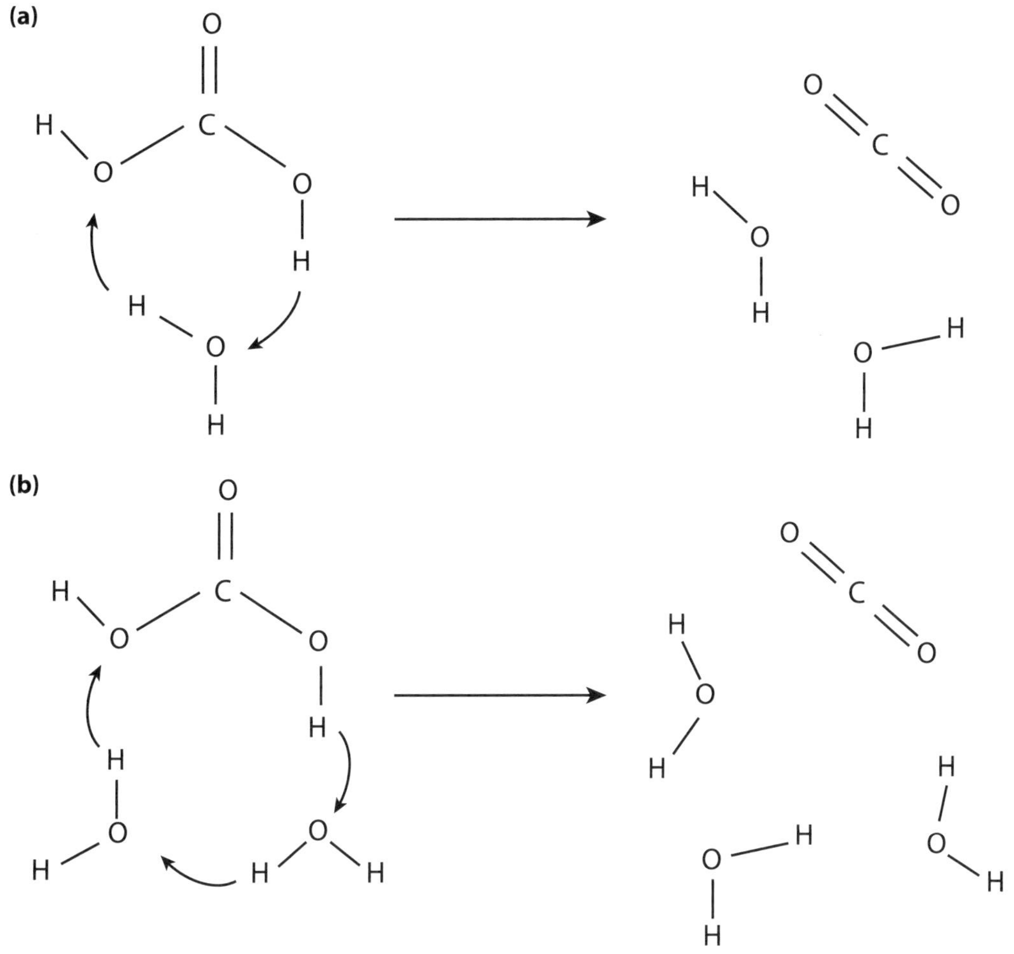

(a)

(b)

FIGURE IX.22. Water-molecule catalyzed decomposition of carbonic acid involving (a) a single added water molecule or (b) a pair of added water molecules.

These ions can diffuse apart and then be stabilized by strong interaction with the surrounding liquid water, details of which for the proton were discussed in Section IX.G. Although it has low probability unless the liquid has been brought to an alkaline (high pH) state by addition of a third component, the bicarbonate ion HCO_3^- can release a second proton to become a carbonate ion:

$$HCO_3^- \leftrightarrow H^+ + CO_3^{2-}. \tag{IX.46}$$

These ionization processes cause the pH = 7 of pure water to decline as it incorporates CO_2. At very low concentrations, chemical equilibrium forces most of the CO_2 added to the liquid to end up as HCO_3^-. However, as increasing pressure causes more and more CO_2 to be forced into the liquid, an increasing fraction remains in this unreacted molecular form, solvated by, but not chemically reacting with, surrounding waters. It should be noted in passing that at low temperature, water and carbon dioxide form a "Class I" clathrate hydrate in which each CO_2 molecule remains intact and is enclosed in a polyhedral cage of water molecules hydrogen-bonded to one another [Davidson, 1973, p. 131]. The qualitative implication of these facts for the energy or enthalpy landscape representation is that the inherent structures for the "binary" system realistically require identification of at least six distinguishable species: H_2O, CO_2, H_2CO_3, HCO_3^-, H^+, and CO_3^{2-}.

The quantitative details of the carbon dioxide solution process in liquid water are specified by a set of measurable chemical "constants." These quantities, and their values that are applicable at 25°C and 1 atm pressure of the gaseous carbon dioxide are the following:

(1) The total solubility of CO_2 in the liquid solution, regardless of subsequent reactions [Carroll et al., 1991]

$$c[\text{"}CO_2\text{", total}] \cong 3.38 \times 10^{-2} \text{ mol/L.} \qquad \text{(IX.47)}$$

(2) The equilibrium concentration ratio K_h in the aqueous solution for the unreacted and reacted forms of carbon dioxide subject to the chemical reaction (IX.44) [Housecroft and Sharpe, 2008, p. 410]:

$$K_h = \frac{c[H_2CO_3]}{c[CO_2]} \qquad \text{(IX.48)}$$

$$\cong 1.70 \times 10^{-3}.$$

(3) The first ionization constant K_1 for carbonic acid described by Eq. (IX.45) [Jolly, 1991, p. 226]:

$$K_1 = \frac{c[H^+]c(HCO_3^-)}{c[H_2CO_3]} \qquad \text{(IX.49)}$$

$$\cong 2.50 \times 10^{-4} \text{ mol/L.}$$

(4) The second ionization constant K_2 for carbonic acid taking the bicarbonate anion to the carbonate anion, reaction (IX.46) [Jolly, 1991, p. 227]:

$$K_2 = \frac{c[H^+]c[CO_3^{2-}]}{c[HCO_3^-]} \qquad \text{(IX.50)}$$

$$\cong 4.84 \times 10^{-11} \text{ mol/L.}$$

These values permit calculation of the equilibrium concentrations of the various species in the aqueous solution at 25°C and 1 atm. The results in mol/L are the following:

$$c[CO_2] \cong 3.36 \times 10^{-2},$$

$$c[H_2CO_3] \cong 5.72 \times 10^{-5},$$

$$c[HCO_3^-] \cong 1.20 \times 10^{-4}, \qquad \text{(IX.51)}$$

$$c[\text{CO}_3{}^{2-}] \cong 4.84 \times 10^{-11},$$

$$c[\text{H}^+] \cong 1.20 \times 10^{-4}.$$

Rates for the forward and reverse reactions in the liquid-phase versions of hydration/dehydration processes, Eq. (IX.44), are quite slow when measured on the typical molecular time scale [Welch et al., 1969; Jolly, 1991, pp. 226–227]. This indicates a kinetic bottleneck in the energy landscape for the aqueous solution, specifically a high transition state or states that must be surmounted for the reactions to proceed. This is consistent with remarks relevant to the gas phase following reaction (IX.44). In the biological regime, living organisms find it advantageous to speed these reactions with an enzyme carbonic anhydrase [Jolly, 1991, p. 567].

In addition to the case of carbon dioxide, there are several other gaseous substances that dissolve in liquid water and can then react chemically with that medium to produce acids with varying ionization strengths. Two examples are nitrogen pentoxide (N_2O_5), which leads to nitric acid (HNO_3), and sulfur trioxide (SO_3), which yields sulfuric acid (H_2SO_4). Many other soluble gases do not react by adding water but are by themselves acids that ionize to produce solvated protons and respective anions. This latter group includes the hydrogen halides ($\text{H}X$, $X = \text{F, Cl,}$ Br, I), as well as hydrogen cyanide (HCN) and some low-molecular-weight organic compounds such as formic acid (HCOOH) and acetic acid (CH_3COOH).

(4) HYDROPHOBICITY

Many nonpolar or weakly polar substances whose molecules are sparingly soluble in liquid water exhibit properties conventionally labeled as hydrophobic effects. In the case of relatively small solute molecules, this involves interactions with surrounding water molecules that are dominated by geometric exclusion effects. Any attractive interactions that may be present are relatively weak and less directional than the hydrogen bonds that operate among the surrounding water molecules. The range of small-particle hydrophobic solutes includes the heavier noble gases (Ne, Ar, Kr, Xe), as well as the diatomics that dominate the terrestrial atmosphere (N_2, O_2). It also includes low-molecular-weight hydrocarbons such as methane (CH_4), ethane (C_2H_6), and cyclopropane (C_3H_6), as well as their perfluorinated analogs (CF_4, C_2F_6, C_3F_6). Sulfur hexafluoride (SF_6) also qualifies for inclusion in this group. Typically, substances that are sparingly soluble in a liquid solvent exhibit increasing solubility as temperature increases. However, hydrophobic solutes in water around room temperature and pressure tend to show a solubility decrease upon heating, though as temperature rises toward the boiling point this trend reverses, thus producing a solubility minimum [Franks, 1973].

The low solubility and its unusual nonmonotonic temperature dependence for these small hydrophobes can be usefully analyzed in terms of solution chemical potentials [Tanford, 1980, Chapters 2 and 4]. In particular, the chemical potential of solute A at mole fraction $x_{A,W}$ in water (W) can be expressed in the following way:

$$\mu_{A,W} = \mu_{A,W}{}^0 + k_B T \ln x_{A,W} + k_B T \ln f_{A,W}. \tag{IX.52}$$

Here $\mu_{A,W}{}^0$ is the standard chemical potential (an additive constant that is required) for the mole-fraction concentration convention. The concentration-dependent activity coefficient $f_{A,W}$ contains the effects of solute–solute interactions in the aqueous solution at $x_{A,W} > 0$, but by definition this quantity approaches unity as $x_{A,W} \to 0$. For another solvent Y that is nonaqueous and nonhydrogen-bonding, such as the higher molecular-weight hydrocarbons that are liquid at room temperature and immiscible with water, the analog of Eq. (IX.52) is

$$\mu_{A,Y} = \mu_{A,Y}{}^0 + k_B T \ln x_{A,Y} + k_B T \ln f_{A,Y}. \tag{IX.53}$$

The distribution of species A between these two solvent media W and Y at thermodynamic equilibrium rests upon equality of the two solute chemical potential expressions. In the low concentration limit for A where both activity coefficients are unity, this equality requires

$$\mu_{A,W}{}^0 - \mu_{A,Y}{}^0 = k_B T \ln(x_{A,W}/x_{A,Y}). \tag{IX.54}$$

The difference in standard chemical potentials for the two liquid solvents, therefore, can be determined by the equilibrium distribution of the solute between the two solution phases in the low concentration limit. It is an experimental fact that this last expression is relatively insensitive to the choice of nonhydrogen-bonding solvent Y, but the concentration ratio $x_{A,W}/x_{A,Y}$ is substantially smaller than unity, verifying that water is a relatively poor solvent for small nonpolar or weakly polar solutes. This amounts to observing that at a given (small) mole fraction of the hydrophobic solute A, its chemical potential is higher in water than in the nonaqueous solvent.

Thermodynamic identities permit the difference of standard chemical potentials to be expressed in terms of infinite-dilution partial molar enthalpies ($\bar{H}_{A,W}{}^0$, $\bar{H}_{A,Y}{}^0$) and partial molar entropies ($\bar{S}_{A,W}{}^0$, $\bar{S}_{A,Y}{}^0$) of A in the two solvents [Tanford, 1980, Chapter 4]:

$$\mu_{A,W}{}^0 - \mu_{A,Y}{}^0 = \bar{H}_{A,W}{}^0 - \bar{H}_{A,Y}{}^0 - T(\bar{S}_{A,W}{}^0 - \bar{S}_{A,Y}{}^0). \tag{IX.55}$$

The enthalpy and the entropy differences can be separately evaluated by virtue of the additional identity:

$$\frac{d[(\mu_{A,W}{}^0 - \mu_{A,Y}{}^0)/T]}{d(1/T)} = \bar{H}_{A,W}{}^0 - \bar{H}_{A,Y}{}^0. \tag{IX.56}$$

In the vicinity of room temperature, this partial molar enthalpy difference has a relatively small magnitude for the small apolar solutes under consideration and is relatively insensitive to the choice of nonaqueous solvent. However, the partial molar entropy difference is distinctly negative and is the reason that solubility is substantially lower in water. The fact that this term in Eq. (IX.55) is multiplied by absolute temperature causes that low solubility to decrease with increasing T near room temperature. Nevertheless, this situation reverses upon increasing the temperature toward the boiling point of water, as mentioned earlier, because of a large positive value measured for the difference in isobaric partial molar heat capacities:

$$\frac{d(\bar{H}_{A,W}{}^0 - \bar{H}_{A,Y}{}^0)}{dT} = (\bar{C}_P{}^0)_{A,W} - (\bar{C}_P{}^0)_{A,Y}. \tag{IX.57}$$

These measurable effects do not appear when pairs of nonaqueous liquid solvents are compared, but they reveal the singular character of liquid water. They can be logically connected to the

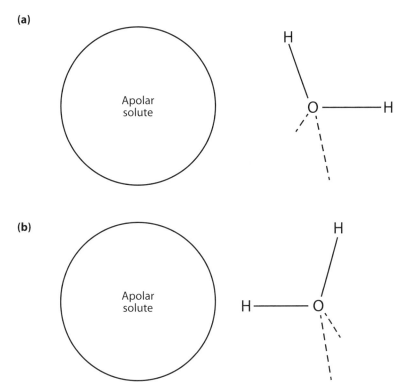

FIGURE IX.23. Orientational preference for water molecules next to a small nonpolar or weakly polar solute. In (a), the possible hydrogen-bonding directions at a first-solvation shell water molecule, in a roughly tetrahedral arrangement, avoid the solute particle and are available to bond to other similarly disposed water molecules. In (b), a different water molecule orientation "wastes" one of the four possible hydrogen-bonding directions as it points toward the solute.

hydrogen-bond patterns that water tends to form in the immediate vicinity of the small nonpolar or weakly polar solutes considered. In particular, around room temperature and below, this involves a strong orientational preference for those water molecules in the first solvent shell to avoid "wasting" a hydrogen-bond possibility that would occur if one of the tetrahedral directions emanating from a neighboring water molecule were pointed toward the relatively inert solute particle with which it cannot hydrogen-bond. This is illustrated schematically in Figure IX.23. The existence of clathrate hydrates and their diverse three-dimensional hydrogen-bond networks illustrates the energetic advantage that such orientational restrictions can yield [Davidson, 1973]. Of course, the typical arrangements of water molecules around the small solute in the liquid solution would not display complete and undistorted polyhedra of hydrogen-bonded water molecules, but identifiable portions of those polyhedra with some distorted and broken bonds are to be expected. Indeed, computer simulations for water surrounding small apolar solutes verify this expectation [Geiger et al., 1979; Chandler, 2005, Figure 1], and it has received indirect support from EXAFS spectroscopy [Bowron et al., 1998]. Furthermore, femtosecond infrared spectrosopy demonstrates that these orientation-restricted water molecules next to apolar molecular groups experience substantial reduction in orientational mobility compared to those in pure bulk water [Rezus and

BOX IX.2. Henry's Law Constants

An alternative experimental approach to the behavior of low-molecular-weight apolar solutes in liquid water can be based on their distribution between the solution and a coexisting gas phase. Henry's law expresses the equilibrium partition between these phases [Guggenheim, 1950, pp. 240–243]. In the low solute concentration limit for both phases, it takes the following form:

$$\lim_{x_{A,W} \to 0} (p_A / x_{A,W}) = h_{A,W}^{(0)}(T).$$

Here p_A is the partial pressure of solute A in the gas phase, $x_{A,W}$ is its mole fraction in the aqueous solution, and the corresponding Henry's law constant has been denoted by $h_{A,W}^{(0)}(T)$. For a nonaqueous liquid solvent Y that is immiscible with water, the analogous Henry's law constant is

$$\lim_{x_{A,Y} \to 0} (p_A / x_{A,Y}) = h_{A,Y}^{(0)}(T).$$

Suppose that the vapor phase containing solute A is simultaneously in thermal equilibrium with both the W and Y liquids (which are therefore in thermal equilibrium with each other). Consequently, in the $p_A \to 0$ limit one has

$$h_{A,W}^{(0)}(T) / h_{A,Y}^{(0)}(T) = x_{A,Y} / x_{A,W},$$

the combination required in Eq. (IX.54) to evaluate the difference between standard chemical potentials for A in the two liquid solvents. Thus, measured Henry's law constants can serve as the input for documenting hydrophobic solvation of small apolar solutes.

Bakker, 2007]. Of course, any clathratelike orientational preference theoretically would be rendered into clearer structural form by mapping the solution configurations onto their underlying inherent structures.

The orientational preferences at room temperature and below explain the anomalous entropy of solution and resulting low solubility. But upon heating the liquid solution toward its boiling point, the clathratelike solvation patterns melt away and are then replaced by higher enthalpy and higher entropy orientational preferences. This produces the observed high partial molar heat capacity and the turnaround in temperature dependence of solubility. As Box IX.2 indicates, experimentally determined Henry's law constants provide access to the solvation thermodynamic properties of some low-molecular-weight solutes.

Convex polyhedra composed of well-hydrogen-bonded water molecules necessarily have an upper limit on their size. When the enclosed solute is less than a nanometer in diameter, polyhedra (or their somewhat defective versions) can form in principle and adhere closely to that solute, especially near the upper size limit. For substantially larger solutes, though, enclosing polyhedra that exhibit tangential hydrogen-bond directions in nearly the tetrahedral directions at each oxygen vertex are no longer available. Therefore, with respect to solute size, at low temperature a qualitative shift in the mode of solvation must occur, from convex clathratelike cage enclosure, to

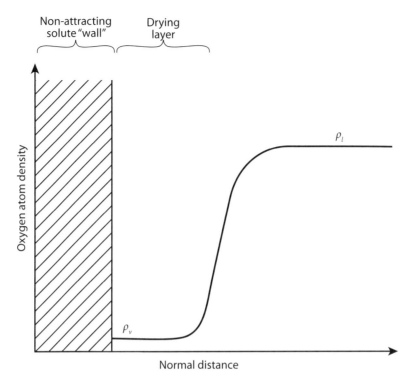

FIGURE IX.24. Surface drying phenomenon next to a near-planar hard wall, i.e., a hypothetical nonattracting large-solute particle.

a less confining interface with a substantial density of severely distorted or broken hydrogen bonds. In the latter circumstance, the neighboring water molecules have no geometric benefit in clustering close to the solute, and so if the ambient pressure is sufficiently low, they can pull away from the solute surface, possibly producing a drying effect [Chandler, 2005; Poynor et al., 2006]. In the hypothetical limit of a very large spherical solute whose surface locally is nearly planar and only weakly attracting for water molecules, the retreating solvent would leave behind essentially a vapor film and a liquid–vapor interface. This possibility was first pointed out in connection with the "scaled particle theory," where the solute particle was modeled simply as a hard sphere with variable radius [Stillinger, 1973]. That situation is illustrated qualitatively in Figure IX.24. The interface is partly driven from the hard particle surface by its natural tendency to minimize free energy by establishing an approximation to the equilibrium liquid–vapor interface profile, including capillary waves, as described in Section V.H. Referring to Eqs. (IX.38) and (IX.39) and the measured temperature dependence of the water liquid–vapor surface tension [Floriano and Angell, 1990], one concludes that this interface should contribute a positive energy and a positive entropy to the large-solute chemical potential.

If the pressure of the solvating water were substantially higher than its vapor pressure at the given temperature, then the vapor film schematically indicated in Figure IX.24 would be squeezed out and the density of water molecules in contact with the large solute would revert to a value comparable to that of the bulk liquid. Similarly, if a substantial attraction (e.g., London dispersion interaction) were operating between the large-solute particle and the water molecules, the vapor film would also be reduced or eliminated [Chandler, 2005]. However, it is significant that computer simulation studies indicate that close approach of two nearly planar hydrophobic surfaces that in isolation have little or no drying layer caused by elevated pressure or attractive forces can in fact spontaneously develop a mutual dry region between them as they approach one another in a parallel arrangement [Walquist and Berne, 1995; Giovambattista et al., 2006]. This mutual drying region generates a large attractive force between the hydrophobic units that can extend over separation distances that are many water molecule diameters [Lum et al., 1999].

At low to moderate temperatures, pairs or larger sets of hydrophobic solutes exhibit water-mediated effective attractions for one another [Ben-Naim, 1980, Chapter 3; Chandler, 2005]. These attractions are simply a reflection of the unfavorable free energy associated with the water molecule structures at the surface of those solutes. The costly interfacial zone around hydrophobic solutes is minimized when they are clustered together, thus replacing at least some water–solute contacts with solute–solute contacts. This phenomenon also operates on single hydrophobic molecules such as long-chain normal alkanes in water to induce intramolecular collapse from an extended linear conformation into a compact folded form [Miller et al., 2007].

Many substances ("amphiphiles") involve molecules whose structures include both hydrophobic and hydrophilic portions. Surfactacts are a significant family with this attribute [Kresheck, 1975]. Examples often consist of a normal alkane "tail" attached at one of its ends to a polar or ionic group. A specific case of the latter is sodium dodecyl sulfate (often abbreviated "SDS"): $CH_3(CH_2)_{11}SO_4^-Na^+$. Because of its hydrophobic tail, its solutions in water display a strong tendency to form micelles, compact clusters comprising roughly 10^2 molecules, in which the hydrophobic alkane tails reside in the interior and the ionic groups gather at the exterior in contact with water [Tanford, 1980, Chapter 6]. These micelles suddenly appear at a "critical micelle concentration" (CMC), which is approximately 0.008 mol/L of surfactant for SDS at 25°C [Kresheck, 1975, p. 97]. Because of the characteristic temperature dependence of hydrophobic solvation that produces the solubility minimum for small apolar solutes that was discussed earlier, critical micelle concentrations also exhibit a minimum with respect to temperature [Kresheck, 1975, pp. 103–105]. In addition to micelle formation, surfactants of course also form a monolayer at the water liquid–vapor interface, with their ionic groups embedded in the outer liquid layer and their hydrophobic tails oriented out of the liquid water, with a consequent reduction in the solution surface tension [Davies and Rideal, 1963].

Molecular biology presents many phenomena and structures made possible by hydrophobic interactions. The formation of bilayer membranes and vesicles involves amphiphilic molecules in the aqueous medium of living organisms. Polypeptides and proteins are linear sequences of chemically linked amino acids that present a significant range of hydrophilic and hydrophobic characteristics. The important subject of protein folding properties and how they depend on hydrophobic–hydrophilic aspects of amino acid sequences constitutes part of the detailed analysis in Chapter XI.

The ranking of a planar solid surface on the hydrophobicity–hydrophilicity scale can be determined experimentally by measuring the equilibrated contact angle θ of a liquid water droplet resting upon that surface. This is illustrated in Figure IX.25. The vapor phase contains the density of water molecules determined by the vapor pressure at the ambient temperature. Three surface free energies per unit area (surface tensions) are involved in this system, to be denoted by γ_{sl}, γ_{lv},

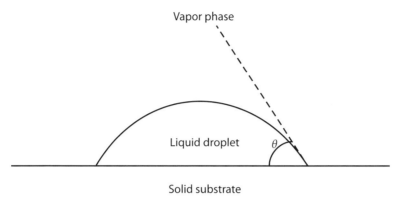

FIGURE IX.25. Liquid droplet on a solid substrate. The tangent line at the droplet edge (three-phase confluence) defines the contact angle θ.

and γ_{sv}, referring respectively to the solid–liquid, liquid–vapor, and solid–vapor interfaces. The contact angle is established by minimizing free energy for the system, which leads to Young's formula [Young, 1805; Rowlinson and Widom, 1982, p. 9]:

$$\gamma_{sv} = \gamma_{sl} + \gamma_{lv} \cos\theta. \qquad (IX.58)$$

In the case of a hydrophilic surface, γ_{sl} is substantially less than the other two surface free energy quantities, and it may even be negative. Consequently, the water tends to spread on the hospitable solid substrate, producing a small positive (or even vanishing) contact angle. But a hydrophobic surface gives rise to a relatively large positive γ_{sl}, so that to maintain equality in Young's formula the contact angle must be large so that $\cos\theta$ becomes small ($\theta \approx \pi/2$) or even becomes negative ($\theta > \pi/2$).

The concept of "superhydrophobicity" has arisen from observations of certain plant species whose leaves are extraordinarily adept at avoiding wettening by water droplets [Neinhuis and Barthlott, 1997]. A similar avoidance of wetting is exhibited by the legs of water strider insects [Gao and Jiang, 2004]. From the standpoint of Young's formula, these biological examples exhibit very large contact angles, i.e., the contact area between the water droplet and the locally planar leaf surface or insect leg is very small. The explanation is that these species have dense arrays of hydrophobic protrusions at the submillimeter length scale, so that the water droplets sit atop those arrays and do not come in full contact with the underlying leaf or leg surface. These empirical observations have stimulated synthetic surface fabrication activity that has been able to produce a wide variety of superhydrophobic surfaces from nonbiological materials, using well-established technologies for small-length-scale manipulation [Quéré, 2005]. The superhydrophobicity phenomenon has also become the object of computer simulation study [Koishi et al., 2009].

J. Amorphous Solid Water

Macroscopic samples of water can be converted into noncrystalline solid form, "amorphous ice," by means of several distinct experimental procedures. The resulting preparations must be kept at low temperature unless the external pressure is high, but in any case there is no reason to believe that their basins are a significant subset of those sampled in any portion of the equilibrium-phase diagram. These solid samples are metastable states of water. At low temperature, they are kinetically trapped in a small set of potential energy or potential enthalpy basins, with small vibrational excursions from the inherent structures at the basin bottoms. The occupancy probabilities for these basins depend on the formation procedure used to produce the amorphous ice. An important question concerns whether the occupied basins for at least some amorphous ice preparations are those that underlie some metastable states of liquid water.

The following are laboratory procedures that have been used to create amorphous solid water:

(a) Slow deposition of water vapor onto cold solid substrates [Burton and Oliver, 1935; Sceats and Rice, 1982]. A variant utilizes deposition from a molecular beam [Smith et al., 1996].
(b) Hyperquenching liquid water by rapid injection into cryofluids [Mayer and Brüggeller, 1982].
(c) Pressurizing ice Ih samples at low temperature ($T \leq 77$ K) to crush their crystalline structure into amorphous form [Mishima et al., 1984].
(d) Exposing low-temperature ice Ih samples to heavy radiation dosage to destroy crystalline order, a process of importance primarily in an astrophysical context [Hansen and McCord, 2004].

The amorphous solid water samples produced by the low-pressure procedures (a) above tend to have properties dependent on the specific details of the deposition method used. However, those differences tend to disappear when those samples are annealed above 77 K. The resulting samples are similar to the amorphous solid water samples produced by (b), ambient-pressure hyperquenching. These amorphous solid water materials conventionally are denoted by the abbreviation "LDA" for "low-density amorphous" water ice. The measured mass density of LDA in its low-temperature range of thermal stability is 0.94 g/cm³ [Debenedetti, 2003, Table 7]. This is the same mass density as that reported for ice Ih at −175°C [Lonsdale, 1958]. However, X-ray and neutron diffraction experiments on LDA indeed clearly show that the molecular-scale structure is noncrystalline [Bellissent-Funel et al., 1992].

When the applied pressure lies in the range of 10 kbar or higher, procedure (c) above produces amorphous solid water with substantially higher mass densities. For example, ice Ih pressurized to 10 kbar at 77 K, then measured at that p, T state has mass density 1.31 g/cm³, but isothermally releasing the pressure to 1 bar leads to a reduction to 1.17 g/cm³ [Mishima et al., 1984]. This alternative type of solid amorphous water is designated "HDA," an acronym for "high-density amorphous" water ice. When it is stored at low temperature, the decompressed form remains indefinitely stable.

A so-called "VHDA" (very high density amorphous) ice has been formed by first heating an HDA sample isobarically at 11 kbar from 77 K to 165 K, or alternatively at 19 kbar from 77 K to 177 K. These pressure-restructured samples are then cooled back to 77 K and decompressed to 1 bar. The mass densities measured for the final products were 1.25 g/cm³, substantially above that of HDA [Loerting et al., 2001a]. As with HDA, this decompressed VHDA state remains stable if kept at low enough temperature. However, heating VHDA under various pressures below that at which it was formed allows it kinetically to relax to HDA or LDA form. Further heating eventually leads to nucleation of either ice Ic or ice Ih.

Interconversions between LDA and HDA samples, whether induced by temperature or pressure rises, occur rather suddenly, with substantial jumps in properties such as density and enthalpy, accompanied by hysteresis effcts [Debenedetti, 2003, Section 6.2]. These features are characteristic of first-order phase transitions. Because this applies to metastable states of water within a portion of the T,p plane that is occupied by crystalline ice phases at equilibrium, a description in terms of potential energy or potential enthalpy landscapes must rest upon application of a configurational constraint, as described in Section III.G, as well as in Chapter VI. In principle, the correspondingly restricted statistical mechanical description should be able to verify whether a strict first-order transition between LDA and HDA is present. In an isochoric representation, this would require specifying the basin enumeration and vibrational quantities $\sigma_a(\varphi, \rho)$ and $f_{vib,a}(\varphi, \beta, \rho)$, respectively, from which Helmholtz free energy and phase-transition behavior could be obtained. The isobaric representation would require the analogous quantities $\hat{\sigma}_a(\hat{\varphi}, p)$ and $\hat{f}_{vib,a}(\hat{\varphi}, \beta, p)$ and would lead to possible phase transitions via the Gibbs free energy for the constrained metastable system.

If it were possible strictly to impose the metastability constraint while heating an LDA sample isochorically at its low-temperature mass density at 0.94 g/cm³, molecular motions should become those of a viscous liquid, with rotational and translational diffusion. If an HDA sample were similarly constrained while being isochorically heated at its high-pressure formation mass density 1.31 g/cm³, it should likewise begin to act kinetically as a viscous liquid. The presumption of a first-order phase transition between LDA and HDA would then amount to "liquid polymorphism." As mentioned in Chapter V, this phenomenon of coexisting dense liquid phases for a pure substance has two known examples in their equilibrium phase behavior: (1) elemental phosphorus (P) well above its conventional liquid–vapor critical temperature and pressure [Katayama et al., 2004] and (2) triphenyl phosphite [P(OC$_6$H$_5$)$_3$] [Kurita and Tanaka, 2004]. The second of these reveals a second critical point in addition to its liquid–vapor critical point, at which a diverging correlation length for density fluctuations arises.

It has been proposed that the apparent LDA–HDA first-order phase transition, occurring along a liquid–liquid transition curve in the metastable T,p plane also terminates at a second critical point for water, a hypothesis supported by molecular dynamics simulations using the "ST2" water effective pair potential [Poole et al., 1992]. A tentative location for this second critical point has been assigned for D$_2$O based on the experimentally observed shapes of its melting curves for ices IV and V [Mishima, 2000]:

$$T_{c2} \approx 230 \text{ K}, \tag{IX.59}$$

$$p_{c2} \approx 0.5 \text{ kbar.}$$

With respect to the equilibrium-phase diagram, this point falls within the ice Ih region of stability for both light and heavy water. Figure IX.26 graphically indicates the location of the presumed LDA–HDA phase transition line and its terminating second critical point.

Because of the experimental difficulty, if not impossibility, of enforcing the required configurational constraint while ensuring thermal equilibrium under that constraint, direct observation

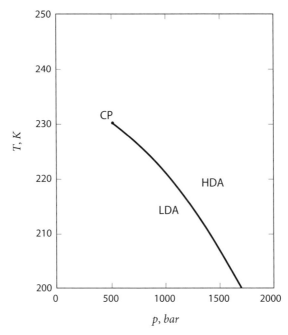

FIGURE IX.26. Inferred location in the *p,T* plane of a first-order phase transition line and its terminating critical point (CP) for the amorphous metastable LDA and HDA phases. Adapted from Mishima, 2000, Figure 2.

of the putative second critical point for water remains beyond reach, as does the extension of the transition line to absolute zero. It should be pointed out that the measurable thermodynamic properties of water can alternatively be explained without the existence of a strict first-order LDA–HDA transition line and second critical point [Rebelo et al., 1998]. Further clarification of this issue may have to await development and analysis of increasingly accurate simulation models for water.

X.

Polymeric Substances

Although many cases of polyatomic molecules have already been considered in the preceding chapters, substances composed of high-molecular-weight molecules each containing far higher numbers of atoms present a very wide range of distinctive physical properties, both in and out of thermal equilibrium. That situation merits separate theoretical consideration, and so the present chapter is devoted to developing connections to the basics of the potential energy landscape/inherent structure formalism for this class of polymeric substances. Because polymer science is such a complex subject, only a few basic aspects of which can be covered here, the reader may wish to consult specialized texts for a more comprehensive view [e.g., Flory, 1953; Stevens, 1990; Matsuoka, 1992; Painter and Coleman, 1997; Rubinstein and Colby, 2003]. Although they formally fall within this polymer category, protein biopolymers constitute a special case with their own distinctive characteristics, so their discussion is deferred to Chapter XI.

A. Polymer Chemical Structures and Classes

The chemical structural possibilities that exist within the polymer category are huge, and as a result they have given rise to a large lexicon of terms to help distinguish them, at least within broad categories. The conceptually simplest class of polymers, the "linear homopolymers," collects members consisting of a single chemical monomeric repeat unit linked by covalent bonds into a linear chain of controllably variable length. The prototypical example arises basically from polymerization of the unsaturated hydrocarbon ethylene (C_2H_4), with final addition of terminating hydrogen atoms. The result of this process is the well-known substance polyethylene (PE), whose linear chemical structure with a backbone consisting of n carbon atoms for a large integer n can be represented schematically as $CH_3[CH_2]_{n-2}CH_3$. This amounts to a high-molecular-weight extension of the normal alkane family, the diminutive initial members of which are methane (CH_4), ethane (C_2H_6), propane (C_3H_8), and butane (C_4H_{10}). Although there is no precise criterion for when the term "polymer" applies, a rough rule for linear homopolymers is that the number n of repeat units along the backbone ($-CH_2-$ for PE) should roughly obey $n \geq 50$. At the high-molecular-weight extreme, PE polymers can be synthesized with n rising into the range 10^5 to 10^6. Although simple ethylene polymerization causes n to be an even integer, PE molecules with odd n are included in Box X.1 for completeness.

The perfluorinated analog of polyethylene, $CF_3[CF_2]_{n-2}CF_3$, is conventionally abbreviated PTFE (for polytetrafluoroethylene). It is another industrially and commercially important linear homopolymer given the trademarked name "Teflon." The fluorine atoms are larger and more

---

BOX X.1. "Polyethylene" Terminology

For the primary purposes of this chapter, the term "polyethylene" refers strictly to the class of linear (un-branched) hydrocarbon polymers that constitute the large-n regime of the normal alkane series C_nH_{2n+2}. But because of the widespread industrial and commercial importance of this and closely related polymeric materials, the term "polyethylene" has been given a wider colloquial meaning that has required modifier acronyms to convey more chemical and physical specificity. It should be kept in mind in this context that distributions of molecular weights are almost invariably involved, not monodisperse substances. Further-more, some deviation from linearity of structure in the form of backbone branching is often present in commercial "polyethylene" materials. Some of the frequently encountered acronyms are the following.

> **ULMWPE, HMWPE, UHMWPE:** Ultra-low, high, and ultra-high molecular weight polyethylene, respectively. The first of these refers to alkanes usually in the $n < 100$ range, the second to molecular weights in the thousands, and the third to molecular weights in the millions. Carbon backbone branching may be involved in varying degrees.
>
> **VLDPE, LDPE, MDPE, HDPE:** Very low density, low-density, medium-density, and high-density polyethylene, referring, respectively, to mass densities under ambient conditions. These acronyms specify variable density ranges due to incorporation of correspondingly varying amounts of hydrocarbon side chains that branch off of the carbon backbone and whose presence interferes with geometrically "efficient" packing of the polymer molecules, thus reducing density.
>
> **XLPE, HDXLPE:** Cross-linked polyethylene, and high-density cross-linked polyethylene. These cat-egories refer to polymeric media that have side chains linking different primary backbones to produce elastomeric materials. The chemical bonds present presumably form a network above the connectivity percolation threshold.
>
> **LLDPE:** Linear low density polyethylene. This category involves polyethylene molecules contain-ing a significant occurrence frequency of short side-chains along the primary backbone, so that the coarse-grained shape feature is still that of an extended linear molecule.

---

electronegative than the hydrogens in PE that they replace, and the result is a distinctly different (and important) set of chemical and physical properties. As in the analogous case of normal al-kanes providing conceptually useful low-molecular-weight precursors for PE, the corresponding perfluorinated alkanes have also been extensively investigated experimentally [e.g., Starkweather, 1986; Schwickert et al., 1991; Albrecht et al., 1991] and are helpful in the interpretation of the behavior of PTFE.

Internal degrees of freedom for a linear polymer include the relatively stiff vibrational mo-tions that only permit covalent bond lengths and the angles between neighboring pairs of those bonds at multivalent atoms to vary just by small amounts at ambient conditions. But especially important for both equilibrium and time-dependent properties of linear polymers are the dihe-dral angles that describe the backbone conformation in terms of the directions of four sequential backbone atoms. Specifically, the dihedral angle measures the rotation between the planes respec-tively containing the first three backbone atoms of the four, and the latter three backbone atoms. For PE and PTFE, the backbone is the covalently bonded carbon chain, with $n - 3$ dihedral angles. Figure X.1 illustrates how these dihedral angles are measured. Many choices of the dihedral angle set are effectively excluded on physical grounds because they would produce encounters close

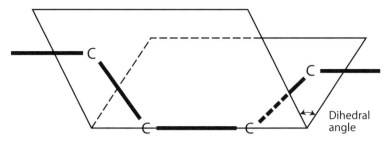

FIGURE X.1. Dihedral angle along the carbon–carbon backbone of polyethylene or one of its homopolymeric derivatives.

enough to cause overlaps between atoms that are remote neighbors as measured along the backbone. These possibilities would automatically be described as large positive intramolecular potential energy configurations in any physically realistic potential energy landscape for the polymer, whether in isolation or as part of a condensed phase, and so would not be present within inherent structures.

Linear homopolymers based on more complex monomeric units can exhibit covalently linked backbones consisting of more than a single atomic species. A simple case is provided by polyethylene oxide (PEO). Its chemical structure is indicated by $H[-O-CH_2-CH_2-]_nOH$. Three dihedral angles per interior repeat unit now need to be specified, in contrast to just one per backbone carbon atom for PE or PTFE. Another example is poly(dimethylsiloxane), denoted briefly by PDMS, possessing the structure $H[-O-Si(CH_3)_2-]_nOH$, with pairs of pendant methyl groups attached covalently to the backbone silicon atoms.

In addition to PDMS, there are many other important polymers with monomeric repeat units that contain chemical groups not directly part of the polymer backbone, but present as pendant side groups attached chemically to the backbone. Polypropylene (PP) provides a simple example. Its backbone consists entirely of carbon atoms as in PE and PTFE, but these carbons alternate between $-CH_2-$ and $-CH(CH_3)-$ types, where the latter has one of its pendant hydrogen atoms $-H$ replaced by a pendant methyl group $-CH_3$. Another frequently encountered case is polystyrene (PS), in which the pendant chemical group attached to every other backbone carbon atom is a benzene ring $-C_6H_5$.

The four covalent single bonds emanating from a carbon atom are rather rigidly arranged relative to one another at approximately the ideal tetrahedral angle $\theta_t = \cos^{-1}(-1/3) \cong 109.5°$. Under usual temperature and pressure conditions, this spatial arrangement of chemical bonds is essentially a permanent feature once synthesis incorporates it into a polymer. This leads to binary choices for attachments of distinguishable side groups to the backbone, which in turn significantly affects the properties of the resulting polymer. The chemical reaction conditions present during polymerization synthesis in some cases are able to exercise control over these choices. To illustrate graphically the distinct possibilities that these alternative choices involve, it is visually clearest if the dihedral angles along the backbone have all been set to 180°. This convention places an all-carbon backbone into a planar zig-zag "all *trans*" configuration. Then if all of the pendant side groups lie on the same side of the backbone plane, the result is called an "isotactic polymer." If instead the side groups strictly alternate from one side of the backbone plane to the other side, the result is called a "syndiotactic polymer." Finally, if the sequence of attachment choices along

(a)

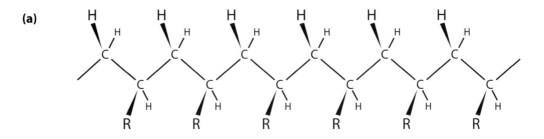

(b)

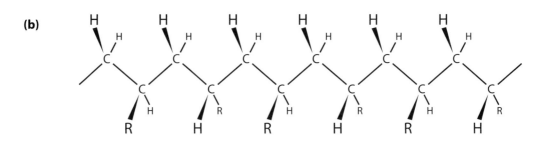

(c)

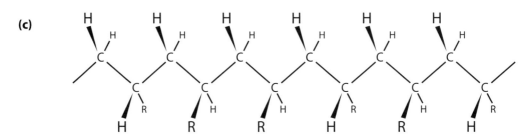

FIGURE X.2. Spatial patterns of side group (R) attachments. For clarity, the carbon backbone has been placed in an "all *trans*" configuration, and at each carbon atom the four covalent bonds are disposed radially at nearly tetrahedral directions: (a) isotactic, (b) syndiotactic, and (c) atactic alternatives.

the backbone is irregular or even completely random, the outcome is classified as an "atactic polymer." These three categories are illustrated in Figure X.2. Although this figure presents the three categories with a pure-carbon backbone, they are also applicable to polymers possessing other backbone atomic compositions. These three distinctions apply as well to the more general case in which the remaining hydrogen atoms indicated in Figure X.2 have been replaced by another type of pendant group R' that is distinguishable from R.

Linear polymers can also be prepared using two (or more) distinct monomer substances, the generic term for which is "copolymer." The specific sequence of the monomers along the copolymer backbone can often be controlled chemically during synthesis. The "nylon" polymer family contains examples with regular alternation of monomer units; an example is "nylon 6,6" formed

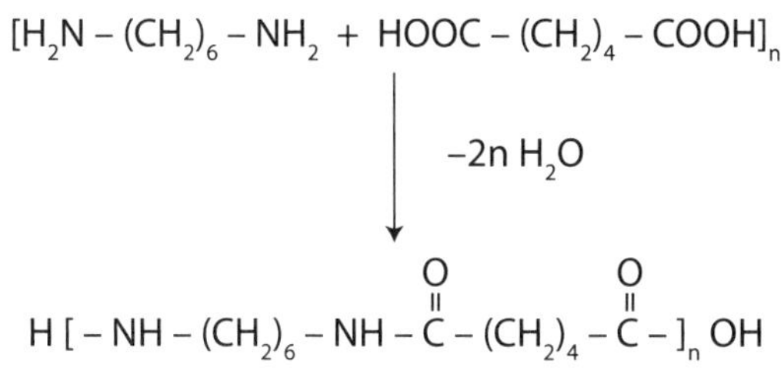

$$[H_2N - (CH_2)_6 - NH_2 + HOOC - (CH_2)_4 - COOH]_n$$

$$-2n \ H_2O$$

$$H \ [- NH - (CH_2)_6 - NH - \overset{\displaystyle O}{\overset{\|}{C}} - (CH_2)_4 - \overset{\displaystyle O}{\overset{\|}{C}} -]_n \ OH$$

FIGURE X.3. Water elimination to create amide linkages that produce the backbone of the alternating block copolymer nylon 6,6.

by H_2O elimination from a 6-carbon diamine and a 6-carbon carboxylic acid, as illustrated in Figure X.3. More generally, "block copolymers" possess backbones with long sequences of just one monomeric unit, alternating with long sequences comprising just another monomeric unit. For two monomer types, this would be schematically indicated as $....(A)_{l_1}(B)_{l_2}(A)_{l_3}(B)_{l_4}....$ Alternatively, the sequence of monomers could be irregular, a situation deserving the obvious name "random copolymer."

Polymer backbones are not limited to simple linear sequences of covalent chemical bonds but can also exhibit more complex connectivities. In particular, the backbone can contain one or more branching points to segments larger than simple side groups. A carbon atom serving as a backbone element can contribute two, three, or four of its bonds to the covalent backbone and thus can serve as the initiation point for two, three, or four emanating backbone directions. This situation is directly analogous to the chemical structures represented in the lower molecular-weight extended family of saturated hydrocarbons, including not only the linear normal alkanes with a simple unbranched chain structure, but also the other saturated hydrocarbons with one or more branching points. In the event that just a single branching point is present, at which linear chains of roughly equal length are attached, the result is called a "star polymer." If the branches radiating outward from the central point themselves undergo repeated branching, the structure is a "dendritic polymer." These structural alternatives are illustrated in highly schematic fashion in Figure X.4. It should be noted that backbone branching points need not be centered on single atoms, but more complex multifunctional units can be chemically synthesized that in principle could be incorporated within a polymer and act as attachment sites for more than four branches.

Unbranched polymers can exist in cyclic form, i.e., possessing no free ends. The examples of the cyclic form of polyethylene would have the chemical formula $(CH_2)_n$ and are just an extension to large n of the low-molecular-weight saturated cyclic hydrocarbons (cyclopropane, cyclobutane, cyclopentane, …). With large n, the resulting loop formed by the chemical bonds could in principle be knotted and would be configurationally trapped in that situation unless a backbone chemical bond were to break, thus permitting escape from that knotted class of polymer spatial configurations. Also possible with large n is the purely spatial interlinking of two or more cyclic polymers, again leading to confined spatial configurations. These latter nonbonded arrangements are analogs of short elements of a linked chain and are called "catenanes." Figure X.5 provides schematic diagrams of these last three possibilities. It should be noted that high-yield synthetic routes have been devised and exploited to form relatively low-molecular-weight examples both of knotted molecules [Ponnuswami et al., 2012] and of catenanes [Weck et al., 1999].

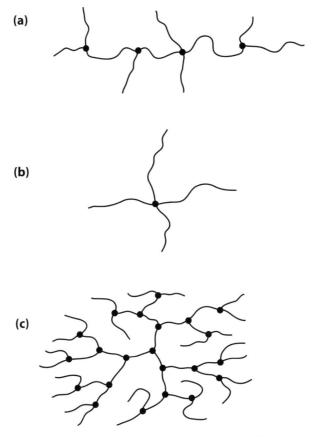

FIGURE X.4. Branched polymer backbone classes: (a) multiple branching points along a primary backbone sequence, (b) a star polymer with a single central branching point, and (c) a dendritic branched polymer.

A detailed statistical mechanical description of a polymeric medium of course requires a chemically and physically accurate specification of the potential energy functions for the collection of species that are present. The intramolecular portion of the overall potential energy normally preserves the chemical bonding structures of the species, but it must describe energy costs involved in molecular shape changes produced by bond stretch and bend degrees of freedom, as well as conformational changes created by dihedral angle variations. The intermolecular interactions naturally are coupled to the intramolecular degrees of freedom. Furthermore, the available regions of configuration space to be included in the statistical mechanical analysis of the polymer or polymers in a system of interest may be severely constrained if knotted or catenated polymers as schematically illustrated in Figure X.5 are present.

Most applications of the theory involve predicting the properties of chemically stable polymers, their mixtures with one another, and their solutions in low-molecular-weight solvents. However, there is yet another class of polymers that violates this stability characteristic but that deserves at least passing attention. This is the so-called "living polymer" category. Specifically, this concerns substances that under prevailing temperature and pressure conditions exhibit continuing chemical reactions at measurable rates involving either or both polymerization and depoly-

(a)

(b)

(c)

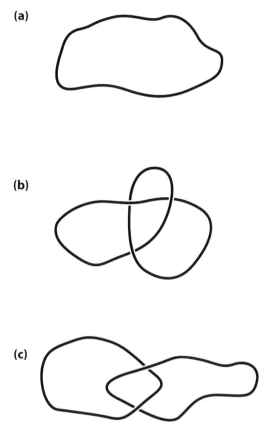

FIGURE X.5. Cyclic polymer possibilities: (a) a simple unencumbered cyclic structure, (b) a cyclic polymer forming a trefoil knot (which is chiral, i.e., inequivalent to its mirror image), and (c) a pair of chemically unbonded but interlinked cyclic polymers forming a catenane structure.

merization processes for the molecules present. This category includes both systems at dynamic thermal equilibrium as well as those undergoing irreversible sets of reactions (e.g., overall time-dependent increase in degree of polymerization). Elemental sulfur above its melting temperature is a well-documented example of a living polymeric system (mentioned in Section I.E and in Box V.6) for which extensive experimental studies have been published [MacKnight and Tobolsky, 1965; Steudel, 1984; Bellissent et al., 1990; Zheng and Greer, 1992]. Modeling the many-body potential energy function for such systems requires a realistic description of these chemical reactions that form and break covalent chemical bonds. The inherent structures and their encompassing basins for these systems could be classified by the populations of polymers present with distinct molecular weights.

Elasticity and rigidity of polymeric materials can be controlled by introducing suitable covalent cross links. These can be viewed as side branches bonded at both ends to what were precross-linking backbones. Rubber vulcanization is one example that utilizes short sulfur chains as the cross links between hydrocarbon backbones to increase rigidity and impact-resisting toughness. An important type of synthetic rubber involves polymerization of 1,3-butadiene

(CH_2=CH–CH=CH_2), which, because of its pair of double bonds, is able to create its own cross links upon polymerization. Thermosetting plastic materials also rely on inclusion of cross links to resist deformation under various stress conditions. In each of these cases, the net result of the cross-linking process is to produce a condensed matter medium that is well above the percolation threshold for covalent bond connectivity. In other words, most if not all of the material medium consists of a single "molecule" that occupies the available macroscopic volume.

Chemical syntheses of high-molecular-weight polymers have somewhat limited capacity to control the degree of polymerization of the final product. Specifically, this can result in a distribution of backbone chain lengths, or more generally a distribution of end-product molecular weights. Two frequently used measures of the distribution are the "number average molecular weight," M_{na}, and the "weight average molecular weight," M_{wa}. If the system under consideration contains N_i polymers whose molecular weight is M_i, these averages are defined as follows:

$$M_{na} = \frac{\sum N_i M_i}{\sum N_i},$$
(X.1)

$$M_{wa} = \frac{\sum N_i M_i^2}{\sum N_i M_i}.$$

The former can be extracted from polymer solution thermodynamic properties, the latter from light scattering [Flory, 1953, Chapter VII]. The extent to which $M_{na} \neq M_{wa}$ provides a measure of the breadth of the molecular weight distribution. Well-designed experiments often produce narrow molecular weight distributions, in which case $M_{na} \cong M_{wa}$, and the (nearly) common value is then often expressed merely as "M" for simplicity. It should be noted in passing that in the averages (X.1), two or more cyclic molecules that are topologically linked in catenane fashion would be counted as a single large molecule possessing the total molecular weight of the included cyclic units.

Atactic linear polymers naturally provide another form of molecular diversity. As illustrated in Figure X.2, there are two independent choices for the attachment direction of a side group at each of the l qualifying backbone locations. Consequently, the number of independent molecular structures that could be formed is 2^l (or 2^{l-1} if the backbone has end-to-end symmetry). Even if the degree of polymerization is quite modest, presenting say $l = 80$ attachment locations, for which

$$2^{80} \cong 1.2 \times 10^{24},$$
(X.2)

the number of distinct structural alternatives for a single polymer molecule is already comparable to Avogadro's number $N_{Av} \cong 6.022045 \times 10^{23}$. For even larger l, this raises the basic question about whether in a macroscopic system any two atactic polymer molecules that are nominally the same substance indeed have identical molecular structures; see Box X.2 below.

B. General Landscape Features

The great chemical diversity presented by polymeric substances leads to a corresponding diversity of potential energy and potential enthalpy landscapes. In general, one must consider an interacting set of N large molecules, not necessarily identical in chemical bonding structure, whose individual spatial configurations can be symbolized by vectors $\mathbf{R}_i$, $1 \le i \le N$. Because of the possible bonding-structure differences and molecular-weight differences, these vectors can contain different numbers of components, specifically, $3a_i$ if polymer molecule i contains a_i atoms, so that the isochoric (constant-volume) configuration space would have dimension

$$D_V = 3\sum_{i=1}^{N} a_i. \tag{X.3}$$

In most applications, N and a_i would be time-invariant, but in the special case of living polymer systems, these would fluctuate as time progresses whether or not thermal equilibrium obtained.

The electronic ground-state potential energy for such an N-molecule system can be expressed in the following obvious way:

$$\Phi(\mathbf{R}_1 \ldots \mathbf{R}_N) = \sum_{i=1}^{N} v_i^{(1)}(\mathbf{R}_i) + \Delta(\mathbf{R}_1 \ldots \mathbf{R}_N), \tag{X.4}$$

where the $v_i^{(1)}(\mathbf{R}_i)$ are the intramolecular potential energies for the individual molecules in isolation (including molecule-wall interaction, if appropriate), and $\Delta(\mathbf{R}_1 \ldots \mathbf{R}_N)$ represents the entirety of intermolecular interactions. Although Δ may be dominated by molecular pair contributions, higher order contributions could also be significant, especially if ionic or strongly polar side groups are present. As usual, the potential enthalpy function corresponding to Φ in Eq. (X.4) is obtained by simply appending the pressure–volume product:

$$\hat{\Phi}(\mathbf{R}_1 \ldots \mathbf{R}_N, V) = \Phi(\mathbf{R}_1 \ldots \mathbf{R}_N) + pV, \tag{X.5}$$

where the associated isobaric (constant-pressure) condition requires inclusion of V as an additional configurational variable, thus extending the configurational space dimension to

$$D_p = D_V + 1. \tag{X.6}$$

Within the usual temperature and pressure ranges that are applicable to polymeric systems, the electron density distributions along intramolecular chemical bonds are sufficiently repelling toward one another that those bonds are not able to cross each other spatially. This is an effective geometric constraint that has fundamental consequences for the dynamical processes which can occur within the polymeric system, as well as for the system's thermodynamic behavior. As a simple preliminary to analysis of the full Φ or $\hat{\Phi}$ landscapes for polymeric systems, it is useful to consider first just the character of this bond-noncrossing constraint when it is expressed in a very stripped-down version. For this purpose, suppose that straight line segments have been drawn between every intramolecular pair of atoms that are directly chemically bonded to one another, thus creating a kind of "stick model" cartoon of each polymer molecule present in the system. Figure X.6 schematically illustrates this prototypical construction. Although the hypothetical line segments cannot cross one another (whether they belong to the same or to different molecules), at this basic level of description they could in principle approach one another arbitrarily closely. The resulting implication for the multidimensional configuration space is that impenetrable but infinitesimal-width barriers have been erected in that space. In other words, the configuration space has been converted to a thin-walled labyrinth, through which the system must find its way dynamically to explore alternative molecular arrangements. Nevertheless, the barriers themselves

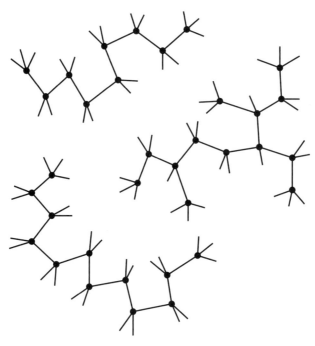

FIGURE X.6. Schematic illustration of the use of straight line segment indicators for intramolecular chemical bonds, to enforce noncrossing constraints in polymeric systems. The example shown can be interpreted as representing neighboring alkane molecules of modest size, but the approach has general applicability to other molecular species, in particular to any high-molecular-weight polymeric species.

have vanishing measure, so the configuration space content overall has not been diminished by their presence.

In the event that all of the molecules in the system have chemical structures devoid of ring closures, as is the case in the schematic illustration in Figure X.6, then all pairs of system configurations are accessible from one another by continuous paths that avoid the thin-walled labyrinth barriers. Of course, many of these connecting pathways might necessarily be very long and tortuous. However, when ring closures are present, the hypothetical barriers partition the full configuration space into mutually inaccessible portions. In the case of simple ring compounds such as the cyclic alkanes $[(CH_2)_n, n \geq 3]$, the individual portions are determined primarily by which, if any, of the ring compounds are catenated [i.e., chain-linked, see Figure X.5(c)], and with which neighbor molecules they share that relationship. Furthermore, if any cyclic molecules present in the system are sufficiently large that they could be intramolecularly knotted as schematically indicated in Figure X.5(b), this becomes a topological feature that the noncrossing constraint also preserves. In other words, intramolecular knotting involves another kind of configuration-limiting barrier in the full configuration space, whether isochoric or isobaric conditions apply.

The next step conceptually is to restore the full potential energy Φ or potential enthalpy $\hat{\Phi}$ to the configuration space. One of the effects is to replace the infinitesimal-width noncrossing barriers with positive-width repelling barriers that would continue to act as configurational constraints. In particular, they would continue the partitioning of the full D_V-dimensional or D_p-dimensional space into distinct nonoverlapping regions when molecular ring closures, intramolecular knots, or catenated rings are present. But whether chemical bonding circumstances allow the full configuration space or only a portion to be available for the molecules to explore kinetically, the complete set of intramolecular and intermolecular interactions produces almost everywhere a continuous and differentiable potential energy or potential enthalpy landscape. These landscapes exhibit inherent structure minima located along the corridors that pass between the labyrinth's noncrossing barriers. As usual, the inherent structures can be located by steepest descent mapping, which also serves to divide the available configuration space into nonoverlapping basins, one surrounding each inherent structure. Transition-state saddle points located in shared basin boundaries of course cannot violate the noncrossing constraints.

As indicated above, let N denote the total number of polymer molecules present, and suppose that this total is resolvable into separate contributions N_α from each of v distinguishable species present:

$$N = \sum_{\alpha=1}^{v} N_\alpha. \qquad (X.7)$$

Regardless of whether isochoric or isobaric conditions apply, each inherent structure and its surrounding basin for this system belongs to a permutational equivalence class of geometrically identical replicas whose number E_p can generally be expressed as follows:

$$E_P = \prod_{\alpha=1}^{v} \gamma_\alpha^{N_\alpha} N_\alpha!. \qquad (X.8)$$

Here each factor $N_\alpha!$ accounts for the permutation possibilities among distinct molecules of the same species α, while for each of those molecules, γ_α stands for the number of intramolecular permutational displacements leading to equivalent spatial configurations without violating the chemical bonding. In the case of a cyclic polyethylene $(CH_2)_n$, the possibility of discrete circumferential displacements requires for that species $\gamma_\alpha = n$. A linear polymer with just end-to-end symmetry would permit an end-for-end flip and thus would have $\gamma_\alpha = 2$. Any polymer molecule containing some number q of pendant methyl groups $(-CH_3)$ arrayed along, or at the ends of, its structure would have a factor 3^q contained within its γ_α because of the corresponding methyl rotational possibilities by $\pm120°$. A structurally asymmetrical polymer molecule devoid of pendant methyl groups (or other rotation-symmetric groups) would possess only the identity intramolecular permutation, so for such molecules, one would have $\gamma_\alpha = 1$.

If v is a fixed finite number, and if the molecular weight of all polymeric species has a fixed upper bound, then under isochoric conditions there would exist sufficiently large macroscopic system sizes N, V such that the total number of inherent structures Ω and the number of structurally distinguishable inherent structures (equivalence classes) Ω_d should have the same asymptotic forms as inferred for systems of lower molecular weight substances, namely:

$$\Omega(\{N_\alpha\}, V) = E_P \Omega_d(\{N_\alpha\}, V), \qquad (X.9)$$

$$\ln \Omega_d(\{N_\alpha\}, V) \sim N \sigma_\infty(\{\rho_\alpha\}),$$

where $\sigma_\infty > 0$ is an intensive quantity depending only on the number densities ρ_α of the species present. In this large-system regime, the leading-order exponential rise indicated here for Ω_d relies on the essentially independent rearrangement possibilities within macroscopic subvolumes (see Section III.C). The corresponding isobaric results are the following:

$$\hat{\Omega}(\{N_\alpha\}, p) = E_p \hat{\Omega}_d(\{N_\alpha\}, p),$$ (X.10)

$$\ln \hat{\Omega}_d(\{N_\alpha\}, p) \sim N\hat{\sigma}_\infty(\{x_\alpha\}, p),$$

in which the species mole fractions are denoted by x_α, and $\hat{\sigma}_\infty > 0$ likewise is an intensive quantity.

The case of atactic polymers formally presents a basic question about enumeration of inherent structures and their basins. It was pointed out at the end of Section X.A that in a system of conventional macroscopic size the huge number of distinct atactic molecules of even modest degree of polymerization could in principle equal or exceed the number of molecules present. If indeed it were the case that all polymer molecules present had distinguishable structures, then the combinatorial factor E_p in Eqs. (X.9) and (X.10) would reduce simply to the product of intramolecular factors γ_α. This situation would then lead to the question whether the essential independence argument for rearrangement of macroscopic subvolumes invoked in asymptotically estimating Ω_d or $\hat{\Omega}_d$ is valid if those subvolumes do not contain any pairs or larger numbers of molecules of the same species. In this connection, Box X.2 presents an elementary calculation specifying the probability that randomly synthesized atactic polymer molecules, all with the same degree of polymerization, are indeed structurally distinct. The implication of the calculation is that even in a sample of typical macroscopic size, atactic polymers of moderate molecular weight could indeed turn out to be all structurally unique. It has been pointed out earlier that huge time intervals would be required for classical dynamics in macroscopic systems of low-molecular-weight species to visit at least one member of all of the distinct inherent-structure basin equivalence classes (Section V.B). This timing situation becomes enormously more extended in the $E_p = \prod \gamma_\alpha^{N_\alpha}$ case of unsymmetrical atactic polymer media, where systemwide diffusive displacements would be required to sample all possible distinct molecule mixings.

If the number of distinct atactic molecular species v is held fixed as the overall size of the system is allowed to grow without bound, there would be a size range above which the enumeration subsystems would contain large numbers of each of those species. Then because each of the subsystems would have substantially a common composition, this would lead to the prior conclusion for monodisperse systems that the number of distinct rearrangements in separate subvolumes would be effectively independent of one another in leading order. As before, this leads in turn to the conclusion that the total number of distinct inherent structures for the whole system would rise exponentially with N, the total number of atactic polymer molecules present. The analysis presented in Box X.2 indicates that the system size range for this argument to hold could be very large, even on the usual macroscopic size scale. However, it should be kept in mind that a mixture containing many distinct molecular species necessarily has a large mixing entropy, and this constitutes a large positive additive constant for the enumeration functions σ_∞ and $\hat{\sigma}_\infty$ in Eqs. (X.9) and (X.10), respectively. These mixing entropy contributions would not be properly included until the enumeration process was carried out in the large-system regime for which each atactic species number N_α, $1 \le \alpha \le v$, was substantially larger than unity.

Thermodynamic equilibrium states in macroscopic systems comprising large numbers N of polymeric molecules can be represented in the same overall format that has been developed earlier

BOX X.2. Atactic Polymer Structural Inequivalence

Consider the case of N atactic linear polymer molecules, all of which have the same degree of polymerization n_a, with one binary choice per monomer for side-group attachment. Each of these n_a attachment directions on the linear backbone is assumed to be independent [see Figure X.2(c)], creating $\Gamma = 2^{n_a}$ distinguishable geometric structures. Suppose that $N \leq \Gamma$. Imagine blindly selecting the atactic polymer molecules one by one from a randomly synthesized reservoir of molecules. At each stage of this selection sequence, one can ask about the probability that the last selection does not possess a geometric pattern that had been previously chosen.

Suppose that $j-1$ molecules have already been selected, with no geometric duplication. The a priori probability that the jth selection also does not duplicate a predecessor is

$$\theta_j = 1 - (j-1)/\Gamma,$$

so that overall success after N selections becomes

$$s(N, n_a) = \prod_{j=1}^{N} \theta_j.$$

Because Γ is large, incrementing j by unity causes little change in the factor θ_j. Consequently, it is feasible to apply logarithms to both members of the last equation and to represent the resulting summation in the right member as an integral:

$$\ln s(N, n_a) \cong \int_1^N \ln[1 - (j-1)/\Gamma]\,dj$$

$$\cong \Gamma\left[\int_1^{N/\Gamma} \ln(1-x)\,dx\right]$$

$$= -(\Gamma - N)\ln[1 - (N/\Gamma)] - N.$$

Equivalently one has

$$s(N, n_a) \cong [1 - (N/\Gamma)]^{-(\Gamma - N)} \exp(-N).$$

If interest centered about a total of one mole of polymer, so that $N = N_{Av} \cong 6.022045 \times 10^{23}$, then using these expressions it is easy to see how large Γ and thus n_a must be so that $s(N, n_a) \geq 0.999$, in other words to ensure with high probability that all N atactic polymers are structurally distinct. The required arithmetic yields the following numbers:

$$\Gamma \geq 1.812 \times 10^{50},$$

$$n_a \geq 167,$$

indicating that only a moderate degree of polymerization is necessary effectively to fulfill the criterion of polymer molecule uniqueness.

for small-molecule systems. However, there are some notable effects on the basic descriptive quantities stemming from the high molecular weights involved. For that reason, it is useful to retrace the earlier development as adapted for the present polymer context. In particular, consider first the classical canonical partition function for a monodisperse macroscopic system containing N identical molecules each composed of n_a atoms. After carrying out all of the momentum integrations, one is left with the following appropriately normalized integral spanning the $3n_aN$-dimensional configuration space for system volume V:

$$Q_N(\beta, V) = \{[\Lambda(\beta)^N N!]\}^{-1} \int \ldots \int \exp[-\beta \Phi(\mathbf{R}_1 \ldots \mathbf{R}_N)] d\mathbf{R}_1 \ldots d\mathbf{R}_N. \qquad \text{(X.11)}$$

Here $\beta = 1/k_B T$ as usual, and $\Lambda(\beta)$ is the product of the cubes of all the individual-atom mean thermal de Broglie wavelengths [Eq. (III.22)] contained in one of the polymer molecules. The integral spans the relevant portion of the bond-noncrossing landscape labyrinth described above, and it can be resolved into separate contributions from each correspondingly accessible basin in the potential energy hypersurface that is defined by interaction potential Φ. Because the permutational equivalence class size for each distinct type of basin is $\gamma^N N!$, only one member of each class needs to be included in the right side of Eq. (X.11). Consequently, Eq. (X.11) is equivalent to the following:

$$Q_N(\beta, V) = [\gamma/\Lambda(\beta)]^N \sum_\zeta \exp(-\beta \Phi_\zeta) \int_{B_\zeta} \ldots \int \exp[-\beta \Delta_\zeta \Phi(\mathbf{R}_1 \ldots \mathbf{R}_N)] d\mathbf{R}_1 \ldots d\mathbf{R}_N. \qquad \text{(X.12)}$$

The summation index ζ covers all distinguishable basin types within the accessible portion of the labyrinth, Φ_ζ is the inherent structure value of the potential energy in basin type ζ, $\Delta_\zeta \Phi$ is the intrabasin rise in potential energy above that of the included inherent structure, and B_ζ stands for the region in the multidimensional configuration space occupied by the one chosen basin of type ζ.

Because the total count of distinct basin types rises exponentially with N, and because the span of inherent structure energies scales as N, it is possible to represent the inherent structure energy distribution itself as an exponentially rising function of N multiplied by an N-independent enumeration function $\sigma(\varphi, \rho)$ of the intensive quantities $\varphi = \Phi_\zeta/N$ and $\rho = N/V$. Specifically, we can set

$$\gamma^N {\sum_\zeta}' 1 \approx C \exp[N\sigma(\varphi, \rho)]\delta\varphi, \qquad \text{(X.13)}$$

where C is a positive constant with dimension $(energy)^{-1}$ and where the primed summation of unit contributions includes only those basins whose inherent structure energies lie in the small interval $N(\varphi \pm \delta\varphi/2)$ The average intrabasin vibrational free energy Nf_{vib} for the same small interval may be expressed in the following way:

$$\exp[-N\beta f_{vib}(\varphi, \beta, \rho)] = [\Lambda(\beta)]^{-N} \left\langle \int_{B_\zeta} \ldots \int \exp[-\beta \Delta_\zeta \Phi(\mathbf{R}_1 \ldots \mathbf{R}_N)] d\mathbf{R}_1 \ldots d\mathbf{R}_N \right\rangle_{\varphi \pm \delta\varphi/2}, \qquad \text{(X.14)}$$

where the average indicated spans the same small φ interval. These definitions lead to the collapse of the multidimensional integral form of the canonical partition function to a simple quadrature over the intensive quantity φ:

$$Q_N(\beta, V) \approx C \int \exp\{N[\sigma(\varphi, \rho) - \beta\varphi - \beta f_{vib}(\varphi, \beta, \rho)]\}d\varphi. \qquad \text{(X.15)}$$

This is a mathematical format previously encountered in Eq. (III.28) for low-molecular-weight systems at equilibrium.

Because of the high molecular weights exhibited by polymers, and the fact that φ is defined as a per-molecule parameter, the difference between the lower and upper limits of the φ integration in Eq. (X.15), determined respectively by the lowest lying and highest lying inherent structures for the system, is correspondingly larger than one expects to encounter with low-molecular-weight substances. Similarly, the per-molecule intrabasin vibrational quantity f_{vib} is magnified compared to its values for low-molecular-weight substances because of the larger number of coordinates involved per molecule, $3n_a$. The isochoric enumeration function $\sigma(\varphi, \rho)$ is also defined on a per-molecule basis and so can be expected to rise with the degree of polymerization (i.e., molecular weight) of the polymer under consideration.

Under the assumption that the N-molecule system can achieve thermal equilibrium, the logarithm of the canonical partition function yields the Helmholtz free energy F_N. When examined on a per-molecule basis, F_N/N is entirely dominated in the large-system asymptotic limit by the maximum of the logarithm of the integrand in Eq. (X.15), which is located at $\varphi = \varphi^*(\beta, \rho)$:

$$\beta F_N/N \sim \beta\varphi^* + \beta f_{vib}(\varphi^*, \beta, \rho) - \sigma(\varphi^*, \rho), \tag{X.16}$$

where $\varphi^*(\beta, \rho)$ is defined by the extremum condition:

$$\{(\partial/\partial\varphi)[\sigma(\varphi, \rho) - \beta f_{vib}(\varphi, \beta, \rho)]\}_{\varphi=\varphi^*} = \beta. \tag{X.17}$$

Isobaric versions of the last two relations (X.16) and (X.17) follow in similar fashion, starting with the isothermal–isobaric partition function $\hat{Q}_N(\beta, p)$:

$$\hat{Q}_N(\beta, p) = [\lambda_V(\beta)\Lambda^N(\beta)N!]^{-1}\int\ldots\int\exp[-\beta\hat{\Phi}(\mathbf{R}_1\ldots\mathbf{R}_N, V)]d\mathbf{R}_1\ldots d\mathbf{R}_N dV, \tag{X.18}$$

in which $\lambda_V(\beta)$ is the formal mean thermal de Broglie wavelength arising from integration over the momentum conjugate to the configurational coordinate V. This multidimensional integral form then can finally be re-expressed as a simple quadrature over the intensive parameter $\hat{\varphi}$, the potential enthalpy per polymer molecule of the isobaric inherent structures:

$$\hat{Q}_N(\beta, p) = \hat{C}\int\exp\{N[\hat{\sigma}(\hat{\varphi}, p) - \beta\hat{\varphi} - \beta\hat{f}_{vib}(\hat{\varphi}, \beta, p)]\}d\hat{\varphi}, \tag{X.19}$$

where $\hat{C} > 0$ is the isobaric analog of the isochoric multiplier C in Eq. (X.13), and where isobaric enumeration function $\hat{\sigma}$ and vibrational free energy function $\hat{f}_{vib}$ are direct extensions of the isochoric definitions upon inclusion of volume V as an additional configurational variable. Utilizing the same asymptotic integrand-maximum approach as before in Eq. (X.19), one finds that the isobaric analog of the Helmholtz free energy per molecule in Eq. (X.16) is the Gibbs free energy per molecule:

$$\beta G_N/N \sim \beta\hat{\varphi}^* + \beta\hat{f}_{vib}(\hat{\varphi}^*, \beta, p) - \hat{\sigma}(\hat{\varphi}^*, p). \tag{X.20}$$

The relevant enthalpy depth $\hat{\varphi}^*(\beta, p)$ of the dominating inherent structures at any given temperature and pressure is determined by the isobaric analog of the isochoric condition Eq. (X.17), specifically

$$\{(\partial/\partial\hat{\varphi})[\hat{\sigma}(\hat{\varphi}, p) - \beta\hat{f}_{vib}(\hat{\varphi}, \beta, p)]\}_{\hat{\varphi}=\hat{\varphi}^*} = \beta. \tag{X.21}$$

In parallel to the previous remark for the isochoric circumstance, the range expected for intensive parameter $\hat{\varphi}$ and the magnitudes of the enumeration function $\hat{\sigma}$ and of the mean vibrational free energy per polymer molecule $\hat{f}_{vib}$ are all expected to rise roughly in proportion to increasing polymer molecular weight.

The definitions of the isochoric and isobaric enumeration functions $\sigma(\varphi, \rho)$ and $\hat{\sigma}(\hat{\varphi}, p)$ can be directly extended to systems containing polymers with molecular weight distributions, as well as to systems containing more general classes of polymer mixtures. This follows simply by requiring that they describe the asymptotic exponential rise with total molecule population N of the numbers of distinct inherent structures in the Φ and $\hat{\Phi}$ landscapes, respectively, keeping in mind that atactic polymer systems may involve inclusion of large additive mixing entropies. Analogously, the definitions of per-molecule vibrational free energy functions $f_{vib}(\varphi, \rho)$ and $\hat{f}_{vib}(\hat{\varphi}, p)$ can immediately be extended to these mixture cases as averages over the basins of depths specified by the intensive parameters φ and $\hat{\varphi}$. In these generalizations, it is important to remember that mean thermal de Broglie wavelength factor $[\Lambda(\beta)]^{-N}$ must be replaced by a product over species $\prod_a [\Lambda_a(\beta)]^{-N_a}$. With these formal interpretive changes, the expressions for Helmholtz and Gibbs free energy per molecule, Eqs. (X.16) and (X.20), continue to apply.

Although the preceding partition function and free energy expressions have been presented within the context of classical statistical mechanics, that is a convenient simplification but not a fundamental necessity nor even strictly realistic in all respects. In fact, the typical occurrence of many light hydrogen atoms chemically confined by strong bonds within the polymer molecules of frequent interest suggests that consideration of quantum effects is inevitable. The steepest descent mapping that underlies analysis of Φ and $\hat{\Phi}$ landscapes in terms of inherent structures and their basins generally operates on a quantum mechanical distribution function for the system as a whole, i.e., the diagonal elements of the density matrix (Section III.D). In particular, this means that quantum mechanical versions of the intrabasin vibrational quantities f_{vib} and $\hat{f}_{vib}$ need to be calculated subject to appropriate boundary conditions at the basin boundaries. So far as the hydrogen atom degrees of freedom are concerned, this means accounting for the lowest few quantized vibrational bound states in each of their chemical bonding circumstances, as well as methyl group rotations. Because the hydrogen vibrational motions are typically so confined and relatively little influenced by other degrees of freedom in the system, they would have little effect on the outcomes of the steepest decent mappings. This situation leads favorably to the so-called "united atom" simplification for molecules, specifically for polymers, in which a heavy atom and its attached hydrogens are treated as a single rigid force center possessing the total mass of those atoms; an example appears in Box X.3 below. In this kind of contracted description, the united atom units can then be treated with classical mechanics and statistical mechanics with reduced need for quantum corrections, provided that realistic choices of effective interaction potentials have been assigned.

A macroscopic system filled entirely with a densely cross-linked polymer amounts to a single huge molecule. Every atom present would be connected to every other atom by one or more paths along chemical bonds, which for present purposes can be regarded as permanent structural features. In this circumstance, the above analyses of isochoric and isobaric thermal equilibrium are not directly applicable. The relevant potential energy and potential enthalpy landscapes for such systems would occupy a severely truncated region of configuration space that is limited not only by bond noncrossing constraints but also by bond length and angle stiffness characteristics along cross links. Whereas any member of an ensemble of N chemically discrete polymers can diffuse across a volume-V system (given sufficient time) and occupy any location within that volume, the cross-linked material has no such freedom. A related additional distinction between the two cases arises from the fact that the preparation procedure for the heavily cross-linked system could have resulted in spatial heterogeneity of one or more properties, such as local mass density, average length of cross links, and stress tensor components, among other possibilities. Absence of diffusive and/or relaxational processes caused by the cross-linking simply means that these inhomogeneities are essentially frozen-in permanent features. While it is formally possible to develop free energy expressions of the types displayed in Eqs. (X.16) and (X.20) for cross-linked materials, to do so would require careful redefinition of the included quantities. As an example, one could use the number N_{nh} of nonhydrogen atoms present in the material in place of the previous number N of distinct polymer molecules. Furthermore, a cross-linked medium exhibits no extensive symmetries or any permutations of identical units (barring rotations of pendant methyl groups), so all equivalence classes of basins would contain only a single element ($E_p = 1$). Nevertheless, at least for homogeneous cross-linked macroscopic systems, the number of inherent structures and basins should rise exponentially with a size measure such as N_{nh}, leading to proper identification of the asymptotic limit enumeration functions σ, $\hat{\sigma}$ and intrabasin vibrational free energy functions f_{vib}, $\hat{f}_{vib}$. Such an approach could serve as a basis for analysis of elastic properties of the cross-linked medium.

C. Crystal Structures and Melting Transitions

A basic aspect of polymer statistical mechanics is how the molecules might be able to arrange themselves in a crystal phase so as to minimize the total potential energy or potential enthalpy. Experimental observations are a valuable guide to theory, but it is important to keep in mind that crystallization phenomena in experimental polymer systems can present significant complications that stem from molecular weight distributions, atactic structural variations, inadvertent backbone branching, and the presence of relaxation processes with time scales that may far exceed the times available for experimental observation. A detailed description of the crystallization behavior of many real polymers is available in a review article by Khoury and Passaglia (1976). In principle, it is possible to avoid some of these difficulties by examining and interpreting equilibrium measurements on pure samples for oligomeric sequences of low-molecular-weight polymer-precursor molecules. Extrapolation of the observed trends for such a sequence as the molecular weight increases should allow one to infer at least some of the basic features of the corresponding equilibrium phase behavior in the high-molecular-weight polymer regime.

The following analysis starts by considering the relatively simple linear polymer, polyethylene (PE). But even in this case, it is conceptually useful to break the discussion into two parts. The first focuses on "prototypical polyethylene," that is, the ideal equilibrium behavior for structurally pure, monodisperse polyethylene, as can be inferred from extrapolation of the normal alkane series properties into the high-molecular-weight regime. This is followed by a discussion of "realistic polyethylene," the crystallization and melting behavior inferred from finite-time experiments for typically encountered samples that include polydispersity. Then using these various polyethylene properties as a starting point, brief discussions are finally presented in this section for the crystallization and melting behaviors of three other linear polymers, polytetrafluoroethylene (PTFE), polystyrene (PS), and polyethylene oxide (PEO) to illustrate how their respective structural modifications affect observable phase-change properties.

(1) MONODISPERSE POLYETHYLENE

The obvious and most easily interpreted starting point involves the series of normal alkane hydrocarbons C_nH_{2n+2}, namely methane ($n = 1$), ethane ($n = 2$), propane ($n = 3$), …, that are the diminutive precursors of strictly linear polyethylene. For these linear hydrocarbons, the conventional melting temperatures T_m (i.e., at $p = 1$ atm) for $1 \leq n \leq 20$ have been plotted against n in Figure X.7. Although the overall trend is for T_m to rise with increasing chain length n, the plot shows a distinctive even–odd alternation, diminishing rapidly in amplitude with increasing n. Furthermore, these T_m values suggest that they may level off in the large-n limit. A rough fit to the 1 atm melting temperatures of large-n alkanes [Lide, 2003, Section 3] beyond those represented in Figure X.7, hence beyond the significant even–odd alternation feature, leads to the following formula:

$$T_m(°C) \approx 141.9 - 2490/n. \qquad (X.22)$$

It is worth noting that the melting temperature measured for high-molecular-weight polyethylene (with a molecular weight distribution) has been reported as 137.5°C [Matsuoka, 1992, p. 200], which agrees reasonably well with the constant term in this last equation. The melting temperature for any n is thermodynamically equal to the ratio $\Delta H/\Delta S$ of melting enthalpy to melting entropy, both of which are expected to rise roughly in proportion to the molecular size n as it becomes large.

For contrast, the 1-atm boiling temperatures T_b have also been plotted versus n in Figure X.8 for $1 \leq n \leq 20$. No corresponding even–odd alternation appears, and the trend shown is consistent with an unbounded rise in T_b as n increases (although thermal decomposition would eventually intervene). That the boiling points would increase with n should not be surprising, since attractive intermolecular interactions that provide binding in the liquid must be overcome for a molecule to

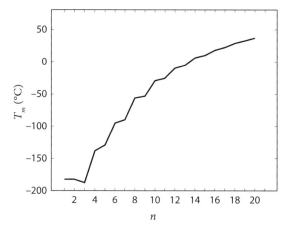

FIGURE X.7. Melting temperatures at 1 atm pressure for the normal alkanes C_nH_{2n+2} plotted versus n. The data have been taken from Lide, 2003, Section 3. See also Ubbelohde, 1978, Figure 7.3.

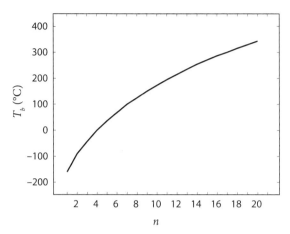

FIGURE X.8. Boiling temperatures at 1 atm pressure for the normal alkanes C_nH_{2n+2}, plotted against carbon number n. The data are taken from Lide, 2003, Section 3.

evaporate and become essentially isolated in the vapor phase, and this binding energy would increase with the length of that molecule. However, it should be kept in mind that the rate of rise of T_b with n would be somewhat diminished as n increases by a decreasing molecular packing fraction because of increasing thermal expansion in the liquid, an effect tending to reduce somewhat the mean attractive binding energy in that liquid phase.

The presence of an n alternation effect in normal alkane melting temperatures but not their boiling temperatures rests upon the crystal structures of those substances. The even-n alkanes tend to have an enhanced T_m that correlates with the fact that their crystal-phase packing as measured by mass densities also displays an enhancement compared to the odd-n members, though, like the melting temperatures, this distinction diminishes in amplitude with increasing chain length [Boese et al., 1999]. The higher packing densities imply that the even-n molecules are able to take increased advantage of attractive intermolecular interactions acting between neighbors in the crystal structure, thereby effectively locking them into a periodic crystal structure to higher temperature.

The crystal structures adopted by the $n \geq 4$ normal alkanes at low temperatures show that the molecules' carbon backbones are present in an all-*trans* planar zig-zag conformation. That is, all $n - 3$ dihedral angles along the backbone, as represented in Figure X.1, are $180°$ [Bunn, 1939; Nyburg and Lüth, 1972; Boese et al., 1999; Bond and Davies, 2002]. Furthermore, the long axes of these all-*trans* molecules are arranged parallel to one another. Nevertheless, a basic difference exists between the low-temperature crystal structures for even and for odd n. When n is even, the crystals exhibit unit cells containing just one molecule, but for n odd the unit cells contain two molecules [Boese et al., 1999]. These represent the inherent structures for the normal alkanes at low temperatures and pressures. By extrapolation into the very large n regime, the inference is that the lowest energy and/or enthalpy inherent structure for monodisperse polyethylene would also display extended all-*trans* backbones with parallel molecular long-axis alignments, presumably with the same even–odd alternating unit cell occupancies.

As temperature rises above absolute zero in the normal alkane crystals, vibrational and librational motions naturally display increasing amplitudes and anharmonicities. This can lead to interbasin transitions, the simplest of which involve terminal methyl group rotations by $\pm 120°$ about their C–C attachments to the linear backbone, as mentioned in Section X.B. Such transitions entail no net energy change and result in an equivalent inherent structure. More consequential are the transitions that create crystal defects, with an associated rise in inherent structure energy. Spectroscopic experiments with crystal phases of alkanes over a range of n values identify defects that involve backbone conformational deformations, specifically *trans* to *gauche* switches whereby dihedral angles change by approximately $\pm 120°$ [Naidu and Smith, 1994]. These structural defects appear to be least costly in potential energy when the dihedral angles involved are closest to the backbone ends, or when they are paired. But as temperature rises, these conformational defects can not only become more numerous but also can occur farther in toward the chain center, in spite of requiring larger excitation energies.

A complicating feature presented by the larger-n alkanes is the occurrence of crystal–crystal phase transitions as T rises toward their higher melting temperatures $T_m(n)$. The combination of increasing density of conformational defects and of increasingly anharmonic intrabasin vibrations produces driving forces toward alternative molecular packing patterns. The normal alkane $C_{33}H_{68}$ provides a clear example, with its ground-state structure giving way successively to three other crystal structures (at 54.5°C, 65.5°C, and 68.0°C) before reaching its melting point at 71.8°C [Ewen et al., 1980]. It is currently unknown what pattern of crystal–crystal transitions, if any, would persist in the large-n limit applicable to monodisperse and properly equilibrated polyethylene. Nevertheless, the available information suggests that the overall backbone geometry remains essentially linear, with aligned molecules in crystalline monodisperse polyethylene up to its melting temperature, in spite of possible crystal-phase transitions.

Semi-empirical potential energy functions have been constructed to represent both intramolecular and intermolecular interactions in the normal alkanes and polyethylene. One example using the simplifying "united atom" approximation appears in Box X.3 [Errington and Panagiotopoulos, 1998, 1999]. This and similar examples have provided the basis for many simulational studies [e.g., Martin and Siepmann, 1998; Nath et al., 1998; Shen et al., 2002].

The all-*trans* conformation that normal alkanes exhibit in their crystalline ground states is also the lowest energy shape available to the individual molecules in isolation, provided that their carbon number n is not too large. However, it should be noted in passing that this must change as n increases. The potential energy cost incurred by a backbone "U-turn" can be overcome by attractive dispersion forces acting between the now-neighboring sections of the backbone. These competing interactions are represented in the united-atom model potential described in Box X.3, respectively by the dihedral angle terms and by the attractive tails of the Buckingham pair interaction terms. For a single, isolated, very long polyethylene molecule *in vacuo*, the lowest lying inherent structure would involve many U-turns, with essentially *trans* sections between them packed together into a compact unit that minimizes surface energy.

At least at room temperature and below, it is believed that the thermodynamically stable crystal form for polyethylene at 1 atm is orthorhombic [Bunn, 1939; Swan, 1962]. This conclusion is based upon X-ray diffraction studies of large-n normal alkanes and of imperfectly crystallized samples of polydisperse polyethylene with focus on the domains with high degree of local backbone packing order. The results are consistent with inferences from extrapolation of the normal alkane series crystallography. The effective "unit cell" for the polymer refers to the local structural

BOX X.3. Approximate Potential Function for the Normal Alkanes

For many purposes, it suffices to describe condensed phases of the normal alkanes and their mixtures with a type of potential function that uses a contracted representation, specifically a "united atom" description. This approach treats methane molecules (CH_4), terminal methyl groups (CH_3-), and the internal methylene groups ($-CH_2-$) each as single spherically symmetric force centers, in the latter two cases arrayed along the molecular backbones. These force centers provide both intermolecular interactions and a subset of the intramolecular interactions. In terms of the generic potential function exhibited in Eq. (X.4), the configurational variables $\mathbf{R}_i$ for each molecule i comprise the locations of the united atom force centers in that molecule. The specific version to be displayed here was derived by Errington and Panagiotopoulos (1998, 1999).

Pair potentials acting between the united atoms of types λ and μ in different molecules, or between united atoms separated by four or more backbone bonds in the same molecule, are assigned the historically prominent "exp-6" Buckingham form [Buckingham, 1938] with appropriate energy and distance scaling:

$$v_{\lambda\mu}(r) = \varepsilon_{\lambda\mu} v_B[x_0(a)r/\sigma_{\lambda\mu}, a_{\lambda\mu}],$$

where

$$v_B(x, a) = +\infty \qquad (0 \leq x < x_{max}(a)),$$

$$= \left(\frac{a}{a-6}\right)\{(6/a)\exp[a(1-x)] - x^{-6}\} \qquad (x_{max}(a) \leq x).$$

The quantity $x_0(a) < 1$ is the root of the "exp-6" function. Also $0 < x_{max}(a) < x_0(a)$ locates the small-x value at which that "exp-6" form passes through a high maximum, and inside of which that function would plunge unphysically to $-\infty$, a catastrophe avoided by the assigned narrow hard core. Note that the reduced interaction $v_B(x, a)$ as a function of x passes through a unit-depth minimum at $x = 1$. For pairs of united atoms of the same type, the distance and energy scaling parameters are assigned the values

	CH_4	$-CH_3$	$-CH_2-$
$\sigma_{\lambda\lambda}$(Å)	3.741	3.679	4.000
$\varepsilon_{\lambda\lambda}/k_B$(K)	160.3	129.6	73.5
$a_{\lambda\lambda}$	15	16	22

For unlike pairs of united atoms, the model utilizes the combining rules:

$$\sigma_{\lambda\mu} = (\sigma_{\lambda\lambda} + \sigma_{\mu\mu})/2,$$

$$\varepsilon_{\lambda\mu} = (\varepsilon_{\lambda\lambda}\varepsilon_{\mu\mu})^{1/2},$$

$$a_{\lambda\mu} = (a_{\lambda\lambda}a_{\mu\mu})^{1/2}.$$

The united-atom force centers are placed at the ends and at the internal nodes of each molecule's backbone. The intramolecular bonds connecting these points are subject to quadratic stretch (u_{str}) and bend (u_{bnd}) potentials with the generic forms

$$u_{str}(r) = (K_r/2)(r - r_{eq})^2,$$

$$u_{bnd}(\theta) = (K_\theta/2)(\theta - \theta_{eq})^2,$$

where

$$K_r/k_B = 96{,}500 \ \mathrm{K} \ \text{\AA}^{-2},$$

$$K_\theta/k_B = 62{,}500 \ \mathrm{K} \ rad^{-2},$$

$$\theta_{eq} = 114° = 1.990 \ rad,$$

and r_{eq} is set equal to 1.839 Å for CH_3–CH_3, to 1.687 Å for CH_3–CH_2, and to 1.535 Å for CH_2–CH_2 bonds. The first two of these distances are larger than the corresponding actual C–C covalent bonds, reflecting the spatial extension of the methyl groups.

An intramolecular torsion potential depending on the dihedral angle ϕ [illustrated in Figure X.1] is assigned to each three-bond sequence along a molecule's backbone. Its form is the following [Smit et al., 1995]:

$$u_{tor}(\phi) = (1/2)\{V_1[1 + \cos\phi] + V_2[1 - \cos(2\phi)] + V_3[1 + \cos(3\phi)]\},$$

$$V_1/k_B = 355.03 \ \mathrm{K},$$

$$V_2/k_B = -68.19 \ \mathrm{K},$$

$$V_3/k_B = 791.32 \ \mathrm{K}.$$

This torsional potential possesses three minima, the lowest at $\phi = 180°$ (the *trans* configuration), and two relative minima at $\phi = \pm 63.45°$ (the two *gauche* configurations). In terms of u_{tor}/k_B at these minima, the *trans* → *gauche* change involves an excitation of 208.7 K, and in order for that to occur as ϕ changes continuously, it is necessary to surmount a torsional barrier of 830.3 K.

repeat unit and does not consider molecule terminal groups. This unit cell incorporates four methylenes (–CH_2–), two from each of two neighboring molecules, with all-*trans* backbones parallel to the unit cell's c dimension. Over the temperature range from –196°C to +130°C, this c dimension remains essentially invariant, but the perpendicular a and b dimensions exhibit thermal expansion [Swan, 1962]. These results imply that the 1 atm mass densities for perfect polyethylene crystals at –196°C, +30°C, and +138°C are 1.043 g/cm³, 0.998 g/cm³, and 0.961 g/cm³, respectively. Figure X.9 illustrates a projection of the effective orthorhombic unit cell perpendicular to the direction of the all-*trans* extended backbones.

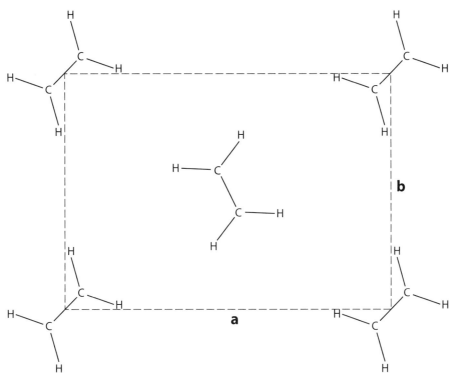

FIGURE X.9. Projection of the orthorhombic effective unit cell for the polyethylene crystal, outlined by the dashed boundary. The *a,b* plane is perpendicular to the long axis of the fully extended carbon backbones. Adapted from Painter and Coleman, 1997, Figure 7.22.

(2) "TYPICAL" POLYETHYLENE BEHAVIOR

In the case of the pure normal alkanes at constant pressure, the observed phase transitions in the crystalline state and the melting transition that occur as temperature increases are sharply defined under isobaric conditions, provided that chain length n is not excessively large. However, one needs to keep in mind that these transitions are at least somewhat obscured in real polyethylene samples by the three previously mentioned complications (keeping in mind that polymer synthesis tends to yield noncrystalline samples). One of these is the fact that some small extent of irregular molecular branching tends to occur in real samples. A second is that even if only linear molecules were present they would have a significant molecular weight distribution, which alone would make it ambiguous as to what the true ground-state crystal structure would be. The third is that the large molecular lengths involved in the polymer would require long times for the torsionally disordered and entangled configurations typically present in the melt to move through the labyrinthine potential energy/enthalpy landscape toward optimally crystalline inherent structures. This last feature amounts to a major kinetic roadblock for configurational conversion of polyethylene to the predominantly extended backbone configurations needed for crystallization, compared to the times available in the real world for sample observation under cooling conditions for a liquid phase.

With respect to the inhibition of crystallization kinetics caused by backbone entanglement, a method for at least partial relief is available. This involves suspending the polyethylene molecules as a dilute solution in a low-molecular-weight solvent, in which the polymer molecules are sparingly soluble but are able to diffuse and to reconfigure much more rapidly than they could in the melt at the same temperature. Subsequent cooling and/or evaporating solvent from the hot solution takes the system to the point where polymer crystalline structures separate from the solution. Specifically, it was discovered that using xylene $[C_6H_4(CH_3)_2]$ as the solvent led to the formation of thin single-crystal lamellae [Fischer, 1957; Keller, 1957; Till, 1957]. As illustrated schematically in Figure X.10, these lamellae consist of polyethylene backbones multiply folded at the top and bottom surfaces, with parallel orientation between neighbor chain segments that are substantially perpendicular to those top and bottom surfaces. Typical dimension ranges observed for such lamellar structures are thicknesses around 150 Å, and lateral widths roughly three orders of magnitude larger [Painter and Coleman, 1997, p. 255], and in this respect they are geometrically similar to sheets of paper. A polyethylene molecule with degree of polymerization approximately equal to 1,000 (i.e., approximately 2,000 carbon atoms along its backbone) has a fully extended length of about 2,500 Å; thus it would contribute 15–20 thickness-spanning sections and U-turns to the lamella in which it resides. A single lamella would incorporate many individual molecules from the dilute starting solution. Subsequent to the initial observations, it has been established that lamellar thickness can be varied somewhat by choice of solvent and of crystallization temperature [Hoffman et al., 1976, Fig. 34(a)].

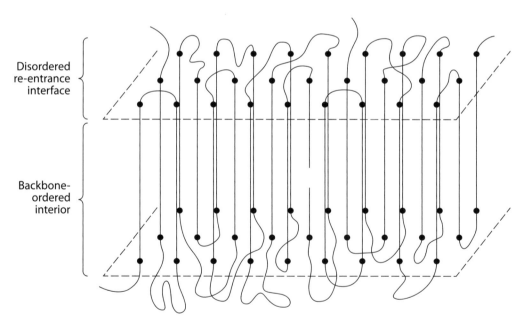

Disordered re-entrance interface

Backbone-ordered interior

FIGURE X.10. Qualitative representation of molecular backbone packing arrangement in a single-crystal polyethylene lamella, formed from dilute solution. The dashed-edge upper and lower planes that separate ordered interior from disordered interface zones and the bold crossing points are shown only for ease of visualization.

The methyl end groups of linear polyethylene molecules can either be embedded within the lamellar crystal structure or could extend into the less ordered surface zones of the lamella. These diverse options permit inclusion of molecules with a distribution of molecular weights. Although it may be impractical to check in the laboratory, this suggests that if a strictly monodisperse sample of high-molecular-weight polyethylene were available, it would also form lamellar structures upon precipitation from a dilute solution. Whether monodisperse or polydisperse, the configurational constraints on participating polymer molecules are far less stringent in forming lamellae by multiple folding than would be the case requiring a fully extended backbone conformation to fit into the ideal crystal structure that has been inferred above by extrapolating the normal alkane series crystal structures. The lamellar structures amount to higher lying inherent structures for the monodisperse polymer, but they are the only ones that are kinetically accessible under conventional conditions.

Cooling a polyethylene melt (no solvent present) at a realistic rate produces a solid medium with partial crystallinity. This also involves backbone ordering with multiple folding, but in a process that includes the nucleation and growth of "spherulites." These patterns become clearly visible under optical microscopy and grow to encompass the entire system [Bunn and Alcock, 1945]. During the initial stages of their appearance, the spherulites are indeed spherical, but their radial growth eventually causes them to come into contact with one another, creating stationary common boundaries and finally space-filling polyhedral shapes. Examination of these spherulite domains with a variety of experimental probes (polarized light, scanning electron microscopy, atomic force microscopy) indicates that the polymer backbones tend to be multiply folded into lamellar ribbons that radiate outward, with helical twisting and branching [Toda et al., 2008; Hatwalne and Muthukumar, 2010]. The fold direction of the backbones is preferentially tangential to the spherulite radial growth direction.

Although the formation of spherulites upon cooling a melt takes the polymeric system to a significantly lower portion of the multidimensional potential-energy/potential-enthalpy landscape, it amounts to a substantially incomplete descent toward the lowest inherent structure basins. The completed spherulitic structure contains both backbone-ordered and backbone-amorphous sections, the latter primarily sandwiched between the twisted and branched lamellar ribbons. These amorphous sandwiched regions contain backbone portions connecting, and thus stitching together, neighboring lamellar regions.

This structural heterogeneity in spherulitic systems has important consequences for what is colloquially identified as the "melting transition" of the solidified material. In fact, it creates a significant "premelting phenomenon." This can be illustrated in terms of the isobaric system volume measured as temperature rises (dilatometry) from a low value where the system is unquestionably solid to a substantially higher temperature where it is unquestionably a homogeneous liquid. If strict thermodynamic equilibrium for a monodisperse sample obtained, the resulting first-order phase transition would produce an unambiguous jump discontinuity in volume, as the well-packed elongated polymer backbones suddenly gave way to the geometrically inefficient irregular and entangled arrangements in the liquid. By contrast, the nonequilibrium transition of the spheroidal solid medium exhibits an asymmetrically broadened transition [Chiang and Flory, 1961]. These distinct scenarios are qualitatively illustrated by Figures X.11(a) and X.11(b), respectively.

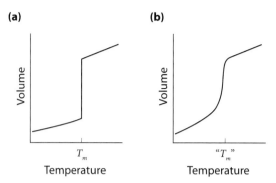

FIGURE X.11. (a) Isobaric volume discontinuity for an equilibrium first-order melting transition in a monodisperse sample at temperature T_m. (b) Asymmetrically broadened melting transition ("premelting phenomenon"), monitored by isobaric volume change, for a spheroidal-texture semicrystalline sample of polyethylene. The effective melting temperature "T_m" may be identified as the location of the maximum downward curvature for volume vs. temperature.

It must be kept in mind that the quantitative details of this broadened nonequilibrium transition depend on the polymer molecular weight distribution, the cooling schedule for producing the initial solid, and the warming rate through the transition. But even with these complicating features, it seems clear that the broadening is a direct consequence of the spatial heterogeneity of crystallinity that is inevitably present in the semicrystalline initial state. Locations including the least well-ordered lamellar arrangements are least able to resist thermal agitation as temperature rises and are the first to melt and to cause volume to begin to rise. Further temperature rise disorders somewhat larger and better arranged lamellar portions with greater influence on measured volume. Finally, the largest well-ordered locales suddenly melt and cause a final steep rise in the volume curve, as shown in Figure X.11(b). Thereafter, only the uniform liquid's thermal expansion determines the isobaric volume behavior.

The asymmetric melting phenomenon is not limited merely to high-molecular-weight polyethylene samples with a distribution of those molecular weights. Monodisperse normal alkanes with moderately long carbon backbones behave somewhat similarly, indicating the presence of substantial crystal-phase defects. This phenomenon has been documented by dilatometry for $C_{94}H_{190}$ [Mandelkern, 1989] and by calorimetry for $C_{192}H_{386}$ [Stack et al., 1989].

An elementary measure of the fraction of crystallinity $0 \leq f_{crys} \leq 1$ present in solidified polyethylene can be based on volumetric measurements. Suppose that at the temperature $T < T_m$ of interest d_c and d_l are the mass densities of the ideal polyethylene orthorhombic crystal and of the temperature-extrapolated supercooled liquid, respectively. Then if d is the measured mass density of the solidified sample at that temperature, the basic interpolation expression accounting for relative sizes of the amorphous and crystalline subregions is the following:

$$f_{crys} = \frac{d_c(d - d_l)}{d(d_c - d_l)}. \tag{X.23}$$

Of course, this rough measure is not sensitive to the spatial distribution of crystallinity, whether homogeneous or heterogeneous on various length scales. An analogous f_{crys} measure can be deduced from calorimetric determination of the heat of transition for a given melting or freezing transition, compared to that for a perfect crystal (if that value is available).

(3) POLY(TETRAFLUOROETHYLENE)

Replacement of all hydrogens in normal alkane molecules with fluorines, producing perfluoroalkanes (C_nF_{2n+2}), introduces substantial property changes. These changes extend into the large-n polymeric limit and significantly influence the behavior of polytetrafluoroethylene (PTFE). A principal source of these changes can be identified as the increase in size and net electrostatic charge of the electron clouds surrounding the backbone-attached atoms, as each H is replaced by a larger and more electronegative F. This replacement gives rise to substantial overlap and coulombic repulsion between pairs of nearby F atoms along the carbon backbone [Jang et al., 2003]. The configurational result is that the planar all-*trans* backbone shape, which serves as an inherent structure for the normal alkane molecules in isolation, is no longer a mechanically stable configuration for the perfluorinated analog molecules. Instead, the perfluorocarbon molecules spontaneously exhibit a gentle twist deformation away from the all-*trans* arrangement, which can occur in either of two equivalent twist directions, thus leading to a pair of equivalent chiral configurations [Bunn and Howells, 1954]. This chiral deformation is schematically illustrated in Figure X.12. In the case of perfluorohexane, the angular twist of the backbone orientation between every other $-CF_2-$ unit is ±27°, which in a longer perfluoroalkane would amount to approximately a full turn (±360°) for every 13 monomer units ($-CF_2-CF_2-$) [Jang et al., 2003, Figure 5].

The tendency for fluorines to avoid one another because of coulombic repulsion, supplemented at short separations by enhanced steric repulsion because of their larger size, should have the additional effect of raising the energy of a backbone "U-turn" compared to that for the unsubstituted alkanes. This implies that the lowest lying inherent structures for isolated C_nF_{2n+2}

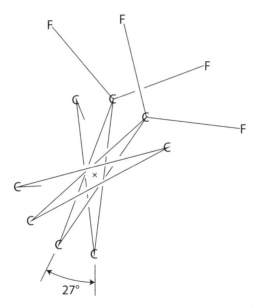

FIGURE X.12. Mechanically stable intrinsically twisted configuration of a short portion of the carbon backbone exhibited by perfluoroalkanes. This schematic view is along the helical axis indicated by "×." As a result of the twist, pairs of fluorine atoms attached to every other backbone carbon are able to reduce their repulsive interactions. In this view, bond lengths are not drawn accurately to scale.

molecules should retain essentially linear backbone geometries to larger n than for the unfluori-nated alkanes. But with sufficiently large n, the lowest energy structure would again involve the backbone folding back on itself under the influence of dispersion attractions.

The initial formation of PTFE by polymerization of the monomer produces a highly crystal-line material ("virgin PTFE") in which the polymer backbones are primarily linear and aligned [Starkweather, 1986]. However, after melting such samples, subsequent freezing produces only partial crystallinity, and the crystalline regions have a lamellar geometry with kinetically imposed backbone folding. The lamellar thickness has been observed to depend on cooling history but is considerably larger than for PE and has been observed to fall in the 500–1,500 Å range [Ferry et al., 1995, Figure 11]. This comparison is to be expected in terms of relative folding energies for PTFE and PE.

Two polymorphic phase transitions have been identified as occurring within the crystalline state of PTFE. Below 19°C, the crystal consists of aligned molecules, each exhibiting an undis-turbed helical conformation. Between 19°C and 30°C, an intermediate slightly defective crystal exists, then to be replaced above 30°C with a more defective structure persisting up to the PTFE melting point that occurs in the neighborhood of 340°C [Starkweather, 1986]. Molecular dynam-ics computer simulation suggests that the energetically least costly defect process involves intra-molecular helicity disruption (localized twist reversals), followed in increasing energy cost by longitudinal chain displacement, and finally by *gauche* deformations of the backbone chain [Sprik et al., 1997]. It should be noted that this crystal polymorphism is not just an attribute of the large-n polymer regime; crystal-phase transitions have also been observed in the normal perfluo-roalkanes C_nF_{2n+2} in the range $7 \leq n \leq 24$ [Starkweather, 1986].

(4) POLYSTYRENE

The geometric placement of side groups along a polymer backbone, illustrated schematically in Figure X.2, has a strong influence on that polymer's crystallization behavior. This is clearly illus-trated by the phenyl side-group isotactic, syndiotactic, and atactic variants of polystyrene (PS). The first of these variants displays a crystal phase that has a 1 atm melting temperature of 237°C [Karasz et al., 1965]. The second similarly has a crystal phase, but with the somewhat higher 1 atm melting temperature of 270°C [Ishihara et al., 1986]. In contrast, atactic polystyrene resists crys-tallization because of the inability of its random pattern of phenyl side groups to pack effectively and consistently next to one another. Consequently, liquid-phase atactic polystyrene remains amorphous upon cooling, eventually passing through a glass transition at $T \approx 100°C$ [Painter and Coleman, 1997, p. 300]. Both the syndiotactic and atactic forms of polystyrene have wide indus-trial applications, in contrast to isostatic polystyrene.

Isotactic polystyrene melts can be slowly crystallized, although typically the extent of crystal-linity is only on the order of 50%. Nevertheless, the resulting solid material suffices for determina-tion of the structure within the crystallized domains by X-ray and electron diffraction [Natta et al., 1960]. In order to avoid phenyl side-group interferences while maximizing interchain dispersion attractions, the macromolecules adopt a backbone twist that repeats after every third phenyl group. Beyond this local helicity, the backbones overall are linear and parallel. The backbone twist direc-tion exhibits regular alternation throughout the crystal, which therefore exhibits no net chirality.

Crystallization of syndiotactic polystyrene produces a complex set of crystallographic structures, depending on the experimental conditions [De Rosa et al., 1992]. These alternatives may reflect underlying crystal transitions in the strict thermodynamic equilibrium behavior of a monodisperse high-molecular-weight system. The most ordered structure has been designated β'' and is orthorhombic. In terms of available information, this appears to be the most likely candidate for the lowest lying inherent structure at 1 atm. The polystyrene molecule backbones are present in an all-*trans* arrangement, with four parallel molecular chains penetrating the unit cell [De Rosa et al., 1992, Figure 5].

Atactic polystyrene appears in many household items as a mechanically tough molded material. It is also very commonly seen as polystyrene foam, used as disposable drinking cups, thermal insulation, and packaging materials to protect fragile items in transit. These foams are produced by dissolving low-molecular-weight gases in the melt under pressure, which is then released as the combination is cooled. The resulting expanded medium typically has a mass density one tenth or less of that before foaming.

Polystyrene in each of its three forms is substantially nonbiodegradable, thus creating concerns over its long-term environmental impact.

(5) POLYETHYLENE OXIDE

A simple example illustrating the effects of chemically modifying the backbone in linear polymers is provided by polyethylene oxide (PEO). This macromolecule is formed by catalyzed polymerization of the strained-ring ethylene oxide molecule $(CH_2)_2O$. The degree of polymerization n (number of monomers) in PEO samples typically lies in the wide range $500 \leq n \leq 250,000$. Unlike the previously discussed polymers PE, TFPE, and PS, PEO is very soluble in water under ambient conditions. This feature stems largely from the availability of backbone oxygen atoms for hydrogen bonding by neighboring water molecules. It underlies the widespread use of PEO in pharmaceutical and other industrial products. An interesting phase behavior for this PEO–water system emerges at elevated temperature and pressure, specifically, a closed-loop phase separation region with lower and upper consolute points [Saeki et al., 1976].

The crystal structure of PEO has been determined by X-ray diffraction [Takahashi and Tadokoro, 1973]. The effective unit cell is monoclinic, containing four helically twisted backbone portions. The twist geometry within a unit cell involves seven monomeric units exhibiting two full rotations. The fact that the backbone exhibits a preference for twist, compared with the planar all-*trans* geometry in polyethylene crystals, evidently reflects the greater dihedral rotational freedom possible around the many –C–O– bonds that are present. This additional configurational freedom also enhances the entropy of the melt and is at least partly responsible for the relatively low reported melting temperature 66°C [Wang et al., 2009], compared to polyethylene.

A technique has been demonstrated for the production of single-molecule PEO crystals [Bu et al., 1996]. This involves spreading a thin layer of a dilute PEO solution in benzene (C_6H_6) on a hot liquid water surface. The benzene solution spreads quickly because of surface tension, thus separating the individual PEO molecules, followed immediately by evaporation of the benzene. The single-molecule crystals that form at the surface are then collected before they can dissolve in the water. The distribution of observed crystal sizes reflects the distribution of molecular weights in the PEO sample. This crystallization scenario contrasts strongly with partial crystallization of PEO melts, which proceeds via spherulite formation [Khoury and Passaglia, 1976, Figure 49(b)].

D. Polymer Melts and Glasses

To the extent that they can remain in thermal equilibrium, atactic, randomly branched, and random copolymer systems continue to display amorphous configurations as they are cooled from hot liquid states. In practice, of course, cooling such materials eventually encounters a glass transition temperature range below which thermal equilibration becomes kinetically unattainable. But above that range, the thermodynamic properties can be inferred from the isochoric enumeration and mean vibrational quantities $\sigma(\varphi, \{\rho_a\})$ and $\beta f_{vib}(\varphi, \beta, \{\rho_a\})$, or from their isobaric analogs $\hat{\sigma}(\hat{\varphi}, \{x_a\}, p)$ and $\beta \hat{f}_{vib}(\hat{\varphi}, \beta, \{x_a\}, p)$, by minimizing the corresponding Helmholtz or Gibbs free energy expressions with respect to φ or $\hat{\varphi}$ [e.g., Eqs. (X.16) or (X.20)].

However, for those polymer systems that are able to produce at least partially crystalline structures, a usable description and analysis of metastable supercooled liquid states requires projecting out sets of inherent structures exhibiting local crystalline configurations that could act as nucleation sites. The inherent structures surviving this projection operation involve disordered backbone conformations, consistent with the volume and entropy increases per monomer that have been measured at melting [Mandelkern, 1989, 2002]. This descriptive approach for supercooling simply adapts to polymeric systems the strategy introduced in Section III.G for low-molecular-weight substances. In view of the tendency for crystallizable polymers to produce locally ordered laminar structures below their melting temperatures, the inherent structures to be eliminated must be those exhibiting portions of such laminae. The projection operation applies both to monodisperse as well as to polydisperse systems of linear polymers. Following previously introduced terminology for the basins and inherent structures remaining after the elimination process, the modified intensive enumeration and vibrational quantities are denoted by $\sigma_a(\varphi, \{\rho_a\})$ and $\beta f_{vib,a}(\varphi, \beta, \{\rho_a\})$ for isochoric conditions, and by $\hat{\sigma}_a(\hat{\varphi}, \{x_a\}, p)$ and $\beta \hat{f}_{vib,a}(\hat{\varphi}, \beta, \{x_a\}, p)$ for isobaric conditions. But whether or not the molecular structure requires a projection operation to describe supercooling, the intensive properties of the resulting amorphous polymeric solid below a glass transition depend on the cooling rate, just as described in Chapter VI for low-molecular-weight substances.

Because of the high molecular weights typically presented by polymers, vector $\mathbf{R}_i$ specifying the full instantaneous configurational details for polymer i in a macroscopic liquid sample includes a huge amount of information. For practical purposes, it is often useful to cite a set of simple geometric measures that amount to contractions of $\mathbf{R}_i$. First, let $\mathbf{s}_j^{(i)}$ denote the position of the center of monomer unit j in that polymer molecule i, which we shall suppose is composed of n monomers. The geometric centroid of the polymer then is located at

$$\mathbf{S}_0^{(i)} = n^{-1} \sum_{j=1}^{n} \mathbf{s}_j^{(i)}. \tag{X.24}$$

Note that for simplicity all monomers have been given equal weights in this expression, although they might differ in chemical structure and mass. Alternatively, end groups, monomers acting as backbone branching units, and distinguishable monomers in copolymers could have been weighted differently. But in the simple case of equal weighting, the radius of gyration $R_g^{(i)}$ for the molecule is defined by the following expression:

$$(R_g^{(i)})^2 = n^{-1} \sum_{j=1}^{n} [\mathbf{s}_j^{(i)} - \mathbf{S}_0^{(i)}]^2. \tag{X.25}$$

This can be transformed into an alternative expression that involves a sum over all pairs of distinct monomers within the molecule [Rubinstein and Colby, 2003, p. 62]:

$$(R_g^{(i)})^2 = n^{-2} \sum_{j=2}^{n} \sum_{k=1}^{j-1} [\mathbf{s}_j^{(i)} - \mathbf{s}_k^{(i)}]^2. \tag{X.26}$$

In the event that molecule i is a linear polymer, another relevant length is the end-to-end distance:

$$R_e^{(i)} = | \mathbf{s}_n^{(i)} - \mathbf{s}_1^{(i)} |. \tag{X.27}$$

In the liquid phase, this latter length for linear polymers is typically significantly less than the contour length for the same molecule, defined as follows:

$$R_c^{(i)} = \sum_{j=1}^{n-1} | \mathbf{s}_{j+1}^{(i)} - \mathbf{s}_j^{(i)} |. \tag{X.28}$$

Each of these three lengths $R_g^{(i)}$, $R_e^{(i)}$, and $R_c^{(i)}$ possesses distribution functions for liquid states at thermal equilibrium, as well as for supercooled-state extensions of those distribution functions in cases of crystallizable polymers by invoking the basin projection procedure mentioned above. The contour length $R_c^{(i)}$ should be relatively insensitive to temperature change, and in particular would be the least sensitive of the three lengths to a melting transition that induces qualitative changes in the overall backbone geometry of a linear polymer. By contrast, the radius of gyration and end-to-end distance would exhibit major changes as a fully extended crystal configuration, or even a lamellar backbone ordering, converted to a disordered and entangled arrangement of polymer backbones.

A basic feature of a polymer melt is how the configurations in an equilibrium ensemble respond under mapping to inherent structures. Because of the pervasive molecular entanglements that exist in the liquid, and the corresponding labyrinthine nature of the configuration space that is accessible to the molecules, the constant-volume steepest descent mapping should cause relatively little change in the lengths $R_g^{(i)}$ and $R_e^{(i)}$. In other words, the distributions of these properties should undergo only minor changes upon mapping to the corresponding sets of inherent structures. However, it might be noted in passing that if the same single-molecule configuration sets were subjected to their own steepest descent mappings for just the intramolecular interaction (i.e., no entangling intermolecular interactions), the resulting distributions could be quite different. In particular, such single-polymer-molecule steepest descent mappings could involve substantial collapse of initially extended polymer conformations, driven by the unblocked intramolecular attractions operating between noncontiguous monomer units.

With regard to high-molecular-weight linear polymers, it is useful to have in hand the implications of some simple geometric idealizations of how the backbones could be arranged within the entangled liquid medium. The most basic of these idealizations is the random walk picture. This viewpoint supposes that the monomers constitute a freely jointed chain, with fixed separation

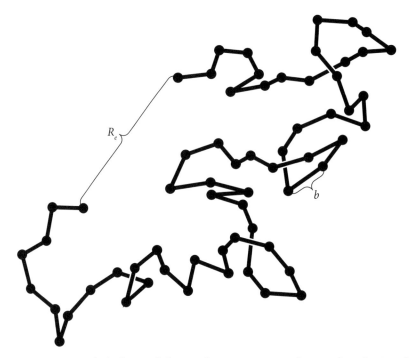

FIGURE X.13. Representation of a freely jointed chain configuration, consisting for ease of visualization of the modest number $n = 63$ of monomer units (dots), with fixed bond length b.

b between neighbors along the chain. Furthermore, non-nearest neighbors are assumed not to interact, and the bond angle defined by three successive monomers along the backbone is entirely free to vary without constraint or potential energy cost. Figure X.13 provides a sketch of such a freely jointed chain configuration. The contour length obviously has the fixed value

$$R_c = (n - 1)b. \tag{X.29}$$

The root-mean-square value of the end-to-end distance R_e is simple to calculate for the freely jointed chain containing n monomer nodes because of the fact that the directions of the bond vectors are all independent:

$$
\begin{aligned}
\langle (R_e)^2 \rangle^{1/2} &= \left\langle \left[\sum_{j=1}^{n-1} (\mathbf{s}_{j+1} - \mathbf{s}_j) \right]^2 \right\rangle^{1/2} \\
&= \left\langle \sum_{j=1}^{n-1} (\mathbf{s}_{j+1} - \mathbf{s}_j) \cdot (\mathbf{s}_{j+1} - \mathbf{s}_j) \right\rangle^{1/2} \\
&= \langle (n-1)b^2 \rangle^{1/2} \\
&\equiv (n-1)^{1/2} b.
\end{aligned}
\tag{X.30}
$$

Here, starting from one end of the chain, we have indicated the jth monomer node position along the chain by $\mathbf{s}_j$, and the averaging operation $\langle .... \rangle$ covers all of the equally probable independent bond orientations. A similar calculation can be carried out for the root-mean-square radius of gyration R_g of the freely jointed chain, using the fact that any subchain consisting of $1 \leq l \leq n - 1$ contiguous bonds is itself a freely jointed chain. The result is found to be

$$\langle (R_g)^2 \rangle^{1/2} = \left[\frac{(n^2 - 1)}{6n} \right]^{1/2} b. \tag{X.31}$$

The oversimplifications involved in the freely jointed chain picture are clear. Real polymer backbones are obviously subject to local angle preferences and energy penalties for deviations from those preferences imposed by the quantum mechanics of chemical bonding. If this feature alone were at issue, one might argue that combining some number > 1 of contiguous monomer units into an effective smaller number of "bond" units whose relative orientations would be close to independent and free might constitute a useful description for high-molecular-weight linear polymers. Corresponding versions of Eqs. (X.29)–(X.31) could then be displayed with n and b replaced empirically with effective values $n_{eff} < n$ and $b_{eff} > b$. This modification does not alter the prediction for melts that in the asymptotic large-molecular-weight limit, both $\langle (R_e)^2 \rangle^{1/2}$ and $\langle (R_g)^2 \rangle^{1/2}$ would be proportional to $n^{1/2}$. That sets a standard against which determinations based on more complete and realistic modeling can be compared. In particular, one subsequently must account for the nonoverlap property of the constituent monomer units, involving both intramolecular and intermolecular interactions.

Small-angle neutron scattering (SANS) is a possible way to determine $\langle (R_g)^2 \rangle^{1/2}$ in polymer melts, after a controlled fraction of the molecules have undergone replacement of their hydrogen atoms with deuterium [Stein and Han, 1985]. This isotopic replacement should have a negligible effect on the entangled conformations in the amorphous melt or glass medium. The measurements indicate that the radius of gyration for linear polymers adheres closely to the freely jointed chain behavior, namely, that its root-mean-square (rms) value scales as $n^{1/2}$ in the asymptotic limit of high degree of polymerization n [e.g., Kirste et al., 1975, for the case of amorphous poly(methylmethacrylate), PMMA]. This conclusion indicates qualitatively that the necessity for any polymer to avoid intersecting itself is largely compensated in melts by the "solvating" effect of neighboring polymer molecules.

In addition to neutron scattering measurements, some insight into the statistical geometry of polymers in the molten state has become available from computer simulations. Although these investigations go beyond the elementary freely jointed chain idealization, they usually involve considerable simplification of chemical structure. Furthermore, even for simplified models, limitations on computer capabilities restrict simulations to considerably smaller chain lengths than those involved in typical experimental investigations of polymer melts. Nevertheless, the limited results appear to be consistent with the experimental conclusion that linear polymers in the melt behave as freely jointed chains with respect to chain length dependence of average end-to-end distance and radius of gyration [e.g., Kremer and Grest, 1990, Figure 1, for $10 \leq n \leq 200$]. It is important to note that simulation of melts composed of noncatenated ring polymers exhibit a distinctly different behavior: The average radius of gyration appears to increase asymptotically as $n^{1/3}$ for large n as opposed to the $n^{1/2}$ behavior of freely jointed chains [Vettorel et al., 2009].

Many of the properties described in Chapter VI for low-molecular-weight substances undergoing supercooling and glass formation are qualitatively applicable to polymer melts. In particular, Figure VI.3 generically represents the isobaric temperature dependence of macroscopic system volume under a range of cooling rates. As was the case for media consisting of small molecules, this cooling rate dependence leads to identification for polymer melts of a glass transition temperature range, not a unique T_g. The rather sudden change in slope of the $V(T, p)$ curve upon passing through the T_g range signifies the separation of isobaric thermal expansion $\alpha_p(T) = (\partial \ln V/\partial T)_p$ into intrabasin and interbasin contributions [Section V.C(4)]. Below the T_g range, the observed value is dominated by the former (with minor involvement from localized "2-level" transitions). Above the T_g range, the slope increases because both classes of contributions are present within the projected portion of the labyrinthine configuration space. Analogous remarks are applicable to the isobaric heat capacity C_p of a polymer melt and its glass, where the change in functional form across the T_g range provides the possibility of intrabasin/interbasin separation of contributions [Section V.C(3)].

The responses to external perturbations by polymer melts and by the glasses they form upon cooling have high technological importance. This provides a substantial impetus for scientific analysis of the molecular kinetic processes following mechanical, electrical, or thermal impulses applied to macroscopic samples, and how those processes relate to the polymer topological structure (linear, branched, cyclic, …) and to the distribution of molecular weights present. These kinetic processes of course also underlie self-diffusion. For high molecular weights typically involved in polymeric materials, the central issue is how those large molecules are able to cope kinetically with the severe entanglement restrictions presented by the potential energy or potential enthalpy landscape labyrinth.

Measurements of melt shear viscosity η (in the vanishing shear rate limit) as a function of molecular weight provides some basic insight [Berry and Fox, 1968, Figures 1 and 2; Painter and Coleman, 1997, p. 414; Rubinstein and Colby, 2003, p. 341]. For many linear polymers, it has been determined that in the low-molecular-weight regime, shear viscosity at a constant temperature increment above the glass transition temperature is approximately proportional to the weight-averaged molecular weight of the medium:

$$\eta \propto M_{wa}. \tag{X.32}$$

However, as M_{wa} rises, this simple behavior is superseded empirically by a nonlinear dependence with a substantially larger positive exponent:

$$\eta \propto (M_{wa})^{3.4}. \tag{X.33}$$

Note that in the large-M_{wa} regime, glass transition temperatures T_g become independent of M_{wa} [e.g., Fox and Flory, 1950, Figure 3]. The crossover between these two regimes (X.32) and (X.33) occurs in the range of 300 to 500 backbone chain atoms. Evidently, this range is where chain entanglements begin to exert a dominating influence on the molecular motions that determine the macroscopic property η. Below this range, the landscape can be viewed qualitatively as exhibiting an incomplete labyrinth, missing a large portion of the impenetrable walls that would be present to produce entanglements at large M_{wa}. Although the results (X.32) and (X.33) are based on experiments that involve samples with variable molecular weight distributions, there is no evidence that the nonzero width of those distributions affects the exponents shown.

In the first of these ranges, Eq. (X.32), the observed viscosities have been rationalized utilizing the "Rouse model" [Rouse, 1953]. Although this model was originally proposed to explain viscoelastic properties of dilute polymer solutions, it is amenable to reinterpretation for dense polymer melts. It views each linear polymer molecule as a sequence of relatively compact structural units that are connected by harmonic springs. The structural units are individually and independently subject to localized hydrodynamic drag because of the surrounding medium; the connecting springs themselves, however, experience no such hydrodynamic interaction. No explicit effect arising from entanglements between neighboring molecules plays any role in the Rouse model. From a thorough analysis of the relaxational modes present in the Rouse model, it is possible to conclude that Eq. (X.32) is satisfied [Rubinstein and Colby, 2003, p. 321].

Self-diffusion constants D have been measured for a substantial variety of linear polymers with high molecular weights [Lodge, 1999]. The isothermal variation with M_{wa} for any one of these substances accurately follows a power law:

$$D \propto (M_{wa})^{-2/3}. \tag{X.34}$$

It is useful to compare Eqs. (X.33) and (X.34) in the context of the previously mentioned elementary Stokes-Einstein relation, Eq. (V.143), which was introduced long ago to describe Brownian motion in simple (nonpolymeric) liquids:

$$D\eta = k_B T / C R_{eff}, \tag{X.35}$$

where C is a dimensionless constant and R_{eff} is an effective hydrodynamic radius. If the polymer η and D expressions (X.33) and (X.34) were to be formally inserted into this Stokes-Einstein relation, the result would imply

$$R_{eff} \propto (\eta D)^{-1} \tag{X.36}$$

$$\propto (M_{wa})^{-1.1}.$$

Considering the possibility of slight numerical imprecisions in the experimental fits in Eqs. (X.33) and (X.34), one perhaps should not distinguish this last exponent from minus unity. This result leads to the obviously unphysical conclusion that the effective hydrodynamic radius of a linear polymer in its melt is inversely proportional to the polymer contour length R_c. In view of the experimental evidence such as SANS that polymer chains in the melt are conformationally similar to freely jointed chains, thus displaying rms radii of gyration proportional to $(M_{wa})^{1/2}$, this last inference points to an important feature of polymer melts. Specifically, it indicates that while entanglements have the effect of retarding viscous flow at the macroscopic length scale, polymer backbone motions at the molecular length scale manage to circumvent some significant aspects of entanglement. Therefore, the effective value of a local shear viscosity η formally required by a Stokes-Einstein relation is much less at the polymer $\langle (R_g)^2 \rangle^{1/2}$ length scale than its value at the macroscopic length scale, thus substantially increasing R_{eff} to a physically reasonable value.

A "tube model" has been created to provide a theoretical treatment of chain entanglement in the high M_{wa} regime, and a "reptation model" has been proposed to describe polymer chain motions within the tube model viewpoint [de Gennes, 1971; Doi and Edwards, 1986]. Figure X.14 provides a simplified view of the local entanglement situation, wherein a single linear polymer is confined in its possible motions by nearby impenetrable barriers. In the figure, these constraints are shown merely as black dots, as though the backbones involved were perpendicular to the plane

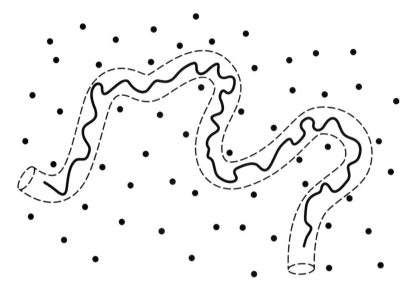

FIGURE X.14. Schematic representation of the tube model for a linear polymer (solid curve) restricted by the presence of neighboring molecules in the melt, whose covalent backbone bonds cannot be crossed. For simplicity, the confining restraints are shown as black dots. Portions of the polymer chain are able to move relatively freely within the momentarily occupied tube indicated by dashed lines.

of the diagram. However, these neighboring polymer molecules can be present in a wide variety of conformations, including being wrapped to some extent around the central molecule. As a result of the noncrossing constraints for covalently bonded backbones, the polymer under consideration is effectively confined for short times to the interior of an occupancy tube (dashed lines in Figure X.14).

Because of its thermal energy, a linear polymer is able to move stepwise along, and to a limited extent across, its momentary occupancy tube. These motions presumably comprise combinations of backbone kink formation and decay, and elastic stretch–compression fluctuations along the backbone contour length. After one end of the polymer molecule has retracted into the tube interior as a result of these thermal motions, it can re-extend either along the previous tube direction or it can wend its way between the existing barriers in a new direction, thus modifying the geometry of the momentary encasing tube. As time proceeds, the tube can increasingly wander away from its original placement within the medium, i.e., the polymer molecule under consideration undergoes self-diffusion. The wormlike or snakelike motions of the entangled polymer have been traditionally called "reptation."

Under the assumption that the impenetrable barriers affecting a test molecule are static features, a scaling analysis for large molecular weight M of that test molecule [de Gennes, 1971] concludes that the self-diffusion constant D for a monodisperse melt should exhibit the following behavior:

$$D \propto M^{-2}. \tag{X.37}$$

A closely related result can be derived for the shear viscosity η [Rubinstein and Colby, 2003, Section 9.2.2]:

$$\eta \propto M^3. \tag{X.38}$$

The latter of these results is based on the presumption that shear viscosity is proportional to the mean time required for a polymer molecule to completely evacuate one occupancy tube in favor of an independent alternative. These scaling relations differ non-negligibly from the experimental results in Eqs. (X.33) and (X.34), indicating that the fixed-barrier assumption has reduced the magnitudes of the exponents involved. Of course, the neighboring molecules providing entanglement constraints also undergo reptation, thus generating finite lifetimes for the local barriers indicated naively as dots in Figure X.14. Presumably, a self-consistent treatment of molecular reptation in the presence of reptating barrier molecules would drive up the exponent magnitudes somewhat to attain better agreement with Eqs. (X.33) and (X.34).

In principle, it should be possible to describe reptation in terms of potential energy or potential enthalpy basins for the melt, the chronological sequence of occupied basins, and the corresponding inherent structures. The elementary transitions involved as the system configuration wanders through the labyrinth presumably are localized at the molecular length scale just as they are for condensed phases of low-molecular-weight substances [Section III.F]. These localized transitions could be generated numerically during the course of molecular dynamics simulations for various polymer models that have well-defined intramolecular and intermolecular interactions. In particular, it would be valuable to classify the geometric differences between successively visited inherent structures with respect to their elementary contributions to reptation, specifically, whether the geometric change primarily involved a single molecule or whether a concerted displacement of two or more tightly entangled molecules was involved.

The discussion in this section has thus far been focused on linear polymers. This is the subset most amenable to analysis, although still scientifically challenging. One other case deserves brief mention in passing with respect to the occupancy-tube and reptation approach, namely high-molecular-weight star polymers. Because these molecules consist of separate linear arms chemically connected together at a central node, the reptation possibilities for each arm are extremely limited. As a result, the self-diffusion constant D of a star polymer melt declines essentially exponentially with molecular weight, while the shear viscosity η rises essentially exponentially with molecular weight [Rubinstein and Colby, 2003, Section 9.4.2]. This constitutes a distinctive departure from the simple algebraic power laws in Eqs. (X.37) and (X.38) for linear polymers.

Entanglements in the melt can cause severe retardation in the time-dependent response of the liquid to external perturbations. This is true even well above a glass transition temperature range. An important aspect of this kinetic retardation is the viscoelastic response of the melt to imposed shear. Consider the case of a macroscopic sample contained between, and in full contact with, a pair of parallel plates. To avoid spurious edge effects, assume that periodic boundary conditions apply in directions parallel to the plate surfaces. These plates are then given a small-amplitude relative shearing motion (plate separation fixed) with sinusoidal time dependence that formally can be represented as proportional to $\exp(i\omega t)$. It is assumed that the melt adheres sufficiently strongly to the two plates that slippage does not occur. In order to maintain a constant amplitude sinusoidal shearing displacement, sinusoidal forces with the same frequency ω must be applied to the plates. This applied force generally involves both in-phase and out-of-phase components, relative to the plate displacements. The former is a nondissipative elastic response that would be present even if the sample were a nonflowing perfect crystal. The latter includes dissipative processes present in the melt. The full linear response of the melt can be represented by a complex function of frequency $G(\omega)$ that is the ratio of shear stress to shear strain acting on the melt. This ratio comprises a "storage modulus" G' and a "loss modulus" G'' [Rubinstein and Colby, 2003, Section 7.6.5]:

$$G(\omega) = G'(\omega) + iG''(\omega). \tag{X.39}$$

Here G' and G'' are real functions of ω (the primes in this conventional notation should not be confused with derivative operations). As ω approaches zero, both G' and G'' for liquids approach zero. The former is proportional to ω^2 in that limit, while the latter is proportional to ω with a multiplier that is the shear viscosity of the liquid:

$$\lim_{\omega \to 0} \left[\frac{G''(\omega)}{\omega} \right] = \eta(T). \tag{X.40}$$

The temporal lag of mechanical response of the entangled polymer melt in response to the driving force applied by the plate shear defines a lag angle $0 \le \delta \le \pi/2$ given by the ratio

$$\tan \delta(\omega) = \frac{G''(\omega)}{G'(\omega)}. \tag{X.41}$$

Because $G(\omega)$ describes the linear response regime, it can be attributed formally to a linear combination of independent responses applied to the sample at all preceding times. Consequently, it is useful to define a stress relaxation function $\Gamma(t)$ to be the stress remaining in the sample at a time increment $t \ge 0$ later than sudden application of a fixed unit strain. In an entangled polymer melt, Γ eventually decays to zero with increasing t, but at a rate strongly dependent on the molecular weight distribution and the temperature. The linearity of response then leads to the following relations:

$$G'(\omega) = \omega \int_0^\infty \Gamma(t) \sin(\omega t) dt, \tag{X.42}$$

$$G''(\omega) = \omega \int_0^\infty \Gamma(t) \cos(\omega t) dt.$$

In view of Eq. (X.40), one has

$$\eta(T) = \int_0^\infty \Gamma(t) dt. \tag{X.43}$$

By invoking an inverse Fourier transform, relations (X.42) lead to an expression for the stress relaxation function ($t \ge 0$):

$$\Gamma(t) = \pi^{-1} \int_0^\infty [\sin(\omega t) G'(\omega) + \cos(\omega t) G''(\omega)] \omega^{-1} d\omega. \tag{X.44}$$

For many polymer melts, it has been observed empirically that the stress relaxation function (equivalently G' and G'') approximately satisfies a "time–temperature superposition principle" [e.g., Catliff and Tobolsky, 1955, Figure 2; Rubinstein and Colby, 2003, Figure 8.15]. This principle implies that for a given polymeric system under isobaric conditions, $\Gamma(t)$ obeys a simple scaling behavior. If this stress relaxation function has been determined at temperature T_0, then its form at another temperature T_1 is specified by

$$\Gamma(t \mid T_1) = b(T_1, T_0)\Gamma[t/a(T_1, T_0) \mid T_0]. \tag{X.45}$$

Consequently, if Γ at these two temperatures were displayed together on a log–log plot (i.e., log Γ vs. log t), the two curves would have identical shapes and orientations but would be displaced relative to one another horizontally and vertically by log $a(T_1, T_0)$ and log $b(T_1, T_0)$, respectively. The relation between Γ and η stated in Eq. (X.43) requires that the scaling variables a and b satisfy a constraint imposed by the temperature variation of the shear viscosity:

$$a(T_1, T_0)b(T_1, T_0) = \eta(T_1)/\eta(T_0). \tag{X.46}$$

As can be expected from the behavior of most glass-forming liquids with low molecular weight, the temperature dependence of shear viscosity for polymer melts deviates significantly from a simple Arrhenius form. In the range from the glass transition temperature T_g to roughly $T_g + 100$ K, the isobaric shear viscosities empirically can be fitted with the Vogel-Tammann-Fulcher (VTF) equation [Wang and Porter, 1995]:

$$\eta(T) \approx \eta_0 \exp\left[\frac{A}{T - T_0}\right], \tag{X.47}$$

which previously appeared as Eq. (VI.4). Here η_0, A, and T_0 are temperature-independent positive fitting parameters, but their values reflect the molecular weight distribution that is present to be in accord with Eqs. (X.32) and (X.33). This VTF form is substantially equivalent to the so-called Williams-Landel-Ferry fitting formula [Williams et al., 1955]. No fundamental significance should be attached to the divergence that the VTF expression exhibits at $T_0 < T_g$. However, the fact that lowering the temperature in the VTF expression involves an increasingly non-Arrhenius behavior arises from the system configuration point tending to inhibit more and more deeply fissured regions of the potential energy/enthalpy landscape within the accessible labyrinth pathways.

The stress relaxation function $\Gamma(t)$ for high-molecular-weight polymers above T_g is most conveniently displayed as a log–log plot. Figure X.15 provides a qualitative representation of typical results when presented in this fashion [Matsuoka, 1992, Chapter 4]. The times involved for a given high-molecular-weight polymer sample often run from microseconds to many hours. The shape shown can be divided qualitatively into three time-delay stages. Stage 1, the earliest, contains relaxation processes that are relatively localized within the liquid medium and are not substantially affected by entanglement constraints (i.e., "intratube" relaxations). Consequently, these relaxations are similar to those present in melts of short-chain, low-molecular-weight substances. As indicated in Figure X.15, this stage 1 can be approximated by a Kohlrausch-Williams-Watts (KWW) stretched exponential function that has previously been discussed in application to nonpolymeric glass formers [cf. Eq. (VI.9)].

Stage 2 is a transition time interval during which entanglement has become a significant retarding phenomenon. Stage 3 is a plateau interval during which entanglement constraints cause the partly relaxed melt to act as an elastic medium; sufficient time has not yet elapsed for the polymer molecules to "reptate" fully away from their initial entanglement constraints, which is

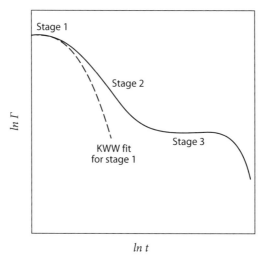

FIGURE X.15. Generic representation of the stress relaxation function $\Gamma(t)$ for a molten polymer above its glass transition temperature T_g, displayed as a log–log plot. Numerical details for a given material depend on its monomer chemical structure, molecular weight distribution, as well as the temperature and pressure.

necessary to complete the stress relaxation process. The length of this stage 3 plateau increases with increasing average molecular weight for a given polymer with approximately the same dependence as indicated in Eq. (X.33) for η [Matsuoka, 1992, Chapter 4]. In the case of a globally cross-linked medium, the stress would never relax entirely to zero, so in principle the plateau would extend to infinite time with a $\Gamma(\infty)$ value corresponding to that medium's rubber elasticity.

The unbounded extension of the stage 3 plateau with increasing molecular weight needs to be viewed in the context of observed glass transition temperatures T_g. It was pointed out earlier that at least for linear polymer samples of a given local chemical structure, a given thermal history (e.g., cooling or heating rate) causes the measured T_g to approach a limiting value as average molecular weight increases without bound [Fox and Flory, 1950, Figure 3]. This implies that the slowest relaxation processes remaining at the long-time end of the stage 3 plateau for those high-molecular-weight polymers have little, if any, influence on the glass transition phenomenon. In other words, the slow disentanglements that long backbones must undergo in order to dissipate remaining shear stress are not processes that control observed T_g values. The sudden changes in heat capacity and volume that typically are used to monitor glass transitions are primarily influenced by local polymer "in tube" configurational transitions.

E. Polymer Solutions

The severe entanglement property that profoundly influences the kinetics of polymer melts can be reduced or even eliminated by dispersing the polymer molecules in a suitable low-molecular-weight solvent. Furthermore, the conformational statistics of individual polymer molecules generally change upon dispersing in a solvent and depend upon the chemical nature of that solvent. These structural attributes and the kinetic characteristics that emerge in these solutions form the focus of this section.

A basic property of polymer solutions is the relative extent to which dilution by solvent has separated the polymers in comparison to their size, specifically measured by their root-mean-square radius of gyration. In an undiluted melt containing N high-molecular-weight polymer molecules in volume V, subject therefore to maximum entanglement, one has

$$\langle (R_g)^2 \rangle^{1/2} \gg (V/N)^{1/3}. \tag{X.48}$$

But with N fixed, when sufficient dilution by solvent causes the volume to increase so that

$$\langle (R_g)^2 \rangle^{1/2} \approx (V/N)^{1/3}, \tag{X.49}$$

the interactions between neighboring polymer molecules will have been substantially reduced but are still present. In this situation, the volume fraction of polymer is very small. Most of the interior of V is occupied by solvent molecules. This is called the "semidilute" polymer solution condition. Finally, the extreme dilution regime with

$$\langle (R_g)^2 \rangle^{1/2} \ll (V/N)^{1/3} \tag{X.50}$$

has the solvated polymers well separated and only rarely interacting with one another.

In the case of homogeneous solution phases, it is useful to classify solvents according to the nature of their interactions with the polymers, and in particular how the variations in interactions between different solvents influence the statistics of polymer backbone configurations. With no solvent present, Eq. (X.4) formally represented the totality of interactions present, including intramolecular, intermolecular, and (if present) polymer-wall interactions. In the solution case, one can in principle average over all solvent degrees of freedom for a given thermodynamic state, holding all polymer configurational degrees of freedom fixed, to obtain a solvent-mediated version of Eq. (X.4) that is distinguished by asterisks:

$$\Phi^*(\mathbf{R}_1 \ldots \mathbf{R}_N) = \sum_{i=1}^{N} v_i^{(1)*}(\mathbf{R}_i) + \Delta^*(\mathbf{R}_1 \ldots \mathbf{R}_N). \tag{X.51}$$

The leading terms $v_i^{(1)*}$ represent the solvent-averaged potentials of mean force for independent polymer molecules, including wall interactions, if any are present. Although not explicitly indicated in this form (X.51) for notational simplicity, the functions appearing here obviously depend on temperature and the solvent composition and its chemical potential(s), or equivalently the osmotic pressure if a membrane permeable only to solvents constrains the polymers to a fixed volume. It should be noted in passing that Δ and Δ^* might differ significantly in the extent to which they can be represented accurately just by a sum of polymer-pair interaction terms.

The solution function Φ^* is a potential of mean force for the collection of N polymer molecules. Its corresponding normalized Boltzmann factor,

$$P(\mathbf{R}_1 \ldots \mathbf{R}_N) = C^{-1} \exp[-\beta \Phi^*(\mathbf{R}_1 \ldots \mathbf{R}_N)], \tag{X.52}$$

$$C = \int d\mathbf{R}_1 \ldots \int d\mathbf{R}_N \exp[-\beta \Phi^*(\mathbf{R}_1 \ldots \mathbf{R}_N)],$$

is the configurational probability density for the set of solvated polymer molecules. It can be used in principle to calculate quantities, such as the average values of the radius of gyration R_g and the end-to-end distance R_e, and to reveal how these attributes depend on solvent composition and polymer concentration, as well as temperature and pressure. By averaging over solvent degrees of freedom, the dimension of the configurational space has been drastically reduced, but it is still a large number. The multidimensional landscape defined by $\Phi^*(\mathbf{R}_1 \ldots \mathbf{R}_N)$ includes bond noncrossing constraints and exhibits multiple minima (inherent structures appearing after solvent averaging) that locate preferred polymer arrangements in the solution medium. Just as for the preaveraged landscape, the solvated polymer configuration space is tiled by basins, each containing a single solvent-averaged inherent structure, defined by a steepest descent mapping on the Φ^* hypersurface. However, the number of basins, their shapes, and their distribution by depth all depend on the solvent composition, temperature, and pressure conditions over which the averaging has been carried out.

Not all polymer–solvent combinations produce homogeneous solutions that are thermodynamically stable over the entire concentration range indicated by expressions (X.48)–(X.50). More typically, there are temperature ranges over which phase separation exists, with a dilute solution coexisting with a concentrated solution. In these cases, the solvent-averaged polymer interaction functions $\Phi^*(\mathbf{R}_1 \ldots \mathbf{R}_N)$ differ between the coexisting phases and in particular yield different results for the polymer extension measures $\langle (R_g)^2 \rangle^{1/2}$ and $\langle (R_e)^2 \rangle^{1/2}$. Furthermore, the distribution of molecular weights for the polymer molecules (assuming that they are not strictly monodisperse) can be expected to differ between the coexisting dilute and concentrated solutions.

Because of the diversity and complexity of polymer systems, there is currently no comprehensive predictive theory for polymer solutions. However, the Flory-Huggins theory has provided a historically important approximate description for solutions of linear polymers [Huggins, 1941; Flory, 1941, 1942]. As illustrated in Figure X.16, it is based upon a regular lattice view of the configurations present in the solution. In the simplest interpretation, each lattice site can be occupied either by a single solvent molecule or by a monomer contained within a polymer molecule. The lattice structure surrounds each site with z nearest neighbor sites, and successive monomers along the backbone of a polymer molecule reside on nearest neighbor sites. The various polymer configurations on the lattice are assumed to have a common value of the intramolecular potential function $v_i^{(1)*}$. The presence of the lattice permits a rough estimate of the number of configurations adopted by a polymer–solvent mixture subject to random mixing. Furthermore, intermolecular interactions are assumed to act only between nearest neighbor site pairs, leading to a mean field approximation for the energies of solvation compared to the intermolecular interaction energies present in the pure liquid components. The resulting form for the Helmholtz free energy of mixing for a solvent and a polymer, each molecule of which consists of n monomers, presented on a per-lattice-site basis, is the following [Rubinstein and Colby, 2003, Section 4.2]:

$$\Delta F_{mix} = k_B T [(\phi/n) \ln \phi + (1 - \phi) \ln (1 - \phi) + \chi \phi (1 - \phi)]. \tag{X.53}$$

Here ϕ stands for the volume fraction of the polymer (i.e., the fraction of lattice sites occupied by monomer units), and the first two terms in the right side estimate the mixing entropy. A linear combination of interaction energies u_{mm}, u_{ss}, and u_{ms} operating between nearest neighbor lattice sites (subscripts m and s denote monomer and solvent species, respectively) forms the Flory interaction parameter χ:

$$\chi = (z/2k_B T)(2u_{ms} - u_{mm} - u_{ss}). \tag{X.54}$$

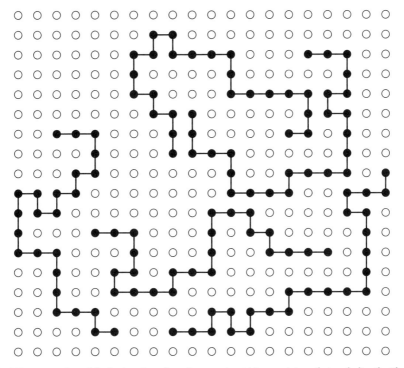

FIGURE X.16. Representation of the lattice view of a polymer–solvent binary mixture that underlies the Flory-Huggins approximation. In this picture, the lattice coordination number $z = 4$ and the number of monomers in each polymer is $n = 24$. The polymer volume fraction as shown is $\phi = 1/3$.

Note that u_{mm} interactions can arise either from intermolecular contacts or from intramolecular contacts between monomers that are not directly bonded to one another but are more remotely located along the same polymer backbone.

The nominal assumption of the Flory-Huggins approximation that monomers and solvent molecules occupy the same volume regardless of which species are involved is unrealistic for most binary mixtures. However, a more liberal interpretation of the terms in Eqs. (X.53) and (X.54) is often invoked. Specifically, this conceptually permits an "effective" monomer unit along the linear polymer backbone to be the occupant of a single lattice site. In that case, n represents the number of such effective units in one polymer molecule rather than its degree of polymerization. Furthermore, the straightforward interpretation of the lattice picture underlying the Flory-Huggins approximation is that the volume of mixing vanishes, in which case the Helmholtz (ΔF_{mix}) and Gibbs (ΔG_{mix}) mixing free energies would be identical. This may be an adequate assumption for some mixtures, especially at low pressure. However, if the volume of mixing deviates significantly from zero and if the pressure p is large, then it may be reasonable to interpret u_{mm}, u_{ss}, and u_{ms} as nearest neighbor interaction enthalpies and to use the corresponding χ from Eq. (X.54) in expression (X.53) to represent ΔG_{mix}.

By using a variety of measurements on the solutions, it is possible to infer values of their χ parameters. Extensive tabulations of this quantity are available for a wide variety of polymer–solvent pairs [e.g., Orwoll and Arnold, 1996]. Although the formal expression (X.54) indicates that χ should be inversely proportional to absolute temperature, the measurements indicate that a better empirical representation is the following:

$$\chi \cong A + B/T. \tag{X.55}$$

Under some circumstances, it can happen that $\chi = 0$. The solvent then is called "athermal." When this is the case, only the mixing entropy terms in the mixing free energy expression (X.53) survive. For any $T > 0$, the free energy of mixing in this athermal solvent is a convex function of ϕ (i.e., positive second derivative) for all $0 < \phi < 1$. Consequently, the solution would be stable against phase separation. This unmixing stability also applies for "good" solvents with $\chi < 0$, in which ms pair contacts are energetically favored over the average of mm and ss contacts. The corresponding interaction term in the Flory-Huggins free energy expression increases the magnitude of the positive second ϕ derivative.

By contrast, a "poor" solvent would have $\chi > 0$. In such a case, the last term in the right side of Eq. (X.53) contributes a negative curvature to $\Delta F_{mix}(\phi)$, and if χ is sufficiently large in magnitude, phase separation is implied. This situation is illustrated qualitatively in Figure X.17, where now the mixing free energy curve is no longer convex but rather possesses a double tangent construction with a straight-line segment that falls below the curve defined by Eq. (X.53). Coexisting solution compositions are located by the points of tangency and are indicated by "I" and "II" in Figure X.17.

The χ value above which phase separation applies corresponds to coincidence of the tangency pair. This occurs at a solution critical point, with a vanishing of the curvature:

$$\partial^2 \Delta F_{mix}/\partial \phi^2 = 0. \tag{X.56}$$

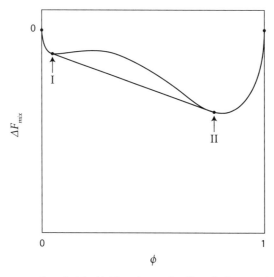

FIGURE X.17. Schematic representation of a $\Delta F_{mix}(\phi, T)$ vs. ϕ curve for a "poor" solvent. In this case, the Flory interaction parameter $\chi > 0$ is sufficiently large to create a double tangent construction. The two points of tangency locate the compositions of the coexisting solution phases, denoted by "I" and "II."

In most cases, this would appear as temperature is lowered, yielding an upper consolute point (also designated as UCST for "upper critical solution temperature"). In terms of the fit function (X.55) for χ, this would arise from $B > 0$. But if $B < 0$, a lower consolute point (LCST, "lower critical solution temperature") could emerge.

Assuming that parameter χ is indeed independent of ϕ, twice differentiating the Flory-Huggins expression (X.53) and setting the result equal to zero leads to the following consolute (critical) point condition:

$$\frac{1}{n\phi} + \frac{1}{1-\phi} - 2\chi = 0. \tag{X.57}$$

This is equivalent to the quadratic equation:

$$2\chi\phi^2 + \left(1 - 2\chi - \frac{1}{n}\right)\phi + \frac{1}{n} = 0, \tag{X.58}$$

the roots for which are given by

$$\phi = \frac{1}{4\chi}\left\{-\left(1 - 2\chi - \frac{1}{n}\right) \pm \left[\left(1 - 2\chi - \frac{1}{n}\right)^2 - \frac{8\chi}{n}\right]^{1/2}\right\}. \tag{X.59}$$

Well into the phase coexistence regime where χ is positive and large, this last expression delivers two real roots. These locate the ϕ values at which the $\Delta F_{mix}(\phi)$ curve exhibits points of inflection within the coexistence range of ϕ. By reducing the magnitude of χ, these inflection points move closer to one another, finally coalescing at the consolute point when $\chi = \chi_c$. At this stage, the two real roots in Eq. (X.59) are degenerate, which means that the term $[....]^{1/2}$ in that equation vanishes. This leads in turn to a quadratic equation that must be satisfied by χ_c:

$$\chi_c^2 - \left(1 + \frac{1}{n}\right)\chi_c + \frac{1}{4}\left(1 - \frac{1}{n}\right)^2 = 0. \tag{X.60}$$

Equation (X.57) requires that $\chi_c > 1/2$, so that the physically relevant solution to Eq. (X.60) is

$$\chi_c = \frac{1}{2}\left(1 + \frac{1}{n^{1/2}}\right)^2. \tag{X.61}$$

Finally, Eq. (X.59) leads to evaluation of the polymer volume fraction at the consolute point:

$$\phi_c = (2\chi_c n)^{-1/2}$$
$$= (n^{1/2} + 1)^{-1} \tag{X.62}$$
$$= n^{-1/2} - n^{-1} + n^{-3/2} - .....$$

Consequently, increasing the length l of the polymer chains creates increasing asymmetry wherein the location of the consolute point converges toward a vanishing polymer volume fraction. In spite of the imprecision contained within the Flory-Huggins formalism, this is a qualitatively correct prediction.

Assuming that a polymer–solvent binary system can be described at least approximately by a Flory-Huggins mixing free energy with an interaction χ conforming to the simple expression (X.55) with constants A and B that are independent of T and ϕ, either an upper or a lower consolute point could in principle arise as already noted, but not both in the same mixture. However, examples are known experimentally that indeed exhibit both. Two such cases involve low-molecular-weight polystyrene (presumably atactic) dissolved respectively in acetone and in diethyl ether, for which miscibility obtains between a low-temperature UCST and a high-temperature LCST [Siow et al., 1972]. It is also worth noting that the inverse situation has been observed experimentally for aqueous solutions of polyoxyethylene, exhibiting a closed-loop immiscibility region in the ϕ, T plane, with an LCST at the bottom and a UCST at the top of that region [Malcolm and Rowlinson, 1957].

The mean field approach incorporated in the Flory-Huggins formalism, specifically its introduction of the Flory interaction parameter χ, becomes unrealistic in the very dilute solution regime, Eq. (X.50). When the polymer molecules are widely separated and thus essentially independent of one another, each of their monomers is embedded in a local concentration of other monomers connected to them by the common backbone. This is a much higher monomer concentration than the systemwide average, computed mostly over pure solvent regions, that is used by the mean field estimate in χ.

In this very dilute solution regime, the system's solvent-averaged interaction function can be consistently represented just by the sum of its single polymer terms:

$$\Phi^*(\mathbf{R}_1 \dots \mathbf{R}_N, \beta) \cong \sum_{i=1}^{N} v_i^{(1)*}(\mathbf{R}_i, \beta), \qquad (X.63)$$

where now the temperature dependence has been explicitly indicated. Various measures of the folding statistics for any chosen molecule i in these isolating circumstances then in principle can be calculated using the corresponding normalized Boltzmann factor:

$$P_i(\mathbf{R}_i, \beta) = C_i^{-1} \exp[-\beta v_i^{(1)*}(\mathbf{R}_i, \beta)], \qquad (X.64)$$

$$C_i = \int \exp[-\beta v_i^{(1)*}(\mathbf{R}_i, \beta)] d\mathbf{R}_i.$$

A polymer in a "poor" solvent would tend to adopt configurations that minimize monomer–solvent contacts, which in the extreme would occur if the polymer backbone were coiled compactly to exclude solvent from the interior of its occupied region. In such a case for linear polymers, the mean square radius of gyration and end-to-end distance would increase less rapidly with degree of polymerization n than for a corresponding ideal freely jointed chain as this size became large. Specifically, $\langle (R_g)^2 \rangle^{1/2}$ would only rise asymptotically as $n^{1/3}$ for the compact coil, as opposed to $n^{1/2}$ for the freely jointed chain. If instead a "good" solvent were involved, the polymer coiling statistics revealed by $P_i(\mathbf{R}_i, \beta)$ would show more extended characteristics than those of the ideal freely jointed chain, with $\langle (R_g)^2 \rangle^{1/2}$ possibly varying as n^μ with $\mu > 1/2$ in the large-n asymptotic regime [Painter and Coleman, 1997, pp. 334–335]. The transition between these regimes occurs in a "theta solvent," or stated equivalently, in a solvent at its "theta temperature" T_θ. In this situation, the mutual exclusions operating between monomer units on the same backbone are canceled by mean attractive forces acting between neighboring monomers, presumably causing the backbone to behave overall as a freely jointed chain with its characteristic $n^{1/2}$ spatial extension measures.

Because theta solvent conditions cause monomers within the same polymer chain to behave toward one another as though they had no exclusion volume, the same situation should apply to pairs of monomers located in different polymer molecules. This implies that in a coarse-grained view, pairs of polymer molecules should be configurationally "invisible" to one another. The result is that osmotic pressure measurements for the binary solution provide a way to identify theta solvent conditions. Specifically, this amounts to identifying solvent composition and/or temperature conditions for which the osmotic pressure equation of state is that of an ideal solution, at least through the second osmotic virial coefficient. In this respect, the theta solvent condition is analogous to the Boyle temperature for gases, at which the standard second virial coefficient vanishes.

In contrast to the theta solvent condition, polymers suspended in athermal solvents experience significant monomer–monomer volume exclusions. Typically, this would occur in a given solvent at temperatures T substantially above T_θ for which the effects of neighbor attractions are thermally insignificant. The volume exclusion dominance was evident in the Flory-Huggins approximation discussed above at $\chi = 0$, where as illustrated in Figure X.16, only single monomer units in the same or different molecules can occupy a lattice site, while mm, ss, and ms neighbor attractions cancel one another. The influence that excluded volume exerts on the distribution of configurations exhibited by a polymeric substance under athermal conditions, and how those distributions vary with degree of polymerization n, have presented considerable scientific challenges. As at least a starting point, it has been helpful to analyze simplified models, specifically those involving polymer chain occupancy on regular lattices where all nonoverlapping configurations are given equal probability. For this reason, considerable attention has been devoted to "self-avoiding walks" (SAWs) on lattices [Madras and Slade, 1993]. The focus of such studies has been on the properties of single SAWs on the square lattice in two dimensions and the simple cubic lattice in three dimensions.

The number of steps taken by a SAW on a lattice is one less than the number of sites visited, $n - 1$ in the notation used above. Basic questions to ask are how the values of $\langle (R_g)^2 \rangle^{1/2}$ and $\langle (R_e)^2 \rangle^{1/2}$ depend on $n - 1$. Computer simulation is able to generate ensembles of isolated SAWs of rather large lengths, so that numerical fits to such data can supply tentative answers to these questions for single polymers in the extreme dilution regime. In the case of the simple cubic lattice in three dimensions, one such simulational investigation has generated SAWs of up to 3.3×10^7 steps [Clisby, 2010], with results for mean square distances satisfying the following form ($x = g, e$):

$$\langle (R_x)^2 \rangle = D_x(n - 1)^{2\tilde{\nu}}\{1 + O[(n - 1)^{-1}]\}. \tag{X.65}$$

Here the distance unit is taken to be the nearest neighbor separation (the step length) in the lattice. Numerical values that were obtained are

$$\tilde{\nu} = 0.587597(7),$$

$$D_g = 0.19514(4), \tag{X.66}$$

$$D_e = 0.122035(25).$$

The exponent $\tilde{\nu}$ is the same for both size measures, exceeding the value 1/2 for freely jointed chains as self-avoidance causes overall expansion of the chain configuration. As might be expected, the numerical coefficients D_x are not quite equal for the two size measures.

Although SAW exponents such as $\tilde{\nu}$ depend on the dimension of the lattice space, they are otherwise believed to be universal. That is, for SAWs generated by short-range steps (but not necessarily confined to nearest neighbor sites), the large-SAW size measures would be described by exponents that are independent of the lattice. Consequently, the numerical value for $\tilde{\nu}$ shown in Eq. (X.66) should equally well describe SAWs on face-centered cubic, hexagonal close-packed, body-centered cubic, and diamond lattices. Furthermore, this universality may also extend beyond lattice SAWs, applying equally well to walks in continuous three-dimensional Euclidean space that are more realistic representations of polymer molecule solution configurations in the very dilute limit, Eq. (X.50).

The inherent structures that underlie a polymer solution are accessed as usual by a steepest descent mapping on the applicable potential energy or potential enthalpy hypersurface. For either an isochoric or isobaric ensemble of initial conditions, a choice of two procedures is available for polymer solutions. One choice is to carry out the mapping process for the full polymer plus solvent system that for isochoric conditions is described by potential function Φ, Eq. (X.4). The alternative, as indicated earlier, is to apply the mapping to the solvent-averaged interaction Φ^*, Eq. (X.51), in the lower dimension space of only polymer configurational degrees of freedom. Because of a dense solvent medium being present in the former mapping scenario, the polymer molecules would undergo only local readjustments as a result of the mapping, and so the exponent $\tilde{\nu}$ should continue to describe the large-n asymptotic size of isolated polymer chains in their inherent structure configurations. It is not clear that the same can be claimed for the mapping results for the solvent-averaged case with mapping on Φ^*.

A full resolution of the solvent-averaged interaction Φ^* into intrinsic single, pair, triplet, ... polymer-molecule contributions has the form

$$\Phi^*(\mathbf{R}_1\ldots\mathbf{R}_N, \beta) = \sum_{i=1}^{N} \mathrm{v}_i^{(1)*}(\mathbf{R}_i, \beta) + \sum_{i=2}^{N}\sum_{j=1}^{i-1} \mathrm{v}_{ij}^{(2)*}(\mathbf{R}_i, \mathbf{R}_j, \beta) + \sum_{i=3}^{N}\sum_{j=2}^{i-1}\sum_{k=1}^{i-1} \mathrm{v}_{ijk}^{(3)*}(\mathbf{R}_i, \mathbf{R}_j, \mathbf{R}_k, \beta)$$

$$+\ldots+ \mathrm{v}_{1\ldots N}^{(N)*}(\mathbf{R}_1\ldots\mathbf{R}_N, \beta). \tag{X.67}$$

The subscripts on the component functions $\mathrm{v}^{(m)*}$ are necessary to distinguish polymer molecules by chemical and/or molecular weight differences, if those differences are present. Upon increasing the polymer concentration from the extreme dilution regime (X.50) into the semidilute regime (X.49), the pair and higher order contributions in Φ^* come into play (if the solvent does not behave strictly as a theta solvent). When the solution is still sufficiently dilute that polymer intermolecular encounters are essentially only pairwise, the resulting configurational characteristics can be extracted from a probability distribution function amounting to a Boltzmann factor of the form

$$P_{ij}(\mathbf{R}_i, \mathbf{R}_j, \beta) = C_{ij}^{-1} \exp\{-\beta[\mathrm{v}_i^{(1)*}(\mathbf{R}_i, \beta) + \mathrm{v}_j^{(1)*}(\mathbf{R}_j, \beta) + \mathrm{v}_{ij}^{(2)*}(\mathbf{R}_i, \mathbf{R}_j, \beta)]\}. \tag{X.68}$$

In particular, this polymer pair probability could be used to compute the distribution of the distance between the centroids of the two molecules. This distribution was estimated long ago for athermal or near-athermal solution conditions, with the conclusion that the effective polymer–polymer pair interaction U_{ij} was a positive Gaussian function of centroid–centroid separation r_{cc} [Flory, 1949, Eq. (5); Flory and Krigbaum, 1950, Eq. (12)]:

$$U_{ij}(r_{cc}, \beta) = \varepsilon_{ij}(\beta) \exp\{-[r_{cc}/\sigma_{ij}(\beta)]^2\}. \tag{X.69}$$

Monte Carlo simulations of freely jointed chains of tangent hard spheres agree qualitatively with this spatial form [Dautenhahn and Hall, 1994]. A renormalization group analysis also produces the repelling Gaussian form for the effective interaction [Krüger et al., 1989].

Although it was not intended to represent polymer solutions, the Gaussian core model defined in Section I.D is constructed around pair potentials superficially similar in spatial form to effective pair interaction (X.69). No three-body or higher order interactions are present in that model, in contrast to the expectation for polymer solutions. The conceptually simple Gaussian core model involves only a single particle species with no internal degrees of freedom, and the energy and distance scaling parameters ε and σ in its pair potential have fixed values (i.e., no temperature dependence) [Stillinger, 1976].

As a polymer species is added under fixed T,p conditions to an initially pure liquid solvent with low molecular weight, the shear viscosity η rises. While still in the very dilute regime, each added polymer contributes independently to that increase. This effect is conventionally represented as follows:

$$\eta/\eta_s = 1 + [\eta]c + O(c^2), \tag{X.70}$$

where η_s is the shear viscosity of the pure solvent, and c is the concentration of the added polymer. The coefficient $[\eta]$ of the linear term is called the intrinsic viscosity. The higher the degree of polymerization of the polymer species, the greater is the viscosity increment per molecule added. For linear polymers, the dependence of $[\eta]$ on degree of polymerization, or equivalently on the molecular weight M, is estimated by the empirical Mark-Houwink equation [Houwink, 1940; Mark, 1950]:

$$[\eta] = KM^a. \tag{X.71}$$

The exponent is found experimentally to lie in the range $0.5 \leq a \leq 0.8$, and the coefficient K depends on the solvent and polymer species involved [Rubinstein and Colby, 2003, Table 1.4].

By definition, η is a linear response measure for the solution under a constant applied shear strain rate. This linearity requires that the perturbation to polymer coil distributions resulting from that shear flow must necessarily be small. Any imposed shear rate that causes a substantial change in the average coil configuration distribution necessarily exceeds the linear response regime. The observation of "shear thinning" for strongly sheared solutions results from hydrodynamically induced elongation of polymer coils, resulting in a reduction in apparent shear viscosity. It is worth pointing out that the higher the degree of polymerization, with consequently larger coil size, the more susceptible the dissolved polymer is to substantial deformation. Therefore, the higher the molecular weight, the lower must be the shear rate in order to remain in the strict linear response regime. In experimentally testing an empirical law such as the Mark-Houwink equation (X.71) for the effect of varying M, it is important to observe this constraint so as to avoid inadvertent intrusion of shear thinning.

The focus thus far in this section has been on solutions of non-ionic polymers. However, solutions of ionizable polymers, "polyelectrolytes," constitute an important additional category. The attainable degree of polymerization for polyelectrolytes is very broad and is comparable to that of non-ionic polymers. Ionizable groups along the backbones of polyelectrolytes can undergo

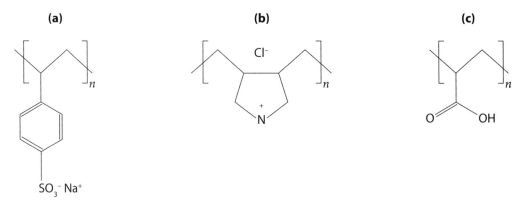

FIGURE X.18. Monomer structures present in three linear polyelectrolytes. Each of these has a covalently bound carbon backbone: (a) poly(sodium styrene sulfonate); (b) poly(diallylmethylammonium chloride); and (c) polyacrylic acid. The degree of polymerization is indicated by n.

charge separation that is aided by immersion in a polar solvent possessing a high dielectric constant. Consequently, many polyelectrolytes are observed as components of aqueous solutions.

Some examples of linear polyelectrolytes are illustrated in Figure X.18. These include a polyanion [example (a)], a polycation [example (b)], and a polymer consisting of weakly ionizing organic acid monomers [example (c), a polyanion following ionization]. Copolymerization can produce "polyampholytes," which are polyelectrolytes whose monomers include both anionic and cationic types. Many biopolymers (proteins, DNA, ...) are polyampholytes. For more detail about the properties of polyelectrolytes than this brief overview can present, the reader can consult several available comprehensive reviews [e.g., Radeva, 2001; Dobrynin and Rubinstein, 2005; Schanze and Shelton, 2009].

The long-range nature of the Coulomb interactions acting between ions produces configurational effects beyond those operating in solutions of non-ionic polymers. Leaving aside for the moment the case of polyampholytes, consider conceptually the limit of extreme dilution in a high-dielectric-constant nonelectrolytic solvent. In this circumstance, the small counterions predominantly diffuse far away from each polyelectrolyte molecule into the intervening pure solvent region. This would be the case even for those polyelectrolytes composed of weakly ionizing monomers, so that in the extreme dilution limit, even they would be fully ionized and largely devoid of nearby counterions. Consequently, each polyelectrolyte molecule would bear a large uncompensated charge because of the included ionized monomers distributed along its backbone, and the Coulomb repulsion acting between these unshielded charges would tend on average to expand the molecule's configuration. If the polyelectrolyte were a linear polymer, this repulsion would drive it toward its fully extended configuration. As the polyelectrolyte solution concentration increases, however, that entropy-driven charge separation abates, and the counterions cluster more and more within the polymer coils. In the semidilute regime, the counterion concentration within the coil regions is high enough to effectively shield the charges arrayed along the backbone, and the radii of gyration and end-to-end distance measures decline substantially.

If the solvent were not a pure electrically nonconducting dielectric liquid, but instead consisted of a moderately concentrated electrolytic solution of a simple salt (e.g., Na^+Cl^-), the extension effect of course would not be present in the extreme dilution regime for the polyelectrolyte. The shielding of the backbone charges would return the polyelectrolyte to spatial extensions more in accord with those of nonelectrolyte polymers of comparable molecular weight. In the cases of weakly ionizing monomers, the same extension reduction effect would be produced by using a solvent with added acid (e.g., H^+Cl^- for weakly acidic monomers) or added base (e.g., Na^+OH^- for weakly basic monomers) to shift the monomer equilibrium state away from ionization.

The Debye inverse-distance parameter κ has a basic role in the description of simple (non-polymeric) electrolytes, but its relevance to polyelectrolytes requires an obvious qualification. Its formal definition appeared in Chapter V [Eq. (V.101)] and is repeated here:

$$\kappa^2 = \frac{4\pi e^2}{k_B T \varepsilon_0(0)} \sum_{\alpha=1}^{\nu} Z_\alpha^2 \rho_\alpha. \tag{X.72}$$

Here the index α runs over the distinct ionic species whose electrostatic charges are $Z_\alpha e$, and number densities are ρ_α, respectively. The static, infinite-spatial-wavelength dielectric constant for the solution medium has been denoted by $\varepsilon_0(0)$. In the case of a fully ionized simple uni-univalent electrolyte such as Na^+Cl^- that is present in solution at number density $\rho = \rho(Na^+) = \rho(Cl^-)$, the definition yields

$$\kappa^2 = \frac{8\pi \rho e^2}{k_B T \varepsilon_0(0)}. \tag{X.73}$$

However, consider the situation where all of the ions of one charge, say the cations, were to be covalently linked into linear polyelectrolytes with polymerization degree n while still leaving the anions unbound, as in example (b) of Figure X.18. Then formally the cation charge and number density would become ne and ρ/n, respectively. As a consequence, the Debye inverse length defined formally in Eq. (X.72) would instead be specified by

$$(\kappa_{poly})^2 = \frac{4(n+1)\pi \rho e^2}{k_B T \varepsilon_0(0)}. \tag{X.74}$$

The resulting predicted shielding distance $1/\kappa_{poly}$ converges to zero in the $n \to +\infty$ limit, normally a physical irrelevancy. In most solution circumstances of interest, the spatial separation of the monomer charges along the backbone and the penetration of counterions into the coil region produce statistical charge distributions with shielding over distances qualitatively described by the $1/\kappa$ for monomer charges treated as individual particles.

Weakly ionizable molecules of low molecular weight in solution, such as acetic acid [CH_3COOH] in water, are conventionally described at the macroscopic level by temperature-dependent dissociation constants $K(T)$. These quantities are defined by the ratio of activities for the ionized product species to that of the un-ionized species. However, a comprehensive statistical mechanical analysis of ionization equilibrium demands a much more detailed description that is precise and unambiguous at the atomic level. Given an arbitrary configuration of the solution system, the description must involve a decision procedure that is able to identify which solute molecules are indeed ionized versus those that are not. Furthermore, in the case of proton-containing solvent molecules like water, the configurational criterion should also include a clear identification of the excess solvated protons supplied by acid dissociation [see discussion in Section IX.G].

BOX X.4. Moment Condition Invariance Properties

Polyelectrolyte solutions are subject to the pair correlation function zeroth and second-moment conditions that were introduced earlier in connection with nonpolymeric electrolytes, Eqs. (V. 98) and (V.100), respectively. These refer to the spatial patterns of charge distributions in the infinite-system limit, and they are a consequence of the fact that the solution's electrical conductivity has the property of shielding the solution interior from any externally applied static electric field. There is a zeroth-moment condition for each ionic species $1 \leq a \leq v$:

$$-Z_a e = \sum_{\gamma=1}^{v} Z_\gamma e \rho_\gamma \int [g_{a\gamma}(s) - 1] d\mathbf{s},$$

specifying that the net surrounding charge of the average electrostatic charge distribution centered on any chosen ion species exactly cancels the charge borne by that central ion. The $g_{a\gamma}(s)$ are the radial pair correlation functions for ion pairs of the subscripted species, with the conventional normalization to unity as scalar pair distance approaches infinity.

If one of the ionic species is a polyelectrolyte, there is a choice either to regard that macroion as a single ionic species with very large net charge or to regard each of its charged monomers as an "individual" ion with modest charge. With the former choice, the correlation functions involving the polyion as one or both of the subscripted species would most naturally use the centroids of the polymers as "positions" of the macroions. With the latter choice, the set of monomer centroids would suffice. But the zeroth-moment conditions remain valid in either convention, although the magnitudes of the central and the shielding charges involved are hugely different.

There is just a single second-moment condition that involves all ionic species present in the solution:

$$-6/\kappa^2 = \left(\sum_{a=1}^{v} Z_a^2 \rho_a \right)^{-1} \sum_{a,\gamma=1}^{v} Z_a Z_\gamma \rho_a \rho_\gamma \int [g_{a\gamma}(s) - 1] s^2 d\mathbf{s}.$$

In contrast to the set of zeroth-moment conditions, it exhibits an explicit appearance of the Debye inverse-length parameter κ, Eq. (X.72). The validity of this second-moment condition is also unaffected by which choice (macroion vs. monomer ion) is applied, although as for the zeroth moments, the magnitudes of the individual terms are strongly affected.

There is another type of invariance that the moment conditions possess. Specifically, this concerns weakly ionizing polyelectrolytes and the freedom in choice of the configurational criterion for deciding when ionization has formally occurred. Although information from structural chemistry may suggest rather tight limits on what local geometries should be consistent with ionization, the statistical mechanical formalism in principle permits wide freedom in that assignment. For example, if one imagines increasing a cutoff distance for oxygen–hydrogen pairs from small to large in determination of neutral vs. ionized state for a monomer, the ion concentrations ρ_a would display corresponding continuous decreases, but both zeroth and second-moment conditions would continue to hold. Note that any formally neutral monomer identified as such, even with a chemically unrealistic cutoff distance, would become a polar monomer contributing to the dielectric constant $\varepsilon_0(0)$ that helps to determine the Debye inverse distance κ that occurs in the second-moment equation.

Given such a species-identification tool, the statistical mechanical theory in principle could then quantitatively evaluate the dissociation constant but could also report much more atomic-level detail.

A similar requirement is also obligatory for a complete statistical mechanical theory for solutions of polyelectrolytes that are synthesized from weakly ionizing monomers. Given the multidimensional potential energy function for such a system, a precise and unambiguous criterion needs to be available to distinguish which monomers have undergone ionization and which have not, regardless of where they may be located along a polymer backbone. Considering the fact that inherent structures present enhanced images of local order in liquids, it would be reasonable to examine a representative sample of those inherent structures for an equilibrated polyelectrolyte solution of interest to see if the local order they present suggests a simple configurational criterion for ionization. Polyacrylic acid [Figure X.18, example (c)] in aqueous solution provides a relevant case. Its inherent structures would provide a basis for deciding how close a proton would be to one of the two monomer oxygen atoms in order for that monomer to be classified as un-ionized; if none were that close, the monomer would then be defined as ionized. Structural chemistry for organic acids suggests that this inherent structure approach would lead to an oxygen–hydrogen cutoff distance approximately equal to 1.5 Å. Box X.4 discusses technical issues that arise formally when the electrolyte spatial correlation moment conditions (Section V.G) are applied to polyelectrolytes.

XI.

Protein Folding Phenomena

The complex molecular properties of naturally occurring proteins and of related synthetic polypeptides pose major challenges to the physical and biological sciences. Not surprisingly, these challenges have created an enormous scientific literature that has been motivated both by intellectual curiosity as well as by the obvious demands of practical applications in biology and medicine. It is certainly not the purpose nor is it feasible here to survey and analyze the entire vast subject of protein characteristics, including their diverse roles in living organisms. Many texts specializing in this subject are available [e.g., Creighton, 1993; Mathews et al., 2000; Alberts, 2008]. But in view of the complexity of this scientific area, and the fact that it continues to present many unsolved problems, it is reasonable to search for novel insights that might result from applying the formalism of the energy landscape/inherent structure viewpoint to these intriguing molecular systems. Of course, proteins constitute only a subset of all biopolymers (e.g., DNA, RNA, carbohydrates, lipids), but any useful insights that emerge from the restricted analysis presented here may have utility in that broader domain.

As they perform their biological tasks in vivo or as they are examined in the laboratory in vitro, proteins are immersed in solvents that play an important role in determining their spatial configurations and kinetic phenomena. The most frequently encountered situations involve water and aqueous solutions as the solvent media, although some notable exceptions exist. This chapter includes a brief examination of how solvents interact with proteins in both liquid and crystalline phases, including the effects produced by temperature, pressure, and solvent composition changes.

A. Protein Chemical Structure—An Overview

Naturally occurring proteins are linear polymers. When they are initially synthesized in living organisms, protein molecules contain monomeric units that are selected from a list of 20 chemically distinct amino acids. The linear sequence order of these amino acid monomers is specified by the genetic code contained in DNA, and that information is conveyed by messenger RNA from the DNA to ribosomes for protein synthesis. Figure XI.1 illustrates these monomers in generic isolated form, showing both the un-ionized molecules, as well as possible high-dipole-moment zwitterionic forms. In the latter case, a proton has ionized from the carboxylic acid group at one location of the molecule and has attached to a nitrogen atom at another location. The 20 amino acids are listed in Table XI.1, which identifies their distinguishing chemical side groups and provides the three-letter and one-letter abbreviations that are conventionally used in the scientific literature to designate them. It should be noted that the last entry in Table XI.1, proline, differs

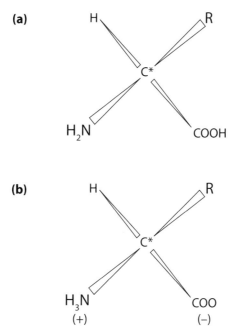

FIGURE XI.1. Generic structure of amino acids. The side group symbolized by –R distinguishes the amino acids. (a) The un-ionized form, with –R and –NH$_2$ displaced toward the viewer, –COOH and –H displaced away from the viewer; (b) a zwitterionic form resulting from dissociation and reattachment of the acidic H$^+$. The representations shown illustrate the chiral form "L" that predominates in the terrestrial biological context, with an asterisk identifying the "α" carbon atom whose stable bonding pattern determines that chiral choice.

structurally from the others by the fact that its side group is attached covalently not only to the usual "α" carbon atom, but to the nitrogen atom as well, thus forming a closed ring of five covalent bonds. Of course, synthetic chemistry has expanded the list of known amino acids far beyond the 20 shown in Table XI.1 by substituting a wide variety of alternative side-group structures.

With the exception of the simplest amino acid glycine, the remaining 19 amino acids are chiral and in principle can exist in either of two distinguishable mirror image forms that are separately stable as such, except under unusual physical or chemical conditions that could lead to distortional conversion of one chiral form to the other. This chirality stems from the normal tetrahedral directionality of the four covalent single bonds emanating from the α carbon that has been identified in Figure XI.1 by an asterisk (C*). Glycine is exempt from this chirality because of the fact that it alone has two nominally indistinguishable hydrogen atoms occupying a pair of the four α carbon covalent bonds. For the other 19 amino acids, interconversion between the two chiral forms *in vacuo* requires passing through a high-energy transition state where the four covalent bonds at the α carbon are coplanar. In the case of alanine, the barrier height has been estimated from quantum calculations for the isolated molecule to be 45,443 cm^{-1} ≅ 130 kcal/mol ≅ 5.63 eV [Lee et al., 2004].

With only rare exceptions [e.g., Lam et al., 2009], the naturally occurring amino acids each exhibit a fixed chirality, conventionally denoted as "L" (the opposite "unnatural" amino acid chirality is denoted as "D"). In fact, these rare D occurrences are usually associated with modifications

TABLE XI.1. The naturally occurring amino acids

	Distinguishing Side Group (–R)	Abbreviation	One-Letter Code
Glycine	–H	Gly	G
Alanine	–CH_3	Ala	A
Valine	–$CH(CH_3)_2$	Val	V
Leucine	–$CH_2CH(CH_3)_2$	Leu	L
Isoleucine	–$CH(CH_3)(CH_2CH_3)$	Ile	I
Serine	–CH_2OH	Ser	S
Threonine	–$CH(CH_3)(OH)$	Thr	T
Aspartic acid	–CH_2COOH	Asp	D
Asparagine	–$CH_2(C=O)NH_2$	Asn	N
Lysine	–$(CH_2)_4NH_2$	Lys	K
Glutamic acid	–CH_2CH_2COOH	Glu	E
Glutamine	–$CH_2CH_2(C=O)NH_2$	Gln	Q
Arginine	–$(CH_2)_3NH(C=NH)NH_2$	Arg	R
Histidine	–$CH_2(C_3N_2H_3)$	His	H
Phenylalanine	–$CH_2(C_6H_5)$	Phe	F
Cysteine	–CH_2SH	Cys	C
Tryptophan	–$CH_2(C_8NH_6)$	Trp	W
Tyrosine	–$CH_2(C_6H_4)OH$	Tyr	Y
Methionine	–$CH_2CH_2SCH_3$	Met	M
Proline	–$CH_2CH_2CH_2$–	Pro	P

of L amino acids after the initial formation of the protein that contains them. Chiral inversions become feasible because of the existence of specific catalytic agents that can be present in the aqueous biological solvent medium [Kreil, 1994; Shaw et al., 1997; Park et al., 2007], the effect of which is to lower substantially the isolated molecule's α carbon inversion barrier. A compilation of the rare D occurrences in biological proteins shows that alanine is by far the most likely amino acid to exhibit this modification, with serine the next most likely case [Khouri et al., 2011, Table 1].

Historically, intense scientific attention and creative literature have been devoted to the fundamental issues surrounding the D vs. L chiral bias for amino acids in the terrestrial biosphere [e.g., Frank, 1953; Mislow, 2003; Blackmond, 2004; Breslow and Cheng, 2009]. As a result, a variety of statistical models have been introduced and investigated to provide possibly relevant many-body phenomena that exhibit such chiral symmetry breaking [e.g., Kondepudi and Asakura, 2001; Cartwright et al., 2007; Hochberg, 2009; Lombardo et al., 2009]. Currently it remains unclear what prebiotic chemical and physical conditions produced this molecular-geometry chiral preference on the Earth and whether chiral asymmetry is an absolute prerequisite for the appearance and subsequent evolution of life as conventionally defined, or merely a historical accident.

A basic chemical property of the amino acids is their ability to link to one another by elimination of a water molecule. This dehydration reaction is illustrated in Figure XI.2. The resulting combination of two linked amino acids is called a "dipeptide." Repetition of the water-elimination reaction at the dipeptide's remaining terminal –COOH or H_2N– groups with a third amino acid

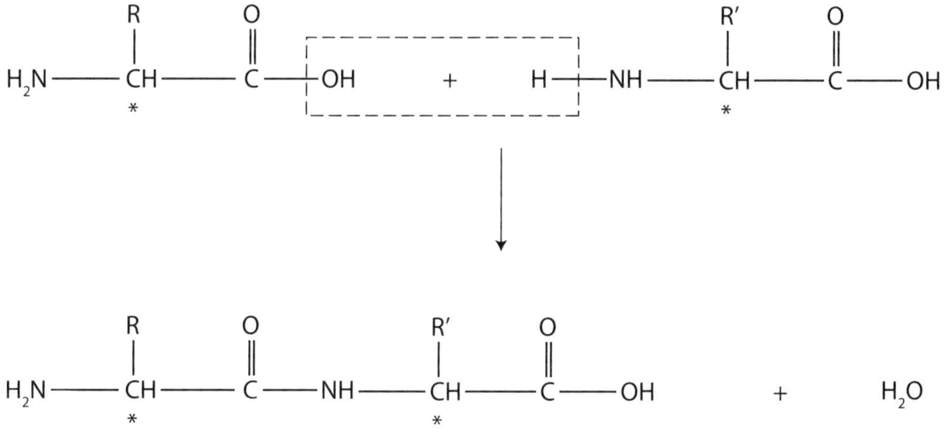

$$\text{H}_2\text{N}\underset{*}{-}\text{CH}\overset{R}{-}\text{C}\overset{O}{=}\text{O}-\text{OH} \quad + \quad \text{H}-\text{NH}-\underset{*}{\text{CH}}\overset{R'}{-}\text{C}\overset{O}{=}\text{O}-\text{OH}$$

$$\downarrow$$

$$\text{H}_2\text{N}-\underset{*}{\text{CH}}\overset{R}{-}\text{C}\overset{O}{=}\text{O}-\text{NH}-\underset{*}{\text{CH}}\overset{R'}{-}\text{C}\overset{O}{=}\text{O}-\text{OH} \quad + \quad \text{H}_2\text{O}$$

FIGURE XI.2. Water elimination reaction to form a peptide linkage.

yields a "tripeptide," a process which then can continue. In this manner, much longer sequences of interlinked amino acids can be created, including of course the naturally occurring proteins. In principle, the resulting sequences of chemically bonded amino acids could include arbitrary selections of L and D monomers. The term "polypeptide" usually applies to linear molecules comprising roughly 25 or more amino acids and can include both biologically occurring proteins as well as synthetically produced amino acid polymers with no natural biological role. Although the linking dehydration reaction that creates these various substances can be reversed, under normal biological circumstances for most proteins, the rate of that reversal can be very low, contributing to the stability required for proteins' effective role in life processes. Terminal –COOH and H$_2$N– groups can be blocked from further dehydration reactions by replacing their hydrogen atoms H– with less reactive units such as methyl groups CH$_3$–. The amino acid subunits included in a peptide or protein are commonly referred to as amino acid "residues."

The linear backbone in peptides and proteins has a definite directionality. This direction can be reckoned as running from a C*–H to the immediately neighboring C=O in the same amino acid monomer, and then in the same direction to the immediately subsequent NH in the next amino acid monomer. The dipeptide dehydration product illustrated in Figure XI.2 has this basic directionality running from left to right. A consequence of this directionality feature is that in general a polypeptide or protein with a given amino acid sequence (known as the "primary structure") does not exhibit the same properties as the polypeptide or protein with the reversed sequence. When proteins are encoded by a horizontal string of one-letter amino-acid indicators (Table XI.1, last column), it should be understood that this left-to-right convention applies to that sequence as presented. The amino acid monomer at the left end possesses an unlinked amino group ("N-terminus" or "amino terminus"), whereas the amino acid monomer at the right end possesses an unlinked carboxyl group ("C-terminus" or "carboxy terminus"). In order for proteins to be assembled stepwise at ribosomes by translation of the genetic code embedded in the messenger RNA structure, individual amino acid "building blocks" are transported to those ribosomes by transfer RNA molecules [Alberts, 2008]. In this stepwise formation process, the chronological sequence of amino acid additions building any given protein proceeds from the N-terminus to the C-terminus.

By way of concrete illustration, the properly ordered letter sequence for the small 26-amino-acid protein melittin is the following [Terwilliger and Eisenberg, 1982]:

GIGAVLKVLTTGLPALISWIKRKRQQ

(backbone direction →).

Its biological assembly and emergence from a ribosome thus proceed from left to right as shown, from G to Q. This naturally occurring substance forms the main toxic component of bee venom. With the exception of its 3 achiral glycine (G) units, the remaining 23 constituent amino acids all possess L chirality. Other proteins can be similarly represented by their own ordered letter sequences, which in most cases are significantly longer than that for melittin.

One should note in passing that the unique letter code specifying the amino acid sequence in any given protein emphasizes a distinction compared to the nonbiological polymers discussed in Chapter X. In that case, the polymer synthesis methods typically produce a distribution of molecular weights, i.e., a distribution of backbone lengths. By contrast, the biological task required of a protein in a living organism normally eliminates the possibility of an analogous backbone length variability. But the fixed backbone length constraint for a specific protein contrasts starkly with the length range exhibited by the full set of distinct proteins in living organisms. An extreme upper limit is illustrated by the titins, a protein component of myocardial muscle tissue, exhibiting backbones that contain approximately 3×10^4 amino acid residues [Labeit and Kolmerer, 1995; Opitz et al., 2003].

It is important to keep in mind that after completion of its assembly at a ribosome, a protein molecule in a living organism is often subjected to chemical modification. As indicated earlier, this modification can involve addition of stabilizing chemical groups at the N-terminus and/or C-terminus and rarely involves cases in which the chirality of an amino acid residue is catalytically switched from L to D. But also in vivo modification of the amino acid side groups along the protein backbone can occur to extend beyond the list of residues of the 20 primary amino acids in Table XI.1. Specific examples of these side-group modifications are carboxylation of glutamate [Vermeer, 1990], hydroxylation of proline [Bhattacharjee and Bansal, 2005], and lysine residue modification [Park, 2006]. Each of these kinds of postassembly changes usually is necessary for the protein to fulfill its biological role effectively.

After creation of its linear sequence of amino acid units, it is possible for the N-terminus and C-terminus of a sufficiently large polypeptide or protein to encounter one another and to link up by the usual water elimination process. This produces a class of cyclic polypeptides and proteins [Trabi and Craik, 2002; Craik, 2006]. If the linear sequence is sufficiently long to include substantial overall conformational freedom, the result of this end-to-end linkage could in principle generate a knotted protein. These topological possibilities were illustrated qualitatively for general polymeric molecules in Chapter X, Figure X.5(a) and (b). However, it should be noted that the term "knotted protein" conventionally does not refer to a backbone topology such as the trefoil knot illustrated in Figure X.5(b) but instead refers to protein conformations where a two-ended backbone has inserted one of its ends through a backbone loop, or where the backbone has formed a "slip knot" [King et al., 2010; Sułkowska et al., 2012a, 2012b].

Polypeptide or protein structures containing two or more cysteine residues may produce disulfide linkages. This normally involves an oxidation reaction stemming from atmospheric oxygen dissolved in a surrounding liquid medium. It can schematically be illustrated as follows:

$$\frac{1}{2} O_2 + 2(\ldots - CH_2 - SH) \rightarrow \ldots - CH_2 - S - S - CH_2 - \ldots + H_2O.$$

The formation of such additional chemical bonds of course requires that the backbone chain be able to contort without excessive free energy cost to bring the thiol (–SH) groups close together. Consequently, the cysteine residues forming a disulfide bond would not be immediate neighbors along the backbone. Once formed, these disulfide linkages can be stabilizing contributors to the biologically required protein conformation. Note that pairs of initially unconnected polypeptide or protein molecules can also be linked by intermolecular disulfide bonds if they contain the required cysteine residues.

Molecular modeling of proteins and other peptides under conditions where they remain chemically stable must account for the basic structural characteristics just covered. Any assignment of an intramolecular potential energy function for those molecules needs to incorporate the necessary L or D chirality at each incorporated amino acid residue, the conformational degrees of freedom of the linear backbone, as well as the degrees of freedom and interactions associated with the distinguishing side groups. The molecular modeling should also respect the fundamental backbone directionality of the primary structure and should be capable of predicting what changes in properties would result from reversal of that directionality. Some technical details of these requirements are examined in Section XI.C.

As a result of unusual radiation or foreign-chemical damage, DNA sequences can undergo mutations. Some of these mutations can amount to modifications of the originally encoded amino acid residue sequence, changing one or more of the residues from the native form of the protein involved. Alternatively, the number of residues contained in the protein might be altered. Most mutations would degrade the biological role of the protein, even turning it into a toxic substance. However, infrequently a mutation improves the contribution of the modified protein to the living organism that contains it, an event that can be viewed as evolutionary progress.

B. Solvent Properties and Effects

Experimental investigations of proteins seldom, if ever, examine these substances as isolated molecules or in pure bulk form. Instead, proteins normally are found residing in a compatible solvent medium that typically is an aqueous solution, but with substantial composition variability. Proteins have also been studied experimentally when suspended in pure water. A notable example of an aqueous solution that can serve as a solvent medium for proteins is "normal saline solution." This liquid is used in the medical context for human intravenous infusion but also has been used in molecular biology and biochemistry laboratory experiments because of its relatively close approximation to the aqueous environment in which many biomolecules reside and operate in a variety of living organisms. The composition of normal saline solution includes only water and sodium chloride at the standard concentration 0.154 mol/L.

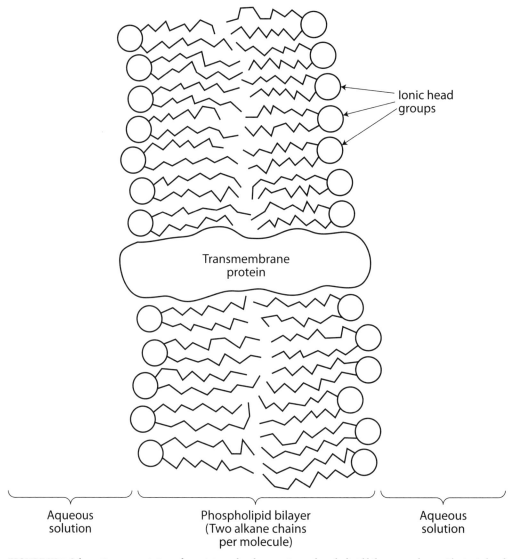

FIGURE XI.3. Schematic representation of a protein molecule spanning a phospholipid bilayer membrane. The ionic head groups of the phospholipids and hydrophilic portions of the protein molecule contact a surrounding aqueous medium. The pairs of alkane chains in each phospholipid molecule are oriented to the interior of the bilayer membrane and present a local nonaqueous medium for the central portion of the protein molecule.

A biologically important alternative solvation situation involves transmembrane proteins that reside at least partially within, and in some cases span, phospholipid bilayer membranes. This combination normally would itself be found suspended in a surrounding aqueous solution medium [Bowie, 2005]. Figure XI.3 provides a simple graphical representation of this circumstance. A basic issue is to determine quantitatively the relevant driving forces that cause any specific protein to arrange itself conformationally in either its simple aqueous solvent or its membrane environment so as to carry out its biological role, i.e., to adopt the configurational "native state" for that protein. It is also important to determine what temperature and pressure changes as well as chemical composition alterations in the protein's surrounding medium would cause modifications in those native-state conformations that are severe enough to result in a "denatured state," that is, severe enough to produce interruptions in the biological role assigned to that individual protein. Some details arising from these considerations are examined in Section XI.E.

In order to generate a proper statistical mechanical formalism for proteins in their supporting media, one needs in principle an appropriate many-molecule potential energy function Φ. For notational simplicity, let $\mathbf{R}$ denote the full set of configurational coordinates necessary to describe the protein molecule or molecules in a system of interest, and let $\mathbf{S}$ be the configurational coordinates required to specify the instantaneous configurations of all of the solvent particles (including those forming a bilayer membrane, if present). Then for a protein–solvent system contained in a volume V, one can resolve Φ into the following obvious contributions:

$$\Phi(\mathbf{R}, \mathbf{S}) = \Phi_1(\mathbf{R}) + \Phi_2(\mathbf{S}) + \Phi_{12}(\mathbf{R}, \mathbf{S}). \tag{XI.1}$$

Here the intramolecular and possible intermolecular interactions involving just the chemically intact protein molecule or molecules in the absence of solvent are entirely represented by the potential energy function Φ_1. Similarly, Φ_2 contains all interaction potentials operating among solvent particles in the absence of the protein(s). All cross interactions between protein(s) and solvent particles appear in Φ_{12}. If repelling walls are present to define system volume V (as opposed to periodic boundary conditions), Φ_1 and Φ_2 both include corresponding position-limiting contributions for the protein(s) and for the solvent particles. As usual, the enthalpy function $\hat{\Phi}(\mathbf{R}, \mathbf{S})$ to describe isobaric conditions can be obtained simply by adding the pressure–volume product pV to $\Phi(\mathbf{R}, \mathbf{S})$:

$$\hat{\Phi}(\mathbf{R}, \mathbf{S}, V) = \Phi(\mathbf{R}, \mathbf{S}) + pV. \tag{XI.2}$$

The occurrence of hydrogen atoms as essential components both of proteins and of their supporting solvents implies that quantum mechanical effects of nontrivial magnitude are present under ambient conditions, as well as under lower temperature conditions. Therefore, use of the isochoric canonical ensemble to describe protein–solvent systems at thermal equilibrium should formally involve quantum statistics that starts with the Hamiltonian operator for the full system:

$$\mathbf{H} = \mathbf{K}_1 + \mathbf{K}_2 + \Phi(\mathbf{R}, \mathbf{S}). \tag{XI.3}$$

In this expression, $\mathbf{K}_1$ and $\mathbf{K}_2$, respectively, represent the kinetic energy operators for the protein and for the solvent configurational degrees of freedom. The canonical partition function Q then has the usual form ($\beta = 1/k_B T$):

$$Q(\beta, V) = Tr[\exp(-\beta \mathbf{H})], \tag{XI.4}$$

where the trace operation includes a complete set of eigenstates for the nuclear motions on the potential energy hypersurface Φ of the combined system of given chemical structures.

The joint probability function for all configurational degrees of freedom is provided by the diagonal elements of the quantum mechanical density matrix in the configurational representation:

$$P(\mathbf{R}, \mathbf{S} \,|\, \beta, V) = [Q(\beta, V)]^{-1} \langle \mathbf{R}, \mathbf{S} \,|\, \exp(-\beta \mathbf{H}) \,|\, \mathbf{R}, \mathbf{S} \rangle. \tag{XI.5}$$

Integrating this joint probability over the solvent coordinates yields a probability density for just coordinate degrees of freedom of the protein(s) in the thermodynamic state specified by β and V:

$$P_1(\mathbf{R} \mid \beta, V) = \int P(\mathbf{R}, \mathbf{S} \mid \beta, V) d\mathbf{S} \tag{XI.6}$$

$$\equiv C(\beta, V) \exp[-\beta F_1(\mathbf{R} \mid \beta, V)],$$

where

$$C(\beta, V) = [\int P_1(\mathbf{R} \mid \beta, V) d\mathbf{R}]^{-1} \tag{XI.7}$$

is a normalizing constant. This serves to introduce a Helmholtz free energy landscape function F_1 for just the protein degrees of freedom, containing not only the contribution of the direct protein interaction potential Φ_1 but also the important $\mathbf{R}$-dependent average solvent effects on the protein(s). Because of the latter contribution, this landscape now varies topographically upon changing the externally imposed temperature, and of course it depends on the solvent chemical composition and density. It should be mentioned that the $\mathbf{R}$ dependence of the free energy landscape defined by F_1 is often colloquially referred to as the protein "energy landscape" [e.g., Onuchic et al., 1997; Dill and MacCallum, 2012, Figure 3], a term that is dimensionally correct but which may obscure the solvent-averaged free energy character of F_1.

In principle, it is possible to submit the full potential energy function $\Phi(\mathbf{R}, \mathbf{S})$ to steepest descent mapping as described in earlier chapters, thus providing a detailed characterization of its multidimensional topography in terms of basins and inherent structures. But for practical purposes, a lower dimensional, and therefore simpler, version of the steepest descent mapping can be carried out to characterize the landscape of the protein free energy function $F_1(\mathbf{R} \mid \beta, V)$. Of course, that simpler topography depends upon temperature and volume at fixed system composition; however, for any specific protein the primary interest focuses on the biologically relevant conditions. Specifically, vector $\mathbf{R}$ may comprise roughly 10^3–10^5 components for a single-protein molecule, whereas at least an order of magnitude larger number of $\mathbf{S}$ components would have to be involved additionally to represent explicitly a local solvation environment for a low-concentration protein solution. It should be kept in mind that if a free energy landscape $F_1(\mathbf{R} \mid \beta, V)$ for a single isolated protein is to be examined, its basin structure is quite different from that presented by many-body systems of low-molecular-weight substances where many permutation-equivalent basins are present. In particular, with just a single-protein molecule in the system of interest there are only limited sets of permutation-equivalent basins, primarily associated with intramolecular methyl group rotations. In this altered circumstance, a basic objective is to identify what portions of the single-protein F_1 landscape correspond respectively to native and to denatured protein states, what their occupancy probabilities are, and what kinds of transition states and kinetics might exist between them. These issues are considered in more detail in Section XI.E.

Isobaric analogs of the isochoric Eqs. (XI.4)–(XI.7) amount to elementary modifications. Volume V gets included in the set of configurational variables with its own kinetic energy operator in the augmented Hamiltonian $\hat{\mathbf{H}}$ that contains interaction function $\hat{\Phi}(\mathbf{R}, \mathbf{S}, V)$. One thus has the following isobaric versions of Eqs. (XI.4)–(XI.7):

$$\hat{Q}(\beta, p) = Tr[\exp(-\beta \hat{\mathbf{H}})], \tag{XI.8}$$

$$\hat{P}(\mathbf{R}, \mathbf{S}, V \mid \beta, p) = [\hat{Q}(\beta, p)]^{-1} \langle \mathbf{R}, \mathbf{S}, V \mid \exp(-\beta \hat{\mathbf{H}}) \mid \mathbf{R}, \mathbf{S}, V \rangle, \tag{XI.9}$$

and

$$\hat{P}_1(\mathbf{R} \mid \beta, p) = \int \hat{P}(\mathbf{R}, \mathbf{S}, V \mid \beta, p) d\mathbf{S} dV \tag{XI.10}$$

$$\equiv \hat{C}(\beta, p) \exp[-\beta G_1(\mathbf{R} \mid \beta, p)],$$

$$\hat{C}(\beta, p) = \left[\int \hat{P}_1(\mathbf{R} \mid \beta, p) d\mathbf{R} \right]^{-1}. \tag{XI.11}$$

Here $G_1(\mathbf{R} \mid \beta, p)$ is the Gibbs free energy landscape function for the protein(s) subject to the temperature and pressure effects of solvation.

The side groups that distinguish the amino acids involve chemical structures that interact differently with a neighboring liquid solvent or membrane medium, as well as differently with one another within the same protein molecule. At a coarse level of distinction, the 20 amino acids listed in Table XI.1 conventionally are divided into two equal subsets, informally designated "nonpolar" and "polar" amino acids. These subsets are identified in Table XI.2.

Reference to Table XI.1 verifies that the nonpolar subset possesses side groups that are entirely, or predominantly, hydrocarbon structures (glycine's H can be regarded as a degenerate case). The two cases of sulfur-containing side groups (cysteine, methionine) are also included in the nonpolar subset. In analogy with simple hydrocarbons, these nonpolar cases can also be classified as hydrophobic, so that a protein containing them that is suspended in an aqueous solvent would have a free energy driving force to adopt conformations that minimize their contact with water. On the other hand, a protein bound within a bilayer membrane would exploit nonpolar side groups to form stabilizing contacts with the hydrocarbon-tail interior of that occupied host membrane.

Polar amino acid side groups can interact relatively strongly and stably with surrounding water molecules. Typically, this occurs by forming hydrogen bonds. For some of the polar cases, this arises from the presence of exposed –OH groups (serine, threonine, tyrosine), carboxylic acid groups –COOH (aspartic acid, glutamic acid), amide groups (asparagine, glutamine), and other structures with associated substantial dipole moments that present exposed nitrogen and/or carboxylic oxygen atoms to the solvent (arginine, histidine, lysine).

TABLE XI.2. Separation of amino acids into "nonpolar" and "polar" subsets

Nonpolar	Polar
Alanine (Ala, A)	Arginine (Arg, R)
Cysteine (Cys, C)	Asparagine (Asn, N)
Glycine (Gly, G)	Aspartic acid (Asp, D)
Isoleucine (Ile, I)	Glutamic acid (Glu, E)
Leucine (Leu, L)	Glutamine (Gln, Q)
Methionine (Met, M)	Histidine (His, H)
Phenylalanine (Phe, F)	Lysine (Lys, K)
Proline (Pro, P)	Serine (Ser, S)
Tryptophan (Try, W)	Threonine (Thr, T)
Valine (Val, V)	Tyrosine (Tyr, Y)

In view of the wide pH range that aqueous solvents can possess, an important additional characteristic of at least some of the polar side groups is that they could add or lose a proton to become ionic, possibly including zwitterion formation analogous to that indicated in Figure XI.1(b). Depending on whether this involved a net gain or loss of protons by the protein, this could alter the dimensionality of the configurational vector $\mathbf{R}$ (with a compensating change in the dimensionality of $\mathbf{S}$) and would influence the resulting free energy landscape functions $F_1(\mathbf{R}\,|\,\beta,\,V)$ and $G_1(\mathbf{R}\,|\,\beta,\,p)$. A related technical issue is what realistic configurational criteria should be used to decide when a hydrogen ion qualifies to be regarded as part of the protein, rather than simply belonging to the solvent. To provide a background context for these protein ionic effects, it is worth considering first the behavior of single isolated molecules of the 20 amino acids before polymerization as they respond to pH variations in a surrounding aqueous environment.

The possibility of an intramolecular proton transfer for an isolated amino acid *in vacuo* to produce a zwitterionic state, Figure XI.1(b), would create a high-potential-energy charge-separated state in spite of remaining electrically neutral overall. However, its large dipole moment would be substantially stabilized by the high dielectric constant of a surrounding aqueous solvent medium. The side groups of the polar amino acids as individual molecules can also develop their own charges by adding or losing a proton from or to a surrounding solvent. In acidic media with pH substantially lower than 7, the dominant state of both nonpolar and polar amino acids would involve a net positive charge, with un-ionized carboxyl groups (–COOH) but with at least one proton added, for example, to a pendant amine group (–NH$_3^+$) or hydroxyl group (–OH$_2^+$). By contrast, with pH well above 7, the situation would reverse, producing a net negative charge on the amino acid as a result of carboxyl group ionization (–COO$^-$) but no added protons elsewhere in the molecule.

In principle, the charge state of an amino acid at equilibrium with a surrounding solvent can fluctuate over time, with probabilities of those charge states strongly dependent on the pH. These probabilities define an average coulombic charge borne by the amino acid of interest, which would be a continuous and differentiable function of pH. This average charge can be accessed experimentally by electrophoresis, that is, by observing the direction and speed of drift of the dissolved amino acid under the influence of an applied static electric field. The drift velocity function of the average charge declines monotonically from positive to negative values as the solvent pH changes continuously from strongly acidic to strongly basic, passing through zero at the "isoelectric point" for that amino acid. The pH value at this isoelectric point of vanishing electrophoretic drift is denoted conventionally by pI. The pH values at which the neutral un-ionized state and the neutral zwitterionic state have their respective probability maxima need not exactly equal pI but are expected to be close.

Table XI.3 presents pI values for the 20 principal amino acids. These results are applicable to room-temperature conditions at atmospheric pressure in suitably buffered aqueous solutions. A notable feature of the entries is that the 10 nonpolar amino acids identified in Table XI.2 have pI values confined to the narrow range $5.15 \le \text{pI} \le 6.30$, whereas the 10 polar amino acids in Table XI.2 present the much broader range $2.98 \le \text{pI} \le 10.76$. The lower and upper pI limits for the polar group are due, respectively, to the strong acid behavior of the aspartic acid side group and the strong proton binding property of the side-group nitrogens in arginine.

Referring to the listed pI values, one can infer that for an aqueous solvent whose pH is close to the neutral value 7, both aspartic acid and glutamic acid are almost always found to be ionized

TABLE XI.3. Isoelectric points for amino acids[a]

Amino Acid	pI
Alanine (Ala, A)	6.11
Arginine (Arg, R)	10.76
Asparagine (Asn, N)	5.43
Aspartic acid (Asp, D)	2.98
Cysteine (Cys, C)	5.15
Glutamic acid (Glu, E)	3.08
Glutamine (Gln, Q)	5.65
Glycine (Gly, G)	6.06
Histidine (His, H)	7.64
Isoleucine (Ile, I)	6.04
Leucine (Leu, L)	6.04
Lysine (Lys, K)	9.47
Methionine (Met, M)	5.71
Phenylalanine (Phe, F)	5.76
Proline (Pro, P)	6.30
Serine (Ser, S)	5.70
Threonine (Thr, T)	5.60
Tryptophan (Try, W)	5.88
Tyrosine (Tyr, Y)	5.63
Valine (Val, V)	6.02

[a]Taken from Liu et al., 2004.

by proton loss and thus to bear a net charge $-e$. On the other hand, arginine and lysine molecules experiencing the same neutral pH range almost always have an added proton attached to an exposed side-group nitrogen, thus conferring a net charge $+e$ to that group.

Formation of the dehydrating peptide linkage between two amino acids, Figure XI.2, causes loss of a carboxyl group and an amine group that otherwise could have ionized by loss or gain of a proton, respectively. The terminal amino residues in a linear polypeptide or protein remain available, unless they have linked together to form a cyclic (simple or topologically knotted) polymer or have been rendered unreactive by substitution of methyl or other stabilizing chemical units. However, in aqueous solvents the ionization behaviors of side groups in protein residues corresponding to the polar subset of amino acids remain relatively immune to these linkage or end-stabilizing processes, at least if the resulting polymeric backbone is in an extended configuration that permits solvent contact. On the other hand, compact backbone configurations that bring side groups into near contact with one another and out of contact with solvent can have a strong influence on side-group ionization. Realistic modeling of polypeptide and protein properties needs to account for these features and in particular needs to accommodate time-varying dimension changes of the configurational vector $\mathbf{R}$ as the evolving protein configuration and solvation influence the extent of ionization.

While the free energy landscape functions $F_1(\mathbf{R} \mid \beta, V)$ and $G_1(\mathbf{R} \mid \beta, p)$ focus attention on the protein conformation under the average influence of the solvent medium, there are circumstances in which a role reversal is a natural viewpoint. In particular, this is the case where "antifreeze" proteins are involved. Specifically, these proteins occur in organisms whose environments present temperatures below the normal freezing point of their aqueous body fluids and whose survival requires avoiding that freezing (i.e., ice crystal production) internally. Antifreeze proteins are found in fish that live in salty ocean waters whose equilibrium freezing temperature is below that of the fishes' internal fluids [Knight et al., 1984; Sun et al., 2014]. Other antifreeze proteins have been found in insects [Graham et al., 1997; Meister et al., 2013]. The role of these proteins does not involve phenomena within the thermal equilibrium regime. Rather, it is to inhibit kinetically the growth of ice crystallites at subfreezing temperatures that would otherwise serve as nuclei for initiation of crystal growth into the macroscopic regime with resulting damage to the tissues of the living organism. This is accomplished by the capacity of the antifreeze proteins to bind to the surfaces of the nanometer-scale ice crystallites when they are well below the critical nucleus size and thus to prevent addition of more water molecules from the surrounding aqueous solution [Garnham et al., 2011]. Viewed from the perspective of the full-system potential energy landscape $\Phi(\mathbf{R}, \mathbf{S})$ or enthalpy landscape $\hat{\Phi}(\mathbf{R}, \mathbf{S}, V)$, the antifreeze proteins block or at least raise the free energy of transition states that exist between the fully fluid basin set in the multidimensional configuration space and the basin set exhibiting substantial ice crystals. Consequently, it should come as no surprise that antifreeze proteins also permit superheating of ice crystals in solutions of those proteins [Celik et al., 2010].

Long-term storage, including storage at low temperature, is an important necessity for many medical and pharmaceutical substances, including proteins. The objective is to ensure long-term stability against irreversible chemical or physical changes in bulk samples. Freeze-drying, or "lyophilization," can be an effective strategy for this purpose, though not without problems [Wang, 2000]. At the beginning, this typically involves adding substantial amounts of organic glass formers such as carbohydrates to the aqueous solvent medium in which the protein is suspended. The system with the correspondingly modified solvent composition can then be cooled quickly to low temperature without experiencing freezing out of large and mechanically disruptive ice crystals, a process that could produce damaging local stresses on the proteins. But for long-term storage, it is often desirable to remove most of the remaining water, whether it is present as ice or not, by sublimation from the glassy matrix. If that is properly done, the system can often be brought back safely to higher temperature with a nonaqueous solvent medium that may still be in a solid amorphous (and thus protective) state. A similar natural process allows certain primitive plant and animal organisms to survive extreme low temperatures in a dehydrated state by replacing their water with sugars [Crowe et al., 1998]. Needless to say, the addition of glass formers and the removal of water by sublimation substantially modify the landscape functions $\Phi(\mathbf{R}, \mathbf{S})$ and $\hat{\Phi}(\mathbf{R}, \mathbf{S}, V)$, the free energy functions $F_1(\mathbf{R} \,|\, \beta, V)$ and $G_1(\mathbf{R} \,|\, \beta, p)$, and their multidimensional tilings by steepest descent basins.

The most effective experimental technique for determining the folding structure of protein molecules has been X-ray diffraction performed on crystals of the substance of interest. Forming high-quality crystals from protein solutions is usually a nontrivial task, but when it succeeds, the result typically contains a large fraction of solvent by weight, often exceeding 50%. An important result is that this abundance of solvent effectively shields each protein molecule from interacting strongly with its neighbors, a phenomenon that relies on the fact that the crystallizable proteins tend to possess compact globular conformations, as opposed to extended backbone shapes. Consequently, the geometric structure adopted by a protein molecule in the crystal is close to that for an isolated single-protein molecule in the same solvent at the same temperature and pressure. A substantial portion of the large quantity of low-molecular-weight solvent molecules in the crystal may possess liquidlike diffusional mobilities, thus causing the solvent content of any specific nominal unit cell in the crystal to fluctuate in geometry and number of molecules with the passage of time. These local solvation fluctuations would create small uncertainties in the protein molecule position in a unit cell, which in principle would be reflected in the free energy functions $F_1(\mathbf{R}\,|\,\beta, V)$ and $G_1(\mathbf{R}\,|\,\beta, p)$. This protein configurational variability would produce a small broadening effect in an X-ray diffraction pattern for the crystal.

C. Interaction Approximations

The remarkable growth of computing power available for scientific research has opened the possibility of carrying out realistic molecular level simulations for at least some biologically important proteins. Such efforts utilizing Monte Carlo or molecular dynamics techniques can yield valuable insights into preferred folding structures and how they are attained kinetically under varying temperature, pressure, and solvent composition conditions. In principle, they can also characterize overall topographic features of a protein's multidimensional free energy landscapes $F_1(\mathbf{R}\,|\,\beta, V)$ and $G_1(\mathbf{R}\,|\,\beta, p)$, as well as revealing the role of consistent amino acid residue chirality and the consequences of its violation. In order to pursue these objectives, the simulations require descriptions for the interactions that are involved at least at a basic level of biochemical reality. Obviously, it is impossible to carry out accurate quantum mechanical calculations for the electronic ground state of an isolated protein or polypeptide molecule over its entire conformation space, let alone for such a molecule with surrounding explicit solvent molecules. Instead, practical prescriptions based on simple approximations that capture essential features of the "real" interactions can be devised. The level of detail and the numerical precision of these approximations can be allowed to vary widely, depending on the scientific objective involved. This situation is analogous to that considered earlier for selection of effective pair potentials [Box IX.1, Eq. (IX.6)].

Although chemical reactions involving the protein or polypeptide can be important phenomena to investigate, one basic simplification in the interest of computational feasibility involves confining attention to the molecular system while temporarily halting relevant reaction kinetics. For many applications that one might envision, attention would focus on cases where those halted reactions had already approached or attained their chemical equilibrium states. In particular, this could involve assignment of ionization states for those residues that add or lose a proton, or transfer a proton between alternative binding sites within the same residue. The pH of the surrounding solvent, as well as its other composition characteristics, would be central to such assignments. In some cases, the prior formation of intramolecular disulfide linkages might also be a legitimate presumption.

The configurational variable $\mathbf{R}$ appearing in the potential energy and enthalpy functions $\Phi(\mathbf{R}, \mathbf{S})$ and $\hat{\Phi}(\mathbf{R}, \mathbf{S}, V)$ nominally refers to the full set of positions for all nuclei included in the polymeric molecule or molecules under consideration. However, a substantial number of intramolecular degrees of freedom are "stiff": Covalent chemical bond lengths as well as the bond angles between successive chemical bonds provide examples. In many applications, it may be reasonable to freeze these degrees of freedom at their mechanically stable values, thus leading to a reduced dimension for the remaining "flexible" intramolecular degrees of freedom included in **R**. A similar dimensional reduction may also be feasible for solvent degrees of freedom comprised in **S**. An obvious example of this latter possibility for aqueous media is to treat each water molecule as a rigid rotor with C_{2v} symmetry, possessing the average OH bond length and HOH bond angle of an isolated molecule, or alternatively the slightly different average length and angle for the interacting molecules in pure liquid water.

Rotational degrees of freedom around covalent bonds are generally less constrained and play a primary role in determining the overall conformation of a protein or polypeptide molecule and how it changes with external conditions of temperature, pressure, and solvent composition. These rotations are described naturally by dihedral angles defined by the directions of three sequential covalent bonds; this geometry was illustrated in Figure X.1 for carbon–carbon bond sequences. The distinguishing side groups listed in Table XI.1 for the amino acid residues themselves present many such rotational degrees of freedom. But at least equally important are the rotational degrees of freedom of the backbone of the protein or polypeptide. Three dihedral angles per residue are present, illustrated in Figure XI.4. For residue j, these have been denoted as φ_j, ψ_j, and ω_j. Although the first two of these dihedral angles are relatively free to vary, it has often been supposed that each ω_j is chemically constrained to the immediate vicinity of its "*trans*" configuration $\omega_j = \pm 180°$ [Creighton, 1993, p. 5]. The reason for this presumption is illustrated in Figure XI.5, specifically the possibility of a potential-energy-lowering resonance involving a partial double bond character of the backbone amide (N–C) linkage, which requires the presence of the *trans* geometry. If this *trans* constraint could be applied to all residues along the backbone, the configurational problem would be significantly simplified. However, analysis of measured protein structures at atomic-level resolution reveals that as much as 25° deviations of the ω_j from the planar *trans* configuration are not extremely uncommon, and the ideal *trans* configuration can be driven out of planarity by the requirements of overall folding geometry for a protein containing that residue [Berkholtz et al., 2012].

The diversity of feasible backbone and side-group conformations can bring residues that are non-neighbors along the backbone into close proximity. This obviously requires $\Phi_1(\mathbf{R})$ in Eq. (XI.1) to include the corresponding interactions. These are composed of short-range repulsions, intermediate-range electrostatic and dispersion effects, and hydrogen bond formation interactions between intramolecular hydrogen donors and acceptors. All are contributions that are involved in determining the free energy landscape functions $F_1(\mathbf{R}\,|\,\beta, V)$, Eq. (XI.6), and $G_1(\mathbf{R}\,|\,\beta, p)$, Eq. (XI.10). Aspects of the dielectric properties of solvents, insofar as they influence protein and polypeptide conformational behavior, reside in principle within the potential component $\Phi_{12}(\mathbf{R}, \mathbf{S})$ in Eq. (XI.1). At fixed configuration **S** for the entire solvent medium, the dielectric response of that solvent to the presence of ionic charges involves many-molecule polarization contributions. In principle, these are not resolvable simply into pairwise-additive terms and therefore could present a computational difficulty. However, under some circumstances, it may be an appropriate approximation simply to rescale coulombic interactions by a scalar high-frequency dielectric constant. In the case of widely separated small ions dissolved in the solvent medium, a macroscopic dielectric constant (equivalent to the square of a refractive index for the solvent) may suffice. However, ionized parts of residue side groups would normally be found in an inhomogeneous location because of proximity both to the macromolecular backbone as well as to the

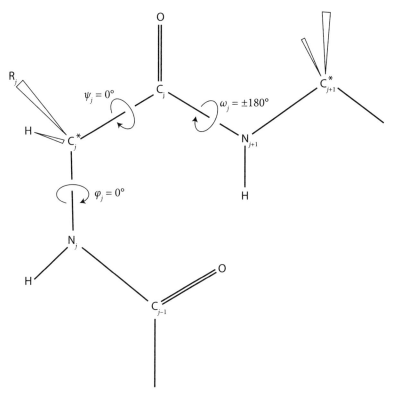

FIGURE XI.4. Assignment of the three backbone dihedral angles for generic residue j. As illustrated, two of the angles (ϕ_j, ψ_j) have been set to the *cis* configuration value 0°, while the third (ω_j) has been set to the opposite extreme *trans* value ±180°. The curved arrows for each of these dihedral angles indicate the rotation direction of the last of the three bonds involved about the middle bond to give rise to a positive increase in the corresponding dihedral angle.

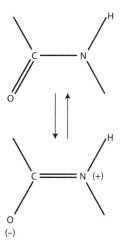

FIGURE XI.5. Chemically resonant structures contributing to partial double bond character and planarity of a backbone amide bond.

fluid solvent, so that a locally relevant position-dependent dielectric tensor might have to be invoked.

The computational complexity of realistic protein and polypeptide simulations has stimulated development of elaborate computational software packages to provide both generic molecular dynamics simulation procedures, as well as applicable model interactions. These have been made widely available to the research community to systematize such investigations, and thus to minimize the investigator's necessary time expenditure. Among many that are now offered, five examples of such packages are identified by the following acronyms:

(1) GROMOS [Scott et al., 1999];
(2) AMBER [Case et al., 2005];
(3) NAMD [Phillips et al., 2005];
(4) GROMACS [Hess et al., 2008];
(5) CHARMM [Brooks et al., 2009].

The possible applications of these resources typically extend beyond the protein and polypeptide regime to include wider families of biologically relevant molecules. Each package has undergone several updates and extensions, including user choices for the level of molecular detail and precision of the description. In some cases, it is possible to import a model potential function from one package into a molecular dynamics procedure from another package. At the most detailed level ("all-atom" description), the prescribed interaction function $\Phi(\mathbf{R}, \mathbf{S})$ is typically composed of component functions derived from direct quantum mechanical energy calculations (in the Born-Oppenheimer approximation) on small molecules related to the residues and the solvent components, as well as from the extensive published literature on computer simulations of these small molecules alone in condensed phases. In particular, the latter relies on available effective pair potentials for water that have been developed via classical simulations (Section IX.A). The alternative class of user choices eliminates the $\mathbf{S}$ coordinates by treating the solvent medium implicitly for a given volume or pressure, which involves assigning ion and polar group solvation free energies, inclusion of low-frequency dielectric constants in Coulomb interactions between charged groups, and configuration-dependent hydrophobic interactions between nonpolar residues.

The available software packages can predict some distinctly different potential energy hypersurface details when applied to a given molecule. One documented example of the difference involves "tetra alanine" with terminating methyl groups, residing alone in vacuum, consisting of 52 atoms [Somani and Wales, 2013]. The computed numbers of potential energy minima were 732 for AMBER (version AMBER99SB), and 13,732 for CHARMM (version CHARM27). Such an enumeration discrepancy may reflect prediction differences primarily for high free energy (i.e., low occupancy probability) portions of the free energy hypersurface, rather than for the more physically relevant configurational vicinity of the absolute free energy minimum. It is the latter that is naturally the focus of fitting procedures used for development of model interactions.

D. Secondary, Tertiary, and Quaternary Structures

A vast number of folded protein three-dimensional structures have been determined experimentally and are collected in the Protein Data Bank (see Box XI.1). As a rational first step toward

BOX XI.1. The Protein Data Bank (PDB)

As rapidly increasing numbers of reliable experimental results that determined protein spatial structures were generated, it became clear that having a reliable repository with that information readily and widely available in standardized form would be an important scientific resource. This realization motivated the initial effort to create the Protein Data Bank (PDB) in 1971 [Berman, 2008]. As a result of drastic improvements in experimental techniques and information processing capacity, the PDB has grown over the subsequent years from its modest beginning to include more than 80,000 entries at the end of 2012, and its content is now accessible over the Internet. At first, the structures placed in this archive were individual proteins of relatively low molecular weight. However, the geometric structure database subsequently expanded to include nucleic acids, protein complexes (quaternary structures), virus particles, and ribosomes, among other biological entities. The primary experimental methods that have contributed data are X-ray crystallography and nuclear magnetic resonance (NMR). Structures submitted for inclusion in the PDB are subjected to consistency testing before acceptance.

Each structure entry is assigned a four-character alphanumeric Protein Data Bank identification code, the "PDB ID." These identify the protein, nucleic acid, or other complex structure involved, but they also distinguish spatial structures for a given biomolecule that have been determined under distinctly different conditions, such as solvent composition or variable quaternary (multiple-protein-complex) circumstances.

---

understanding the daunting structural complexity presented by this data, it is useful to identify frequently occurring geometrical motifs exhibited by subsets of the residues that are present in protein primary structures. These motifs constitute "secondary structures," and their localized patterns amount to important geometric units contained within the overall "tertiary structures" exhibited by each of the full protein molecules. Because these secondary structures are relatively robust, their presence helps to define and maintain the biologically important features of the native tertiary structures. The two most prominent secondary structure types observed in protein native states are "alpha helices" and "beta sheets."

In order to help characterize experimentally observed secondary structures, it is useful to display aspects of their backbone dihedral angles in "Ramachandran plots" [Ramachandran et al., 1963; Ramachandran and Sasisekharan, 1968]. These plots concentrate on the angles φ_j and ψ_j for a residue j that was illustrated schematically in Figure XI.4. Those angle pairs fall in the square region:

$$-180° \leq \varphi_j, \psi_j < 180°. \tag{XI.12}$$

This region is subject to periodic boundary conditions since adding or subtracting 360° to either angle returns to the same backbone geometry. For the purposes of interpreting Ramachandran plots, the third dihedral angle ω_j for residue j is presumed to lie sufficiently close to its *trans* values ±180° to be temporarily disregarded. Because of interactions between pendant atoms and side groups attached to the backbone, the dihedral angle pairs are subject to significant constraints as they occur in native protein structures.

Figure XI.6 presents a Ramachandran plot indicating the observed high-probability locations for backbone dihedral angle pairs φ_j, ψ_j for those residues involved in alpha-helical and in beta-sheet secondary structures. This is based on experimental structure determinations for a large

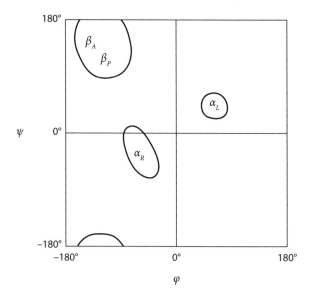

FIGURE XI.6. Observed preferential locations of backbone dihedral angles for residues in alpha-helical and beta-sheet secondary structures. Each motif has two subsets: the right-hand (α_R) and left-hand (α_L) helices, and the parallel (β_P) and antiparallel (β_A) sheets. The closed contours indicate roughly the locations of the majority of angular variations observed for each of these motifs. Periodicity applies to both horizontal (φ) and vertical (ψ) directions.

database of proteins [Lovell et al., 2003]. The four sites α_R, α_L, β_P, β_A indicated in the Ramachandran plot are the most likely locations for the secondary structure variants, but because of diversity in other structural and solvation effects, the corresponding distributions exhibit significant breadths whose examples occasionally drift outside the closed loops shown. Notice, however, that the lower right-hand quadrant

$$0 < \varphi < 180°, \tag{XI.13}$$

$$-180° < \psi < 0°,$$

is virtually unoccupied. The locations indicated in Figure XI.6 refer to biological proteins whose included chiral amino acids all have L form. If instead, analogous mirror-image proteins including only D form amino acids were to be examined, the corresponding population maxima would occur at locations that represent inversions through the Ramachandran plot origin, i.e., $\varphi, \psi \rightarrow -\varphi, -\psi$.

Right-handed alpha helices α_R are the most frequently encountered secondary structure. Figure XI.7 indicates the backbone geometry involved. A key feature is that the C=O group of residue j is hydrogen-bonded to the N–H group of residue $j + 4$. These essentially linear hydrogen bonds are nearly parallel to the helix axis. Consequently, identifiable alpha helices must contain at least five neighboring residues along the backbone. The average pitch of the helix is such that 3.6 residues occur for each full turn; the axial-direction translation is 1.50 Å per residue, which amounts to 5.41 Å per full turn [Creighton, 1984, p. 171]. The helix axis is surrounded coaxially by an empty core region, which is too narrow to accommodate any solvent.

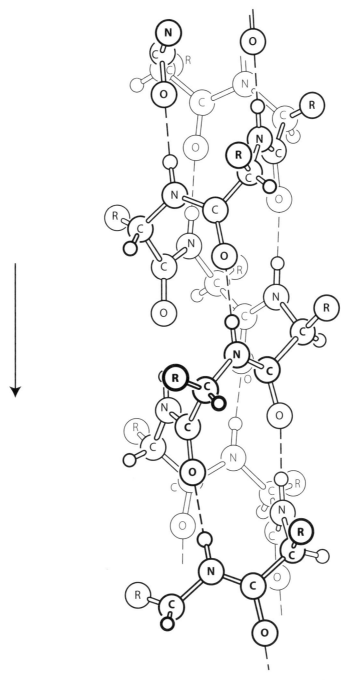

FIGURE XI.7. Geometrical arrangement of a protein backbone in a right-handed alpha-helical secondary structure. In this view, the basic direction of the backbone from the N terminus to the C terminus is downward along the helix axis. Dashed lines indicate hydrogen bonds. Adapted with permission of the publisher from Pauling, 1960, Figure 12-18. Copyright © 1960 by Cornell University.

In principle, a single alpha helix could encompass an entire protein. However, the majority of structures observed experimentally tend to show helix lengths with fewer than 50 residues serving as incorporated secondary structures. At the ends of an alpha helix, the protein can exhibit a sequence of residues that are not arranged in a simply recognizable secondary structure but serve as connecting portions between distinct alpha helices or other secondary structures, such as the beta-sheet sections described below. The amino acid proline presents a special case because of its backbone-constraining pentagonal heterocycle. This constraint prevents it from occurring within the interior of an alpha helix. However, it can be included in a nonhelical connecting portion of the protein backbone between alpha-helix portions, and indeed its presence there may stabilize that interval of residues that are not part of secondary structures.

Residue side groups in an alpha helix all radiate outward from the central axis, as shown in Figure XI.7. Consequently, these residues are all exposed to contact with surrounding solvent, to other portions of the same protein's tertiary structure, or to other protein molecules present in the system under examination. The polar vs. nonpolar distinctions listed in Table XI.2 for side groups are important in this respect. The specific linear sequence of amino acids along an alpha-helical portion of the backbone thus determines a polar and/or nonpolar helix-periphery pattern, and that is a significant controlling feature in how that secondary structure contributes to the biological role of the protein.

Left-handed alpha helices formed from L amino acids are only rarely observed. As with the common right-handed version, they also involve hydrogen bonds between the C=O group of a residue j and the N–H group of residue $j + 4$. However, the different dihedral angles involved, as shown in Figure XI.6, typically entail a higher intramolecular energy than for the right-hand version. Consequently, a special compensating free energy contribution would be necessary to stabilize its incorporation in a protein native state.

In contrast to the geometry of alpha helices, beta sheets consist of neighboring stretches of extended backbones ("beta strands") that link to one another by lateral N–H $\cdots$ O=C hydrogen bonds. This distinction is reflected in the separate locations of backbone dihedral angle distributions for the alpha and beta secondary structures, shown in the Ramachandran plot of Figure XI.6. Neighboring strands in a beta sheet that are connected by hydrogen bonds can have either parallel or antiparallel directions of their respective N-terminus to C-terminus orientations. These are illustrated in Figure XI.8. The lateral connecting hydrogen bonds in both cases alternate in direction. For antiparallel strands, the hydrogen bonds are arranged in close pairs; for parallel strands, the hydrogen bonds are not clustered in pairs. The antiparallel variant can form as a result of a few-residue tight turn at the end of one strand that then forms the beginning of its neighbor strand. A longer intervening sequence of residues would be required to allow emplacement of a parallel strand within the same molecule, and that intervening sequence could itself be involved in some other secondary structure before returning to the beta sheet of interest. In analogy to the α helix situation, proline residues do not fit naturally within beta-sheet structures and are not able to participate in the lateral connecting hydrogen bonds; however, they frequently occur within the tight turns between the ends of neighboring antiparallel strands.

Residue-identifying side groups in beta-sheet secondary structures protrude alternately from one side of the beta sheet to the other. These protrusions are amplified somewhat because the strands, and thus the sheets themselves, have a pleated texture. As a result of these side-group

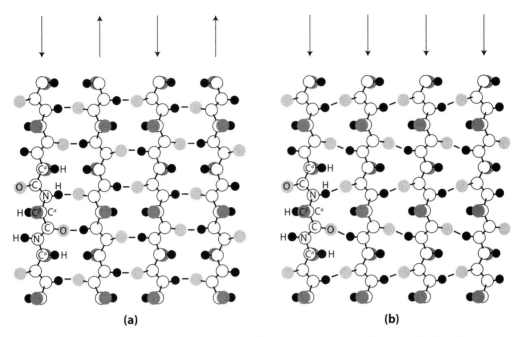

FIGURE XI.8. Beta-sheet patterns of lateral hydrogen bonds between neighboring backbone strands whose directions are indicated by arrows. (a) Close-by pairs of oppositely directed hydrogen bonds connecting antiparallel strands. (b) Sequence of hydrogen bonds between parallel strands. Adapted with permission of the publisher from Pauling, 1960, Figures 12-19 and 12-20. Copyright © 1960 by Cornell University.

locations, it is possible for all residues on one side of the sheet to be polar while those on the other side are nonpolar (Table XI.2). Consequently, the beta sheet would simultaneously possess both hydrophilic and hydrophobic surfaces. This is a feature that can play an important role in establishing the native structure of the protein involved.

The hydrogen bond patterns for antiparallel or parallel beta strands do not require the overall shape of a beta sheet (beyond its pleated texture) to be close to planar. Various distortions from overall planarity are possible and are the norm in experimentally observed native structures. Presumably, such flexibility incurs relatively little energy or free energy cost and can be an important contributor to the ability of the protein to conduct its biological role.

A protein's "tertiary structure" specifies the overall spatial deployment of its constituent backbone secondary structures and of the residues along the backbone that connect those secondary structures. Conventionally, this phrase refers to the three-dimensional geometry of the molecule in its biologically active native state, and this can vary widely from a compact globular form that mostly excludes interior solvent to more open or even linearly extended shapes. In any case, the stability of these tertiary structures depends on the composition of the surrounding solvent medium and its thermodynamic state, which in some circumstances may involve contributing one or more ions that are bound within the folded protein. The majority of proteins whose tertiary structures have been determined each include both alpha-helix and beta-sheet secondary structure portions. However, there are significant numbers of proteins that exhibit primarily just one of those two types of secondary structures. Examples containing primarily alpha helices are the following:

(α1) the transmembrane sodium–calcium ion exchanger NCX_Mj with 10 helical segments, comprising more than 350 amino acid residues [Liao et al., 2012];

(α2) a de novo designed three-helix protein comprising 73 amino acid residues [Walsh et al., 1999];

(α3) members of a large family of "ankyrin repeat proteins," containing variable numbers of alpha-helical portions, each of which comprises about 33 amino acid residues [Kohl et al., 2003].

By contrast, the following examples have tertiary structures dominated structurally by beta sheets:

(β1) hollow "beta barrel" proteins, named "porins," that present channels for transfer of various substances across bacterial outer membranes [Schirmer, 1998];

(β2) human carbonic anhydrase II, containing a bound zinc ion from the surrounding aqueous solvent [Eriksson et al., 1988];

(β3) transthyretin (previously named "prealbumin") monomers that consist of 127 amino acid residues, which in tetrameric form binds the thyroid hormone thyroxine [Blake et al., 1974; Blake, 1981].

In many cases, protein molecules need to be included in cooperative quaternary structural arrangements of several protein molecules in order to carry out their biological roles. Often this involves aggregation of a specific small number of identical or similar proteins, held together in a definite spatial shape by some combination of intermolecular interactions that may include hydrogen bonds, hydrophobic interactions between nonpolar residues, disulfide linkages, and/or ion bridges between ionized residue side groups. The individual protein molecules remain identifiable by their own backbones, which in quaternary structures do not become chemically fused to one another by amide linkages to create a larger backbone. Case (β1) presents an example, wherein three beta-barrel porin molecules bound together laterally by hydrophilic residues create a membrane-spanning channel for passage of small molecules. Case (β3) presents another example, whose tetrameric quaternary geometry consists of two identical dimers, each one of which includes hydrogen bonding between side-by-side beta sheets connecting separate protein monomers (a situation that is known as a "cross-β conformation"). A third example is the bacterial enzyme phosphofructokinase (from *Bacillus stearothermophilus*), consisting of four identical protein units, each containing both alpha-helical and beta-sheet secondary structures [Evans et al., 1981].

In addition to these quaternary combinations of a small number of individual protein molecules, much larger protein aggregates of indefinite size also play vital roles in living organisms. Structural proteins such as actin [Otterbein et al., 2001], the collagens [Di Lullo et al., 2002], keratins [Kreplak et al., 2004], tubulins [Nogales et al., 1998], and silk fibroins [Sashina et al., 2006] are proteins that fall into this category. Although structural details vary, the individual molecules in these substances tend to be organized to constitute extended fibers rather than globular tertiary structures, a basic feature that contributes to the necessary mechanical properties demanded by their biological roles.

Unfortunately, there are also pathological cases of quaternary aggregation of large numbers of "misfolded" protein molecules [Dobson, 2003; Selkoe, 2003; Chiti and Dobson, 2006]. This situation underlies the family of "amyloid diseases," also known as "amyloidoses." Besides "mad cow" disease, this category includes several serious human disorders such as Alzheimer's disease, Parkinson's disease, Creutzfeldt-Jakob disease, Huntington's disease, and Type II diabetes. The anatomical reasons for occurrence of these pathologies, and how they might best be treated and prevented, remain important research objectives.

It is important to keep in mind that the biologically occurring proteins now available for study are the outcome of millions of years of evolutionary selection. Although they may seem large in number and in chemical diversity, they constitute an extremely small fraction of all conceivable primary structures. The 20 available amino acids in Table XI.1, even when confined to the native L chirality, could in principle be linked in 20^N distinct ways to form an N-residue polypeptide. For just the modest length $N = 50$, this amounts to approximately 10^{65} possibilities. Evidently virtually all of these, and of those for other lengths N, are biologically useless, if not lethal.

E. Native Versus Denatured States

The biologically important native configurations of proteins that are present in healthy living organisms usually occur at, or in the neighborhood of, the absolute minima of the corresponding isochoric and isobaric free energy functions $F_1(\mathbf{R} \,|\, \beta, V)$ and $G_1(\mathbf{R} \,|\, \beta, p)$. These functions include the basic effects of biologically natural solvation conditions of temperature, pressure, and solvent composition. This is an interpretation of Anfinsen's thermodynamic hypothesis for globular proteins [Anfinsen, 1973], as re-expressed in terms of the present formalism. However, broader interest concerns how various changes in solvation conditions drive the preferred protein configurations away from the native state into a distribution of denatured configurations that are sufficiently different to eliminate biological activity.

In general, the concept of "native state" may refer either to single-protein molecules that carry out their biological role in monomeric form, or it may refer to a specific number of proteins bound together in an operationally necessary quaternary state. As indicated in Section XI.D, the sodium–calcium exchanger listed as (α1) is an example of the former, and an example of the latter is the tetramer transthyretin listed as (β3).

For simplicity, assume that the system contains only a single protein or a single quaternary complex of interest. Let $\mathbf{R}$ denote the multidimensional space of configurations $\mathbf{R}$ for that single protein or protein complex under consideration. A basic assumption now to be invoked is that $\mathbf{R}$ can be divided exhaustively into two nonoverlapping subsets, $\mathbf{N}$ and $\mathbf{D}$, which represent, respectively, the configurations of native and of denatured states. In set-theoretic notation, one thus has

$$\mathbf{R} = \mathbf{N} \cup \mathbf{D}. \qquad (XI.14)$$

It should be stressed that this $\mathbf{R}$-space division does not depend on whether isochoric or isobaric conditions apply, and it is independent of solvent thermodynamic conditions; of course, those conditions affect the occupancy probabilities of the $\mathbf{N}$ and $\mathbf{D}$ subsets.

Introduction of the native-state configuration subset $\mathbf{N}$ at least superficially recalls the configuration space projection operation introduced in Section III.G to isolate metastable phases.

However, that earlier strategy involved inclusion or exclusion of entire steepest descent basins in a potential energy or potential enthalpy landscape according to the geometric character of the inherent structures within those basins. But now the $F_1(\mathbf{R}\,|\,\beta,\,V)$ and $G_1(\mathbf{R}\,|\,\beta,\,p)$ landscapes depend on temperature, pressure, and solvent composition because of the averaging over solvent degrees of freedom. Consequently, it is generally inappropriate for the present protein native-state identification problem to rely upon the changeable basin topography. Instead, $\mathbf{N}$ is to be defined simply as a domain of at-most modest configurational deformation from the ideal native configuration for the single-protein molecule or protein molecule complex under consideration. The number of free energy minima, global or relative, and their surrounding basins defined by steepest descent that the landscape places within $\mathbf{N}$ can vary widely. In principle, it is possible that $\mathbf{N}$ could contain no interior minima under ambient T and p conditions if the surrounding solvent contained a high concentration of powerful denaturant.

A basic mathematical attribute of the native-state region $\mathbf{N}$ concerns its connectivity. If a monomeric protein molecule were involved, and if that protein molecule consisted of a small number of residues (e.g., ≤ 50), then it would be reasonable to expect $\mathbf{N}$ to be a singly connected domain. The several free energy minima that might then be contained in $\mathbf{N}$ could correspond to minor configurational readjustments that would not be so large as to interfere with biological function. However, if the monomeric protein under consideration consisted of a much larger number of residues (e.g., $>1,000$), it might be possible for a substantial portion of the molecule not including its biologically active site to adopt two or more distinct folding structures without interfering with the geometry and hence the operation of that active site. But in order to switch between those alternatives, the active site at the transition state might temporarily have to be substantially disrupted. Such a scenario could indeed require $\mathbf{N}$ to be disconnected. In those cases involving quaternary binding of several identical molecules (the transthyretin and phosphofructokinase tetramers mentioned in Section XI.D are examples), $\mathbf{N}$ disconnects into separate portions that represent the distinct permutational options for the monomers in that quaternary structure (that are not interconvertible by simple overall rotation).

Given specific temperature and volume conditions for an isochoric ensemble, or alternatively specific temperature and pressure conditions for an isobaric ensemble, the respective probabilities $v(\beta,\,V)$ and $\hat{v}(\beta,\,p)$ that the protein or protein multimer is within its native state are given formally by integrals of the respective probability functions P_1 and $\hat{P}_1$ over the configurational domain $\mathbf{N}$:

$$v(\beta,\,V) = \int_{\mathbf{N}} P_1(\mathbf{R}\,|\,\beta,\,V)d\mathbf{R}, \tag{XI.15}$$

$$\hat{v}(\beta,\,p) = \int_{\mathbf{N}} \hat{P}_1(\mathbf{R}\,|\,\beta,\,p)d\mathbf{R}. \tag{XI.16}$$

These expressions depend implicitly on the chemical composition of the solvent medium in which the proteins are suspended. Under the experimentally conventional isobaric conditions, the latter of these expressions could be used in principle to construct contours in the T,p plane along which $\hat{v}$ is constant. In particular, for a solvent composition close to that in which the protein monomer or multimer operates in a living organism, there would exist a specific contour along which

$$\hat{v}(\beta = 1/k_B T,\,p) = 0.50, \tag{XI.17}$$

which would serve as a natural boundary between native and denatured regions in the T,p plane.

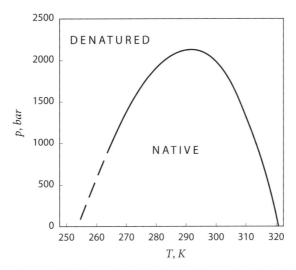

FIGURE XI.9. Approximate location of the $\hat{v} = 0.50$ boundary for staphylococcal nuclease in the T,p plane. This corresponds to the behavior of the protein as a dilute aqueous solution maintained at pH 5.5 by a phosphate buffer. Adapted from Ravindra and Winter, 2003, Figure 1.

It must be noted in passing that if the native quaternary state of a single-protein multimer is under consideration, the surrounding solvent medium must be limited in size. If not, the individual protein monomers for that multimer would spend most of their time widely separated from one another in the extended solvent and would therefore be unavailable to cooperatively produce the desired native quaternary structure. If that were the case, native-state probability $\hat{v}$ would be much smaller than 0.50 throughout the T,p plane.

The constant-$\hat{v}$ contours in the T,p plane are closely analogous to the constant elevation curves on a geographic map. Closely spaced curves on the latter indicate steep changes in elevation. Closely spaced constant-$\hat{v}$ curves indicate rapid free energy changes with respect to T,p location. If the native vs. denatured state of a protein was highly cooperative, thus presenting a rough approximation to a true phase transition, the $\hat{v} = 0.50$ curve would be densely crowded on both sides with neighboring curves of smaller and larger $\hat{v}$.

Changes in solvent composition cause shifts of the constant-$\hat{v}$ contours in the T,p plane. For a homogeneous aqueous medium, these shifts could result from a change of the pH or ionic strength, or from addition of any of several water-soluble denaturants such as urea $[(NH_2)_2C=O]$ or guanidinium salts $[(H_2N)_2C=NH_2{}^+X^-$, where $X^- = Cl^-$ or $SCN^-]$. If the denaturing effect is sufficiently strong, the result for a protein initially exhibiting a native-state region in the T,p plane could be the complete disappearance of that region.

The protein staphylococcal nuclease consists of 149 amino acid residues. It is monomeric in its native state, exhibiting a globular, rather than extended, form in its natural aquatic medium. By applying a combination of experimental probes, it has been possible to estimate its native-denatured boundary $\hat{v} = 0.50$ in the T,p plane [Ravindra and Winter, 2003]. The result is shown in Figure XI.9. The native states are confined to the interior of the $p \geq 0$ portion of a roughly inverted parabola region. Similarly shaped native-state regions have been observed for other proteins in suitably buffered aqueous media, though those regions vary considerably in size, shape, and orientation in the T,p plane [Smeller, 2002].

With a region qualitatively similar to that presented in Figure XI.9, there are three empirically distinguished routes for protein denaturation, at least in an aqueous solvent of fixed composition. Starting from a native-state location, these are

(a) heat denaturation, resulting from a sufficiently large temperature increase at constant pressure to cross the $\hat{v} = 0.50$ boundary;

(b) cold denaturation, by lowering the temperature at constant pressure so as to cross the boundary in the opposite direction from (a); and

(c) pressure denaturation, resulting from a substantial pressure increase at fixed temperature.

The first of these (a) is simply driven by the increasing entropic dominance as temperature rises of unfolded and spatially extended structures over more compact globular structures that have stabilizing interactions. The cold denaturation phenomenon (b) is generally believed to result from the fact that globular proteins in aqueous media have their nonpolar side groups collected within the interior of the native state because of hydrophobic attractive interactions; however, around room temperature and below, these hydrophobic interactions decrease in strength as temperature decreases [Section IX.I(4)], eventually failing to maintain the protein's globular native structure. The pressure denaturation phenomenon (c) involves forced insertion of water molecules into the hydrophobic cores of globular proteins, resulting in water distribution roughly analogous to clathrate formation [Section IX.C]; this disrupts the native structure enough to render it biologically inactive but not nearly so much as to produce a random coil [Hummer et al., 1998; Sarupria et al., 2010]. The configurational distributions exhibited by a protein that has respectively undergone each of the denaturing processes (a), (b), and (c) above may be qualitatively different from one another, a behavior that has been suggested by simple lattice models that involve molecularly explicit water solvent [Castrillón et al., 2012a, 2012b].

Evidence has been presented that some proteins possess essentially elliptical native-state regions in the T,p plane [Lesch et al., 2004; Wiedersich et al., 2008]. The resulting $\hat{v} = 0.50$ locus can dip into the metastable negative-pressure (tension) regime, which nevertheless could be experimentally accessible, at least in part. The existence of such a closed-loop native-state region requires adding a tension-induced denaturation path (d) to the list (a)–(c) above, whereby an isothermal pressure reduction causes disruption of the native state.

The highest pressure environment on the Earth in which living organisms have been observed is the Mariana Trench in the Pacific Ocean. This portion of the ocean floor exhibits a maximum depth below the ocean surface of approximately 11 km. The pressure there is approximately 1.1 kbar, well below the maximum pressure indicated in Figure XI.9 for at least the one protein staphylococcus nuclease in its native-state region. However, some other proteins in living organisms occupying more conventional environmental conditions are less resistant to pressure denaturing. In order to produce living species able to withstand pressure extremes, evolution would have had to mutate some protein primary structures into alternatives possessing greater pressure resistance. Similar remarks can be made concerning the high temperatures near hydrothermal vents both above and below ocean surfaces, at which life forms have evolved to include proteins with unusual resistance to heat denaturing [e.g., Takai et al., 2008].

Isothermal kinetics of any polypeptide or specific protein, whether under isochoric or isobaric conditions, can be represented as a classical-mechanics trajectory evolving, respectively, on a free energy surface $F_1(\mathbf{R}|\beta, V)$ or $G_1(\mathbf{R}|\beta, p)$. These trajectories are stochastic because of implicit coupling to solvent. Consequently, any initial configuration and momentum condition of a polypeptide on the free energy surface inevitably spreads out into a distribution of $\mathbf{R}$ positions at subsequent times.

It has been argued that in respect to free energy surfaces biological proteins constitute a special subset of the entire polypeptide family because of their necessity of being able to fold into a useful and stable native state in a reasonable time following ribosome synthesis. This has led to advocating a rough-surfaced "folding funnel" picture of $F_1(\mathbf{R}|\beta, V)$ and $G_1(\mathbf{R}|\beta, p)$ landscapes for globular proteins [Onuchic et al., 1997, Figure 4]. This view represents a native-state domain $\mathbf{N}$ residing at the bottom of the funnel, into which $\mathbf{R}$-space trajectories starting within the dominant high and wide portion of the funnel would eventually fall if the temperature and volume or pressure conditions corresponded to the interior of a native-state domain such as that illustrated in Figure XI.9. Although such a visual image is appealing, its simplicity may be misleading. Computer simulations of small foldable polypeptides indicate that the native-state domain $\mathbf{N}$ can include a significant collection of inequivalent landscape minima (inherent structures) connected by a complex network of intradomain transition paths [Rao and Karplus, 2010]. Furthermore, a simple two- or three-dimensional "cartoon" of a folding funnel can disguise nonobvious protein geometric features underlying a free energy landscape. In particular, the proper landscape topography can involve restrictive pathways from extended denatured protein configurations to the native-state domain. This feature may become especially significant for the minority of proteins with native states that are knotted or that exhibit slip-knot entanglements [Sułkowska et al., 2012b].

For many proteins whose native state is a compact globular configuration, the transition to that native state from an extended denatured configuration involves passing through a region of configuration space that has been called a "molten globule state" [Arai and Kuwajima, 2000]. The molten globule state exhibits most or all of the native-state secondary structure, but its tertiary structure has not yet been properly assembled. The threefold distinction between native, molten globule, and fully denatured states is roughly analogous to the distinction between crystalline, liquid, and vapor phases for a low-molecular-weight substance, but of course an isolated protein molecule cannot strictly exhibit first-order phase transitions.

Recognition of a molten globule state suggests splitting the $\mathbf{R}$-space domain $\mathbf{D}$ into two portions:

$$\mathbf{D} = \mathbf{M} \cup \mathbf{U}, \tag{XI.18}$$

where $\mathbf{M}$ collects all molten globule configurations, and $\mathbf{U}$ collects all remaining denatured configurations. It is currently unclear if $\mathbf{M}$ would always entirely surround the native-state domain $\mathbf{N}$ for any globular protein, thus eliminating any direct kinetic path between $\mathbf{N}$ and $\mathbf{U}$. By analogy with the definitions in Eqs. (XI.15) and (XI.16), the respective occupancy probabilities in the molten globule domain would have the following forms:

$$v_m(\beta, V) = \int_{\mathbf{M}} P_1(\mathbf{R}|\beta, V)d\mathbf{R}, \tag{XI.19}$$

$$\hat{v}_m(\beta, p) = \int_{\mathbf{M}} \hat{P}_1(\mathbf{R}|\beta, p)d\mathbf{R}. \tag{XI.20}$$

F. Rudimentary Models

The structural diversity of proteins and the complexity of interactions that determine protein native states and their regions of stability present formidable challenges to understanding. This makes it attractive to develop rudimentary theoretical models that are designed to illuminate qualitatively at least a restricted set of basic phenomena. With simplicity comes clarity, at least to a limited extent. Comprehensive understanding of such stripped-down versions in principle can help to guide effective theoretical and numerical examination of more detailed realistic descriptions. This section presents two species of simplified models to highlight their modest virtues and restrictions. They represent specific examples from a large number of possibilities that would select specific aspects of the general protein folding problem on which to focus [Chan et al., 2011; Whitford et al., 2012]. Sequentially adding various levels of increased sophistication to these rudimentary models could incrementally bridge the complexity gap toward the full statistical mechanical description outlined in the preceding sections.

(1) HP LATTICE MODELS

Considering how useful simple lattice models have proved to be for representing the chain statistics of nonbiological polymers, it was natural that an analogous view would be developed for the protein folding problem. One of the first of these to be explored was the elementary "HP" model [Lau and Dill, 1989]. In its original version, this approach represented polypeptides and proteins as self-avoiding walks on a two-dimensional square lattice. These walks are nonbranching paths of fixed length consisting of nearest-neighbor lattice-site jumps, with free ends. The symbols "H" and "P" designate two types of vertices occurring along the walk, respectively representing hydrophobic and polar residues. This H vs. P vertex distinction presumably adheres at least roughly to the classification of amino acids, respectively, as "nonpolar" vs. "polar" in Table XI.2. Solvent is assumed implicitly to fill the lattice sites not occupied by the self-avoiding HP "molecules." Figure XI.10 provides some length-15 examples of the HP model's molecules. By confining the molecules to a lattice, the continuum space of configurations $\mathbf{R}$ becomes approximated by the discontinuous space of self-avoiding walks.

The HP model focuses attention on hydrophobic interactions. It assigns a fixed negative free energy $-\varepsilon_{HH}$ to every intramolecular HH pair of nearest neighbor vertices that are not directly connected ("chemically bonded") along the backbone. If more than one molecule is present, the same hydrophobic interaction $-\varepsilon_{HH}$ would be assigned to each nearest neighbor intermolecular HH contact. Because the hydrophobic effect arises from the implicit solvent, ε_{HH} can be assigned a temperature dependence. No interaction is assigned to any of the HP or PP nonbonded pairs of nearest neighbor vertices. Furthermore, no energy is involved distinguishing whether two successive backbone bonds are in a straight or a 90° bent relative orientation. Similarly, no free energy assignment is present to distinguish the two-dimensional versions of a dihedral angle for the relative configuration of three successive bonds. With these simplifications, the corresponding intramolecular free energies for the structures illustrated in Figure XI.10 are (a) $-6\varepsilon_{HH}$; (b) $-3\varepsilon_{HH}$; and (c) $-\varepsilon_{HH}$.

Obviously, the chiral distinctions between D and L amino acid residues that characterize real polypeptides and proteins have been entirely suppressed in this elementary HP model. So too has backbone directionality been lost. Nevertheless, there remain nontrivial statistical mechanical

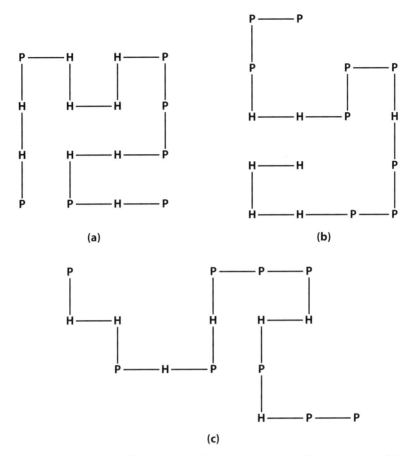

FIGURE XI.10. Examples of length-15 self-avoiding walks (16 vertices) on the two-dimensional square lattice that would be considered by the basic "HP" model.

attributes of the HP model that have received substantial research attention. At least for the moderate backbone lengths that for single molecules could be exhaustively enumerated by computer, "native states" have been identified as those configurations of isolated single walks exhibiting global minimum free energy because of their intramolecular HH contacts. For some HP sequences along the backbone, the global minimum free energy arises from a single folding arrangement (aside from trivial symmetry operations on the lattice). For other HP sequences, the global minimum free energy of a single molecule is degenerate, i.e., attained by two or more distinct folding configurations [Lau and Dill, 1989].

The growth of computing power has permitted study of single isolated molecules on the square lattice with selected HP model sequences in the length range $>10^2$ under variable temperature conditions using Monte Carlo techniques. It has also made it feasible to extend the HP model to a three-dimensional version that resides on the simple cubic lattice [Zhang et al., 2007; Wüst and Landau, 2009], as well as on the face-centered cubic lattice [Backofen, 2004]. In addition, a variant of the HP lattice model that decorates the backbone with side chains has been examined [Hart and Istrail, 1997]. In any of these versions, locating the minimum free energy configurations rises rapidly in difficulty as the size of the molecule increases, and in fact it has been demonstrated that this minimization search is classified by mathematicians as "NP-complete" [Berger and Leighton, 1998].

Let S_n denote a specific ordered HP backbone sequence consisting of n vertices ($n - 1$ bonds). Suppose that this single molecule resides on a portion of a lattice that is subject to periodic boundary conditions. Furthermore, it is assumed that this portion containing M lattice sites is large enough in each direction that even when fully extended the n-vertex molecule cannot encounter its own periodic images. The canonical configurational partition function Z at temperature $T = (k_B T)^{-1}$ for this HP system and the associated free energy F may be expressed in the following manner:

$$Z(\beta \,|\, S_n) = M \sum_{j \geq 0} D(j \,|\, S_n) \exp(\beta \varepsilon_{HH} j), \tag{XI.21}$$

$$\equiv \exp[-\beta F(\beta \,|\, S_n)].$$

Here index j represents the number of intramolecular nearest neighbor hydrophobic contacts, and the degeneracy quantity $D(j \,|\, S_n)$ enumerates the number of distinct configurations with exactly j contacts where one of the backbone ends has been confined to a specific lattice site. The presummation factor M arises from the fact that this enumeration is the same regardless of which lattice site has been chosen for the backbone-end confinement. The heat capacity increment ΔC arising from the presence of the protein molecule can subsequently be extracted from the partition function

$$\Delta C/k_B = \beta^2 (\partial^2 \ln Z/\partial \beta^2)$$

$$= -\beta^2 \left(\frac{\partial^2 \beta F}{\partial \beta^2} \right)$$

$$= \left[\beta \left(\frac{\partial \beta \varepsilon_{HH}}{\partial \beta} \right) \right]^2 [\langle j^2 \rangle - \langle j \rangle^2] + \beta^2 \left(\frac{\partial^2 \beta \varepsilon_{HH}}{\partial \beta^2} \right) \langle j \rangle. \tag{XI.22}$$

This expression involves the following temperature-dependent, degeneracy-weighted moments of the contact number j:

$$\langle j^q \rangle = \left[\sum_{j \geq 0} j^q D(j \,|\, S_n) \exp(\beta \varepsilon_{HH} j) \right] / \left[\sum_{j \geq 0} D(j \,|\, S_n) \exp(\beta \varepsilon_{HH} j) \right]. \tag{XI.23}$$

Beyond study of the thermodynamic properties for any n-site molecule, the HP model offers the possibility of monitoring the geometric characteristics of its molecules as functions of backbone length, HP sequence along the backbone, and temperature. Obviously, such characteristics are the root-mean-square (rms) radius of gyration $\langle (R_g)^2 \rangle^{1/2}$ and the rms end-to-end distance $\langle (R_e)^2 \rangle^{1/2}$, attributes introduced in Section X.D for general polymers. The extent of backbone interpenetration of neighboring molecules in more concentrated systems is also accessible in principle.

(2) "OFF-LATTICE" AB MODELS

The separation of amino acid residues into two coarse categories has also been invoked for some simple polypeptide–protein models for which the space of configurations **R** is indeed continuous,

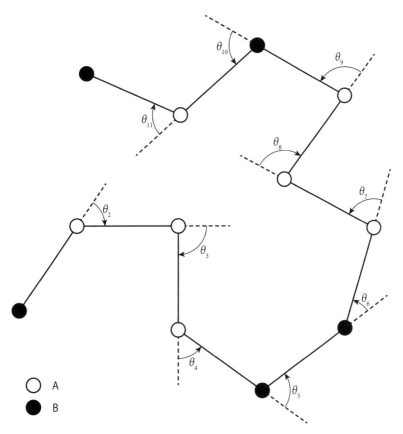

FIGURE XI.11. Representative example of a 12-residue molecule in the two-dimensional version of the AB model. The ten angles $\theta_2, \ldots, \theta_{11}$, positive for clockwise bond rotations, are required to specify the backbone geometry. Open and filled circles represent A and B residues, respectively.

not discrete. Consequently, these latter rudimentary models have been described as "off-lattice." Removal of the lattice constraint allows for inclusion of a wider, more realistic range of backbone geometries. The "residues" are once again limited to two types, now labeled "A" and "B," because they do not necessarily represent exactly the same broad categories of residues as "H" and "P" do in the lattice models described in Section XI.F(1). In addition to relaxing the lattice-occupancy constraint, the off-lattice AB models also assign nonzero pair interactions to all three pair types. As is the case with the HP models, the AB models have been examined in both two- and three-dimensional versions.

In this off-lattice model approach, an n-residue molecule involves a backbone composed of a sequence of $n - 1$ fixed-length bonds. The A and B residue units occupy the $n - 2$ bond connection vertices along the backbone, as well as at the two ends. Figure XI.11 presents a schematic illustration of a molecule in the two-dimensional version of the model. The A and B residue types are illustrated, respectively, by small open and filled circles. Continuous bond bending along the backbone can be specified in two dimensions, as shown in Figure XI.11 by a set of angles:

$$-\pi < \theta_i < +\pi \qquad (2 \leq i \leq n - 1), \qquad (XI.24)$$

where index $1 \leq i \leq n$ identifies all of the residue positions sequentially along the backbone, positive angles represent clockwise rotations as shown in the Figure XI.11, and by definition $\theta_i = 0$ indicates no bend at residue i from one bond to the next.

An n-residue molecule involves an ordered sequence of A and B residues along the backbone, denoted for this model by $\tilde{S}_n$. Any such sequence can be encoded by a set of binary indices $\xi_1, \ldots,$ ξ_n, each of which adopts a value as follows:

$$\xi_i(A) = 1 \text{ (residue } i \text{ is A)}, \tag{XI.25}$$

$$\xi_i(B) = -1 \text{ (residue } i \text{ is B)}.$$

Then in the original two-dimensional version of the AB model, the following potential energy function is postulated for an isolated molecule [Stillinger et al., 1993]:

$$\Phi(\mathbf{R} \,|\, \tilde{S}_n) = \sum_{i=2}^{n-1} V_1(\theta_i) + \sum_{i=1}^{n-2} \sum_{j=i+2}^{N} V_2(r_{ij}, \xi_i, \xi_j). \tag{XI.26}$$

As the summation limits indicate, pair interactions V_2 occur only for residues not directly bonded along the backbone. After choosing appropriate energy units, and the fixed bond length as the length unit, the interaction functions were assigned the following specific forms:

$$V_1(\theta_i) = \left(\frac{1}{4}\right)(1 - \cos\theta_i), \tag{XI.27}$$

which has a minimum at the $\theta_i = 0$ unbent configuration; and

$$V_2(r_{ij}, \xi_i, \xi_j) = 4[r_{ij}^{-12} - f(\xi_i, \xi_j)r_{ij}^{-6}], \tag{XI.28}$$

where the modification parameter f in this Lennard-Jones-type pair interaction is

$$f(\xi_i, \xi_j) = \left(\frac{1}{8}\right)(1 + \xi_i + \xi_j + 5\xi_i\xi_j). \tag{XI.29}$$

This last expression takes on values $+1$, $+1/2$, and $-1/2$ for AA, BB, and AB pairs, respectively, corresponding roughly to strong attraction, weak attraction, and weak repulsion for the pair potentials. These pair interactions can serve equally well for intermolecular interactions, but the studies of the two-dimensional AB model thus far have concentrated on the properties of isolated single molecules. As was pointed out in Section XI.F(1) for HP models, the AB model molecules also fail to represent both residue chirality and intrinsic backbone directionality. Studies of the AB models thus far have not assigned any temperature dependence to the functions V_1 and V_2, although a fluid solvent medium is implicitly assumed to be present.

Global Φ minima in two dimensions have been identified for all sequences $\tilde{S}_n$ in the range $3 \leq n \leq 7$ [Stillinger et al., 1993]. In addition, global Φ minima have also been identified for "center-doped" sequences $(A)_l B(A)_l$, where $4 \leq l \leq 12$ [Stillinger and Head-Gordon, 1995]. For some short AB sequences, Φ attains its global minimum for a rigorously straight backbone (all $\theta_i = 0$), but this attribute disappears for large sequences. Nevertheless, it appears to be true that every AB sequence, regardless of length, has at least a relative Φ minimum when all $\theta_i = 0$.

Although compressing the number of distinguishable residues from the natural amino acid number 20 to merely 2 in the HP and AB models is a drastic simplification, the numbers of distinct

BOX XI.2. Fibonacci Sequence

The Fibonacci sequence, originating in the medieval period, is an infinite sequence of integers F_n. Its mathematical properties are diverse and remain a source of research interest even at present [Hoggatt, 1969]. Given a starting pair of terms

$$F_0 = 0, F_1 = 1,$$

the remainder of the numerical sequence is generated by the following rule:

$$F_{n+2} = F_n + F_{n+1}.$$

The leading members of the sequence are the following:

0, 1, 1, 2, 3, 5, 8, 13, 21, 34, 55, 89, 144, 233, 377, 610, 987, 1597, 2584, 4181, ...

The above generation rule can also be used formally to extend the sequence in the opposite direction ($n < 0$), which results in the same integers but with alternating signs. A basic property of the numerical Fibonacci sequence is its relation to the "golden ratio" (a.k.a. "golden mean"):

$$\lim_{n \to +\infty} (F_{n+1}/F_n) = (1 + 5^{1/2})/2$$

$$\approx 1.618034.$$

For the purposes of protein modeling, it is useful to translate the sequence-generating rule from arithmetic addition to symbolic sequence concatenation. For this version, one can formally set

$$\tilde{S}_0 = A, \tilde{S}_1 = B,$$

and then for $n > 1$,

$$\tilde{S}_{n+2} = \tilde{S}_n * \tilde{S}_{n+1},$$

where the symbol $*$ indicates the concatenation. This protocol generates the following strings:

A, B, AB, BAB, ABBAB, BABABBAB, ABBABBABABBAB,
BABABBABABBABBABABBAB,
ABBABBABABBABBABABBABBABABBAB, ...

This procedure does not have a useful analog to the negative-direction extension indicated above for the numerical Fibonacci sequence. The total number of A and B letters in string $\tilde{S}_n$ for $n \geq 0$ is F_{n+1}. The HP model described in Section XI.F(1) above includes the possibility of similar sequences, where A and B are rewritten as H and P, or even as P and H, respectively.

It is clear from the above alphabetical strings that the Fibonacci algorithm distributes the A and B "residues" rather uniformly along the sequences. In particular, each interior A is immediately preceded and followed by a B, and the interior B residues appear only as isolated single or double units immediately preceded and followed by A residues. In any one of the $\tilde{S}_n$ sequences, the respective numbers of A and B residues are equal to successive integers from the F_n numerical sequence shown above, so that the ratio of B to A concentrations approaches the golden ratio in the long-sequence asymptotic limit.

In view of the fact that the AB model treats the pair interactions, Eq. (XI.28), asymmetrically with respect to A and B, it would be instructive to examine the model's predictions for those other molecules for which notational inversion A, B → B, A had been imposed on the Fibonacci sequence. An obvious result of such an inversion is that the A residues become the dominant species in the longer sequences.

---

sequences S_n and $\tilde{S}_n$ nevertheless still rise exponentially with increasing n. Consequently, it becomes virtually impossible to produce detailed thermodynamic and kinetic characterizations of all AB-model molecules when n exceeds approximately 12. For that reason, it has been sensible to select a specific well-defined family of sequences with unbounded length that could be a common research object. Such a family of sequences should avoid trivial patterns (i.e., all A or all B). In particular, the venerable and well-publicized Fibonacci sequence generates a good example; see Box XI.2 for details. There are published results for the global Φ minima of Fibonacci sequences $\tilde{S}_n$ up to $n = 55$ [Hsu et al., 2003; Elser and Rankenburg, 2006; Liu et al., 2009]. If for no other reason, these specific sequences provide a test bed for the relative effectiveness of numerical search routines that might be applied more generally to the protein configurational problem.

Temperature dependence of the folding and unfolding properties of isolated AB molecules has received some attention [Irbäck and Potthast, 1995; Irbäck et al., 1997a]. Perhaps not surprisingly, a conclusion is that for sequences of a given length n, the folding rate into a "native" state as temperature declines is strongly dependent on sequence details. This is consistent with the generally held opinion that biologically useful polypeptides are far from random in their residue sequences.

The off-lattice AB model can be directly extended from two dimensions to three dimensions, utilizing without change the potential energy function Φ defined in Eqs. (XI.26) to (XI.29) [Elser and Rankenburg, 2006]. The result contains no dihedral angle potential energy contribution from each sequence of three bonds along the backbone. Increasing the dimension for a given sequence $\tilde{S}_n$ tends to increase the number of Φ minima (inherent structures). Although there are exceptions such as the mechanically stable linear structures, when passing to the three-dimensional model extension, many of the inherent structures for the two-dimensional model become planar saddle point configurations of various orders connecting neighboring basins for chiral inherent structures in the resulting configuration space.

Other less direct extensions of the AB model from two to three dimensions have been proposed and examined. One of these includes dihedral angle potentials. It retains the simplification of two distinguishable "residues" but changes the residue pair interactions so that all three become proportional to the standard Lennard-Jones 12,6 potential [Irbäck et al., 1997b].

References

Abascal, J.L.F., and Vega, C. (2005). A General Purpose Model for the Condensed Phases of Water: TIP4P/2005. *J. Chem. Phys.* 123, 234505.

Abel, W.R., Anderson, A.C., and Wheatley, J.B. (1966). Propagation of Zero Sound in Liquid He^3 at Low Temperatures. *Phys. Rev. Lett.* 17, 74–78.

Abraham, F.F. (1974). *Homogeneous Nucleation Theory. The Pretransition Theory of Vapor Condensation.* New York: Academic Press.

Abramowitz, M., and Stegun, I.A. (1964). *Handbook of Mathematical Functions.* National Bureau of Standards Applied Mathematics Series 55. Washington, DC: U.S. Government Printing Office.

Adam, G., and Gibbs, J.H. (1965). On the Temperature Dependence of Cooperative Relaxation Properties in Glass-Forming Liquids. *J. Chem. Phys.* 43, 139–146.

Ahlers, G. (1971). Heat Capacity near the Superfluid Transition in He^4 at Saturated Vapor. *Phys. Rev. A* 3, 696–716.

Ahlers, G. (1973). Thermodynamic and Experimental Tests of Static Scaling and Universality near the Superfluid Transition in He^4 under Pressure. *Phys. Rev. A* 8, 530–568.

Ahlers, G. (1980). Critical Phenomena at Low Temperature. *Rev. Modern Phys.* 52, 489–503.

Ahlrichs, R., Penco, P., and Scoles, G. (1976). Intermolecular Forces in Simple Systems. *Chem. Phys.* 19, 119–130.

Alberts, B. (2008). *Molecular Biology of the Cell*, 5th edition. New York: Garland Science.

Albrecht, T., Elben, H., Jaeger, R., Kimmig, M., Steiner, R., Strobl, G., and Stühn, B. (1991). Molecular Dynamics of Perfluoro-n-Eicosane. II. Components of Disorder. *J. Chem. Phys.* 95, 2807–2816.

Alder, B.J., and Wainwright, T.E. (1957). Phase Transition for a Hard Sphere System. *J. Chem. Phys.* 27, 1208–1209.

Alder, B.J., and Wainwright, T.E. (1970). Decay of the Velocity Autocorrelation Function. *Phys. Rev. A* 1, 18–21.

Aldrich, L.T., and Nier, A.O. (1948). The Occurrence of He^3 in Natural Sources of Helium. *Phys. Rev.* 74, 1590–1594.

Alexander, S., and McTague, J. (1978). Should All Crystals Be bcc? Landau Theory of Solidification and Crystal Nucleation. *Phys. Rev. Lett.* 41, 702–705.

Alvesalo, T.A., Haavasoja, T., and Manninen, M.T. (1981). Specific Heat of Normal and Superfluid ^{3}He. *J. Low Temp. Phys.* 45, 373–405.

Anderson, P.W., Halperin, B.I., and Varma, C.M. (1972). Anomalous Low-Temperature Properties of Glasses and Spin Glasses. *Philos. Mag.* 25, 1–9.

Anderson, V.J., and Lekkerkerker, H.N.W. (2002). Insights into Phase Transition Kinetics from Colloid Science. *Nature* 416, 811–815.

Andreev, A.F. (1964). Singularity of Thermodynamic Quantities at a First Order Phase Transition Point. *Soviet Phys. J.E.T.P.* 18, 1415–1416.

Anfinsen, C.B. (1973). Principles That Govern the Folding of Protein Chains. *Science* 181, 223–230.

Angelani, L., Di Leonardo, R., Ruocco, G., Scala, A., and Sciortino, F. (2002). Quasisaddles as Relevant Points of the Potential Energy Surface in the Dynamics of Supercooled Liquids. *J. Chem. Phys.* 116, 10297–10306.

Angell, C.A. (1982). Supercooled Water. In *Water, a Comprehensive Treatise,* Vol. 7, F. Franks, Editor. New York: Plenum Press, 1–81.

Angell, C.A. (1985). Spectroscopy Simulation and Scattering, and the Medium Range Order Problem in Glass. *J. Non-Cryst. Solids* 73, 1–17.

Angell, C.A. (1988). Structural Instability and Relaxation in Liquid and Glassy Phases near the Fragile Liquid Limit. *J. Non-Cryst. Solids* 102, 205–231.

Angell, C.A. (1990). Dynamic Processes in Ionic Glasses. *Chem. Rev.* 90, 523–542.

Angell, C.A., Ngai, K.L., McKenna, G.B., McMillan, P.F., and Martin, S.W. (2000). Relaxation in Glassforming Liquids and Amorphous Solids. *J. Appl. Phys.* 88, 3113–3157.

Angell, C.A., and Rao, K.J. (1972). Configurational Excitations in Condensed Matter, and the "Bond Lattice" Model for the Liquid–Glass Transition. *J. Chem. Phys.* 57, 470–481.

Angell, C.A., Shuppert, J.C., and Tucker, J.C. (1973). Anomalous Properties of Supercooled Water. Heat Capacity, Expansivity, and Proton Magnetic Resonance Chemical Shift from 0 to –38°. *J. Phys. Chem.* 77, 3092–3099.

Angell, C.A., and Smith, D.L. (1982). Test of the Entropy Basis of the Vogel-Tammann-Fulcher Equation. Dielectric Relaxation of Polyalcohols near T_g. *J. Phys. Chem.* 86, 3845–3852.

Arai, M., and Kuwajima, K. (2000). Role of the Molten Globule State in Protein Folding. *Adv. Protein Chem.* 53, 209–282.

Arkus, N., Manoharan, V.M., and Brenner, M.P. (2009). Minimal Energy Clusters of Hard Spheres with Short Range Attractions. *Phys. Rev. Lett.* 103, 118303.

Arms, D.A., Shah, R.S., and Simmons, R.O. (2003). X-Ray Debye-Waller Factor Measurements of Solid ^{3}He and ^{4}He. *Phys. Rev. B* 67, 094303.

Ashcroft, N.W., and Mermin, N.D. (1976). *Solid State Physics.* London: Thomson Learning.

Asmis, K.R., Santambrogio, G., Zhou, J., Garand, E., Headrick, J., Goebbert, D., Johnson, M.A., and Neumark, D.M. (2007). Vibrational Spectroscopy of Hydrated Electron Clusters $(H_2O)_{15-20}^-$ via Infrared Multiple Photon Dissociation. *J. Chem. Phys.* 126, 191105.

Atkins, K.R. (1959). *Liquid Helium.* Cambridge, U.K.: Cambridge University Press.

Atkins, P.W. (1970). *Molecular Quantum Mechanics, An Introduction to Quantum Chemistry,* Vol. II. Oxford, U.K.: Clarendon Press, 456–460.

Avramov, I,. and Milchev, A. (1988). Effect of Disorder on Diffusion and Viscosity in Condensed Systems. *J. Non-Cryst. Solids* 104, 253–260.

Aziz, R.A., McCourt, F.R.W., and Wong, C.C.K. (1987). A New Determination of the Ground State Interatomic Potential for He_2. *Mol. Phys.* 61, 1487–1511.

Aziz, R.A., Nain, V.P.S., Carley, S., Taylor, W.L., and McConville, G.T. (1979). An Accurate Intermolecular Potential for Helium. *J. Chem. Phys.* 70, 4330–4342.

Aziz, R.A., and Slaman, M.J. (1991). An Examination of Ab Initio Results for the Helium Potential Energy Curve. *J. Chem. Phys.* 94, 8047–8053.

Backofen, R. (2004). A Polynomial Time Upper Bound for the Number of Contacts in the HP-Model on the Face-Centered-Cubic Lattice (FCC). *J. Discrete Algorithms* 2, 161–206.

Band, W. (1939a). Dissociation Treatment of Condensing Systems. *J. Chem. Phys.* 7, 324–326.

Band, W. (1939b). Dissociation Treatment of Condensing Systems. II. *J. Chem. Phys.* 7, 927–931.

Bardeen, J., Cooper, L.N., and Schrieffer, J.R. (1957). Theory of Superconductivity. *Phys. Rev.* 108, 1175–1204.

Bartell, L.S., and Huang, J. (1994). Supercooling of Water below the Anomalous Range near 226 K. *J. Phys. Chem.* 98, 7455–7457.

Baym, G., and Lamb, F.K. (2005). Neutron Stars. In *Encyclopedia of Physics*, 3rd edition, Vol. 2, R.G. Lerner and G.L. Trigg, editors. Weinheim, Germany: Wiley-VCH, 1722.

Beaumont, R.H., Chihara, H., and Morrison, J.A. (1961). Transitions between Different Forms of Ice. *J. Chem. Phys.* 34, 1456–1457.

Becker, O.M., and Karplus, M. (1997). The Topology of Multidimensional Potential Energy Surfaces: Theory and Application to Peptide Structure and Kinetics. *J. Chem. Phys.* 106, 1495–1517.

Bell, R.A., Cristoph, G.G., Fronczek, F.R., and Marsh, R.E. (1975). The Cation $H_{13}O_6^+$: A Short Symmetric Hydrogen Bond. *Science* 190, 151–152.

Bellissent, R., Descotes, L., Boué, F., and Pfeuty, F. (1990). Liquid Sulfur: Local-Order Evidence of a Polymerization Transition. *Phys. Rev. B* 41, 2135–2138.

Bellissent-Funel, M.C., Bosio, L., Hallbrucker, A., Mayer, E., and Sridi-Dorbez, R. (1992). X-Ray and Neutron Scattering Studies of the Structure of Hyperquenched Glassy Water. *J. Chem. Phys.* 97, 1282–1286.

Ben-Naim, A. (1980). *Hydrophobic Interactions.* New York: Plenum Press.

Berendsen, H.J.C., Grigera, J.R., and Straatsma, T.P. (1987). The Missing Term in Effective Potentials. *J. Phys. Chem.* 91, 6269–6271.

Berger, B., and Leighton, T. (1998). Protein Folding in the Hydrophobic–Hydrophilic (HP) Model Is NP-Complete. *J. Comput. Biol.* 5, 27–40.

Berkholtz, D.S., Driggers, C.M., Shapovalov, M.V., Dunbrack, R.L., Jr., and Karplus, P.A. (2012). Nonplanar Peptide Bonds in Proteins Are Common and Conserved But Not Biased toward Active Sites. *Proc. Natl. Acad. Sci. USA* 109, 449–453.

Berman, H.M. (2008). The Protein Data Bank: A Historical Perspective. *Acta Cryst. A* 64, 88–95.

Berman, R., Klemens, P.G., Simon, F.E., and Fry, T.M. (1950). Effect of Neutron Irradiation on the Thermal Conductivity of a Quartz Crystal at Low Temperature. *Nature* 166, 864–866.

Bernal, J.D. (1964). The Bakerian Lecture 1962. The Structure of Liquids. *Proc. R. Soc. London, Ser. A. Math. Phys. Sci.* 280, 299–322.

Bernal, J.D., and Fowler, R.H. (1933). A Theory of Water and Ionic Solution, with Particular Reference to Hydrogen and Hydroxyl Ions. *J. Chem. Phys.* 1, 515–548.

Berry, G.C., and Fox, T.G. (1968). The Viscosity of Polymers and Their Concentrated Solutions. *Adv. Polymer Sci.* 5, 261–357.

Berry, R.S., Davis, H.L., and Beck, T.L. (1988). Finding Saddles on Multidimensional Potential Surfaces. *Chem. Phys. Lett.* 147, 13–17.

Bett, K.E., and Cappi, J.B. (1965). Effect of Pressure on the Viscosity of Water. *Nature* 207, 620–621.

Betts, D.S. (1974). Pomeranchuk Cooling by Adiabatic Solidification of Helium-3. *Contemp. Phys.* 15, 227–247.

Bhattacharjee, A., and Bansal, M. (2005). Collagen Structure: The Madras Triple Helix and the Current Scenario. *IUBMB Life* 57, 161–172.

Bich, E., Millat, J., and Vogel, E. (1988). On the He–He Interatomic Potential and Related Properties. *Z. Phys. Chem.—Leipzig* 269, 917–924.

Biswas, R., and Hamann, D. (1987). New Classical Models for Silicon Structural Energies. *Phys. Rev. B* 36, 6434–6445.

Bjerrum, N. (1952). Structure and Properties of Ice. *Science* 115, 385–390.

Blackmond, D.G. (2004). Asymmetric Autocatalysis and Its Implications for the Origin of Homochirality. *Proc. Natl. Acad. Sci. USA* 101, 5732–5736.

Blake, C.C.F. (1981). Prealbumin and the Thyroid Hormone Nuclear Receptor. *Proc. R. Soc. London [Biol.]* 211, 413–431.

Blake, C.C.F., Geisow, M.J., Swan, I.D.A., Rerat, C., and Rerat, B. (1974). Structure of Human Plasma Prealbumin at 2.5 Å Resolution. A Preliminary Report on the Polypeptide Chain Conformation, Quaternary Structure and Thyroxine Binding. *J. Mol. Biol.* 88, 1–12.

Bloch, F. (1932). Zur Theorie des Austauschproblems und der Remanenzerscheinung der Ferromagnetika. *Zeit. Physik* 74, 295–335.

Bluhm, H., Ogletree, D.F., Fadley, C.S., Hussain, Z., and Salmeron, M. (2002). The Premelting of Ice Studied with Photoelectron Spectroscopy. *J. Phys.: Condens. Matter* 14, L227–L233.

Boese, R., Weiss, H.-C., and Bläser, D. (1999). The Melting Point Alternation in the Short-Chain *n*-Alkanes: Single-Crystal X-Ray Analyses of Propane at 30 K and of *n*-Butane to *n*-Nonane at 90 K. *Angew. Chem. Int. Ed.* 38, 988–992.

Bond, A.D., and Davies, J.E. (2002). *n*-Decane. *Acta Cryst. E* 58, 196–197 (online).

Born, M., Heisenberg, W., and Jordan, P. (1925–1926). Zur Quantenmechanik. II. *Zeit. Physik* 35, 557–615.

Born, M., and Oppenheimer, J.R. (1927). 1. Zur Quantentheorie der Molekeln. *Ann. d. Phys.* 84, 457–484.

Botti, A., Bruni, F., Isopo, A., Ricci, M.A., and Soper, A.K. (2002). Experimental Determination of the Site–Site Radial Distribution Functions of Supercooled Ultrapure Bulk Water. *J. Chem. Phys.* 117, 6196–6199.

Bove, L.E., Klotz, S., Paciaroni, A., and Sacchetti, F. (2009). Anomalous Proton Dynamics in Ice at Low Temperature. *Phys. Rev. Lett.* 103, 165901.

Bowie, J.U. (2005). Solving the Membrane Protein Folding Problem. *Nature* 438, 581–589.

Bowron, D.T., Filiponi, A., Roberts, M.A., and Finney, J.L. (1998). Hydrophobic Hydration and the Formation of a Clathrate Hydrate. *Phys. Rev. Lett.* 81, 4164–4167.

Bragg, W.L., and Williams, E.J. (1934). The Effect of Thermal Agitation on Atomic Arrangement in Alloys. *Proc. R. Soc. (London)* A145, 699–730.

Braier, P.A., Berry, R.S., and Wales, D.J. (1990). How the Range of Pair Interactions Governs Features of Multidimensional Potentials. *J. Chem. Phys.* 93, 8745–8756.

Bramwell, S.T., and Gingras, M.J.P. (2001). Ice State in Frustrated Magnetic Pyrochlore Materials. *Science* 294, 1495–1501.

Breslow, R., and Cheng, Z.-L. (2009). On the Origin of Terrestrial Homochirality for Nucleosides and Amino Acids. *Proc. Natl. Acad. Sci. USA* 106, 9144–9146.

Bridgeman, P.W. (1935). The Pressure–Volume–Temperature Relations of the Liquid and the Phase Diagram of Heavy Water. *J. Chem. Phys.* 3, 597–605.

Briggs, J.M., Nguyen, T.B., and Jorgensen, W.L. (1991). Monte Carlo Simulations of Liquid Acetic Acid and Methyl Acetate with the OPLS Potential Functions. *J. Phys. Chem.* 95, 3315–3322.

Brillouin, L. (1953). *Wave Propagation in Periodic Structures.* New York: Dover.

Broderix, K., Bhattacharya, K.K., Cavagna, A., Zippelius, A., and Giardina, I. (2000). Energy Landscape of a Lennard-Jones Liquid: Statistics of Stationary Points. *Phys. Rev. Lett.* 85, 5360–5363.

Brodskaya, E.N., and Zakharov, V.V. (1995). Computer Simulation Study of the Surface Polarization of Pure Polar Liquids. *J. Chem. Phys.* 102, 4595–4599.

Brooks, B.R., Brooks, C.L., III, Mackerell, A.D., Jr., Nilsson, L., Petrella, R.J., Roux, B., Won, Y., Archontis, G., Bartels, C., Boresch, S., Caflisch, A., Caves, L., Cui, Q., Dinner, A.R., Feig, M., Fischer, S., Gao, J., Hodoscek, M., Im, W., Kuczera, K., Lazaridis, T., Ma, J., Ovchinnikov, V., Paci, E., Pastor, R.W., Post, C.B., Pu, J.Z., Schaefer, M., Tidor, B., Venable, R.M., Woodcock, H.L., Wu, X., Yang, W., York, D.M., and Karplus, M. (2009). CHARMM: The Biomolecular Simulation Program. *J. Comput. Chem.* 30, 1545–1614.

Brown, W.F., Jr. (1950a). Dielectric Constants of Non-Polar Fluids. I. Theory. *J. Chem. Phys.* 18, 1193–1200.

Brown, W.F., Jr. (1950b). Dielectric Constants of Non-Polar Fluids. II. Analysis of Experimental Data. *J. Chem. Phys.* 18, 1200–1206.

Brown, W.F., Jr. (1956). *Handbuch der Physik, Vol. XVII, Dielectrics.* Berlin: Springer-Verlag.

Bruch, L.W., and McGee, I.J. (1970). Semiempirical Helium Intermolecular Potential. II. Dilute Gas Properties. *J. Chem. Phys.* 52, 5884–5895.

Bu, H., Shi, S., Chen, E., Hu, H., Zhang, Z., and Wunderlich, B. (1996). Single-Molecule Single Crystals of Poly(ethylene oxide). *J. Macromol. Sci.—Phys.* B 35, 731–747.

Buch, V., Miley, A., Vácha, R., Jungwirth, P., and Devlin, J.P. (2007). Water Surface Is Acidic. *Science* 104, 7342–7347.

Büchner, S., and Heuer, A. (1999). Potential Energy Landscape of a Model Glass Former: Thermodynamics, Anharmonicities, and Finite Size Effects. *Phys. Rev. E* 60, 6507–6518.

Buckingham, M.J., and Fairbank, W.M. (1961). The Nature of the λ-Transition. In *Progress in Low Temperature Physics,* Vol. 3, C.J. Gorter, editor. Amsterdam: North Holland Publishing, 80–112.

Buckingham, R.A. (1938). The Classical Equation of State of Gaseous Helium, Neon, and Argon. *Proc. R. Soc. (London)* 168A, 264–283.

Buff, F.P., Lovett, R.A., and Stillinger, F.H. (1965). Interfacial Density Profile for Fluids in the Critical Region. *Phys. Rev. Lett.* 15, 621–623.

Buff, F.P., and Stillinger, F.H. (1956). Surface Tensions of Ionic Solutions. *J. Chem. Phys.* 25, 312–318.

Bunn, C.W. (1939). The Crystal Structure of Long-Chain Normal Paraffin Hydrocarbons. The "Shape" of the >CH_2 Group. *Trans. Faraday Soc.* 35, 482–491.

Bunn, C.W., and Alcock, T.C. (1945). The Texture of Polythene. *Trans. Faraday Soc.* 41, 317–325.

Bunn, C.W., and Howells, E.R. (1954). Structures of Molecules and Crystals of Fluorocarbons. *Nature* 174, 549–551.

Burley, J.C., Duer, M.J., Stein, R.S., and Vrcelj, R.M. (2007). Enforcing Ostwald's Rule of Stages: Isolation of Paracetamol Forms III and II. *Euro. J. Pharm. Sci.* 31, 271–276.

Burton, E.F., and Oliver, W.F. (1935). The Crystal Structure of Ice at Low Temperatures. *Proc. R. Soc. A* 153, 166–172.

Callen, H.B., and Greene, R.F. (1952). On a Theorem of Irreversible Thermodynamics. *Phys. Rev.* 86, 702–710.

Callen, H.B., and Welton, T.A. (1951). Irreversibility and Generalized Noise. *Phys. Rev.* 83, 34–40.

Capasso, F., Sirtori, C., Faist, J., Sivco, D.L., Chu, S.-N.G., and Cho, A.Y. (1992). Observation of an Electronic Bound State above a Potential Well. *Nature (London)* 358, 565–567.

Cappelezzo, M., Capelari, C.A., Pezin, S.H., and Coelho, L.A.F. (2007). Stokes-Einstein Relation for Pure Simple Fluids. *J. Chem. Phys.* 126, 224516.

Carignano, M.A., Shepson, P.B., and Szleifer, I. (2005). Molecular Dynamics Simulations of Ice Growth from Supercooled Water. *Mol. Phys.* 103, 2957–2967.

Carroll, J.J., Slupsky, J.D., and Mather, A.E. (1991). The Solubility of Carbon Dioxide in Water at Low Pressure. *J. Phys. Chem. Ref. Data* 20, 1201–1209.

Carruthers, P. (1961). Theory of Thermal Conductivity of Solids at Low Temperatures. *Rev. Modern Phys.* 33, 92–138.

Cartwright, J.H.E., Piro, O., and Tuval, I. (2007). Ostwald Ripening, Chiral Crystallization, and the Common-Ancestor Effect. *Phys. Rev. Lett.* 98, 165501.

Case, D.A., Cheatham, T.E., III, Darden, T., Gohlke, H., Luo, R., Merz, K.M., Jr., Onufriev, A., Simmerling, C., Wang, B., and Woods, R. (2005). The Amber Biomolecular Simulation Programs. *J. Comput. Chem.* 26, 1668–1688.

Castelnovo, C., Moessner, R., and Sondhi, S.L. (2008). Magnetic Monopoles in Spin Ice. *Nature* 451, 42–45.

Castrillón, S.R.-V., Matysiak, S., Stillinger, F.H., Rossky, P.J., and Debenedetti, P.G. (2012a). Phase Behavior of a Lattice Hydrophobic Oligomer in Explicit Water. *J. Phys. Chem. B* 116, 9540–9548.

Castrillón, S.R.-V., Matysiak, S., Stillinger, F.H., Rossky, P.J., and Debenedetti, P.G. (2012b). Thermal Stability of Hydrophobic Helical Oligomers: A Lattice Simulation Study in Explicit Water. *J. Phys. Chem. B* 116, 9963–9970.

Catliff, E., and Tobolsky, A.V. (1955). Stress-Relaxation of Polyisobutylene in the Transition Region (1,2). *J. Colloid Sci.* 10, 375–392.

Cawley, M.F., McGlynn, D., and Mooney, P.A. (2006). Measurement of the Temperature of Density Maximum of Water Solutions Using a Convective Flow Technique. *Int. J. Heat Mass Transfer* 49, 1763–1772.

Celik, Y., Graham, L.A., Mok, Y.-F., Bar, M., Davies, P.L., and Braslavsky, I. (2010). Superheating of Ice Crystals in Antifreeze Protein Solutions. *Proc. Natl. Acad. Sci. USA* 107, 5423–5428.

Ceperley, D.M. (1995). Path Integrals in the Theory of Condensed Helium. *Rev. Modern Phys.* 67, 279–355.

Chacón, E., and Tarazona, P. (2003). Intrinsic Profiles beyond the Capillary Wave Theory: A Monte Carlo Study. *Phys. Rev. Lett.* 91, 166103.

Chakrabarty, R.K., Moosmüller, H., Arnott, W.P., Garro, M.A., Tian, G., Slowik, J.G., Cross, E.S., Han, J.-H., Davidovits, P., Onasch, T.B., and Worsnop, D.R. (2009). Low Fractal Dimension Cluster-Dilute Soot Aggregates from a Premixed Flame. *Phys. Rev. Lett.* 102, 235504.

Chan, H.S., Zhang, Z., Wallin, S., and Liu, Z. (2011). Cooperativity, Local–Nonlocal Coupling, and Nonnative Interactions: Principles of Protein Folding from Coarse-Grained Models. *Annu. Rev. Phys. Chem.* 62, 301–326.

Chandler, D. (2005). Interfaces and the Driving Force of Hydrophobic Assembly. *Nature* 437, 640–647.

Chang, S.S., and Bestul, A.B. (1972). Heat Capacity and Thermodynamic Properties of *o*-Terphenyl Crystal, Glass, and Liquid. *J. Chem. Phys.* 56, 503–516.

Chang, S.S., Horman, J.A., and Bestul, A.B. (1967). Heat Capacities and Related Thermal Data for Diethyl Phthalate Crystal, Glass, and Liquid to 360 °K. *J. Res. NBS* 71A, 293–305.

Chapman, A.J. (1981). Heat Transfer. In *Encyclopedia of Physics*, R.G. Lerner and G.L. Trigg, editors. Reading, MA: Addison-Wesley Publishing, 385–388.

Chester, G.V. (1970). Speculations on Bose-Einstein Condensation and Quantum Crystals. *Phys. Rev. A* 2, 256–258.

Cheung, D.L., and Bon, S.A.F. (2009). Interaction of Nanoparticles with Ideal Liquid–Liquid Interfaces. *Phys. Rev. Lett.* 102, 066103.

Chiang, R., and Flory, P.J. (1961). Equilibrium between Crystalline and Amorphous Phases in Polyethylene. *J. Am. Chem. Soc.* 83, 2857–2862.

Chiti, F., and Dobson, C.M. (2006). Protein Misfolding, Functional Amyloid, and Human Disease. *Annu. Rev. Biochem.* 75, 333–366.

Choi, H., Takahashi, D., Kono, K., and Kim, E. (2010). Evidence of Supersolidity in Rotating Solid Helium. *Science* 330, 1512–1515.

Christy, R.W., and Lawson, A.W. (1951). High Temperature Specific Heat of AgBr. *J. Chem. Phys.* 19, 517.

Ciccotti, G., Ferrario, M., and Ryckaert, J.-P. (1982). Molecular Dynamics of Rigid Systems in Cartesian Coordinates. A General Formulation. *Mol. Phys.* 47, 1253–1264.

Cicerone, M.T., Blackburn, F.R., and Ediger, M.D. (1995). How Do Molecules Move near T_g? Molecular Rotation of Six Probes in *o*-Terphenyl across 14 Decades in Time. *J. Chem. Phys.* 102, 471–479.

Clark, B.K., and Ceperley, D.M. (2006). Off-Diagonal Long-Range Order in Solid ^{4}He. *Phys. Rev. Lett.* 96, 105302.

Clisby, N. (2010). Accurate Estimate of the Critical Exponent ν for Self-Avoiding Walks via a Fast Implementation of the Pivot Algorithm. *Phys. Rev. Lett.* 104, 055702.

Cole, M.W. (1970). Width of the Surface Layer of He^4. *Phys. Rev. A* 1, 1838–1840.

Cole, M.W. (1980). Amplitude of Liquid–Vapor Interfacial Oscillations. *J. Chem. Phys.* 73, 4012–4014.

Connors, K.A. (1990). *Chemical Kinetics*. New York: VCH Publishers.

Covington, A.K., Robinson, R.A., and Bates, R.G. (1966). The Ionization Constant of Deuterium Oxide from 5 to 50°. *J. Phys. Chem.* 70, 3820–3824.

Cowley, R.A., and Woods, A.D.B. (1971). Inelastic Scattering of Thermal Neutrons from Liquid Helium. *Can. J. Phys.* 49, 177–200.

Cox, J.D., and Herington, E.F.G. (1956). The Coexistence Curve in Liquid–Liquid Binary Systems. *Trans. Faraday Soc.* 52, 926–930.

Craik, D.J. (2006). Seamless Proteins Tie Up Their Loose Ends. *Science* 311, 1563–1564.

Creighton, T.E. (1984). *Proteins. Structures and Molecular Principles*. New York: W.H. Freeman and Co.

Creighton, T.E. (1993). *Proteins. Structures and Molecular Properties*, 2nd edition. New York: W.H. Freeman and Co.

Croll, I.M., and Scott, R.L. (1958). Fluorocarbon Solutions at Low Temperatures. III. Phase Equilibria and Volume Changes in the CH_4–CF_4 System. *J. Phys. Chem.* 62, 954–957.

Crowe, J.H., Carpenter, J.F., and Crowe, L.M. (1998). The Role of Vitrification in Anhydrobiosis. *Annu. Rev. Physiol.* 60, 73–103.

Das, S.K., Kim, Y.C., and Fisher, M.E. (2011). When Is a Conductor Not a Conductor? Sum Rules Fail under Critical Fluctuations. *Phys. Rev. Lett.* 107, 215701.

Das, S.K., Kim, Y.C., and Fisher, M.E. (2012). Near Critical Electrolytes: Are the Charge–Charge Sum Rules Obeyed? *J. Chem. Phys.* 137, 074902.

Dautenhahn, J., and Hall, C.K. (1994). Monte Carlo Simulation of Off-Lattice Polymer Chains: Effective Pair Potentials in Dilute Solution. *Macromolecules* 27, 5399–5412.

Davidson, D.W. (1973). Clathrate Hydrates. In *Water, A Comprehensive Treatise,* Vol. 2, F. Franks, editor. New York: Plenum Press, 115–234.

Davies, J.T., and Rideal, E.K. (1963). *Interfacial Phenomena*, 2nd edition. New York: Academic Press.

de Gennes, P.G. (1971). Reptation of a Polymer Chain in the Presence of Fixed Obstacles. *J. Chem. Phys.* 55, 572–579.

de Grotthuss, C.J.T. (1806). Sur la décomposition de l'eau et des corps qu'elle tient en dissolution á l'aide de l'électricité galvanique. *Ann. Chim.* (*Paris*) LVIII, 54–74.

de Koning, M., Antonelli, A., da Silva, A.J.R., and Fazzio, A. (2006). Orientational Defects in Ice Ih: An Interpretation of Electrical Conductivity Measurements. *Phys. Rev. Lett.* 96, 075501.

De Rosa, C., Rapacciuola, M., Guerra, G., Petraccone, V., and Corradini, P. (1992). On the Crystal Structure of the Orthorhombic Form of Syndiotactic Polystyrene. *Polymer* 33, 1423–1428.

de Souza, V.K., and Wales, D.J. (2009). Connectivity in the Potential Energy Landscape for Binary Lennard-Jones Systems. *J. Chem. Phys.* 130, 194508.

De Wette, F.W., and Nijboer, B.R.A. (1965). On the Stability of Close-Packed Crystals with Lennard-Jones Interaction. *Phys. Lett.* 18, 19–20.

Debenedetti, P.G. (1996). *Metastable Liquids. Concepts and Principles*. Princeton, NJ: Princeton University Press.

Debenedetti, P.G. (2003). Supercooled and Glassy Water. *J. Phys.: Condens. Matter* 15, R1669–R1726.

Debenedetti, P.G., and Stillinger, F.H. (2001). Supercooled Liquids and the Glass Transition. *Nature* 410, 259–267.

Debenedetti, P.G., Truskett, T.M., Lewis, C.P., and Stillinger, F.H. (2001). Theory of Supercooled Liquids and Glasses: Energy Landscape and Statistical Geometry Perspectives. *Adv. Chem. Eng.* 28, 21–79.

Debye, P. (1912). Zur Theorie der spezifischen Wärmen. *Ann. Physik, Leipzig* 39, 789–839.

Debye, P., and Hückel, E. (1923). Zur Theorie der Elektrolyte. *Physik. Zeit.* 24, 185–206.

Dhaubhadel, R., Pierce, F., Chakrabarti, A., and Sorensen, C.M. (2006). Hybrid Superaggregate Morphology as a Result of Aggregation in a Cluster-Dense Aerosol. *Phys. Rev. E* 73, 011404.

Di Gennaro, M., Saha, S.K., and Verstraete, M.J. (2013). Role of Dynamical Instability in the *Ab-Initio* Phase Diagram of Calcium. *Phys. Rev. Lett.* 111, 025503.

Di Lullo, G.A., Sweeney, S.M., Körkkö, J., Ala-Kokko, L., and San Antonio, J.D. (2002). Mapping the Ligand-Binding Sites and Disease-Associated Mutations on the Most Abundant Protein in the Human, Type I Collagen. *J. Biol. Chem.* 277, 4223–4231.

Dill, K.A., and MacCallum, J.L. (2012). The Protein-Folding Problem, 50 Years On. *Science* 338, 1042–1046.

DiMarzio, E.A., and Stillinger, F.H. (1964). Residual Entropy of Ice. *J. Chem. Phys.* 40, 1577–1581.

DiMasi, E., Tostmann, H., Ocko, B.M., Pershan, P.S., and Deutsch, M. (1998). X-Ray Reflectivity Study of Temperature-Dependent Surface Layering in Liquid Hg. *Phys. Rev. B* 58, R13419–R13422.

Ding, K., and Andersen, H.C. (1986). Molecular-Dynamics Simulation of Amorphous Germanium. *Phys. Rev. B* 34, 6987–6991.

Dirac, P.A.M. (1947). *The Principles of Quantum Mechanics,* 3rd edition. Oxford, U.K.: Clarendon Press.

Dixon, P.K., and Nagel, S.R. (1988). Frequency-Dependent Specific Heat and Thermal Conductivity at the Glass Transition in *o*-Terphenyl Mixtures. *Phys. Rev. Lett.* 61, 341–344.

Dobbs, E.R. (2000). *Helium Three*. Oxford, U.K.: Oxford University Press.

Dobrynin, A.V., and Rubinstein, M. (2005). Theory of Polyelectrolytes in Solutions and at Surfaces. *Prog. Polym. Sci.* 30, 1049–1118.

Dobson, C.M. (2003). Protein Folding and Misfolding. *Nature* 426, 884–890.

Doi, M., and Edwards, S.F. (1986). *The Theory of Polymer Dynamics*. Oxford, U.K.: Clarendon Press.

Doliwa, B., and Heuer, A. (2003). Finite-Size Effects in a Supercooled Liquid. *J. Phys.: Condens. Matter* 15, S849–S858.

Domb, C. (1960a). On the Theory of Cooperative Phenomena in Crystals. *Advances in Physics* 9, 149–244.

Domb, C. (1960b). On the Theory of Cooperative Phenomena in Crystals. *Advances in Physics* 9, 245–361.

Domcke, W., and Yarkony, D.R. (2012). Role of Conical Intersections in Molecular Spectroscopy and Photoinduced Chemical Dynamics. *Annu. Rev. Phys. Chem.* 63, 325–352.

Donati, C., Douglas, J.F., Kob, W., Poole, P.H., Plimpton, S.J., and Glotzer, S.C. (1998). Stringlike Cooperative Motion in a Supercooled Liquid. *Phys. Rev. Lett.* 80, 2338–2341.

Donev, A., Torquato, S., and Stillinger, F.H. (2005). Pair Correlation Function Characteristics of Nearly Jammed Disordered and Ordered Hard-Sphere Packings. *Phys. Rev. E*, 71, 011105.

Donohue, J. (1982). *The Structures of the Elements*. Malabar, FL: Robert E. Krieger Publishing Co.

Dowell, L.G., and Rinfret, A.P. (1960). Low-Temperature Forms of Ice as Studied by X-Ray Diffraction. *Nature* 188, 1144–1148.

Doye, J.P.K., and Calvo, F. (2002). Entropic Effects on the Structure of Lennard-Jones Clusters. *J. Chem. Phys.* 116, 8307–8317.

Doye, J.P.K., and Wales, D.J. (1996). The Effect of the Range of the Potential on the Structure and Stability of Simple Liquids: From Clusters to Bulk, from Sodium to C_{60}. *J. Phys. B: At. Mol. Opt. Phys.* 29, 4859–4894.

Dubin, D.H.E., and Dewitt, H. (1994). Polymorphic Phase Transition for Inverse-Power-Potential Crystals Keeping the First-Order Anharmonic Correction to the Free Energy. *Phys. Rev. B* 49, 3043–3048.

Duh, D.-M., Perera, D.N., and Haymet, A.D.J. (1995). Structure and Properties of the CF1 Central Force Model of Water: Integral Equation Theory. *J. Chem. Phys.* 102, 3736–3746.

Dyke, T.R., Howard, B.J., and Klemperer, W. (1972). Radiofrequency and Microwave Spectrum of the Hydrogen Fluoride Dimer; a Nonrigid Molecule. *J. Chem. Phys.* 56, 2442–2454.

Dyke, T.R., and Muenter, J.S. (1973). Electric Dipole Moments of Low J States of H_2O and D_2O. *J. Chem. Phys.* 59, 3125–3127.

Ediger, M.D. (2000). Spatially Heterogeneous Dynamics in Supercooled Liquids. *Annu. Rev. Phys. Chem.* 51, 99–128.

Ediger, M.D. (2014). Vapor-Deposited Glasses Provide Clearer View of Two-Level Systems. *Proc. Natl. Acad. Sci. USA* 111, 11232–11233.

Edwards, D.O., and Pandorf, R.C. (1965). Heat Capacity and Other Properties of Hexagonal Close-Packed Helium-4. *Phys. Rev.* 140, A816–A825.

Eigen, M. (1964). Proton Transfer, Acid–Base Catalysis, and Enzymatic Hydrolysis. Part I: Elementary Processes. *Angew. Chem. Int. Ed.* 3, 1–19.

Einstein, A. (1905). Über die von der molekularkinetischen Theorie der Wärme geforderte Bewegung von in ruhenden Flüssigkeiten suspendierten Teilchen. *Ann. Phys. (Leipzig)* 17, 549–560.

Einstein, A. (1906). Zur Theorie der Brownschen Bewegung. *Ann. Phys. (Leipzig)* 19, 371–381.

Einstein, A. (1907). Die Plancksche Theorie der Strahlung und die Theorie der spezifischen Wärme. *Ann. Phys.* 22, 180–190.

Einstein, A. (1911). Eine Beziehung zwischen dem elastischen Verhalten und der spezifischen Wärme bei festen Körpern mit einatomigen Molecül. *Ann. Phys.* 34, 170–174.

Eisenberg, D., and Kauzmann, W. (1969). *The Structure and Properties of Water.* New York: Oxford University Press.

Eisenman, G., editor. (1967). *Glass Electrodes for Hydrogen and Other Cations.* New York: Marcel Dekker.

Elbaum, M., Lipson, S.G., and Dash, J.G. (1993). Optical Study of Surface Melting on Ice. *J. Cryst. Growth* 129, 491–505.

Elmatad, Y.S., Chandler, D., and Garrahan, J.P. (2009). Corresponding States of Structural Glass Formers. *J. Phys. Chem. B* 113, 5563–5567.

Elser, V., and Rankenburg, I. (2006). Deconstructing the Energy Landscape: Constraint-Based Algorithms for Folding Heteropolymers. *Phys. Rev. E* 73, 026702.

Eriksson, A.E., Jones, T.A., and Liljas, A. (1988). Refined Structure of Human Carbonic Anhydrase II at 2.0 Å Resolution. *Proteins* 4, 274–282.

Errea, I., Rousseau, B., and Bergara, A. (2011). Anharmonic Stabilization of the High-Pressure Simple Cubic Phase of Calcium. *Phys. Rev. Lett.* 106, 165501.

Errington, J.R., and Panagiotopoulos, A.Z. (1998). Phase Equilibria of the Modified Buckingham Exponential-6 Potential from Hamiltonian Scaling Grand Canonical Monte Carlo. *J. Chem. Phys.* 109, 1093–1100.

Errington, J.R., and Panagiotopoulos, A.Z. (1999). A New Intermolecular Potential Model for the *n*-Alkane Homologous Series. *J. Phys. Chem. B* 103, 6314–6322.

Essam, J.W., and Fisher, M.E. (1963). Padé Approximant Studies of the Lattice Gas and the Ising Ferromagnet below the Critical Point. *J. Chem. Phys.* 38, 802–812.

Estrada-Torres, R., Iglesias-Silva, G.A., Ramos-Estrada, M., and Hall, K.R. (2007). Boyle Temperatures for Pure Substances. *Fluid Phase Equilib.* 258, 148–154.

Evans, P.R., Farrants, G.W., Hudson, P.J., and Britton, H.G. (1981). Phosphofructokinase: Structure and Control. *Phil. Trans. R. Soc. London, Ser. B, Biol. Sci.* 293, 53–62.

Ewen, B., Strobl, G.R., and Richter, D. (1980). Phase Transitions in Crystals of Chain Molecules. Relation between Defect Structures and Molecular Motion in the Four Modifications of n-$C_{33}H_{68}$. *Faraday Discuss. Chem. Soc.* 69, 19–31.

Fan, Y., Percus, J.K., Stillinger, D.K., and Stillinger, F.H. (1991). Constraints on Collective Density Variables: One Dimension. *Phys. Rev. A* 44, 2394–2402.

Faraday, M. (1860). Note on Regelation. *Proc. R. Soc. London* 10, 440–450.

Fargas, J., de Feraudy, M.F., Raoult, B., and Torchet, J. (1987). Noncrystalline Structure of Argon Clusters. I. Polyicosahedral Structure of Ar_N Clusters, $20 < N < 50$. *J. Chem. Phys.* 78, 5067–5080.

Feder, R., and Charbnau, H.P. (1966). Equilibrium Defect Concentration in Crystalline Sodium. *Phys. Rev.* 149, 464–471.

Feder, R., and Nowick, A.S. (1967). Equilibrium Vacancy Concentration in Pure Pb and Dilute Pb–Tl and Pb–In Alloys. *Phil. Mag.* 15, 805–812.

Feeney, M.R., Debenedetti, P.G., and Stillinger, F.H. (2003). A Statistical Mechanical Model for Inverse Melting. *J. Chem. Phys.* 119, 4582–4591.

Fernández, L.A., Martín-Mayor, V., Seoane, B., and Verrocchio, P. (2012). Equilibrium Fluid–Solid Coexistence of Hard Spheres. *Phys. Rev. Lett.* 108, 165701.

Ferry, L., Vigier, G., Vassaille, R., and Bessede, J.L. (1995). Study of Polytetrafluoroethylene Crystallization. *Acta Polymer.* 46, 300–306.

Feynman, R.P. (1939). Forces in Molecules. *Phys. Rev.* 56, 340–343.

Feynman, R.P. (1954). Atomic Theory of the Two-Fluid Model of Liquid Helium. *Phys. Rev.* 94, 262–277.

Feynman, R.P. (1972). *Statistical Mechanics, a Set of Lectures*, J. Shaham, editor. Reading, MA: W.A. Benjamin, Inc.

Feynman, R.P., and Cohen, M. (1956). Energy Spectrum of the Excitations in Liquid Helium. *Phys. Rev.* 102, 1189–1204.

Fischer, E.W. (1957). Stufenformiges und Spiralformiges Kristalwachstum bei Hochpolymeren. *Zeit. Naturforsch. A* 12(9), 753–754.

Fisher, D.S., and Weeks, J.D. (1983). Shape of Crystals at Low Temperatures: Absence of Quantum Roughening. *Phys. Rev. Lett.* 50, 1077–1080.

Fisher, M.E. (1966). Theory of Critical Fluctuations and Singularities. *Critical Phenomena. Proceedings of a Conference Held in Washington, D.C., April 1965*, M.S. Green and J.V. Sengers, editors. Washington, DC: National Bureau of Standards, 108–115.

Fisher, M.E. (1967a). Theory of Equilibrium Critical Phenomena. *Repts. Prog. Phys.* 30, 615–730.

Fisher, M.E. (1967b). Theory of Condensation and Critical Point. *Physics—New York* 3, 255–283.

Fisher, M.E., and Felderhof, B.U. (1970). Phase Transitions in One-Dimensional Cluster-Interaction Fluids. 1A. Thermodynamics. *Ann. Phys.* 58, 176–216.

Fisher, M.E., and Ruelle, D. (1966). The Stability of Many-Particle Systems. *J. Math. Phys.* 7, 260–270.

Fletcher, N.H. (1970). *The Chemical Physics of Ice*. Cambridge, U.K.: Cambridge University Press.

Floriano, M.A., and Angell, C.A. (1990). Surface Tension and Molar Free Energy and Entropy of Water to −27.2°C. *J. Phys. Chem.* 94, 4199–4202.

Flory, P.J. (1941). Thermodynamics of High Polymer Solutions. *J. Chem. Phys.* 9, 660–661.

Flory, P.J. (1942). Thermodynamics of High Polymer Solutions. *J. Chem. Phys.* 10, 51–61.

Flory, P.J. (1949). Statistical Mechanics of Dilute Polymer Solutions. *J. Chem. Phys.* 17, 1347–1348.

Flory, P.J. (1953). *Principles of Polymer Chemistry*. Ithaca, NY: Cornell University Press, 526, 539.

Flory, P.J., and Krigbaum, W.R. (1950). Statistical Mechanics of Dilute Polymer Solutions. II. *J. Chem. Phys.* 18, 1086–1094.

Fox, J.R., and Andersen, H.C. (1984). Molecular Dynamics Simulations of a Supercooled Monatomic Liquid and Glass. *J. Phys. Chem.* 88, 4019–4027.

Fox, T.G., and Flory, P.J. (1950). Second-Order Transition Temperatures and Related Properties of Polystyrene. I. Influence of Molecular Weight. *J. Appl. Phys.* 21, 581–591.

Frank, F.C. (1953). On Spontaneous Asymmetric Synthesis. *Biochim. Biophys. Acta* 11, 459–463.

Franks, F. (1973). The Solvent Properties of Water. In *Water, a Comprehensive Treatise,* Vol. 2, F. Franks, editor. New York: Plenum Press, 1–54.

Frauenfelder, H., et al., editors. (1997). *Landscape Paradigms in Physics and Biology. Concepts, Structures, and Dynamics*. Amsterdam: Elsevier.

Frenkel, J. (1939a). Statistical Theory of Condensation Phenomena. *J. Chem. Phys.* 7, 200–201.

Frenkel, J. (1939b). A General Theory of Heterophase Fluctuations and Pretransition Phenomena. *J. Chem. Phys.* 7, 538–547.

Frenkel, J. (1955). *Kinetic Theory of Liquids*. New York: Dover Publications.

Fricke, J. (1988). Aerogels. *Sci. Am.* 258(5), 92–97.

Fujara, F., Geil, B., Sillescu, H., and Fleischer, G. (1992). Translational and Rotational Diffusion in Supercooled Orthoterphenyl Close to the Glass Transition. *Zeit. Phys. B—Cond. Matter* 88, 195–204.

Fukuhara, T., Sugawa, S., and Takahashi, Y. (2007). Bose-Einstein Condensation of an Ytterbium Isotope. *Phys. Rev. A* 76, 051604.

Fytas, G., Wang, C.H., Lilge, D., and Dorfmüller, Th. (1981). Homodyne Light Beating Spectroscopy of *o*-Terphenyl in the Supercooled Liquid State. *J. Chem. Phys.* 75, 4247–4255.

Galperin, Y.M., Karpov, V.G., and Kozub, V.I. (1991). Localized States in Glasses. *Adv. Phys.* 38, 669–737.

Gao, X., and Jiang, L. (2004). Biophysics: Water-Repellent Legs of Water Striders. *Nature* 432, 36.

Garland, C.W., and Jones, J.S. (1963). Elastic Constants of Ammonium Chloride near Lambda Point. *J. Chem. Phys.* 39, 2874–2880.

Garland, C.W., and Renard, R. (1966). Order–Disorder Phenomena. III. Effect of Temperature and Pressure on Elastic Constants of Ammonium Chloride. *J. Chem. Phys.* 44, 1130–1139.

Garnham, C.P., Campbell, R.L., and Davies, P.L. (2011). Anchored Clathrate Waters Bind Antifreeze Proteins to Ice. *Proc. Natl. Acad. Sci. USA* 108, 7363–7367.

Gebremichael, Y., Vogel, M., and Glotzer, S.C. (2004). Particle Dynamics and the Development of String-Like Motion in a Simulated Monatomic Supercooled Liquid. *J. Chem. Phys.* 120, 4414–4427.

Geiger, A., Rahman, A., and Stillinger, F.H. (1979). Molecular Dynamics Study of the Hydration of Lennard-Jones Solutes. *J. Chem. Phys.* 70, 263–276.

Gezelter, J.D., Rabani, E., and Berne, B.J. (1997). Can Imaginary Instantaneous Normal Mode Frequencies Predict Barriers to Self-Diffusion? *J. Chem. Phys.* 107, 4618–4627.

Giauque, W.F., and Stout, J.W. (1936). The Entropy of Water and the Third Law of Thermodynamics. The Heat Capacity of Ice from 15 to 273 °K. *J. Am. Chem. Soc.* 58, 1144–1150.

Gibbs, J.W. (1906). *The Scientific Papers of J. Willard Gibbs,* Vol. 1. New York: Longmans, Green; reprinted by Dover Publications, New York, 1961.

Gillis, K.A., Shinder, I.I., and Moldover, M.R. (2005). Bulk Viscosity of Stirred Xenon near the Critical Point. *Phys. Rev. E* 72, 051201.

Giovambattista, N., Rossky, P.J., and Debenedetti, P.G. (2006). Effect of Pressure on the Phase Behavior and Structure of Water Confined between Nanoscale Hydrophobic and Hydrophilic Plates. *Phys. Rev. E* 73, 041604.

Gleim, T., Kob, W., and Binder, K. (1998). How Does the Relaxation of a Supercooled Liquid Depend on Its Microscopic Dynamics? *Phys. Rev. Lett.* 81, 4404–4407.

Glyde, H.R. (1994). *Excitations in Liquid and Solid Helium.* Oxford, U.K.: Clarendon Press.

Goettal, K.A., Eggert, J.H., Silvera, I.F., and Moss, W.C. (1989). Optical Evidence for the Metallization of Xe at 132(5) GPa. *Phys. Rev. Lett.* 62, 665–668.

Goldman, N., Leforestier, C., and Saykally, R.J. (2005). A "First Principles" Potential Energy Surface for Liquid Water from VRT Spectroscopy of Water Clusters. *Phil. Trans. R. Soc. A* 363, 493–508.

Goldstein, H. (1953). *Classical Mechanics.* Cambridge, MA: Addison-Wesley Publishing Co.

Goldstein, M. (1969). Viscous Liquids and the Glass Transition: A Potential Energy Barrier Picture. *J. Chem. Phys.* 51, 3728–3739.

Goldstein, M. (1977). Viscous Liquids and the Glass Transition. VII. Molecular Mechanisms for a Thermodynamic Second Order Transition. *J. Chem. Phys.* 67, 2246–2253.

Gould, H., and Wong, V.K. (1978). Ripplon Damping in Superfluid ^{4}He at Low Temperatures. *Phys. Rev. B* 18, 2124–2126.

Graham, L.A., Liou, Y.-C., Walker, V.K., and Davies, P.L. (1997). Hyperactive Antifreeze Protein from Beetles. *Nature* 388, 727–728.

Graves, R.E., and Agrow, B.M. (1999). Bulk Viscosity: Past to Present. *J. Thermo. Heat Transfer* 13, 337–342.

Green, M.F. (1954). Markoff Random Processes and the Statistical Mechanics of Time-Dependent Phenomena. II. Irreversible Processes in Fluids. *J. Chem. Phys.* 22, 398–413.

Greene, R.F., and Callen, H.B. (1952). On a Theorem of Irreversible Thermodynamics. II. *Phys. Rev.* 88, 1387–1391.

Greet, R.J., and Turnbull, D. (1967). Glass Transition in *o*-Terphenyl. *J. Chem. Phys.* 46, 1243–1251.

Greywall, D.S. (1983). Specific Heat of Normal ^{3}He. *Phys. Rev. B.* 27, 2747–2766.

Greywall, D.S. (1985). ^{3}He Melting-Curve Thermometry at Millikelvin Temperatures. *Phys. Rev B* 31, 2675–2683.

Grigoropoulis, C., Rogers, M., Ko, S.H., Golovin, A.A., and Matkowsky, B.J. (2006). Explosive Crystallization in the Presence of Melting. *Phys. Rev. B* 73, 184125.

Grilly, E.R., and Mills, R.L. (1962). PVT Relations in He^4 near the Melting Curve and the λ-Line. *Ann. Phys.* 18, 250–263.

Gubskaya, A.V., and Kusalik, P. (2002). The Total Molecular Dipole Moment of Water. *J. Chem. Phys.* 117, 5290–5302.

Guggenheim, E.A. (1950). *Thermodynamics.* Amsterdam: North-Holland Publishing Co.

Guillot, B. (2002). A Reappraisal of What We Have Learnt during Three Decades of Computer Simulation on Water. *J. Mol. Liquids* 101, 219–260.

Hagen, M.H.J., Meijer, E.J., Moolj, G.C.A.M., Frenkel, D., and Lekkerkerker, H.N.W. (1993). Does C_{60} Have a Liquid Phase? *Nature* 365, 425–426.

Hales, T.C. (2005). A Proof of the Kepler Conjecture. *Ann. Math.* 162, 1065–1185.

Hansen, G.B., and McCord, T.B. (2004). Amorphous and Crystalline Ice on the Galilean Satellites: A Balance between Thermal and Radiolytic Processes. *J. Geophys. Research. Planets* 109, E01012.

Hansen, J.P., and McDonald, I.R. (1976). *Theory of Simple Liquids.* New York: Academic Press.

Hansen, J.P., and McDonald, I.R. (1986). *Theory of Simple Liquids,* 2nd edition. New York: Academic Press.

Hansen, J. P., and Pollock, E.L. (1972). Ground-State Properties of Solid Helium-4 and -3. *Phys. Rev. A* 5, 2651–2665.

Hansen, J.P., and Verlet, L. (1969). Phase Transitions of the Lennard-Jones System. *Phys. Rev.* 184, 151–161.

Harris, J.G., Gryko, J., and Rice, S.A. (1987). Self-Consistent Monte Carlo Simulations of the Electron and Ion Distributions of Inhomogeneous Liquid Alkali Metals. I. Longitudinal and Transverse Density Distributions in the Liquid–Vapor Interface of a One-Component System. *J. Chem. Phys.* 87, 3069–3081.

Harris, J.G., and Stillinger, F.H. (1990). Fundamental Aspects of Chemical Kinetics in Condensed Phases. *Chemical Physics* 149, 63–80.

Harris, J.G., and Stillinger, F.H. (1991). Isomerization and Inherent Structure in Liquids. A Molecular Dynamics Study of Liquid Cyclohexane. *J. Chem. Phys.* 95, 5953–5965.

Harrison, W.A. (1980). *Electronic Structure and the Properties of Solids.* San Francisco: W.H. Freeman and Co.

Hart, W.E., and Istrail, S. (1997). Lattice and Off-Lattice Side Chain Models of Protein Folding: Linear Time Structure Prediction Better Than 86% of Optimal. *J. Comput. Biol.* 4, 241–259.

Harvey, A.H., and Lemmon, E.W. (2004). Correlation for the Second Virial Coefficient of Water. *J. Phys. Chem. Ref. Data* 33, 369–376.

Hasted, J.B., Ritson, D.M., and Collie, C.H. (1948). Dielectric Properties of Aqueous Ionic Solutions. Parts I and II. *J. Chem. Phys.* 16, 1–21.

Hatwalne, Y., and Muthukumar, M. (2010). Chiral Symmetry Breaking in Crystals of Achiral Polymers. *Phys. Rev. Lett.* 105, 107801.

Hausdorff, F. (1919). Dimension und äusseres Mass. *Math. Annalen* 79, 157–179.

Hecksher, T., Nielsen, A.I., Olsen, N.B., and Dyre, J.C. (2008). Little Evidence for Dynamic Divergences in Ultraviscous Molecular Liquids. *Nature Phys.* 4, 737–741.

Helfand, E. (1983). On Inversion of the Williams-Watts Function for Large Relaxation Times. *J. Chem. Phys.* 78, 1931–1934.

Heller, P. (1967). Experimental Investigations of Equilibrium Critical Phenomena. *Repts. Prog. Phys.* 30, 731–826.

Hellmann, H. (1937). *Einfuhrung in die Quantenchemie*. Leipzig, Germany: Deuticke, 285.

Henderson, S.J., and Speedy, R.J. (1987a). Temperature of Maximum Density in Water at Negative Pressure. *J. Phys. Chem.* 91, 3062–3068.

Henderson, S.J., and Speedy, R.J. (1987b). Melting Temperature of Ice at Positive and Negative Pressures. *J. Phys. Chem.* 91, 3069–3072.

Hepler, L.G., and Woolley, E.M. (1973). Hydration Effects and Acid–Base Equilibria. In *Water, A Comprehensive Treatise*, Vol. 3, F. Franks, editor. New York: Plenum Press, 145–172.

Herrero, C.P. (2002). Isotope Effects in Structural and Thermodynamic Properties of Solid Neon. *Phys. Rev. B* 65, 014112.

Herring, C. (1951). Some Theorems on the Free Energies of Crystal Surfaces. *Phys. Rev.* 82, 87–93.

Hertz, J., Krogh, A., and Palmer, R.G. (1991). *Introduction to the Theory of Neural Computation*. New York: Addison-Wesley Publishing Co., 125.

Herzberg, G. (1950). *Molecular Spectra and Molecular Structure. I. Spectra of Diatomic Molecules*. New York: Van Nostrand Reinhold, 527.

Hess, B., Kutzner, C., van der Spoel, D., and Lindahl, E. (2008). GROMACS 4: Algorithms for Highly Efficient, Load-Balanced, and Scalable Molecular Simulation. *J. Chem. Theory Comput.* 4, 435–447.

Heuer, A. (2008). Exploring the Potential Energy Landscape of Glass-Forming Systems: From Inherent Structures via Metabasins to Macroscopic Transport. *J. Phys.: Condens. Matter* 20, 373101.

Heuer, A., and Büchner, S. (2000). Why Is the Density of Inherent Structures of a Lennard-Jones-Type System Gaussian? *J. Phys.: Condens. Matter* 12, 6535–6541.

Hill, N.E. (1969). Theoretical Treatment of Permittivity and Loss. In *Dielectric Properties & Molecular Behaviour*, N.E. Hill, W.E. Vaughan, A.H. Price, and M. Davies, editors. London: Van Nostrand Reinhold Co., 1–107.

Hill, T.L. (1956). *Statistical Mechanics. Principles and Selected Applications*. New York: McGraw-Hill Book Co.

Hill, T.L. (1960). *Introduction to Statistical Thermodynamics*. Reading, MA: Addison-Wesley Publishing Co.

Hirschfelder, J.O., Curtiss, C.F., and Bird, R.B. (1954). *Molecular Theory of Gases and Liquids*. New York: John Wiley & Sons.

Hoare, M.R. (1979). Structure and Dynamics of Simple Microclusters. *Adv. Chem. Phys.* 40, 49–135.

Hoare, M.R., and McInnes, J. (1976). Statistical Mechanics and Morphology of Very Small Atomic Clusters. *J. Chem. Soc., Faraday Discuss.* 61, 12–24.

Hoare, M.R., and McInnes, J. (1983). Morphology and Statistical Statics of Simple Microclusters. *Adv. Phys.* 32, 791–821.

Hochberg, D. (2009). Mirror Symmetry Breaking and Restoration: The Role of Noise and Chiral Bias. *Phys. Rev. Lett.* 102, 248101.

Hodgdon, J.A., and Stillinger, F.H. (1995). Inherent Structures in the Potential Energy Landscape of Solid ^{4}He. *J. Chem. Phys.* 102, 457–464.

Hoffman, J.D., Davis, T., and Lauritzen, J.I., Jr. (1976). The Rate of Crystallization of Linear Polymers with Chain Folding. Chapter 7 in *Treatise on Solid State Chemistry, Vol. 3, Crystalline and Noncrystalline Solids*, N.B. Hannay, editor. New York: Plenum Press.

Hoggatt, V.E., Jr. (1969). *Fibonacci and Lucas Numbers*. Boston: Houghton Mifflin.

Honeycutt, J.D., and Andersen, H.C. (1987). Molecular Dynamics Study of Melting and Freezing of Small Lennard-Jones Clusters. *J. Phys. Chem.* 91, 4950–4963.

Hoover, W.G., and Ree, F.H. (1968). Melting Transition and Communal Entropy for Hard Spheres. *J. Chem. Phys.* 49, 3609–3617.

Hoover, W.G., Young, D.A., and Grover, R. (1972). Statistical Mechanics of Phase Diagrams. I. Inverse Power Potentials and the Close-Packed to Body-Centered Cubic Transition. *J. Chem. Phys.* 56, 2207–2210.

Housecroft, C.E., and Sharpe, A.G. (2008). *Inorganic Chemistry*, 3rd edition. Harlow, U.K.: Pearson-Prentice-Hall.

Houwink, R. (1940). The Interrelationship between Viscosimetric and Osmotic Identified Degree of Polymerization in High Polymers. *J. für Praktische Chemie-Leipzig* 157, 15–18.

Hsu, H.-P., Mehra, V., and Grassberger, P. (2003). Structure Optimization in an Off-Lattice Protein Model. *Phys. Rev. E* 68, 037703.

Huang, K. (1963). *Statistical Mechanics*. New York: John Wiley & Sons.

Huang, Y., and Chen, G. (2005). Melting-Pressure and Density Equations of ^{3}He at Temperatures from 0.001 to 30 K. *Phys. Rev. B* 72, 184513.

Huber, K.P., and Herzberg, G. (1979). *Molecular Spectra and Molecular Structure. IV. Constants of Diatomic Molecules*. New York: Van Nostrand Reinhold, 214–215.

Hückel, E. (1928). Theorie der Beweglichkeiten des Wasserstoff- und Hydroxylions in Wässriger Lösung. *Z. Elektrochem.* 34, 546–562.

Huggins, M.L. (1941). Solutions of Long-Chain Compounds. *J. Chem. Phys.* 9, 440.

Huggins, M.L. (1951). Transitions in Silver Halides. In *Phase Transformations in Solids*, R. Smoluchowski, J.E. Mayer, and W.A. Weyl, editors. New York: John Wiley & Sons, 238–256.

Hummer, G., Garde, S., Garcia, A.E., Paulaitis, M.E., and Pratt, L.R. (1998). The Pressure Dependence of Hydrophobic Interactions Is Consistent with the Observed Pressure Denaturation of Proteins. *Proc. Natl. Acad. Sci. USA* 95, 1552–1555.

Hunklinger, S. (1997). Tunneling States. In *Amorphous Insulators and Semiconductors*, M.F. Thorpe and M.I. Mitkova, editors. Dordrecht, Netherlands: Kluwer, 469.

Ikeda, A., and Miyazaki, K. (2011). Thermodynamic and Structural Properties of the High Density Gaussian Core Model. *J. Chem. Phys.* 135, 024901.

Irbäck, A., Peterson, C., and Potthast, F. (1997a). Identification of Amino Acid Sequences with Good Folding Properties in an Off-Lattice Model. *Phys. Rev. E* 55, 860–867.

Irbäck, A., Peterson, C., Potthast, F., and Sommelius, O. (1997b). Local Interactions and Protein Folding: A Three-Dimensional Off-Lattice Approach. *J. Chem. Phys.* 107, 273–282.

Irbäck, A., and Potthast, F. (1995). Studies of an Off-Lattice Model for Protein Folding: Sequence Dependence and Improved Sampling at Finite Temperature. *J. Chem. Phys.* 103, 10298–10305.

Ishihara, N., Seimiya, T., Kuramoto, M., and Uoi, M. (1986). Crystalline Syndiotactic Polystyrene. *Macromolecules* 19, 2464–2465.

Jäckle, J. (1986). Models of the Glass Transition. *Repts. Prog. Phys.* 49, 171–231.

Jang, S.S., Blanco, M., Goddard, W.A., III, Caldwell, G., and Ross, R.B. (2003). The Source of Helicity in Perfluorinated *N*-Alkanes. *Macromolecules* 36, 5331–5341.

Jastrow, R. (1955). Many-Body Problem with Strong Forces. *Phys. Rev.* 98, 1479–1484.

Jenkins, A.C., DiPaolo, F.S., and Birdsall, C.M. (1955). The System Ozone–Oxygen. *J. Chem. Phys.* 23, 2049–2054.

Johari, G.P. (1973). Intrinsic Mobility of Molecular Glasses. *J. Chem. Phys.* 58, 1766–1770.

Johari, G.P., and Goldstein, M. (1970). Viscous Liquids and the Glass Transition. II. Secondary Relaxation in Glasses of Rigid Molecules. *J. Chem. Phys.* 53, 2372–2388.

Jolly, W.L. (1991). *Modern Inorganic Chemistry*, 2nd edition. New York: McGraw-Hill, Inc.

Jones, J.E. (1924). On the Determination of Molecular Fields. II. From the Equation of State of a Gas. *Proc. R. Soc. London, Ser. A* 106, 463–477.

Jones, R.E., and Templeton, D.H. (1958). The Crystal Structure of Acetic Acid. *Acta Cryst.* 11, 484–487.

Jordan, K.D. (2004). A Fresh Look at Electron Hydration. *Science* 306, 618–619.

Jorgensen, W.L., Chandrasekhar, J., Madura, J.D., Impey, R.W., and Klein, M.L. (1983). Comparison of Simple Potential Functions for Simulating Liquid Water. *J. Chem. Phys.* 79, 926–935.

Jura, G., Fraga, D., Maki, G., and Hildebrand, J.H. (1953). Phenomena in the Liquid–Liquid Critical Region. *Proc. Natl. Acad. Sci. USA* 39, 19–23.

Kalos, M.H., Lee, M.A., Whitlock, P.A., and Chester, G.V. (1981). Modern Potentials and the Properties of ^{4}He. *Phys. Rev. B* 24, 115–130.

Karasz, F.E., Bair, H.E., and O'Reilly, J.M. (1965). Thermal Properties of Atactic and Isotactic Polystyrene. *J. Phys. Chem.* 69, 2657–2667.

Katayama, Y., Inamura, Y., Mizutani, T., Yamakata, M., Utsumi, W., and Shimomura, O. (2004). Macroscopic Separation of Dense Fluid Phase and Liquid Phase of Phosphorus. *Science* 306, 848–851.

Kaufman, L.C. (1999). Reduced Storage, Quasi-Newton Trust Region Approaches to Function Optimization. *SIAM J. Optim.* 10, 56–69.

Kaufmann, B. (1949). Crystal Statistics. II. Partition Function Evaluated by Spinor Analysis. *Phys. Rev.* 76, 1232–1243.

Kauzmann, W. (1948). The Nature of the Glassy State and the Behavior of Liquids at Low Temperatures. *Chem. Rev.* 43, 219–256.

Kearns, K.L., Swallen, S.F., Ediger, M.D., Sun, Y., and Yu, L. (2009). Calorimetric Evidence for Two Distinct Molecular Packing Arrangements in Stable Glasses of Indomethacin. *J. Phys. Chem. B* 113, 1579–1586.

Kearns, K.L., Swallen, S.F., Ediger, M.D., Wu, T., Sun, Y., and Yu, L. (2008). Hiking Down the Energy Landscape: Progress Toward the Kauzmann Temperature via Vapor Deposition. *J. Phys. Chem. B* 112, 4934–4942.

Keller, A. (1957). A Note on Single Crystals in Polymers: Evidence for a Folded Chain Configuration. *Phil. Mag.* 2(21), 1171–1175.

Kelly, A., and Knowles, K.M. (2012). *Crystallography and Crystal Defects*, 2nd edition. Chichester, Sussex, U.K.: Wiley.

Kestin, I., Knierim, K., Mason, B.A., Najafi, B., Ro, S.T., and Waidman, M. (1984). Equilibrium and Transport Properties of the Noble Gases and Their Mixtures at Low Density. *J. Phys. Chem. Ref. Data* 13, 229–303.

Keutsch, F.N., Braly, L.B., Brown, M.G., Harker, H.A., Petersen, P.B., Leforestier, C., and Saykally, R.J. (2003a). Water Dimer Hydrogen Bond Stretch, Donor Torsion Overtone, and "In-Plane Bend" Vibrations. *J. Chem. Phys.* 119, 8927–8937.

Keutsch, F.N., Cruzan, J.D., and Saykally, R.J. (2003b). The Water Trimer. *Chem. Rev.* 103, 2533–2577.

Keyes, T. (1994). Unstable Modes in Supercooled and Normal Liquids: Density of States, Energy Barriers, and Self-Diffusion. *J. Chem. Phys.* 101, 5081–5092.

Keyes, T. (1997). Instantaneous Normal Mode Approach to Liquid State Dynamics. *J. Phys. Chem. A* 101, 2921–2930.

Keyes, T., and Chowdhary, J. (2002). Potential Energy Landscape and Mechanisms of Diffusion in Liquids. *Phys. Rev. E* 65, 041106.

Khoury, F., and Passaglia, E. (1976). Morphology of Crystalline Synthetic Polymers. Chapter 6 in *Treatise on Solid State Chemistry, Vol. 3, Crystalline and Noncrystalline Solids*, N.B. Hannay, editor. New York: Plenum Press.

Khoury, G.A., Baliban, R.C., and Floudas, C.A. (2011). Proteome-Wide Post-Translational Modification Statistics: Frequency Analysis and Curation of the Swiss-Prot Database. *Sci. Rep.* 1, (paper 90). doi: 10.1038/srep00090.

Kihara, T., and Koba, S. (1952). Crystal Structures and Intermolecular Forces of Rare Gases. *J. Phys. Soc. Japan* 7, 348–354.

Kim, D.Y., and Chan, M.H.W. (2012). Absence of Supersolidity in Solid Helium in Porous Vycor Glass. *Phys. Rev. Lett.* 109, 155301.

Kim, E., and Chan, M.H.W. (2004). Observation of Superflow in Solid Helium. *Science* 305, 1941–1944.

King, N.P., Jacobitz, A.W., Sawaya, M.R., Goldschmidt, L., and Yeates, T.O. (2010). Structure and Folding of a Designed Knotted Protein. *Proc. Natl. Acad. Sci. USA* 107, 20732–20737.

Kirkwood, J.G. (1933). Quantum Statistics of Almost Classical Assemblies. *Phys. Rev.* 44, 31–37. Erratum: Kirkwood, J.G. (1934). *Phys. Rev.* 45, 116–117.

Kirkwood, J.G. (1936). On the Theory of Dielectric Polarization. *J. Chem. Phys.* 4, 592–601.

Kirkwood, J.G. (1939). The Dielectric Polarization of Polar Liquids. *J. Chem. Phys.* 7, 911–919.

Kirkwood, J.G., and Buff, F.P. (1951). The Statistical Mechanical Theory of Solutions. *J. Chem. Phys.* 19, 774–777.

Kirste, R.G., Kruse, W.A., and Ibel, K. (1975). Determination of the Conformation of Polymers in the Amorphous Solid State and in Concentrated Solution by Neutron Diffraction. *Polymer* 16, 120–124.

Kittel, C. (1956). *Introduction to Solid State Physics*, 2nd edition. New York: John Wiley & Sons.

Kittel, C. (1963). *Quantum Theory of Solids*. New York: John Wiley & Sons.

Kittel, C. (1996). *Introduction to Solid State Physics*, 7th edition. New York: John Wiley & Sons.

Knight, C.A., DeVries, A.L., and Oolman, L.D. (1984). Fish Antifreeze Protein and the Freezing and Recrystallization of Ice. *Nature* 308, 295–296.

Kob, W., and Andersen, H.C. (1995). Testing Mode-Coupling Theory for a Supercooled Binary Lennard-Jones Mixture. The van Hove Correlation Function. *Phys. Rev. E* 51, 4626–4641.

Kobayashi, R. (1993). Modeling and Numerical Simulations of Dendritic Crystal Growth. *Physica D* 63, 410–423.

Kohl, A., Binz, H.K., Forrer, P., Stumpp, M.T., Plückthun, A., and Grütter, M.G. (2003). Designed to Be Stable: Crystal Structure of a Consensus Ankyrin Repeat Protein. *Proc. Natl. Acad. Sci. USA* 100, 1700–1705.

Koishi, T., Yasuoka, K., Fujikawa, S., Ebisuzaki, T., and Zeng, X.C. (2009). Coexistence and Transition between Cassie and Wentzel State on Pillared Hydrophobic Surface. *Proc. Natl. Acad. Sci. USA.* 106, 8435–8440.

Kollman, P.A., and Bender, C.F. (1973). Structure of the H_3O^+ (Hydronium) Ion. *Chem. Phys. Lett.* 21, 271–274.

Kołos, W., and Wolniewicz, L. (1965). Potential Energy Curves for the $X^1\Sigma_g^+$, $b^3\Sigma_u^+$, and $C^1\Pi_u$ States of the Hydrogen Molecule. *J. Chem. Phys.* 43, 2429–2441.

Kondepudi, D.K., and Asakura, K. (2001). Chiral Autocatalysis, Spontaneous Symmetry Breaking, and Stochastic Behavior. *Acc. Chem. Res.* 34, 946–954.

Korn, G.A., and Korn, T.M. (1968). *Mathematical Handbook for Scientists and Engineers*, 2nd edition. New York: McGraw-Hill Book Co.

Koza, M.M., Schober, H., Hansen, T., Tölle, A., and Fujara, F. (2000). Ice XII in Its Second Regime of Stability. *Phys. Rev. Lett.* 84, 4112–4115.

Kraft, S., Vogt, F., Appel, O., Riehle, F., and Sterr, U. (2009). Bose-Einstein Condensation of Alkaline Earth Atoms: ^{40}Ca. *Phys. Rev. Lett.* 103, 130401.

Kreil, G. (1994). Conversion of L- to D-Amino Acids: A Posttranslational Reaction. *Science* 266, 996–997.

Kremer, K., and Grest, G.S. (1990). Dynamics of Entangled Linear Polymer Melts: A Molecular-Dynamics Simulation. *J. Chem. Phys.* 92, 5057–5086. Erratum: *J. Chem. Phys.* 94, 4103 (1991).

Kreplak, L., Doucet, J., Dumas, P., and Briki, F. (2004). New Aspects of the Alpha-Helix to Beta-Sheet Transition in Stretched Hard Alpha-Keratin Fibers. *Biophys. J.* 87, 640–647.

Kresheck, G.C. (1975). Surfactants. In *Water, A Comprehensive Treatise,* Vol. 4, F. Franks, editor. New York: Plenum Press, 95–167.

Krüger, B., Schäfer, L., and Baumgärtner, A. (1989). Correlations among Interpenetrating Polymer Coils: The Probing of a Fractal. *J. Physique* 50, 3191–3222.

Kubo, R., Toda, M., and Hashitsume, N. (1991). *Statistical Physics II. Nonequilibrium Statistical Mechanics,* 2nd edition. Berlin: Springer-Verlag.

Kurita, R., and Tanaka, H. (2004). Critical-Like Phenomena Associated with Liquid–Liquid Transition in a Molecular Liquid. *Science* 306, 845–848.

Kurita, R., and Tanaka, H. (2005). On the Abundance and General Nature of the Liquid–Liquid Phase Transition in Molecular Systems. *J. Phys.: Condens. Matter* 17, L293–L302.

La Nave, E., Mossa, S., and Sciortino, F. (2002). Potential Energy Landscape Equation of State. *Phys. Rev. Lett.* 88, 225701.

La Nave, E., Sciortino, F., Tartaglia, P., De Michele, C., and Mossa, S. (2003). Numerical Evaluation of the Statistical Properties of a Potential Energy Landscape. *J. Phys.: Condens. Matter* 15, S1085–S1094.

Labeit, S., and Kolmerer, B. (1995). Titins: Giant Proteins in Charge of Muscle Ultrastructure and Elasticity. *Science* 270, 293–296.

Labowitz, L.C., and Westrum, E.F., Jr. (1961). A Thermodynamic Study of the System Ammonium Fluoride—Water. II. The Solid Solution of Ammonium Fluoride in Ice. *J. Phys. Chem.* 65, 408–414.

Lam, H., Oh, D.-C., Cava, F., Takacs, C.N., Clardy, J., de Pedro, M.A., and Waldor, M.K. (2009). D-Amino Acids Govern Stationary Phase Cell Wall Remodeling in Bacteria. *Science* 325, 1552–1555.

Landau, L.D., and Lifshitz, E.M. (1958a). *Quantum Mechanics. Non-Relativistic Theory,* translated by J.B. Sykes and J.S. Bell. Reading, MA: Addison-Wesley Publishing Co.

Landau, L.D., and Lifshitz, E.M. (1958b). *Statistical Physics,* translated by E. Peierls and R.F. Peierls. Reading, MA: Addison-Wesley Publishing Co.

Landau, L.D., and Lifshitz, E.M. (1959). *Theory of Elasticity,* translated by J.B. Sykes and W.H. Reid. Reading, MA: Addison-Wesley Publishing Co.

Landau, L.D., and Lifshitz, E.M. (1960). *Electrodynamics of Continuous Media,* translated by J.B. Sykes and J.S. Bell. Reading, MA: Addison-Wesley Publishing Co.

Langer, J.S. (1967). Theory of the Condensation Point. *Ann. Phys.* 41, 108–157.

Lau, K.F., and Dill, K.A. (1989). A Lattice Statistical Mechanics Model of the Conformational and Sequence Spaces of Proteins. *Macromolecules* 22, 3986–3997.

Laughlin, W.T., and Uhlmann, D.R. (1972). Viscous Flow in Simple Organic Liquids. *J. Phys. Chem.* 76, 2317–2325.

LaViolette, R.A., Budzien, J.L., and Stillinger, F.H. (2000). Inherent Structure of a Molten Salt. *J. Chem. Phys.* 112, 8072–8078.

LaViolette, R.A., and Stillinger, F.H. (1985). Multidimensional Geometric Aspects of the Solid–Liquid Transition in Simple Substances. *J. Chem. Phys.* 83, 4079–4085.

LaViolette, R.A., and Stillinger, F.H. (1986). Thermal Disruption of the Inherent Structure of Simple Liquids. *J. Chem. Phys.* 85, 6027–6033.

Lawson, A.W. (1940). The Variation of the Adiabatic and Isothermal Elastic Moduli and Coefficient of Thermal Expansion with Temperature through the λ-Point Transition in Ammonium Chloride. *Phys. Rev.* 57, 417–426.

Lawson, A.W. (1950). Thermal Expansion in Silver Halides. *Phys. Rev.* 78, 185.

Leadbetter, A.J., Ward, R.C., Clark, J.W., Tucker, P.A., Matsuo, T., and Suga, H. (1985). The Equilibrium Low-Temperature Structure of Ice. *J. Chem. Phys.* 82, 424–428.

Lee, C., Vanderbilt, D., Laasonen, K., Car, R., and Parrinello, M. (1992). *Ab Initio* Studies on High Pressure Phases of Ice. *Phys. Rev. Lett.* 69, 462–465.

Lee, K., Shin, S., and Ka, J. (2004). Tunneling Dynamics of Amino-Acid: Model for Chiral Evolution? *J. Molec. Struct.: Theochem* 679, 59–63.

Lee, M.A., Schmidt, K.E., Kalos, M.H., and Chester, G.V. (1981). Green's Function Monte Carlo Method for Liquid ^{3}He. *Phys. Rev. Lett.* 46, 728–731.

Leforestier, C., Gatti, F., Fellers, R.S., and Saykally, R.J. (2002). Determination of a Flexible (12D) Water Dimer Potential via Direct Inversion of Spectroscopic Data. *J. Chem. Phys.* 117, 8710–8722.

Leggett, A.J. (1970). Can a Solid be "Superfluid"? *Phys. Rev. Lett.* 25, 1543–1546.

Lemberg, H.L., and Stillinger, F.H. (1975). Central-Force Model for Liquid Water. *J. Chem. Phys.* 62, 1677–1690.

Lesch, H., Hecht, C., and Friedrich, J. (2004). Protein Phase Diagrams: The Physics behind Their Elliptic Shape. *J. Chem. Phys.* 121, 12671–12675.

Lewis, L.J., and Wahnström, G. (1993). Relaxation of a Molecular Glass at Intermediate Times. *Solid State Commun.* 86, 295–299.

Lewis, L.J., and Wahnström, G. (1994). Molecular-Dynamics Study of Supercooled *Ortho*-Terphenyl. *Phys. Rev. E* 50, 3865–3877.

Li, D., and Rice, S.A. (2004). Some Properties of "Madrid" Liquids. *J. Phys. Chem. B* 108, 19640–19646.

Liao, J., Li, H., Zeng, W., Sauer, D.B., Belmares, R., and Jiang, Y. (2012). Structural Insight into the Ion-Exchange Mechanism of the Sodium/Calcium Exchanger. *Science* 335, 686–690.

Libál, A., Reichhardt, C., and Reichhardt, C.J.O. (2006). Realizing Colloidal Artificial Ice on Arrays of Optical Traps. *Phys. Rev. Lett.* 97, 228302.

Lide, D.R., editor-in-chief. (2003). *Handbook of Chemistry and Physics*, 84th edition, 2003–2004. New York: CRC Press.

Lide, D.R., editor-in-chief. (2006). *CRC Handbook of Chemistry and Physics*, 87th edition, 2006–2007. New York: Taylor & Francis.

Lie, G.C., Clementi, E., and Yoshimine, M. (1976). Study of the Structure of Molecular Complexes. XIII. Monte Carlo Simulation of Liquid Water with a Configurational Interaction Pair Potential. *J. Chem. Phys.* 64, 2314–2323.

Lieb, E.H. (1967). Exact Solution of the Problem of the Entropy of Two-Dimensional Ice. *Phys. Rev. Lett.* 18, 692–694.

Lindemann, F.A. (1910). Über die Berechnung molecularer Eigenfrequenzen. *Physik. Z.* 11, 609–612.

Liu, C.-S., Liang, Y.-F., Zhu, Z.-G., and Li, G.-X. (2005). A Molecular Level Study of Liquid Water with a Flexible Water Model. *Chin. Phys.* 14, 785–790.

Liu, D., Zhang, Y., Chen, C.-C., Mou, C.-Y., Poole, P.H., and Chen, S.-H. (2007). Observation of the Density Minimum in Deeply Supercooled Confined Water. *Proc. Natl. Acad. Sci. USA* 104, 9570–9574.

Liu, H.X., Zhang, R.S., Yao, X.J., Liu, M.C., Hu, Z.D., and Fan, B.T. (2004). Prediction of the Isoelectric Point of an Amino Acid Based on GA-PLS and SVMs. *J. Chem. Inf. Comput. Sci.* 44, 161–167.

Liu, J., Xue, S., Chen, D., Geng, H., and Liu, Z. (2009). Structure Optimization of the Two-Dimensional Off-Lattice Hydrophobic–Hydrophilic Model. *J. Biol. Phys.* 35, 245–253.

Lobban, C., Finney, J.L., and Kuhs, W.F. (1998). The Structure of a New Phase of Ice. *Nature* 391, 268–270.

Lodge, T.P. (1999). Reconciliation of the Molecular Weight Dependence of Diffusion and Viscosity in Entangled Polymers. *Phys. Rev. Lett.* 83, 3218–3221.

Loerting, T., Salzmann, C., Kohl, I., Mayer, E., and Hallbrucker, A. (2001a). A Second Structural "State" of High-Density Amorphous Ice at 77 K and 1 bar. *Phys. Chem. Chem. Phys.* 3, 5355–5357.

Loerting, T., Tautermann, C., Kroemer, R.T., Kohl, I., Hallbrucker, A., Meyer, E., and Leidel, K.R. (2001b). On the Surprising Kinetic Stability of Carbonic Acid. *Angew. Chem. Int. Ed.* 39, 891–895.

Lombardo, T.G., Stillinger, F.H., and Debenedetti, P.G. (2009). Thermodynamic Mechanism for Solution Phase Chiral Amplification via a Lattice Model. *Proc. Natl. Acad. Sci. USA* 106, 15131–15135.

Longuet-Higgins, H.C., and Widom, B. (1964). Rigid Sphere Model for Melting of Argon. *Mol. Phys.* 8, 549–556.

Lonsdale, K. (1946). Statistical Structure of Ice and of Ammonium Fluoride. *Nature* 158, 582.

Lonsdale, K. (1958). The Structure of Ice. *Proc. R. Soc. A* 247, 424–434.

Lovell, S.C., Davis, I.W., Arendall, W.B., III, de Bakker, P.I.W., Word, J.M., Prisant, M.G., Richardson, J.S., and Richardson, D.C. (2003). Structure Validation by $C\alpha$ Geometry: φ, ψ and $C\beta$ Deviation. *Proteins* 50, 437–450.

Lubachevsky, B.D., Stillinger, F.H., and Pinson, E.N. (1991). Disks vs. Spheres: Contrasting Properties of Random Packings. *J. Stat. Phys.* 64, 501–524.

Lubchenko, V., and Wolynes, P.G. (2003). The Origin of the Boson Peak and Thermal Conductivity Plateau in Low-Temperature Glasses. *Proc. Natl. Acad. Sci. USA* 100, 1515–1518.

Luijten, E., Fisher, M.E., and Panagiotopoulos, A.Z. (2002). Universality Class of Criticality in the Restricted Primitive Model Electrolyte. *Phys. Rev. Lett.* 88, 185701.

Lum, K., Chandler, D., and Weeks, J.D. (1999). Hydrophobicity at Small and Large Length Scales. *J. Phys. Chem. B* 103, 4570–4577.

Lundgren, J.-O., and Olovsson, I. (1967). Hydrogen Bond Studies. XV. The Crystal Structure of Hydrogen Chloride Dihydrate. *Acta Cryst.* 23, 966–971.

Lundgren, J.-O., and Olovsson, I. (1968). Hydrogen Bond Studies. XXX. The Crystal Structure of Hydrogen Bromide Tetrahydrate $(H_7O_3)^+(H_9O_4)^+2Br^- \cdot H_2O$. *J. Chem. Phys.* 49, 1068–1074.

MacFarland, T., Vitiello, S.A., Reatto, L., Chester, G.V., and Kalos, M.H. (1994). Trial Shadow Wave Function for the Ground State of ^{4}He. *Phys. Rev. B* 50, 13577–13593.

Mackay, A.L. (1962). A Dense Non-Crystallographic Packing of Equal Spheres. *Acta Cryst.* 15, 916–918.

MacKnight, W.J., and Tobolsky, A.V. (1965). In *Elemental Sulfur, Chemistry and Physics,* B. Meyer, editor, Chapter 5 (Properties of Polymeric Sulfur). New York: Wiley-Interscience.

Madras, N., and Slade, G. (1993). *The Self-Avoiding Walk.* Boston: Birkhäuser.

Magnussen, O.M., Ocko, B.M., Regan, M., Penanen, K., Pershan, P.S., and Deutsch, M. (1995). X-Ray Reflectivity Measurements of Surface Layering in Liquid Mercury. *Phys. Rev. Lett.* 74, 4444–4447.

Makin, M.J. (1968). Electron Displacement Damage in Copper and Aluminium in a High Voltage Electron Microscope. *Phil. Mag.* 18, 637–653.

Malandro, D.L., and Lacks, D.J. (1998). Molecular-Level Mechanical Instabilities and Enhanced Self-Diffusion in Flowing Liquids. *Phys. Rev. Lett.* 81, 5576–5579.

Malcolm, G.N.. and Rowlinson, J.S. (1957). The Thermodynamic Properties of Aqueous Solutions of Polyethylene Glycol, Polypropylene Glycol, and Dioxane. *Trans. Faraday Soc.* 53, 921–931.

Mandelbrot, B.B. (1977). *Fractals: Form, Chance and Dimension.* San Francisco: W.H. Freeman and Co.

Mandelkern, L. (1989). Crystallization and Melting. In *Comprehensive Polymer Science, Vol. 2, Polymer Properties,* C. Booth and C. Price, editors. New York: Pergamon Press, Chapter 11.

Mandelkern, L. (2002). *Crystallization of Polymers,* 2nd edition. New York: Cambridge University Press.

Mandelstam, L. (1913). Über die Rauhigkeit freier Flüssigkeitsoberflächen. *Ann. Phys.* 41, 609–624.

Mao, W.L., Koh, C.A., and Sloan, E.D. (2007). Clathrate Hydrates under Pressure. *Phys. Today* 60(10), 42–47.

Mao, W.L., Wang, L., Ding, Y., Yang, W., Liu, W., Kim, D.Y., Luo, W., Ahuja, R., Meng, Y., Sinogeikin, S., Shu, J., and Mao, H.-k. (2010). Distortions and Stabilization of Simple-Cubic Calcium at High Pressure and Low Temperature. *Proc. Natl. Acad. Sci USA* 107, 9965–9968.

Mark, H. (1950). Viskosität und Molekulargewicht macromolekular Lösungen. *Monatshefte für Chemie* 81, 140–150.

Martin, C.H., and Singer, S.J. (1991). Behavior of Point Defects in a Model Crystal near Melting. *Phys. Rev. B* 44, 477–488.

Martin, C.J., and O'Connor, D.A. (1977). An Experimental Test of Lindemann's Melting Law. *J. Phys. C: Solid State Phys.* 10, 3521–3526.

Martin, M.G., and Siepmann, J.I. (1998). Transferable Potentials for Phase Equilibria. 1. United Atom Description of *n*-Alkanes. *J. Phys. Chem. B* 102, 2569–2577.

Mary, T.A., Evans, J.S.O., Vogt, T., and Sleight, A.W. (1996). Negative Thermal Expansion from 0.3 to 1050 Kelvin in ZrW_2O_8. *Science* 272, 90–92.

Mas, E.M., Bukowski, R., and Szalewicz, K. (2003). *Ab Initio* Three-Body Interactions for Water. I. Potential and Structure of Water Trimer. *J. Chem. Phys.* 118, 4386–4403.

Mathews, C.K., van Holde, K.E., and Ahern, K.G. (2000). *Biochemistry,* 3rd edition. San Francisco: Benjamin Cummings.

Matsuoka, S. (1992). *Relaxation Phenomena in Polymers.* Munich, Germany: Hanser Publishers.

Mauro, J.C., Yue, Y., Ellison, A.J., Gupta, P.K., and Allan, D.C. (2009). Viscosity of Glass-Forming Liquids. *Proc. Natl. Acad. Sci. USA* 106, 19780–19784.

Mayer, E., and Brüggeller, P. (1982). Vitrification of Pure Liquid Water by High Pressure Jet Freezing. *Nature* 298, 715–718.

Mayer, J.E. (1947). Integral Equations between Distribution Functions of Molecules. *J. Chem. Phys.* 15, 187–201.

Mayer, J.E., and Mayer, M.G. (1940). *Statistical Mechanics.* New York: John Wiley & Sons.

McMillan, W.L. (1965). Ground State of Liquid He^4. *Phys. Rev.* 138, A442–A463.

McQuarrie, D.A. (1976). *Statistical Mechanics.* New York: Harper & Row.

Meakin, P. (1983). Diffusion-Controlled Cluster Formation in Two, Three, and Four Dimensions. *Phys. Rev. A* 27, 604–607.

Meakin, P. (1988). The Growth of Fractal Aggregates and Their Fractal Measures, in *Phase Transitions and Critical Phenomena,* Vol. 12, C. Domb and J.L. Lebowitz, editors. New York: Academic Press, 335–489.

Meier, K., Laesecke, A., and Kabelac, S. (2004a). Transport Coefficients of the Lennard-Jones Model Fluid. I. Viscosity. *J. Chem. Phys.* 121, 3671–3687.

Meier, K., Laesecke, A., and Kabelac, S. (2004b). Transport Coefficients of the Lennard-Jones Model Fluid. II. Self-Diffusion. *J. Chem. Phys.* 121, 9526–9535.

Meister, K., Ebbinhaus, S., Xu, Y., Duman, J.G., DeVries, A., Gruebele, M., Leitner, D.M., and Havenith, M. (2013). Long-Range Protein–Water Dynamics in Hyperactive Insect Antifreeze Proteins. *Proc. Natl. Acad. Sci. USA* 110, 1617–1622.

Mermin, N.D. (1968). Crystalline Order in Two Dimensions. *Phys. Rev.* 176, 250–254. Errata: *Phys. Rev. B* 20, 4762 (1979); *Phys. Rev. B* 74, 149902 (2006).

Meyer, B., editor. (1965). *Elemental Sulfur, Chemistry and Physics.* New York: Wiley-Interscience.

Miller, T.F., III, Vanden-Eijnden, E., and Chandler, D. (2007). Solvent Coarse-Graining and the String Method Applied to the Hydrophobic Collapse of a Hydrated Chain. *Proc. Natl. Acad. Sci. USA* 104, 14559–14564.

Milne-Thomson, L.M. (1960). *Theoretical Hydrodynamics,* 4th edition. New York: The Macmillan Co.

Mishima, O. (2000). Liquid–Liquid Critical Point in Heavy Water. *Phys. Rev. Lett.* 85, 334–336.

Mishima, O., Calvert, L.D., and Whalley, E. (1984). 'Melting Ice' I at 77 K and 10 kbar: A New Method of Making Amorphous Solids. *Nature* 310, 393–395.

Mislow, K. (2003). Absolute Asymmetric Synthesis: A Commentary. *Collect. Czech. Chem. Comm.* 68, 849–864.

Mo, H., Evmenenko, G., Kewalramani, S., Kim, K., Ehrlich, S.N., and Dutta, P. (2006). Observation of Surface Layering in a Nonmetallic Liquid. *Phys. Rev. Lett.* 96, 096107.

Mohanan, J.L., Arachchige, I.U., and Brock, S.L. (2005). Porous Semiconductor Chalcogenide Aerosols. *Science* 307, 397–400.

Moroni, S., and Senatore, G. (1991). Theory of Freezing for Quantum Fluids: Crystallization of ^{4}He at Zero Temperature. *Europhys. Lett.* 16, 373–378.

Morse, P.M. (1929). Diatomic Molecules According to the Wave Mechanics. II. Vibrational Levels. *Phys. Rev.* 34, 57–64.

Moses, H.E., and Tuan, S.F. (1959). Potentials with Zero Scattering Phase. *Nuovo Cimento (Serie X)* 13, 197–206.

Mossa, S., La Nave, E., Stanley, H.E., Donati, C., Sciortino, F., and Tartaglia, P. (2002). Dynamics and Configurational Entropy in the Lewis-Wahnström Model for Supercooled Orthoterphenyl. *Phys. Rev. E* 65, 041205.

Nagle, J.F. (1966). Lattice Statistics of Hydrogen Bonded Crystals. I. The Residual Entropy of Ice. *J. Math. Phys.* 7, 1484–1491.

Naidu, S.V., and Smith, F.A. (1994). Defects in the Rotator and Liquid Phases of *n*-Alkanes: A Study Using FTIR and Positron Annihilation. *J. Phys.: Condens. Matter* 6, 3865–3878.

Naleway, C.A., and Schwartz, M.E. (1972). *Ab Initio* Studies of the Interactions of an Electron and Two Water Molecules as a Building Block for a Model of the Hydrated Electron. *J. Phys. Chem.* 76, 3905–3908.

Nath, S.K., Escobedo, F.A., and de Pablo, J.J. (1998). On the Simulation of Vapor–Liquid Equilibria for Alkanes. *J. Chem. Phys.* 108, 9905–9911.

Natta, G., Corradini, P., and Bassi, I.W. (1960). Crystal Structure of Isotactic Polystyrene. *Il Nuovo Cimento* 15, 68–82.

Neinhuis, C., and Barthlott, W. (1997). Characterization and Distribution of Water-Repellent Self-Cleaning Plant Surfaces. *Ann. Bot.* 79, 667–677.

Nellis, W.J., van Thiel, M., and Mitchell, A.C. (1982). Shock Compression of Liquid Xenon to 130 GPa (1.3 Mbar). *Phys. Rev. Lett.* 48, 816–818.

Newton, M.D., and Ehrenson, S. (1971). Ab Initio Studies on the Structures and Energetics of Inner- and Outer-Shell Hydrates of the Proton and the Hydroxide Ion. *J. Am. Chem. Soc.* 93, 4971–4990.

Nix, F.C., and Shockley, W. (1938). Order–Disorder Transformations in Alloys. *Rev. Modern Phys.* 10, 1–71.

Nogales, E., Wolf, S.G., and Downing, K.H. (1998). Structure of the Alpha Beta Tubulin Dimer by Electron Crystallography. *Nature* 391, 199–203.

Nyburg, S.C., and Lüth, H. (1972). *n*-Octadecane: A Correction and Refinement of the Structure Given by Hayashida. *Acta Cryst. B* 28, 2992–2995.

Ocko, B.M., Wu, X.Z., Sirota, E.B., Sinha, S.K., Gang, O., and Deutsch, M. (1997). Surface Freezing in Chain Molecules: Normal Alkanes. *Phys. Rev. E* 55, 3164–3182.

Olijnyk, H., and Holzapfel, W.B. (1984). Phase Transitions in Alkaline Earth Metals under Pressure. *Phys. Lett.* 100A, 191–194.

Olovsson, I. (1968). Hydrogen Bond Studies. XXIX. Crystal Structure of Perchloric Acid Dihydrate, $H_5O_2^+ClO_4^-$. *J. Chem. Phys.* 49, 1063–1067.

Onsager, L. (1944). Crystal Statistics. I. A Two-Dimensional Model with an Order–Disorder Transition. *Phys. Rev.* 65, 117–149.

Onsager, L. (1949). (Discussion remarks following C.J. Gorter paper). *Nuovo Cimento* 6, Series 9, Suppl. 2, 249–250.

Onsager, L. (1973). Introductory Lecture. In *Physics and Chemistry of Ice*, E. Whalley, S.J. Jones, and L.W. Gold, editors. Ottawa: Royal Society of Canada, 7–12.

Onsager, L., and Samaras, N.N.T. (1934). The Surface Tension of Debye-Hückel Electrolytes. *J. Chem. Phys.* 2, 528–536.

Onuchic, J.N., Luthey-Schulten, Z., and Wolynes, P.G. (1997). Theory of Protein Folding: The Energy Landscape Perspective. *Annu. Rev. Phys. Chem.* 48, 545–600.

Opitz, C.A., Kulke, M., Leake, M.C., Neagoe, C., Hinson, H., Hajjar, R.J., and Linke, W.A. (2003). Damped Elastic Recoil of the Titin Spring in Myofibrils of Human Myocardium. *Proc. Natl. Acad. Sci. USA* 100, 12688–12693.

Orwoll, R.A., and Arnold, P.A. (1996). Polymer–Solvent Interaction Parameter χ. Chapter 14 in *Physical Properties of Polymers Handbook*, J.E. Mark, editor. New York: AIP Press, 177–196.

Ostwald, W. (1897). Studien über die Bildung und Umwandlung fester Körper. *Z. Phys. Chem.* 22, 289–330.

Otterbein, L.R., Graceffa, P., and Dominguez, R. (2001). The Crystal Structure of Uncomplexed Actin in the ADP State. *Science* 293, 708–711.

Oxtoby, D.W. (1992). Homogeneous Nucleation: Theory and Experiment. *J. Phys.: Condens. Matter* 4, 7627–7650. Erratum: 10, 897 (1998).

Paik, D.H., Lee, I-R., Yang, D.S., Baskin, J.S., and Zewail, A.H. (2004). Electrons in Finite-Sized Water Cavities: Hydration Dynamics Observed in Real Time. *Science* 306, 672–675.

Painter, P.C., and Coleman, M.M. (1997). *Fundamentals of Polymer Science: An Introductory Text*, 2nd edition. Boca Raton, FL: CRC Press.

Palko, J.W., and Kieffer, J. (2004). Inherent Structures in the NaCl–CsCl System at the Boundary between Crystallization and Glass Formation. *J. Phys. Chem. B* 108, 19867–19873.

Pamuk, B., Soler, J.M., Ramírez, R., Herrero, C.P., Stephens, P.W., Allen, P.B., and Fernández-Serra, M.-V. (2012). Anomalous Nuclear Quantum Effects in Ice. *Phys. Rev. Lett.* 108, 193003.

Panoff, R.M., and Carlson, J. (1989). Fermion Monte Carlo Algorithms and Liquid ^{3}He. *Phys. Rev. Lett.* 62, 1130–1133.

Park, H., Kim, K.M., Lee, A., Ham, S., Nam, W., and Chin, J. (2007). Bioinspired Chemical Inversion of L-Amino Acids to D-Amino Acids. *J. Am. Chem. Soc.* 129, 1518–1519.

Park, M.H. (2006). The Post-Translational Synthesis of a Polyamine Derived Amino Acid, Hypusine, in the Eukaryotic Translation Initiation Factor 5A (elF5A). *J. Biochem.* 139, 161–169.

Parr, R.G. (1963). *The Quantum Theory of Molecular Electronic Structure*. New York: W.A. Benjamin, Inc.

Pauli, W. (1925). Über den Zusammenhang des Abschlusses der Elektronengruppen im Atom mit der Komplexstruktur der Spektren. *Zeit. f. Phys.* 31, 765–783.

Pauling, L. (1935). The Structure and Entropy of Ice and of Other Crystals with Some Randomness of Atomic Arrangement. *J. Am. Chem. Soc.* 57, 2680–2684.

Pauling, L. (1960). *The Nature of the Chemical Bond*, 3rd edition. Ithaca, NY: Cornell University Press.

Pauling, L. (1988). *General Chemistry*. New York: Dover Publishers.

Pauling, L., and Wilson, E.B., Jr. (1935). *Introduction to Quantum Mechanics*. New York: McGraw-Hill Book Co.

Penrose, O., and Onsager, L. (1956). Bose-Einstein Condensation and Liquid Helium. *Phys. Rev.* 104, 576–584.

Pérez, C., Muckle, M.T., Zaleski, D.P., Seifert, N.A., Temelso, B., Shields, G.C., Kisiel, Z., and Pate, B.H. (2012). Structures of Cage, Prism, and Book Isomers of Water Hexamer from Broadband Rotational Spectroscopy. *Science* 336, 897–901.

Pérez-Castañeda, T., Rodriguez-Tinoco, C., Rodriguez-Viejo, J., and Ramos, M.A. (2014). Suppression of Tunneling Two-Level Systems in Ultrastable Glasses of Indomethacin. *Proc. Natl. Acad. Sci USA* 111, 11275–11280.

Pfann, W.G. (1966). *Zone Melting*, 2nd edition. New York: Wiley.

Phillips, J.C., Braun, R., Wang, W., Gumbart, J., Tajkhorshid, E., Villa, E., Chipot, C., Skeel, R.D., Kalé, L., and Schulten, K. (2005). Scalable Molecular Dynamics with NAMD. *J. Comput. Chem.* 26, 1781–1802.

Phillips, W.A. (1972). Tunneling States in Amorphous Solids. *Low Temp. Phys.* 7, 351–360.

Polman, A., Mous, D.J.W., Stolk, P.A., Sinke, W.C., Bulle-Lieuma, C.W.T., and Vandenhoudt, D.E.W. (1989). Epitaxial Explosive Crystallization of Amorphous Silicon. *Appl. Phys. Lett.* 55, 1097–1099.

Ponnuswami, N., Cougnon, F.B.L., Clough, J.M., Pantoş, G.D., and Sanders, J.K.M. (2012). Discovery of an Organic Trefoil Knot. *Science* 338, 783–785.

Poole, P.H., Sciortino, F., Essmann, U., and Stanley, H.E. (1992). Phase Behavior of Metastable Water. *Nature* 360, 324–328.

Poyner, A., Hong, L., Robinson, I.K., Granick, S., Zhang, Z., and Fenter, P.A. (2006). How Water Meets a Hydrophobic Surface. *Phys. Rev. Lett.* 97, 266101.

Prasad, S., and Dhinojwala, A. (2005). Rupture of a Two-Dimensional Alkane Crystal. *Phys. Rev. Lett.* 95, 117801.

Prestipino, S., Saija, F., and Giaquinta, P.V. (2005). Phase Diagram of Softly Repulsive Systems: The Gaussian and Inverse-Power-Law Potentials. *J. Chem. Phys.* 123, 144110.

Price, A.H. (1969). The Dielectric Properties of Gases. *Dielectric Properties & Molecular Behaviour*, N.E. Hill, W.E. Vaughn, A.H. Price, and M. Davies, editors. London: Van Nostrand Reinhold Co., Chapter 3.

Qi, D.W., Lu, J., and Wang, S. (1992). Crystallization Properties of a Supercooled Metallic Liquid. *J. Chem. Phys.* 96, 513–516.

Quéré, D. (2005). Non-Sticking Drops. *Repts. Prog. Phys.* 68, 2495–2532.

Radeva, T., editor. (2001). *Physical Chemistry of Polyelectrolytes*. New York: Marcel Dekker.

Rahman, A., Mandell, M.J., and McTague, J.P. (1976). Molecular Dynamics Study of an Amorphous Lennard-Jones System at Low Temperature. *J. Chem. Phys.* 64, 1564–1568.

Rahman, A., and Stillinger, F.H. (1973). Hydrogen-Bond Patterns in Liquid Water. *J. Am. Chem. Soc.* 95, 7943–7948.

Rahman, A., Stillinger, F.H., and Lemberg, H.L. (1975). Study of a Central Force Model for Liquid Water by Molecular Dynamics. *J. Chem. Phys.* 63, 5223–5230.

Ramachandran, G.N., Ramakrishnan, C., and Sasisekharan, V. (1963). Stereochemistry of Polypeptide Chain Configurations. *J. Molec. Biol.* 7, 95–99.

Ramachandran, G.N., and Sasisekharan, V. (1968). Conformation of Polypeptides and Proteins. *Adv. Protein Chem.* 23, 283–437.

Rao, F., and Karplus, M. (2010). Protein Dynamics Investigated by Inherent Structure Analysis. *Proc. Natl. Acad. Sci. USA* 107, 9152–9157.

Ravindra, R., and Winter, R. (2003). On the Temperature–Pressure Free-Energy Landscape of Proteins. *ChemPhysChem* 4, 359–365.

Reatto, L., and Chester, G.V. (1967). Phonons and the Properties of Bose Systems. *Phys. Rev.* 155, 88–100.

Reatto, L., and Masserini, G.L. (1988). Shadow Wave Function for Many-Boson Systems. *Phys. Rev. B* 38, 4516–4522.

Rebelo, L.P.N., Debenedetti, P.G., and Sastry, S. (1998). Singularity-Free Interpretation of Supercooled Water. II. Thermal and Volumetric Behavior. *J. Chem. Phys.* 109, 626–633.

Reiss, H., Frisch, H.L., and Lebowitz, J.L. (1959). Statistical Mechanics of Hard Spheres. *J. Chem. Phys.* 31, 369–380.

Rezus, Y.L.A., and Bakker, H.J. (2007). Observation of Immobilized Water Molecules around Hydrophobic Groups. *Phys. Rev. Lett.* 99, 148301.

Rickardi, J., Fries, P.H., Fischer, R., Rast, R., and Krienke, H. (1998). Liquid Acetone and Chloroform: A Comparison between Monte Carlo Simulation, Molecular Ornstein-Zernike Theory, and Site–Site Ornstein-Zernike Theory. *Mol. Phys.* 93, 925–938.

Roberts, C.J., Debenedetti, P.G., and Stillinger, F.H. (1999). Equation of State of the Energy Landscape of SPC/E Water. *J. Phys. Chem. B* 103, 10258–10265.

Robinson, R.A., and Stokes, R.H. (1959). *Electrolyte Solutions*. London: Butterworths.

Romero, D., Barrón, C., and Gómez, S. (1999). The Optimal Geometry of Lennard-Jones Clusters. *Comput. Phys. Commun.* 123, 87–96.

Rosen, M.J. (2004). *Surfactants and Interfacial Phenomena*, 3rd edition. New York: Wiley.

Rosenberg, R. (2005). Why Is Ice Slippery? *Physics Today* 58(12), 50–55.

Ross, M., and McMahan, A.K. (1980). Condensed Xenon at High Pressure. *Phys. Rev. B* 21, 1658–1664.

Röttger, K., Endriss, A., Ihringer, J., Doyle, S., and Kuhs, W.F. (1994). Lattice Constants and Thermal Expansion of H_2O and D_2O Ice Ih between 10 and 265 K. *Acta Cryst. B* 50, 644–648.

Röttger, K., Endriss, A., Ihringer, J., Doyle, S., and Kuhs, W.F. (2012). Lattice Constants and Thermal Expansion of H_2O and D_2O Ice Ih between 10 and 265 K. Addendum. *Acta Cryst. B* 68, 91.

Rouse, P.E. (1953). A Theory of the Linear Viscoelastic Properties of Dilute Solutions of Coiling Polymers. *J. Chem. Phys.* 21, 1272–1280.

Rowden, R.W., and Rice, O.K. (1951). Critical Phenomena in the Cyclohexane–Aniline System. *J. Chem. Phys.* 19, 1423–1424.

Rowlinson, J.S. (1959). *Liquids and Liquid Mixtures*. London: Butterworths Scientific Publications.

Rowlinson, J.S., and Widom, B. (1982). *Molecular Theory of Capillarity*. Oxford, U.K.: Clarendon Press.

Rubinstein, M., and Colby, R.H. (2003). *Polymer Physics*. New York: Oxford University Press.

Ruelle, D. (1969). *Statistical Mechanics*. New York: W.A. Benjamin.

Rushbrooke, G.S. (1963). On the Thermodynamics of the Critical Region for the Ising Problem. *J. Chem. Phys.* 39, 842–843.

Saeki, S., Kuwahara, N., Nakata, M., and Kaneko, M. (1976). Upper and Lower Critical Solution Temperatures in Poly(ethylene glycol) Solutions. *Polymer* 17, 685–689.

Salzmann, C.G., Radaelli, P.G., Finney, J.L., and Mayer, E. (2008). A Calorimetric Study on the Low Temperature Dynamics of Doped Ice V and Its Reversible Phase Transition to Hydrogen Ordered Ice XIII. *Phys. Chem. Chem. Phys.* 10, 6313–6324.

Salzmann, C.G., Radaelli, P.G., Hallbrucker, A., Mayer, E., and Finney, J.L. (2006). The Preparation and Structure of Hydrogen Ordered Phases of Ice. *Science* 311, 1758–1761.

Salzmann, C.G., Radaelli, P.G., Mayer, E., and Finney, J.L. (2009). Ice XV: A New Thermodynamically Stable Phase of Ice. *Phys. Rev. Lett.* 103, 105701.

Sanyal, M.K., Sinha, S.K., Huang, K.G., and Ocko, B.M. (1991). X-Ray Scattering Study of Capillary-Wave Fluctuations at a Liquid Surface. *Phys. Rev. Lett.* 66, 628–631.

Sanz, E., Vega, C., Abascal, J.L.F., and MacDowell, L.G. (2004). Tracing the Phase Diagram of the Four-Site Water Potential (TIP4P). *J. Chem. Phys.* 121, 1165–1166.

Sarupria, S., Ghosh, T., Garcia, A.E., and Garde, S. (2010). Studying Pressure Denaturation of a Protein by Molecular Dynamics Simulation. *Proteins* 78, 1641–1651.

Sashina, E.S., Bochek, A.M., Novoselov, N.P., and Kirichenko, D.A. (2006). Structure and Solubility of Natural Silk Fibroin. *Russian J. Appl. Chem.* 79, 869–876.

Sastry, S. (2001). The Relationship between Fragility, Configurational Entropy and the Potential Energy Landscape of Glass-Forming Liquids. *Nature* 409, 164–167.

Sastry, S., Debenedetti, P.G., and Stillinger, F.H. (1997). Statistical Geometry of Particle Packings. II. "Weak Spots" in Liquids. *Phys. Rev. E* 56, 5533–5543.

Sastry, S., Debenedetti, P.G., and Stillinger, F.H. (1998). Signatures of Distinct Dynamical Regimes in the Energy Landscape of a Glass-Forming Liquid. *Nature* 393, 554–557.

Sastry, S., Debenedetti, P.G., Stillinger, F.H., Schrøder, T.B., Dyre, J.C., and Glotzer, S.C. (1999). Potential Energy Landscape Signatures of Slow Dynamics in Glass Forming Liquids. *Physica A* 270, 301–308.

Sceats, M.G., and Rice, S.A. (1982). Amorphous Solid Water and Its Relationship to Liquid Water: A Random Network Model for Water. In *Water, A Comprehensive Treatise*, Vol. 7, F. Franks, editor. New York: Plenum Press, 83–214.

Schanze, K.S., and Shelton, A.H. (2009). Functional Polyelectrolytes. *Langmuir* 25, 13698–13702.

Schiff, L.I. (1968). *Quantum Mechanics*, 3rd edition. New York: McGraw-Hill.

Schirmacher, W., Diezemann, G., and Ganter, C. (1998). Harmonic Vibrational Excitations in Disordered Solids and the "Boson Peak." *Phys. Rev. Lett.* 81, 136–139.

Schirmer, T. (1998). General and Specific Porins from Bacterial Outer Membranes. *J. Struct. Biol.* 121, 101–109.

Schmidt, E. (1969). *Properties of Water and Steam in SI-Units*. New York: Springer-Verlag.

Schwartz, D.K., Schlossman, M.L., Kawamoto, E.H., Kellogg, G.J., Pershan, P.S., and Ocko, B.M. (1990). Thermal Diffuse X-Ray Scattering Studies of the Water–Vapor Interface. *Phys. Rev. A* 41, 5687–5690.

Schwartz, P. (1971). Order–Disorder Transition in NH_4Cl. III. Specific Heat. *Phys. Rev. B* 4, 920–928.

Schwickert, H., Strobl, G., and Kimmig, M. (1991). Molecular Dynamics in Perfluoro-n-Eicosane. I. Solid Phase Behavior and Crystal Structure. *J. Chem. Phys.* 95, 2800–2806.

Sciortino, F., Kob, W., and Tartaglia, P. (1999). Inherent Structure Entropy of Supercooled Liquids. *Phys. Rev. Lett.* 83, 3214–3217.

Sciortino, F., Kob, W., and Tartaglia, P. (2000). Thermodynamics of Supercooled Liquids in the Inherent-Structure Formalism: A Case Study. *J. Phys.: Condens. Matter* 12, 6525–6534.

Sciortino, F., and Tartaglia, P. (2001). Extension of the Fluctuation–Dissipation Theorem to the Physical Aging of a Model Glass-Forming Liquid. *Phys. Rev. Lett.* 86, 107–110.

Scott, W.R.P., Hünenberger, P.H., Tironi, I.G., Mark, A.E., Billeter, S.R., Fennen, J., Torda, A.E., Huber, T., Krüger, P., and van Gunsteren, W.F. (1999). The GROMOS Biomolecular Simulation Program Package. *J. Phys. Chem. A* 103, 3596–3607.

Sedlmeier, F., Horinek, D., and Netz, R.R. (2009). Nanoroughness, Intrinsic Density Profile, and Rigidity of the Air–Water Interface. *Phys. Rev. Lett.* 103, 136102.

Seeley, G., and Keyes, T. (1989). Normal-Mode Analysis of Liquid-State Dynamics. *J. Chem. Phys.* 91, 5581–5586.

Seitz, F. (1940). *The Modern Theory of Solids*. New York: McGraw-Hill.

Selkoe, D.J. (2003). Folding Proteins in Fatal Ways. *Nature* 426, 900–904.

Seydel, T., Madsen, A., Tolan, M., Grübel, G., and Press, W. (2001). Capillary Waves in Slow Motion. *Phys. Rev. B* 63, 073409.

Shapiro, S.L., and Teukolsky, S.A. (1983). *Black Holes, White Dwarfs, and Neutron Stars*. New York: Wiley, p. 2, Table 1.1.

Sharma, M., Resta, R., and Car, R. (2007). Dipolar Correlations and the Dielectric Permittivity of Water. *Phys. Rev. Lett.* 98, 247401.

Shaw, J.P., Petsko, G.A., and Ringe, D. (1997). Determination of the Structure of Alanine Racemase from *Bacillus stearothermophilus* at 1.6 Å Resolution. *Biochem.* 36, 1329–1342.

Shell, M.S., Debenedetti, P.G., and Stillinger, F.H. (2004). Inherent-Structure View of Self-Diffusion in Liquids. *J. Phys. Chem. B* 108, 6772–6777.

Shen, V.K., Debenedetti, P.G., and Stillinger, F.H. (2002). Energy Landscape and Isotropic Tensile Strength of n-Alkane Glasses. *J. Phys. Chem. B* 106, 10447–10459.

Sheng, H.W., and Ma, E. (2004). Atomic Packing of the Inherent Structure of Simple Liquids. *Phys. Rev. E* 69, 062202.

Shumake, N.E., and Garland, C.W. (1970). Infrared Investigation of Structural and Ordering Changes in Ammonium Chloride and Bromide. *J. Chem. Phys.* 53, 392–407.

Silence, S.M., Duggal, A.R., Dhar, L., and Nelson, K.A. (1992). Structural and Orientational Relaxation in Supercooled Liquid Triphenylphosphite. *J. Chem. Phys.* 96, 5448–5459.

Silvera, I.F. (1980). The Solid Molecular Hydrogens in the Condensed Phase: Fundamentals and Static Properties. *Rev. Mod. Phys.* 52, 393–452.

Simmons, R.O., and Balluffi, R.W. (1960a). Measurements of Equilibrium Vacancy Concentrations in Aluminum. *Phys. Rev.* 117, 52–61.

Simmons, R.O., and Balluffi, R.W. (1960b). Measurement of the Equilibrium Concentration of Lattice Vacancies in Silver near the Melting Point. *Phys. Rev.* 119, 600–605.

Simmons, R.O., and Balluffi, R.W. (1962). Measurement of Equilibrium Concentrations of Lattice Vacancies in Gold. *Phys. Rev.* 125, 862–872.

Simmons, R.O., and Balluffi, R.W. (1963). Measurement of Equilibrium Concentrations of Vacancies in Copper. *Phys. Rev.* 129, 1533–1544.

Simon, F. (1922). Untersuchungen über die spezifische Wärme bei tiefen Temperaturen. *Ann. Phys.* 68, 241–280.

Simon, F. (1930). Fünfundzwanzig Jahre Nernstscher Wärmesatz. *Ergeb. Exakten Naturwiss.* 9, 222–274.

Simon, F.E., and Glatzel, G. (1929). Bemerkungen zur Schmelzdruckkurve. *Z. Anorg. Allgem. Chem.* 178, 309–316.

Siow, K.S., Delmas, G., and Patterson, D. (1972). Cloud-Point Curves in Polymer Solutions with Adjacent Upper and Lower Critical Solution Temperatures. *Macromolecules* 5, 29–34.

Slater, J.C. (1933). The Virial and Molecular Structure. *J. Chem. Phys.* 1, 687–691.

Smeller, L. (2002). Pressure–Temperature Phase Diagrams of Biomolecules. *Biochim. Biophys. Acta* 1595, 11–29.

Smit, B., Karaborni, S., and Siepmann, J.I. (1995). Computer Simulations of Vapor–Liquid Phase Equilibria of *n*-Alkanes. *J. Chem. Phys.* 102, 2126–2140.

Smith, R.S., Huang, C., Wong, E.K.L., and Kay, B.D. (1996). Desorption and Crystallization Kinetics in Nanoscale Thin Films of Amorphous Water Ice. *Surf. Sci.* 367, L13–L18.

Sokhan, V.P., and Tildesley, D.J. (1997). The Free Surface of Water: Molecular Orientation, Surface Potential and Nonlinear Susceptibility. *Mol. Phys.* 92, 625–640.

Sokolov, A.P., Rössler, E., Kisliuk, A., and Quitmann, D. (1993). Dynamics of Strong and Fragile Glass Formers: Differences and Correlation with Low-Temperature Properties. *Phys. Rev. Lett.* 71, 2062–2065.

Somani, S., and Wales, D.J. (2013). Energy Landscapes and Global Thermodynamics for Alanine Peptides. *J. Chem. Phys.* 139, 121909.

Somasi, S., Khomami, B., and Lovett, R. (2000). Simulation of the Third Law Free Energies of Face-Centered-Cubic and Hexagonal-Close-Packed Lennard-Jones Solids. *J. Chem. Phys.* 113, 4320–4330.

Soper, A.K. (2000). The Radial Distribution Functions of Water and Ice from 220 to 673 K and at Pressures up to 400 MPa. *Chem. Phys.* 258, 121–137.

Souvatzis, P., Eriksson, O., Katsnelson, M.I., and Rudin, S.P. (2008). Entropy Driven Stabilization of Energetically Unstable Crystal Structures Explained from First Principles Theory. *Phys. Rev. Lett.* 100, 095901.

Sprik, M., Röthlisberger, U., and Klein, M.L. (1997). Structure of Solid Poly(tetrafluoroethylene): A Computer Simulation Study of Chain Orientational, Translational, and Conformational Disorder. *J. Phys. Chem. B* 101, 2745–2749.

Stack, G.M., Mandelkern, L., Kröhnke, C., and Wagner, G. (1989). Melting and Crystallization Kinetics of a High Molecular Weight *n*-Alkane: $C_{192}H_{386}$. *Macromolecules* 22, 4351–4361.

Stanley, H.E. (1971). *Introduction to Phase Transitions and Critical Phenomena*. New York: Oxford University Press.

Starkweather, H.W., Jr. (1986). Melting and Crystalline Transitions in Normal Perfluoroalkanes Poly(tetrafluoroethylene). *Macromolecules* 19, 1131–1134.

Stauffer, D., and Aharony, A. (1992). *Introduction to Percolation Theory*, 2nd edition. London: Taylor & Francis.

Stein, R.S., and Han, C.C. (1985). Neutron Scattering from Polymers. *Physics Today* 38, 74–80.

Stern, H.A., and Berne, B.J. (2001). Quantum Effects in Liquid Water: Path-Integral Simulations of a Flexible and Polarizable *Ab Initio* Model. *J. Chem. Phys.* 115, 7622–7628.

Steudel, R. (1984). Elemental Sulfur and Related Homocyclic Compounds and Ions. In *Studies in Inorganic Chemistry 5: Sulfur, Its Significance for Chemistry, for the Geo-, Bio- and Cosmosphere, and Technology*, A. Müller and B. Krebs, editors. New York: Elsevier, 3–37.

Stevens, M.J., and Robbins, M.O. (1993). Melting of Yukawa Systems: A Test of Phenomenological Melting Criteria. *J. Chem. Phys.* 98, 2319–2324.

Stevens, M.P. (1990). *Polymer Chemistry: An Introduction*. New York: Oxford University Press.

Stillinger, F.H. (1963). Rigorous Basis of the Frenkel-Band Theory of Association Equilibrium. *J. Chem. Phys.* 38, 1486–1494.

Stillinger, F.H. (1968). Ion Distribution in Concentrated Electrolytes. *Proc. Natl. Acad. Sci. USA* 60, 1138–1143.

Stillinger, F.H. (1970). Effective Pair Interactions in Liquids. Water. *J. Phys. Chem.* 74, 3677–3687.

Stillinger, F.H. (1972). Density Expansions for Effective Pair Potentials. *J. Chem. Phys.* 57, 1780–1787.

Stillinger, F.H. (1973). Structure in Aqueous Solutions of Nonpolar Solutes from the Standpoint of Scaled-Particle Theory. *J. Solution Chem.* 2, 141–158.

Stillinger, F.H. (1976). Phase Transitions in the Gaussian Core System. *J. Chem. Phys.* 65, 3968–3974.

Stillinger, F.H. (1978). Proton Transfer Reactions and Kinetics in Water, in *Theoretical Chemistry, Advances and Perspectives*, Vol. 3, H. Eyring and D. Henderson, editors. New York: Academic Press, 177–234.

Stillinger, F.H. (1979). Duality Relations for the Gaussian Core Model. *Phys. Rev. B* 20, 299–302.

Stillinger, F.H. (1982). Capillary Waves and the Inherent Density Profile for the Liquid–Vapor Interface. *J. Chem. Phys.* 76, 1087–1091.

Stillinger, F.H. (1988a). Supercooled Liquids, Glass Transitions, and the Kauzmann Paradox. *J. Chem. Phys.* 88, 7818–7825.

Stillinger, F.H. (1988b). Collective Phenomena in Statistical Mechanics and the Geometry of Potential Energy Hypersurfaces. In *Mathematical Frontiers in Computational Chemical Physics*, D.G. Truhlar, editor. New York: Springer-Verlag, 157–173.

Stillinger, F.H. (1990). Relaxation Behavior in Atomic and Molecular Glasses. *Phys. Rev. B* 41, 2409–2416.

Stillinger, F.H. (1998). Enumeration of Isobaric Inherent Structures for the Fragile Glass Former *o*-Terphenyl. *J. Phys. Chem. B* 102, 2807–2810.

Stillinger, F.H. (2000). Inherent Structure Enumeration for Low-Density Materials. *Phys. Rev. E* 63, 011110.

Stillinger, F.H. (2001). Lattice Sums and Their Phase Diagram Implications for the Classical Lennard-Jones Model. *J. Chem. Phys.* 115, 5208–5212.

Stillinger, F.H. (2008). An Inherent Structure View of Liquid–Vapor Interfaces. *J. Chem. Phys.* 128, 204705.

Stillinger, F.H., and Debenedetti, P.G. (1999). Distinguishing Vibrational and Structural Equilibration Contributions to Thermal Expansion. *J. Phys. Chem. B* 103, 4052–4059.

Stillinger, F.H., and Debenedetti, P.G. (2002). Energy Landscape Diversity and Supercooled Liquid Properties. *J. Chem. Phys.* 116, 3353–3361.

Stillinger, F.H., and Debenedetti, P.G. (2003). Phase Transitions, Kauzmann Curves, and Inverse Melting. *Biophys. Chem.* 105, 211–220.

Stillinger, F.H., and Debenedetti, P.G. (2005). Alternative View of Self-Diffusion and Shear Viscosity. *J. Phys. Chem. B* 109, 6604–6609.

Stillinger, F.H., Debenedetti, P.G., and Sastry, S. (1998). Resolving Vibrational and Structural Contributions to Isothermal Compressibility. *J. Chem. Phys.* 109, 3983–3988.

Stillinger, F.H., Debenedetti, P.G., and Truskett, T.M. (2001). The Kauzmann Paradox Revisited. *J. Phys. Chem. B* 105, 11809–11816.

Stillinger, F.H., and Head-Gordon, T. (1993). Perturbational View of Inherent Structures in Water. *Phys. Rev. E* 47, 2484–2490.

Stillinger, F.H., and Head-Gordon, T. (1995). Collective Aspects of Protein Folding Illustrated by a Toy Model. *Phys. Rev. E* 52, 2872–2877.

Stillinger, F.H., Head-Gordon, T., and Hirshfeld, C.L. (1993). Toy Model for Protein Folding. *Phys. Rev. E* 48, 1469–1477.

Stillinger, F.H., and Herrick, D.R. (1975). Bound States in the Continuum. *Phys. Rev. A* 11, 446–454.

Stillinger, F.H., and Hodgdon, J.A. (1994). Translation–Rotation Paradox for Diffusion in Fragile Glass-Forming Liquids. *Phys. Rev. E* 50, 2064–2068.

Stillinger, F.H., and Kirkwood, J.G. (1960). Quantum Statistics of Nonideal Systems. *Phys. Rev.* 118, 361–369.

Stillinger, F.H., and Lemberg, H.L. (1975). Symmetry Breaking in Water Molecule Interactions. *J. Chem. Phys.* 62, 1340–1346.

Stillinger, F.H., and Lovett, R. (1968a). Ion-Pair Theory of Concentrated Electrolytes. I. Basic Concepts. *J. Chem. Phys.* 48, 3858–3868.

Stillinger, F.H., and Lovett, R. (1968b). General Restriction on the Distribution of Ions in Electrolytes. *J. Chem. Phys.* 49, 1991–1994.

Stillinger, F.H., and Rahman, A. (1974). Improved Simulation of Liquid Water by Molecular Dynamics. *J. Chem. Phys.* 60, 1545–1557.

Stillinger, F.H., Sakai, H., and Torquato, S. (2002). Statistical Mechanical Models with Effective Potentials: Definitions, Applications, and Thermodynamic Consequences. *J. Chem. Phys.* 117, 288–296.

Stillinger, F.H., and Stillinger, D.K. (1990). Computational Study of Transition Dynamics in 55-Atom Clusters. *J. Chem. Phys.* 93, 6013–6024.

Stillinger, F.H., and Stillinger, D.K. (1997). Negative Thermal Expansion in the Gaussian Core Model. *Physica A* 244, 358–369.

Stillinger, F.H., and Stillinger, D.K. (2006). Expanded Solid Matter: Two-Dimensional LJ Modeling. *Mech. Materials* 38, 958–968.

Stillinger, F.H., and Weber, T.A. (1981). Structural Aspects of the Melting Transition. *Kinam* 3, 159–171.

Stillinger, F.H., and Weber, T.A. (1982). Hidden Structure in Liquids. *Phys. Rev. A* 25, 978–989.

Stillinger, F.H., and Weber, T.A. (1983a). Inherent Structure in Water. *J. Phys. Chem.* 87, 2833–2840.

Stillinger, F.H., and Weber, T.A. (1983b). Dynamics of Structural Transitions in Liquids. *Phys. Rev. A* 28, 2408–2416.

Stillinger, F.H., and Weber, T.A. (1984a). Inherent Pair Correlation in Simple Liquids. *J. Chem. Phys.* 80, 4434–4437.

Stillinger, F.H., and Weber, T.A. (1984b). Point Defects in BCC Crystals: Structures, Transition Kinetics, and Melting Implications. *J. Chem. Phys.* 81, 5095–5103.

Stillinger, F.H., and Weber, T.A. (1985a). Computer Simulation of Local Order in Condensed Phases of Silicon. *Phys. Rev. B* 31, 5262–5271; Errata: *Phys. Rev. B* 33, 1451 (1986).

Stillinger, F.H., and Weber, T.A. (1985b). Inherent Structure Theory of Liquids in the Hard-Sphere Limit. *J. Chem. Phys.* 83, 4767–4775.

Stillinger, F.H., and Weber, T.A. (1987). Molecular Dynamics Study of Chemical Reactivity in Liquid Sulfur. *J. Phys. Chem.* 91, 4899–4907.

Stillinger, F.H., and Weber, T.A. (1988). Molecular Dynamics Simulation for Chemically Reactive Substances. Fluorine. *J. Chem. Phys.* 88, 5123–5133.

Stillinger, F.H., Weber, T.A., and LaViolette, R.A. (1986). Chemical Reactions in Liquids: Molecular Dynamics Simulation for Sulfur. *J. Chem. Phys.* 85, 6460–6469.

Struik, L.C.E. (1978). *Physical Aging in Amorphous Polymers and Other Materials.* Amsterdam: Elsevier.

Sułkowska, J.I., Noel, J.K., and Onuchic, J.N. (2012a). Energy Landscape of Knotted Protein Folding. *Proc. Natl. Acad. Sci. USA* 109, 17783–17788.

Sułkowska, J.I., Rawdon, E.J., Millett, K.C., Onuchic, J.N., and Stasiak, A. (2012b). Conservation of Complex Knotting and Slipknotting Patterns in Proteins. *Proc. Natl. Acad. Sci. USA* 109, E1715–E1723.

Sun, T., Lin, F.-H., Campbell, R.L., Allingham, J.S., and Davies, P.L. (2014). An Antifreeze Protein Folds with an Interior Network of More Than 400 Semi-Clathrate Waters. *Science* 343, 795–798.

Svedberg, T., and Pedersen, K.O. (1940). *The Ultracentrifuge.* Oxford, U.K.: Clarendon Press; reprinted by Johnson Reprint Corp., New York, 1959.

Svishchev, I.M., and Kusalik, P.G. (1993). Structure in Liquid Water: A Study of Spatial Distribution Functions. *J. Chem. Phys.* 99, 3049–3058.

Swallen, S.F., Bonvallet, P.A., McMahon, R.J., and Ediger, M.D. (2003). Self-Diffusion of *tris*-Naphthylbenzene near the Glass Transition Temperature. *Phys. Rev. Lett.* 90, 015901.

Swallen, S.F., Kearns, K.L., Mapes, M.K., Kim, Y.S., McMahon, R.J., Ediger, M.D., Wu, T., Yu, L., and Satija, S. (2007). Organic Glasses with Exceptional Thermodynamic and Kinetic Stability. *Science* 315, 353–356.

Swan, P.R. (1962). Polyethylene Unit Cell Variations with Temperature. *J. Polymer Sci.* 56, 403–407.

Swanson, J.M.J., Maupin, C.M., Chen, H., Petersen, M.K., Xu, J., Wu, Y., and Voth, G.A. (2007). Proton Solvation and Transport in Aqueous and Biomolecular Systems: Insight from Computer Simulations. *J. Phys. Chem. B* 111, 4300–4314.

Swenson, C.A. (1950). The Liquid–Solid Transformation in Helium near Absolute Zero. *Phys. Rev.* 79, 626–631.

Taesler, I., and Olovsson, I. (1969). Hydrogen Bond Studies. XXXVII. The Crystal Structure of Sulfuric Acid Dihydrate $(H_3O^+)_2SO_4$. *J. Chem. Phys.* 51, 4213–4219.

Tajima, Y., Matsuo, T., and Suga, H. (1982). Phase Transition in KOH-Doped Hexagonal Ice. *Nature* 299, 810–812.

Takahashi, S., and Takatsuka, K. (2006). On the Validity Range of the Born-Oppenheimer Approximation: A Semiclassical Study for All-Particle Quantization of Three-Body Coulomb Systems. *J. Chem. Phys.* 124, 144101.

Takahashi, Y., and Tadokoro, H. (1973). Structural Studies of Polyethers, $(-(CH_2)_m-O-)_n$. X. Crystal Structure of Poly(ethylene oxide). *Macromolecules* 6, 672–675.

Takai, K., Nakamura, K., Toki, T., Tsunogai, U., Miyazaki, M., Miyazaki, J., Hirayama, H., Nakagawa, S., Nunoura, T., and Horikoshi, K. (2008). Cell Proliferation at 122°C and Isotopically Heavy CH_4 Production by a Hyperthermophilic Methanogen under High-Pressure Cultivation. *Proc. Natl. Acad. Sci. USA* 105, 10949–10954.

Takaiwa, D., Hatano, I., Koga, K., and Tanaka, H. (2008). Phase Diagram of Water in Carbon Nanotubes. *Proc. Natl. Acad. Sci. USA* 105, 39–43.

Tanaka, H., and Nakanishi, K. (1991). Hydrophobic Hydration of Inert Gases: Thermodynamic Properties, Inherent Structures, and Normal Mode Analysis. *J. Chem. Phys.* 95, 3719–3727.

Tanford, C. (1980). *The Hydrophobic Effect: Formation of Micelles and Biological Membranes*, 2nd edition. New York: Wiley-Interscience.

Tang, K.T., Toennies, J.P., and Yiu, C.L. (1995). Accurate Analytical He–He van der Waals Potential Based on Perturbation Theory. *Phys. Rev. Lett.* 74, 1546–1549.

Tang, Y., and Lu, B.C.-Y. (1993). A New Solution of the Ornstein-Zernike Equation from the Perturbation Theory. *J. Chem. Phys.* 99, 9828–9835.

ten Wolde, P.R., and Frenkel, D. (1997). Enhancement of Protein Crystal Nucleation by Critical Density Fluctuations. *Science* 277, 1975–1978.

ter Haar, D. (1954). *Elements of Statistical Mechanics.* New York: Rinehart & Co.

Terwilliger, T.C., and Eisenberg, D. (1982). The Structure of Melittin. *J. Biol. Chem.* 257, 6016–6022.

Tessman, J.R., Kahn, A.H., and Shockley, W. (1953). Electronic Polarizabilities of Ions in Crystals. *Phys. Rev.* 92, 890–895.

Thacher, P.D. (1967). Effect of Boundaries and Isotopes on the Thermal Conductivity of LiF. *Phys. Rev.* 156, 975–988.

Thackray, M. (1970). Melting Point Intervals of Sulfur Allotropes. *J. Chem. Eng. Data* 15, 495–497.

Theil, F. (2006). A Proof of Crystallization in Two Dimensions. *Commun. Math. Phys.* 262, 209–236.

Till, P.H. (1957). The Growth of Single Crystals of Linear Polyethylene. *J. Polymer Sci.* 24, 301–306.

Toda, A., Okamura, M., Taguchi, K., Hikosaka, M., and Kagioka, H. (2008). Branching and Higher Order Structure in Banded Polyethylene Spherulites. *Macromolecules* 41, 2484–2493.

Toda, M., Kubo, R., and Saitô, N. (1992). *Statistical Physics I. Equilibrium Statistical Mechanics*, second edition. Berlin: Springer-Verlag.

Tödheide, K. (1972). Water at High Temperatures and Pressures. In *Water, A Comprehensive Treatise,* Vol. 1, F. Franks, editor. New York: Plenum Press, 463–514.

Togeas, J.B. (2005). Acetic Acid Vapor: 2. A Statistical Mechanical Critique of Vapor Density Measurements. *J. Phys. Chem. A* 109, 5438–5444.

Torquato, S., and Stillinger, F.H. (2001). Multiplicity of Generation, Selection, and Classification Procedures for Jammed Hard-Particle Packings. *J. Phys. Chem. B* 105, 11849–11853.

Torquato, S., and Stillinger, F.H. (2007). Toward the Jamming Threshold of Sphere Packings: Tunneled Crystals. *J. Appl. Phys.* 102, 093511. Erratum: *J. Appl. Phys.* 103, 129902 (2008).

Torquato, S., and Stillinger, F.H. (2010). Jammed Hard Particle Packings: From Kepler to Bernal and Beyond. *Rev. Mod. Phys.* 82, 2633–2672.

Torrens, I.M. (1972). *Interatomic Potentials.* New York: Academic Press.

Trabi, M., and Craik, D.J. (2002). Circular Proteins: No End in Sight. *Trends Biochem. Sci.* 27, 132–138.

Trachenko, K., Dove, M.T., and Heine, V. (2002). Origin of the T^{1+a} Dependence of the Heat Capacity of Glasses at Low Temperature. *Phys. Rev. B* 65, 092201.

Tritt, T.M., editor. (2004). *Thermal Conductivity. Theory, Properties, and Applications.* New York: Kluwer Academic/Plenum Publishers.

Tsai, C.J., and Jordan, K.D. (1993a). Use of an Eigenmode Method to Locate the Stationary Points on the Potential Energy Surfaces of Selected Argon and Water Clusters. *J. Phys. Chem.* 97, 11227–11237.

Tsai, C.J., and Jordan, K.D. (1993b). Use of the Histogram and Jump-Walking Methods for Overcoming Slow Barrier Crossing Behavior in Monte Carlo Simulations: Applications to the Phase Transitions in $(Ar)_{13}$ and $(H_2O)_8$ Clusters. *J. Chem. Phys.* 99, 6957–6970.

Tyrrell, H.J.V., and Harris, K.R. (1984). *Diffusion in Liquids: A Theoretical and Experimental Study.* London: Butterworths.

Ubbelohde, A.R. (1978). *The Molten State of Matter.* New York: John Wiley & Sons.

Uche, O.U., Stillinger, F.H., and Torquato, S. (2004). Constraints on Collective Density Variables: Two Dimensions. *Phys. Rev. E* 70, 046122.

Uche, O.U., Torquato, S., and Stillinger, F.H. (2006). Collective Coordinate Control of Density Distributions. *Phys. Rev. E* 74, 031104.

Uhlenbeck, G.E., Hemmer, P.C., and Kac, M. (1963). On the van der Waals Theory of the Liquid–Vapor Equilibrium. II. Discussion of the Distribution Functions. *J. Math. Phys.* 4, 229–247.

Utz, M., Debenedetti, P.G., and Stillinger, F.H. (2000). Atomistic Simulation of Aging and Rejuvenation in Glasses. *Phys. Rev. Lett.* 84, 1471–1474.

Utz, M., Debenedetti, P.G., and Stillinger, F.H. (2001). Isotropic Tensile Strength of Molecular Glasses. *J. Chem. Phys.* 114, 10049–10057.

van Santen, R.A. (1984). The Ostwald Step Rule. *J. Phys. Chem.* 88, 5768–5769.

Vermeer, C. (1990). Gamma-Carboxyglutamate-Containing Proteins and the Vitamin K-Dependent Carboxylase. *Biochem. J.* 266, 625–636.

Vettorel, T., Grosberg, A.Y., and Kremer, K. (2009). Statistics of Polymer Rings in the Melt. *Phys. Biol.* 6, 025013.

Vieira, P.A., and Lacks, D.J. (2003). Particle Packing in Soft- and Hard-Potential Liquids. *J. Chem. Phys.* 119, 9667–9672.

Vitiello, S.A., Runge, K., and Kalos, M.H. (1988). Variational Calculations for Solid and Liquid ^{4}He with a "Shadow" Wave Function. *Phys. Rev. Lett.* 60, 1970–1972.

Vollhardt, D., and Wölfle, P. (1990). *The Superfluid Phases of Helium 3.* New York: Taylor & Francis.

Vollmayr, K., Kob, W., and Binder, K. (1996). How Do the Properties of a Glass Depend on the Cooling Rate? A Computer Simulation Study of a Lennard-Jones System. *J. Chem. Phys.* 105, 4714–4728.

von Neumann, J., and Wigner, E. (1929). Über merkwürdige diskrete Eigenwerte. *Physik. Zeit.* 30, 465–467.

Wagner, H., and Richert, R. (1999). Dielectric Beta Relaxations in the Glassy State of Salol? *J. Chem. Phys.* 110, 11660–11663.

Wahnström, G., and Lewis, L.J. (1993). Molecular Dynamics Simulation of a Molecular Glass at Intermediate Times. *Physica A* 201, 150–156.

Wales, D.J. (2003). *Energy Landscapes.* Cambridge, U.K.: Cambridge University Press.

Wales, D.J., and Berry, R.S. (1990). Melting and Freezing of Small Argon Clusters. *J. Chem. Phys.* 92, 4283–4295.

Wales, D.J., Doye, J.P.K., Miller, M.A., Mortenson, P.N., and Walsh, T.R. (2000). Energy Landscapes: From Clusters to Biomolecules. *Adv. Chem. Phys.* 115, 1–111.

Wales, D.J., and Hodges, M.P. (1998). Global Minima of Water Clusters $(H_2O)_n$, $n \leq 21$, Described by an Empirical Potential. *Chem. Phys. Lett.* 286, 65–72.

Wales, D.J., and Uppenbrink, J. (1994). Rearrangements in Model Face-Centered-Cubic Solids. *Phys. Rev. B* 50, 12342–12361.

Walker, N.A., Lamb, D.M., Adamy, S.T., Jonas, J., and Dare-Edwards, M.P. (1988). Self-Diffusion in the Compressed, Highly Viscous Liquid 2-Ethylhexyl Benzoate. *J. Phys. Chem.* 92, 3675–3679.

Wallace, D.C. (1997). Statistical Mechanics of Monatomic Liquids. *Phys. Rev. E* 56, 4179–4186.

Walquist, A., and Berne, B.J. (1995). Computer Simulation of Hydrophobic Hydration Forces on Stacked Plates at Short Range. *J. Phys. Chem.* 99, 2893–2899.

Walsh, S.T.R., Cheng, H., Bryson, J.W., Roder, H., and DeGrado, W.F. (1999). Solution Structure and Dynamics of a *de Novo* Designed Three-Helix Bundle Protein. *Proc. Natl. Acad. Sci. USA* 96, 5486–5491.

Wang, H., Keum, J.K., Hiltner, A., Baer, E., Freeman, B., Rozanski, A., and Galeski, A. (2009). Confined Crystallization of Polyethylene Oxide in Nanolayer Assemblies. *Science* 323, 757–760.

Wang, J., and Porter, R.S. (1995). On the Viscosity–Temperature Behavior of Polymer Melts. *Rheol. Acta* 34, 496–503.

Wang, R.F., Nisoli, C., Freitas, R.S., Li, J., McConville, W., Cooley, B.J., Lund, M.S., Samarth, N., Leighton, C., Crespi, V.H., and Schiffer, P. (2006). Artificial 'Spin Ice' in a Geometrically Frustrated Lattice of Ferromagnetic Islands. *Nature* 439, 303–306.

Wang, W. (2000). Lyophilization and Development of Solid Protein Pharmaceuticals. *Int. J. Pharmaceutics* 203, 1–60.

Wannier, G.H. (1959). *Elements of Solid State Theory*. Cambridge, U.K.: Cambridge University Press.

Wannier, G.H. (1966). *Statistical Physics*. New York: John Wiley & Sons.

Weast, R.C., editor-in-chief. (1976). *CRC Handbook of Chemistry and Physics*, 57th edition, 1976–1977. Cleveland, OH: CRC Press.

Weast, R.C., editor-in-chief. (1978). *CRC Handbook of Chemistry and Physics*, 59th edition, 1978–1979. West Palm Beach, FL: CRC Press.

Weber, M.D., and Friauf, R.J. (1969). Interstitialcy Motion in the Silver Halides. *J. Phys. Chem. Solids* 30, 407–419.

Weber, T.A., and Stillinger, F.H. (1984). Inherent Structures and Distribution Functions for Liquids That Freeze into BCC Crystals. *J. Chem. Phys.* 81, 5089–5094.

Weber, T.A., and Stillinger, F.H. (1985a). Local Order and Structural Transitions in Amorphous Metal–Metalloid Alloys. *Phys. Rev. B* 31, 1954–1963.

Weber, T.A., and Stillinger, F.H. (1985b). Interactions, Local Order, and Atomic-Rearrangement Kinetics in Amorphous Nickel–Phosphorus Alloys. *Phys. Rev. B* 32, 5402–5411.

Weber, T.A., and Stillinger, F.H. (1993). Melting of Square Crystals in Two Dimensions. *Phys. Rev. E* 48, 4351–4358.

Weck, M., Mohr, B., Sauvage, J.-P., and Grubbs, R.H. (1999). Synthesis of Catenane Structures via Ring-Closing Metathesis. *J. Org. Chem.* 64, 5463–5471.

Weeks, J.D. (1977). Structure and Thermodynamics of the Liquid–Vapor Interface. *J. Chem. Phys.* 67, 3106–3121.

Weidmann, J. (1967). Zur Spektraltheorie von Sturm-Liouville-Operatoren. *Math. Zeit.* 98, 268–302.

Weiner, B.B., and Garland, C.W. (1972). Order–Disorder Phenomena. VII. Critical Variations in Length of NH_4Cl Single Crystals at High Pressures. *J. Chem. Phys.* 56, 155–165.

Weissbluth, M. (1978). *Atoms and Molecules*. New York: Academic Press, 554–555.

Welch, M.J., Lifton, J.F., and Seck, J.A. (1969). Tracer Studies with Radioactive Oxygen-15. Exchange between Carbon Dioxide and Water. *J. Phys. Chem.* 73, 3351–3356.

Whalley, E. (1983). Cubic Ice in Nature. *J. Phys. Chem.* 87, 4174–4179.

Whalley, E., and Davidson, D.W. (1965). Entropy Changes at the Phase Transitions in Ice. *J. Chem. Phys.* 43, 2148–2149.

Whalley, E., Heath, J.B.R., and Davidson, D.W. (1968). Ice IX: An Antiferroelectric Phase Related to Ice III. *J. Chem. Phys.* 48, 2362–2370.

Whitford, P.C., Sanbonmatsu, K.Y., and Onuchic, J. (2012). Biomolecular Dynamics: Order–Disorder Transitions and Energy Landscapes. *Rep. Prog. Phys.* 75, 076601.

Wick, C.D., Dang, L.X., and Jungwirth, P. (2006). Simulated Surface Potentials at the Vapor–Water Interface for the KCl Aqueous Electrolyte Solution. *J. Chem. Phys.* 125, 024706.

Widmer-Cooper, A., and Harrowell, P. (2006). Predicting the Long-Time Dynamic Heterogeneity in a Supercooled Liquid on the Basis of Short-Time Heterogeneities. *Phys. Rev. Lett.* 96, 185701.

Wiedersich, J., Köhler, S., Skerra, A., and Friedrich, J. (2008). Temperature and Pressure Dependence of Protein Stability: The Engineered Fluorescein-Binding Lipocalin FluA Shows an Elliptic Phase Diagram. *Proc. Natl. Acad. Sci. USA* 105, 5756–5761.

Wiewiorowski, T.K., Parthasarathy, A., and Slaten, B.L. (1968). Molten Sulfur Chemistry. V. Kinetics of Chemical Equilibration in Pure Liquid Sulfur. *J. Phys. Chem.* 72, 1890–1892.

Wilks, J., and Betts, D.S. (1987). *An Introduction to Liquid Helium*, 2nd edition. Oxford, U.K.: Clarendon Press.

Williams, D.E. (1966). Nonbonded Potential Parameters Derived from Crystalline Aromatic Hydrocarbons. *J. Chem. Phys.* 45, 3770–3778.

Williams, M.L., Landel, R.F., and Ferry, J.D. (1955). The Temperature Dependence of Relaxation Mechanisms in Amorphous Polymers and Other Glass-Forming Liquids. *J. Am. Chem. Soc.* 77, 3701–3707.

Wilson, K.G. (1971). Renormalization Group and Critical Phenomena. I. Renormalization Group and the Kadanoff Scaling Picture. *Phys. Rev. B* 4, 3174–3183.

Winnick, J., and Cho, S.J. (1971). *PVT* Behavior of Water at Negative Pressure. *J. Chem. Phys.* 55, 2092–2097.

Wirth, F.H., and Hallock, R.B. (1987). X-Ray Determinations of the Liquid-Structure Factor and the Pair-Correlation Function of ^{4}He. *Phys. Rev. B* 35, 89–105.

Witten, T.A., and Sander, L.M. (1981). Diffusion-Limited Aggregation, a Kinetic Critical Phenomenon. *Phys. Rev. Lett.* 47, 1400–1403.

Woodruff, D.P. (1973). The Solid–Liquid Interface. London: Cambridge University Press.

Wu, X.Z., Ocko, B.M., Sirota, E.B., Sinha, S.K., Deutsch, M., Cao, B.H., and Kim, M.W. (1993). Surface Tension Measurements of Surface Freezing in Liquid Normal Alkanes. *Science* 261, 1018–1021.

Wüst, T., and Landau, D.P. (2009). Versatile Approach to Access the Low Temperature Thermodynamics of Lattice Polymers and Proteins. *Phys. Rev. Lett.* 102, 178101.

Xiang, H.W. (2003). Vapor Pressure and Critical Point of Tritium Oxide. *J. Phys. Chem. Ref. Data* 32, 1707–1711.

Xu, N., Wyart, M., Liu, A.J., and Nagel, S.R. (2007). Excess Vibrational Modes and the Boson Peak in Model Glasses. *Phys. Rev. Lett.* 98, 175502.

Yang, C.N. (1962). Concept of Off-Diagonal Long-Range Order and the Quantum Phases of Liquid Helium and of Superconductors. *Rev. Modern Phys.* 34, 694–704.

Yardimci, H., and Leheny, R.L. (2006). Aging of the Johari-Goldstein Relaxation in the Glass-Forming Liquids Sorbitol and Xylitol. *J. Chem. Phys.* 124, 214503.

Yarkony, D.R. (1996). Diabolical Conical Intersections. *Rev. Modern Phys.* 68, 985–1013.

Yarkony, D.R. (2001). Conical Intersections: The New Conventional Wisdom. *J. Phys. Chem. A* 105, 6277–6293.

Yoon, Y.K., and Carpenter, G.B. (1959). The Crystal Structure of Hydrogen Chloride Monohydrate. *Acta Cryst.* 12, 17–20.

Young, D.A., McMahon, A.K., and Ross, M. (1981). Equation of State and Melting Curve of Helium to Very High Pressure. *Phys. Rev. B* 24, 5119–5127.

Young, T. (1805). An Essay on the Cohesion of Fluids. *Phil. Trans. R. Soc. London* 95, 65–87.

Yukawa, H. (1935). On the Interaction of Elementary Particles. I. *Proc. Phys.-Math. Soc. Japan* 17, 48–57.

Yvon, J. (1937). *Actualités Scientifiques et Industrielles*, Nos. 542 and 543. Paris: Herman et Cie.

Zeller, R.C., and Pohl, R.O. (1971). Thermal Conductivity and Specific Heat of Noncrystalline Solids. *Phys. Rev. B* 4, 2029–2041.

Zhang, J., Kou, S.C., and Liu, J.S. (2007). Biopolymer Structure Simulation and Optimization via Fragment Regrowth Monte Carlo. *J. Chem. Phys.* 126, 225101.

Zheng, K.M., and Greer, S.C. (1992). The Density of Liquid Sulfur near the Polymerization Temperature. *J. Chem. Phys.* 96, 2175–2182.

Zheng, Q., Durben, D.J., Wolf, G.H., and Angell, C.A. (1991). Liquids at Large Negative Pressures: Water at the Homogeneous Nucleation Limit. *Science* 254, 829–832.

Ziman, J.M. (2001). *Electrons and Phonons: The Theory of Transport Phenomena in Solids*. New York: Oxford University Press.

Zundel, G. (1976). Easily Polarizable Hydrogen Bonds—Their Interactions with the Environment—IR Continuum and Anomalous Large Proton Conductivity. In *The Hydrogen Bond. Recent Developments in Theory and Experiments. II. Structure and Spectroscopy*, P. Schuster, G. Zundel, and C. Sandorfy, editors. Amsterdam: North-Holland, 683–766.

Zwanzig, R. (1965). Time-Correlation Functions and Transport Coefficients in Statistical Mechanics. *Ann. Rev. Phys. Chem.* 16, 67–102.

Zwanzig, R. (1988). Diffusion in a Rough Potential. *Proc. Natl. Acad. Sci. USA* 85, 2029–2030.

Zwanzig, R. (2001). *Nonequilibrium Statistical Mechanics*. Oxford, U.K.: Oxford University Press.

Zwanzig, R.W. (1954). High-Temperature Equation of State by a Perturbation Method. I. Nonpolar Gases. *J. Chem. Phys.* 22, 1420–1426.

Index